Trends in Logic

Studia Logica Library

Volume 63

The book series Trends in Logic covers essentially the same areas as the journal Studia Logica, that is, contemporary formal logic and its applications and relations to other disciplines. The series aims at publishing monographs and thematically coherent volumes dealing with important developments in logic and presenting significant contributions to logical research.

Volumes of Trends in Logic may range from highly focused studies to presentations that make a subject accessible to a broader scientific community or offer new perspectives for research. The series is open to contributions devoted to topics ranging from algebraic logic, model theory, proof theory, philosophical logic, non-classical logic, and logic in computer science to mathematical linguistics and formal epistemology. This thematic spectrum is also reflected in the editorial board of Trends in Logic. Volumes may be devoted to specific logical systems, particular methods and techniques, fundamental concepts, challenging open problems, different approaches to logical consequence, combinations of logics, classes of algebras or other structures, or interconnections between various logic-related domains.

This book series is indexed in SCOPUS.

Authors interested in proposing a completed book or a manuscript in progress or in conception can contact either chris.wilby@springer.com or one of the Editors of the Series.

Igor Sedlár • Shawn Standefer • Andrew Tedder
Editors

New Directions in Relevant Logic

Springer

Editors
Igor Sedlár
Institute of Computer Science
Czech Academy of Sciences
Prague, Czech Republic

Shawn Standefer
Department of Philosophy
National Taiwan University
Taipei, Taiwan

Andrew Tedder
Department of Philosophy I
Ruhr University Bochum
Bochum, Germany

ISSN 1572-6126 ISSN 2212-7313 (electronic)
Trends in Logic
ISBN 978-3-031-69942-9 ISBN 978-3-031-69940-5 (eBook)
https://doi.org/10.1007/978-3-031-69940-5

This work was supported by American Friends of the Alexander von Humboldt Foundation.

This Springer imprint is published by the registered company Springer Nature Switzerland AG
The registered company address is: Gewerbestrasse 11, 6330 Cham, Switzerland

Contents

Introduction: Directions and New Directions

Igor Sedlár, Shawn Standefer, and Andrew Tedder

Abstract In this chapter, we will provide some background on the volume and the topics of the papers. We begin by presenting the context for the workshop that gave rise to the volume. We then present a short historical overview of relevant logics. We then provide context to situate each of the papers in this volume, organized by the topics of the parts, namely Philosophical Foundations, Model Theory, Proof Theory, and Applications.

In 1974 the first international conference on relevant/relevance logics was held at Southern Illinois University (USA), and a proceedings volume was planned. Because of a range of hurdles and issues, no such volume saw the light of day until 1989 with the publication of *Directions in Relevant Logic* [35]. This volume, edited by Jean Norman and Richard Sylvan of Australian National University (AUS), included some of the papers presented in '74, but included some more recent material as well. Notably, a pair of very long papers were included by Sylvan, as well as a more recent paper by Kit Fine which proved a landmark incompleteness result. The delay in publication was largely due, it seems, to difficulties finding a publisher for work on this often odd, sometimes radical, and thoroughly non-classical style of logic. In addition, the papers in *Directions* [35] run the gamut from cutting-edge mathematical work on those logics called relevant, to philosophical reflections on their value and purpose in the logical landscape. In the decades since its publication, *Directions* has proved to be an invaluable resource for researchers working in relevant, and related, logics. This state of affairs might suggest that further international conferences, and further *Directions* volumes, might have a useful role to play in the contemporary scene.

I. Sedlár
Institute of Computer Science, Czech Academy of Sciences, Prague, Czech Republic

S. Standefer
Department of Philosophy, National Taiwan University, Taipei, Taiwan

A. Tedder (✉)
Ruhr University Bochum, Bochum, Germany

I. Sedlár et al. (eds.), *New Directions in Relevant Logic*, Trends in Logic 63,
https://doi.org/10.1007/978-3-031-69940-5_1

To that end the editors of the present volume organised an online event in late 2022 bringing together a range of researchers in the field. The online aspect of the conference was not only a cost-saving measure, but a good way to open doors to people who might otherwise not have been able to travel the many kilometers needed to get to whatever one physical location we might have chosen. That event also boasted a range of contributions, covering technical/mathematical aspects of relevant logics, philosophical reflections on these, and everything in between. This volume then serves both as a (long overdue) follow up to *Directions* [35], and as a companion piece to our own international conference on relevant logics.

Getting academics to agree on anything is an uphill battle; this is perhaps doubly so for philosophically-minded logicians. Non-classical logicians, in proposing, advocating for, and investigating systems of logic which differ (some in fairly radical ways) from the norm in contemporary logic, are unsurprisingly even more likely to adopt idiosyncratic positions, and so even less likely to come to ready agreement. As a result, even given a number of logicians who reckon that the notion of *relevance* plays a central role in logic, it is quite likely that one will wind up with as many systems as investigators, and perhaps even deeper rifts concerned with logical methodology. As a result of these tendencies, there are relevant logicians doing similar sorts of things all over the place, but with enough differences (deep or superficial) that there are occasional difficulties with 'interfaith dialogue'. Our goal has been to foster such dialogue, and we hope that this conference, and volume, both speak to our success, and will not be the last such. There are logicians all over the globe interested in, persuaded by, or just in possession of interesting things to say about relevant logics. Our goal with the conference was to get as many of these folks into one 'room' to talk with each other, assess the field as it stands, and consider what *new* directions beckon for our investigation.

The goal of fostering increased dialogue on relevant logics was, we think, a success. The workshop featured talks on a range of topics, by logicians at different stages of their careers. Following the workshop, we put out an open call for papers for this volume. Most of the papers presented at the workshop, though not all, appear in some form in this volume. Several of the papers in this volume were not presented at the workshop. All of the papers went through a process of anonymous peer review, and editors who (co-)authored a paper were not involved with the review process of that paper in any way. The resulting volume, we think, is a collection of papers that presents a good view of contemporary work on relevant logics and directions for future research.

1 A Very Brief, Very Partial Historical Overview

There has not, to our knowledge, been a detailed history of relevant logics written, not even relevant logics in the tradition coming out of the work of Anderson and

Belnap.[1] We will not provide such a history here, but we will say a few words about relevant logics and their origins.

Relevant logics mainly come out of a reaction to issues highlighted by C. I. Lewis.[2] Lewis highlighted several so-called paradoxes of implication, of which we will list three.

- $A \to (B \to A)$
- $A \to (B \to B)$
- $(A \wedge \neg A) \to B$

These are all valid on the classical construal of implication as *material implication*. Lewis, in searching for a connective that expresses *entailment*, came up with alternatives on which these are not valid. His strict implications were able to avoid some of the paradoxes, but not all of them.

Ackermann [1] proposed a "strenge Implikation", presenting a logic, Π', that avoided Lewis's paradoxes. Anderson and Belnap built on the work of Ackermann, defining their logics **E** of entailment and **R** of relevance. Both of these logics avoided all of Lewis's paradoxes. These, and related logics, were studied vigorously by Anderson and Belnap, along with their students, who included Dunn, Meyer, Urquhart, and others. The work of what we might call the early days of the Anderson-Belnap approach to relevant logics was put together into *Entailment* volume I [2].

Relevant logics were studied from a proof-theoretic point of view initially. Along with proof systems, algebraic and matrix methods were used and developed to obtain a variety of results, including independence, unprovability, and completeness. Models more like the relational structures used in modal logics came later. A variety of model structures were developed around the same time, prominent examples of which are the operational models of Urquhart [47], the operational-relational models of Fine [21], and, most famously, the ternary relation models of Sylvan (né Routley) and Meyer [41]. There is much to say about the different kinds of models, but we will focus on the ternary relational models of Sylvan and Meyer.

The ternary relational models use a ternary relation to interpret the implication, or conditional, connective. An implication $A \to B$ holds at a point x iff for all points y and z if $Rxyz$ and A holds at y, then B holds at z. In these models, even logical truths can fail at some points. Validity is defined with respect to a distinguished set of points, rather than all the points of the model, so there can be valid formulas that fail at some points in the models. Stronger logics can be obtained from the models by placing conditions on the ternary relations. While Anderson and Belnap focused on the stronger logics **E** and **R**, the ternary relational models naturally highlighted

[1] Mares [31] perhaps comes closest, but his focus is on the logic of entailment, a topic which overlaps with relevant logic but is not coextensive with it.

[2] There were other routes to relevant logics, as illustrated by Orlov [37], Moh [34], and Church [13]. We will be unable to discuss their work here.

weaker logics that required fewer conditions, including the logic **B** that results from no additional conditions beyond the basic ones required by the models.

The International Conference, whose proceedings turned into the *Directions in Relevant Logic* volume, was held in 1974, shortly after the publication of the work of Sylvan and Meyer and shortly before the publication of *Entailment* volume I. Some years after the conference saw the publication of *Relevant Logics and their Rivals* volume 1 [42] by Sylvan, Meyer, Plumwood, and Brady. This volume included extended philosophical discussions of relevant logics, likely authored by Sylvan, as well as systematic development of the ternary relational models and their metatheory.

There is a sense in which there were two different approaches to relevant logics being pursued with differing philosophical viewpoints.[3] The orientation and focus of *Relevant Logics and their Rivals* was very different from that of *Entailment*. While Anderson and Belnap argued that classical logic was mistaken, they tended to be somewhat conservative in how to rectify those mistakes. Sylvan et al., on the other hand, were more radical in their recommendations and their criticisms, arguing that Anderson and Belnap's preferred logics were also flawed.

What we might call the Golden Age of relevant logics comes to a close with the 1992 publication of *Entailment* volume II [3], edited by Anderson, Belnap, and Dunn. This work records much of the work in the field done since the publication of *Entailment* volume I, with a focus on the preferred logics of Anderson and Belnap. This includes the development of different models, as well advances in proof theory, undecidability results, and various applications. Work on relevant logics in the broad tradition of Anderson and Belnap, inclusive of the work of Sylvan, Meyer, and their co-authors, continued after the publication of *Entailment* volume II, and many advances, both formal and philosophical, have been made in the intervening 30 years. We will not say anything more about the history of the tradition in this period. Much of the work of this volume is in that broad tradition, and the reader can get a sense of some of these advances by reading the background provided by these articles.

For the reader interested in reading more about the logics and techniques discussed above, there are many excellent sources to consult. We can recommend the survey articles by Dunn and Restall [19] and Bimbó [10] as well as the books by Read [39], Mares [30], and Logan [29].

So far we have focused on the broad Anderson-Belnap tradition of relevant logics. Tennant takes a different approach to relevance. As with Anderson and Belnap, Tennant thinks that a contradiction, $A \wedge \neg A$, does not entail everything. Rather than require that the *formula* $(A \wedge \neg A) \rightarrow B$ be invalid, he takes it that the *sequent* $A \wedge \neg A : B$ should be invalid. Tennant defines proof systems that obey a relevance constraint at the sequent level, rather than the connective level, as with the logics of Anderson and Belnap. This alternative approach to relevance is pursued

[3] Standefer [45] discusses the different motivations for relevant logics in more detail.

primarily via proof systems, a presentation and defense of which can be found in Tennant's *Core Logic* [46].

2 An Introduction to, and Schematic View of, the Volume's Contents

Let us reflect a bit on where the field of relevant logic is as of writing, and place the volume's contents into that broader context. In the remainder of this introduction, we'll spell out the basics of the various areas the collected papers touch upon, and introduce the contributions in this context.

2.1 Philosophical Foundations

Relevant logic is a diverse field, held together by the idea that arguments in which the premises begun with are irrelevant to the conclusion reached are somehow *defective*, and that defectiveness should be diagonsed in, broadly speaking, *logic* or *semantics*, and not just in *pragmatics*. Classical logicians can admit that irrelevant arguments are faulty in some way, but, like Grice [25], point the finger at pragmatic features as where the fault shows up—on this line, Boolean algebras (or perhaps some subclass of algebras) provide a perfectly adequate model of the logical/semantic features of natural languages, for example. Relevant logicians have, generally speaking, rejected this, saying that classical logic, intuitionistic logic, and other logics permitting *irrelevant* arguments among their validities, are worse off *as logics* for these permissions. So they have, generally speaking, sought to replace such irrelevant logics with better, because relevant, systems.[4]

There are a variety of ways in which researchers in this field have understood *logic*, *argument*, and what-have-you, and for these reasons (along with others), there are a wide range of what the missing *relevance* of the irrelevant logics looks like. More or less broad agreement has coalesced around Nuel Belnap's VARIABLE-SHARING PROPERTY (VSP) as a necessary condition on a logic **L** being relevant:

(VSP) If A entails B (in **L**), then A and B must have a common atomic subformula (or 'a common propositional variable' in the old-school parlance).

One can see that VSP will cut against many well-known logics, such as classical logic and intuitionistic logic, both of which have $p \to (q \to q)$ as theorems. VSP, stated in the above schematic form, is quite broadly agreed upon as, at least, a good indicator of relevance. Indeed, it is not sufficient to reach any of the stronger relevant logics, such as **R**, from classical logic.

[4] We are presenting these logics as genuine rivals, but it is worth noting that there are philosophical alternatives. The logical pluralism of Beall and Restall [5] allows one to see ways in which these different logics are different options rather than rivals.

In his contribution (Part I, Chap. 4), Standefer proposes that more than a mere necessary condition, VSP may be seen as a necessary *and sufficient* condition on logics being relevant—so that VSP constitutes an actual *definition* of relevance for logics. Relevant logics, on this definition, include many logics stronger than **R**, as well as many that are incomparable with it.

While there is general agreement on VSP, there are more or less wide disagreements about how to understand what *entailment* is, and more importantly how it ought to be formally represented, and this leads to some downstream disagreements about what *logics* are. The most standard option, following on work by Anderson and Belnap [2], is to take logics as FMLA systems (in the terminology of Humberstone [26]), and to take *entailment* or *logical consequence* to be expressed in terms of the validity of implication formulas. So the statement of VSP above would have the first line written as something like $\vdash_{\mathbf{L}} A \to B$, for $\to$ a *connective*. A dissenting view which takes *derivability* in a proof system as the correct representation of logical consequence has long been championed by Tennant, who develops a unique take on the issue of how relevant logics, and relevance properties, can be developed.

In his contribution (Part III, Chap. 14), Tennant shows that the rules of his relevant system, of *Core Logic*, can be seen to preserve the important features of classical truth tables, while still preserving relevance in the way his system is known to do.

Brauer's contribution (Part I, Chap. 1) works in a Tennant-inspired setting, cataloging and investigating a range of relevant consequence relations definable in a sequent-style system.

It is possible, however, to define metalanguage-style consequence relations which capture the kind of formalisms presented by Anderson and Belnap.

Øgaard's second contribution (Part III, Chap. 12) investigates a range of consequence relations, building from the axiomatic presentations of some of the key Anderson and Belnap systems. He assesses the extent to which these capture the core intuitions behind relevant logics.

So even without getting into whether something like VSP is *correct*, we can see that there are already choice points available in response to the question of how these logics, and the salient notions of consequence associated with them, ought to be presented.

Furthermore, these decisions are tied into issues of whether relevant logics actually serve the purposes to which they are meant to be put—namely, giving an account of consequence which avoids issues with irrelevance. A range of arguments have been proposed to this effect over the years, though the most famous are likely those presented by Anderson and Belnap throughout *Entailment* [2].

Øgaard's first contribution (Part I, Chap. 3) assesses the two main proposals in Anderson and Belnap's work for how relevance should be understood – VSP and the USE criterion, which requires that in order for an argument to be valid, the premises must be *really used* in deriving the conclusion. Øgaard argues that neither of these provide an adequate motivation for relevant logics, and considers some options for how to proceed.

One potential way forward might come in the form of precisifying what VSP, at least, allows us to capture. An avenue that has gathered a bit of steam recently is that VSP demands shared *topic* between antecedents and consequents of valid arguments.[5]

Ferguson's contribution (Part I, Chap. 2) places relevant logics in the Anderson and Belnap tradition within the broader setting, including the systems of *analytic implication* developed by Parry and Angell, by reading VSP topically. This sees a range of non-classical logics as similarly engaged in a broadly topical project.

So far, this is all by way of reflection on the foundational issues in relevant logics, and the papers in this volume which touch on those issues explicitly. Now let us turn our attention to the remaining three parts of the volume, and what drives the kinds of investigations to be found there.

2.2 Model Theory

The first models of relevant logics were *matrices*, which are structures consisting of a (usually lattice-ordered) algebra with operations interpreting connectives like negation $\neg$ and implication $\rightarrow$, along with a set of *designated values*.[6] Initially, these were *ad hoc* structures, introduced entirely to prove things about some of the existing systems. For example, Ackermann [1] introduced a matrix in order to prove that his logic Π' (closely related to Anderson and Belnap's **E**) satisfied the property that formulas of the form $A \rightarrow (B \rightarrow C)$ were invalid if A had no occurrences of $\rightarrow$. Similarly, Belnap [6] used such a matrix argument to show that **E** satisfied VSP. After the introduction of *De Morgan monoids* by Dunn [16], the matricial/algebraic investigation of relevant logics, and their allies, was off to the races. This continues to be one of the major avenues of investigation for these systems.

Robles and Méndez's contribution (Part II, Chap. 9) investigates VSP in the context of logics which are sound and complete with respect to characteristic three-valued matrices.[7] In particular, they pinpoint one extension of the well-known three valued logic due to Kleene (the "Strong Kleene logic" as it is sometimes called) which includes an implication connective for which VSP holds.

Perhaps the signal development in the model theory of relevant logics is the discovery of a collection of similar classes of *relational* structures which can serve as models. As we noted above, different models were developed by Urquhart, Fine, and Routley and Meyer. As in the case of (classically-based) modal logic, these relational models have probably been the single most significant driver of technical advancement in the field of relevant logics. Of particular importance has been the

[5] Recent examples of papers include those by Øgaard [36] and Ferguson and Logan [20].

[6] For more information about the use of matrices in model theory, one may consult textbooks by Dunn and Hardegree [18] or Font [24].

[7] A matrix is characteristic w.r.t. a logic just in case every theorem of the logic is valid on the matrix, and no non-theorem is valid on the matrix. That is, all and only the theorems of the logic take designated values on every interpretation.

correspondence theory, which categorizes rules and axioms of some logic in terms of what properties a relational model must satisfy in order for those rules to preserve validity, or those axioms to be valid.

Conradie and Goranko's contribution (Part II, Chap. 5) builds on their previous paper [14] in developing the theory of *algorithmic correspondence*. This allows one to compute the frame correspondents for a range of axioms and rules, and provides a very powerful tool for studying relevant logics and their relational semantics.

Lying somewhere between the algebraic/matrix semantics and the relational semantics, *neighbourhood* semantics employs structures where formulas are interpreted at sets of points (as in relational semantics), but where some connectives are interpreted as *operations* on sets of points (as in the matrix semantics). Perhaps the most famous use of neighbourhood semantics in relevant logics is the use it has been put to, by Mares and Goldblatt [32], in interpreting (first order) quantifiers in relevant logics.

Ferenz and Tedder's contribution (Part II, Chap. 6) develops neighbourhood semantics for expansions of relevant logics including both *quantifiers* and *modalities*. This work provides a means for investigating principles which involve the interaction between quantifiers and modal expressions, such as the *Barcan* formulas, which they investigate here.

Ways to coherently handle quantified expressions are significant for applying relevant logics (as we'll see more of in our discussion of Part IV), and to that end having tools around for studying *higher-order* quantification is significant.

Logan's contribution (Part II, Chap. 7) investigates second-order quantification in a relational semantic setting. Building on Fine's [22] *stratified* models of quantifiers, Logan shows how to interpret second-order quantifiers, and provides the key metatheoretical results needed to get such an interpretation off the ground.

Another option for providing relational semantics for relevant logics, and related systems, builds on the *truthmaker* semantics investigated by Fine [23].

Punčochář and Ferguson's contribution (Part II, Chap. 8) investigates a relevant logic, RAC, which combines insights from the usual Anderson-Belnap tradition, as well as from the tradition of Analytic Containment logic, developed by Angell, Parry, and others. They provide a truthmaker semantics which includes, as the title suggests, machinery to interpret *implication* between propositions as well as *containment*, and defines a range of conditional operators in terms of this machinery. The resulting logic has significant connections to Urquhart's semilattice semantics, as well as more recent work on inquisitive logics, which incorporate questions into the usual logical vocabulary.

2.3 *Proof Theory*

In addition to spaces of models, the most common way of presenting logics is in terms of *proof systems*. Relevant logics were originally presented in terms of different proof systems by Anderson and Belnap: Hilbert-style axiomatic systems, sequent systems, and Fitch-style natural deduction. Tree-style natural deduction systems for some relevant logics were presented by Prawitz in his *Natural Deduction*

[38]. The Fitch-style systems of Anderson and Belnap used subscripted indices on formulas, to indicate which assumptions in a proof were used in their derivation.[8]

The systems that Anderson and Belnap used required a separate rule of distribution, from $A \wedge (B \vee C)_\alpha$ to infer $(A \wedge B) \vee (A \wedge C)_\alpha$, in addition to the introduction and elimination rules for conjunction and disjunction. A rule of distribution is not required in standard natural deduction and sequent systems for classical and intuitionistic logics. The reason the rule is needed in the context of Anderson and Belnap's systems is that there is a distinction that must be drawn between the *additive* and *multiplicative* connectives, sometimes called *extensional* and *intensional*, respectively. The standard additive connectives are conjunction and disjunction while the standard multiplicative connectives are implication and fusion.

It is perhaps easiest to illustrate the difference with conjunction and implication. In order to use $\wedge I$ to obtain $A \wedge B_\alpha$, the premises of the rule, A_α and B_α, must depend on the same assumptions, as indicated by their shared indices. On the other hand, no such restriction is placed on $\rightarrow E$, permitting the premises, $A \rightarrow B_\alpha$ and A_β, to depend on different assumptions, yielding the conclusion $B_{\alpha\cup\beta}$. A distinction between the two types of connectives is needed in order to avoid re-introducing irrelevancies into the proof system. Development of proof systems that adequately deal with the distinction between additive and multiplicative connectives while also having proof-theoretically nice features, such as normalization or cut admissibility, has been the source of much research in proof theory for relevant logics.

One of the big breakthrough in the area was the development of sequent systems with two (or more) different kinds of structure. This was first developed by Mints [33] and Belnap et al. [8]. In these kinds of sequent systems, one has different structures that can be built up out of formulas. The different structures obey different structural rules, so that one can have one sort of structure that can obey a structural rule of weakening and another sort of structure that does not obey weakening. The former is connected to additive connectives and the latter to multiplicative connectives. The use of multiple structures has been exploited to obtain illuminating results, by, for example, Slaney [44], Read [39] and Belnap [7], among others. Apart from the preceding, the interested reader should consult Bimbó [11] for some details on sequent systems for relevant logics as well as De Martin Polo's contribution to this volume, in particular §7 which contains a discussion of many alternative proof systems for relevant logics.

Interest in developing proof systems for relevant logics is an ongoing task. New methodologies and techniques have been developed and alternative analyses of existing systems are often fruitful. One technique that has been used to great effect is labelled sequent systems. Labelled sequent systems typically use labelled formulas together with non-formula additional elements that are incorporated into the rules. These additional elements provide additional flexibility in formulating rules that can result in proof systems for a wide range of logics. Kurokawa and Negri [27] apply

[8] While Anderson and Belnap were interested in systems for the logics **R**, **E**, and **T**, their methodology has been generalized to many other relevant logics by Brady [12].

this methodology to relevant logics, using proof systems with ternary-relational statements to provide proof systems for a large range of relevant logics.

De Martin Polo's contribution (Part III, Chap. 11) investigates multiplicative connectives of relevant logics using the framework of labelled sequent systems. He obtains cut admissibility results as well as soundness and completeness results, further demonstrating the utility of labelled sequent systems. His paper closes with an overview of proof-theoretic frameworks to relevant logics.

Labelled proof systems can be adapted to capture the logics of many different sorts of models, and this technique is brought to bear on one of the main approaches to modeling negation in relevant logics. There are two main approaches to negation in relevant logics, the *American* plan and the *Australian* plan. The American plan, recently defended by De and Omori [15], takes two truth values, true and false, as primitive and interprets negation as toggling back and forth between them. The Australian plan, recently defended by Berto and Restall [9], interprets negation using an accessibility relation, or incompatibility relation, on points, building on work of Dunn [17], Restall [40], and others. The incompatibility relation can be seen as a generalization of the Routley star, due to Routley and Routley [43],[9] an operation on points which has often been used to model negation in relevant logics. The American and Australian plans have formal and philosophical differences, and research on them is ongoing.

Restall, in his contribution (Part III, Chap. 13), takes a proof-theoretic approach to the Australian plan. In particular, Restall explores versions of labelled natural deduction proof systems where labels, called contexts, admit the incompatibility relation and the star operation, and where both assertion and denial of statements within proofs is allowed.

One of the difficult questions that arise in the context of first-order relevant logics is how to regiment restricted quantification. In the setting of classical logic, $\mathsf{All}\,x[A, B]$ can be rendered as $\forall x(A \to B)$, and $\mathsf{Some}\,x[A, B]$ can be rendered as $\exists x(A \wedge B)$. Adopting these renderings in a relevant logic leads to the invalidity of many inferences involving restricted quantification that one might want. As Beall et al. [4] show, maintaining all the standard restricted quantifier inferences leads to trouble for the relevant logician. They offer an alternative, but it is not the last word on restricted quantification.

Brady, in his contribution (Part III, Chap. 10) proposes a new option for restricted quantification in relevant logics. After surveying three extant proposals, Brady offers his new proposal in an axiomatic setting and investigates some of its consequences and features.

[9] Later in life, Val Plumwood changed her surname from Routley, under which her early articles were published.

2.4 Applications

From the beginning relevant logicians have been motivated to define notions of consequence which avoid irrelevant consequences not just because they are intrinsically undesirable, but because doing so has been seen as significant for applications of logic. The canonical applications of logic involve *closing extra-logical theories* under consequence, to discover what are the commitments of those theories.[10] In order for this sort of enterprise to work, the theories in question must be *formalised* (or, in a Quinean idiom, *regimented*) so that they can be characterised in the language of the logic as *proper axioms*. The first such theories were mathematical in nature, and the study of such theories in classical logic provided the grist for the most significant applications of logic (e.g., Gödel's incompleteness theorems, the Löwenheim-Skolem theorem, and the independence of the continuum hypothesis).

While applications of relevant logics to mathematical theories such as these has not given rise to *such* spectacular results, there has been a steady output of results in this vein in the field since, at least, the 1970s. Two of the papers in this volume comprise such results. One is very much in line with the sort of work that the author, and collaborators, have been doing in this area for a while.

Slaney's contribution (Part IV, Chap. 17) investigates theories of *rational numbers* in the context of the relevant logic **R**. The resulting systems, extending the natural number arithmetic usually constructed on this logic ($\mathbf{R}^{\#}$), differ in subtle an interesting ways. This indicates an instance of a common theme in relevant logics, and theories constructed thereon: the sensitivity of the relevant logical vocabulary allows for a range of significantly different theories to be constructed by taking subtly different 'wordings' of the key axioms. Slaney investigates these options, and makes some suggestions for going forward.

Setting variations of classical mathematical theories in expansions of relevant logics by first-order quantifiers is the usual way things proceed. There is, however, an alternate way to go, influenced by the *abstractionism* program, inspired by Frege's treatment of mathematical theorizing. The basic plan of abstractionism is to try to derive large fragments of mathematics from logic combined with certain abstraction principles—quantified biconditionals which serve to *reduce* mathematical concepts to logical ones. These biconditionals are usually quantified by second-order quantifiers, allowing for significant expressive resources in the reduction program.

Logan and Boccuni's contribution (Part IV, Chap. 15) investigates abstraction principles in the context of relevant logics expanded by second-order quantifiers. They show that the axioms of second-order Peano Arithmetic are derivable from Frege's Basic Law V in the context of relevant logics. They show this using the non-contractive relevant logic DL.

---

[10] This is such a significant application in relevant logics that some, such as Logan [28], have sought to interpret the relational semantics *entirely* in terms of theories, and operations/relations thereon.

Mathematical theories are, however, not the only important application of relevant logics. The sensitivity of relevant connectives is also significant for applications to operators which draw finer distinctions than are usually available in the extensional idiom of classical logic, or even the intensional idiom of classical logic expanded by modal operators. The *hyperintensional* flavour of relevant logics are useful in areas such as epistemic or deontic logics, where very fine-grained distinctions can matter for assessing the truth values of claims.

Sedlár and Vigiani's contribution (Part IV, Chap. 16) investigates the results of adding strict implication and a classicality constant to the logic of First Degree Entailment (a four-valued, implicationless, logic which underwrites most of the usual systems with implication). The resulting logic allows them to define modal operators representing explicit and implicit belief, respectively. They show that Levesque's FDE-based logic of implicit and explicit belief embeds into their system.

These applications only scratch the surface of what is possible, but provide insights into the kinds of areas where relevant logics can provide a valuable perspective on the sorts of theories to which logics have fruitfully been applied.

During the preparation of this volume, Nuel Belnap passed away. Belnap was a giant in the field of relevant logics, as well as a major figure in philosophical logic more generally. He will be missed by many of us working in the field.

Acknowledgments We would like to thank the two anonymous referees for feedback on this introduction as well as the volume as whole. We would also like to thank Greg Restall for feedback on this introduction. Shawn Standefer's work was supported by research grant 111-2410-H-002–006-MY3 from the National Science and Technology Council of Taiwan.

References

1. Ackermann, W. (1956). Begründung einer Strengen Implikation. *Journal of Symbolic Logic*, *21*(2), 113–128.
2. Anderson, A. R., & Belnap, N. D. (1975). *Entailment: The logic of relevance and necessity* (Vol. 1). Princeton University Press.
3. Anderson, A. R., Belnap, N. D., & Michael Dunn, J. (1992). *Entaillment: The logic of relevance and necessity* (Vol. 2). Princeton University Press.
4. Beall, J., Brady, R. T., Hazen, A. P., Priest, G., & Restall, G. (2006). Relevant restricted quantification. *Journal of Philosophical Logic*, *35*(6), 587–598.
5. Beall, J., & Restall, G. (2006). *Logical pluralism*. Oxford University Press.
6. Belnap, N. D. (1959). *The formalization of entailment*. Ph.D. Dissertation, Yale University.
7. Belnap, N. D. (1982). Display logic. *Journal of Philosophical Logic*, *11*(4), 475–217.
8. Belnap, N. D., Gupta, A., & J. Michael Dunn. (1980). A consecution calculus for positive implication with necessity. *Journal of Philosophical Logic*, *9*, 343–362.
9. Berto, F., & Restall, G. (2019). Negation on the Australian plan. *Journal of Philosophical Logic*, *48*(6), 1119–1144.
10. Bimbó, K. (2007). Relevance logics. In D. Jacquette (Ed.), *Philosophy of logic* (pp. 723–789). Elsevier.
11. Bimbó, K. (2014). *Proof theory: Sequent calculi and related formalisms*. CRC Press.
12. Brady, R. T. (1984). Natural deduction systems for some quantified relevant logics. *Logique Et Analyse*, *27*(8), 355–377.

13. Church, A. (1951). The weak theory of implication. In A. Menne (Ed.), *Kontrolliertes Denken: Untersuchungen zum Logikkalkül und zur Logik der Einzelwissenschaften* (pp. 22–37). K. Alber.
14. Conradie, W., & Goranko, V. (2022). Algorithmic correspondence for relevance logics I. The algorithm PEARL. In I. Düntsch & E. Mares (Eds.), *Alasdair Urquhart on nonclassical and algebraic logic and complexity of proofs* (pp. 163–209). Springer.
15. De, M., & Omori, H. (2018). There is more to negation than modality. *Journal of Philosophical Logic*, *47*(2), 281–299.
16. Dunn, J. M. (1966). *The algebra of intensional logics*. Ph.D. Dissertation, University of Pittsburgh.
17. Dunn, J. M. (1993). Star and Perp: Two treatments of negation. *Philosophical Perspectives*, *7*, 331–357.
18. Dunn, J. M., & Hardegree, G. (2001). *Algebraic methods in philosophical logic*. Oxford University Press.
19. Dunn, J. M., & Restall, G. (2002). Relevance logic. In D. M. Gabbay & F. Guenthner (Eds.), *Handbook of philosophical logic* (2nd ed., pp. 1–136). Kluwer.
20. Ferguson, T. M., & Logan, S. A. (2023). Topic transparency and variable sharing in weak relevant logics. *Erkenntnis*. https://doi.org/10.1007/s10670-023-00748-6
21. Fine, K. (1974). Models for entailment. *Journal of Philosophical Logic*, *3*, 347–372.
22. Fine, K. (1988). Semantics for quantified relevance logic. *Journal of Philosophical Logic*, *17*, 27–59.
23. Fine, K. (1997). Truthmaker semantics. In B. Hale, C. Wright, & A. Miller (Eds.), *A companion to the philosophy of language* (pp. 556–577). Wiley-Blackwell.
24. Font, J. M. (2016). *Abstract algebraic logic: An introductory textbook*. College Publications.
25. Grice, H. P. (1975). Logic and conversation. In D. Davidson (Ed.), *The logic of grammar* (pp. 64–75). Dickinson Pub. Co.
26. Humberstone, L. (2011). *The connectives*. MIT Press.
27. Kurokawa, H., & Negri, S. (2020). Relevant logics: from semantics to proof systems. In A. Rezus (Ed.), *Contemporary logic and computing* (pp. 259–306). College publications.
28. Logan, S. A. (2020). Deep fried logic. *Erkenntnis*, *87*(1), 257–286.
29. Logan, S. A. (2024). *Relevance logic*. Cambridge University Press.
30. Mares, E. D. (2004). *Relevant logic: A philosophical interpretation*. Cambridge University Press.
31. Mares, E. D. (2024). *The logic of entailment and its history*. Cambridge University Press.
32. Mares, E. D., & Goldblatt, R. (2006). An alternative semantics for quantified relevance logic. *Journal of Symbolic Logic*, *71*(1), 163–187.
33. Mints, G. E. (1976). Cut-elimination theorem in relevant logics. *The Journal of Soviet Mathematics*, *6*, 422–428. Translated from Zapiski Nauchnykh Seminarov Leningradskogo Otdeleniya Matematicheskogo Instituta im V. A. Steklova Akademii Nauk SSSR, 32:90–97 (1972).
34. Moh, S.-K. (1950). The deduction theorems and two new logical systems. *Methodos*, *2*, 56–75.
35. Norman, J., & Sylvan, R. (1989). *Directions in relevant logic*. Kluwer Academic Publishers.
36. Øgaard, T. F. (2023). Relevance through topical unconnectedness: Ackermann and Plumwood's motivational ideas on entailment. *Australasian Journal of Logic*, *20*(2), 154–187.
37. Orlov, I. (2024). The calculus of the compatibility of propositions. Translated by Werner Stelzner. *Australasian Journal of Logic*, *21*(1), 20–46.
38. Prawitz, D. (1965). *Natural deduction: A proof-theoretical study*. Almqvist and Wicksell.
39. Read, S. (1988). *Relevant logic: A philosophical examination of inference*. Blackwell.
40. Restall, G. (1999). Negation in relevant logics: How I stopped worrying and learned to love the Routley Star. In D. Gabbay & H. Wansing (Eds.), *What is negation?* (pp. 53–76). Kluwer Academic Publishers.
41. Routley, R., & Meyer, R. K. (1972). The semantics of entailment. In H. Leblanc (Ed.), *Truth, syntax and modality* (pp. 194–243). North Holland.

42. Routley, R., Plumwood, V., Meyer, R. K., & Brady, R. T. (1982). *Relevant logics and their rivals* (Vol. 1). Ridgeview.
43. Routley, R., & Routley, V. (1972). The semantics of first degree entailment. *Noûs*, *6*(4), 335–359.
44. Slaney, J. (1990). A general logic. *Australasian Journal of Philosophy*, *68*(1), 74–88.
45. Standefer, S. (2024). Routes to relevance: Philosophies of relevant logics. *Philosophy Compass*, *19*(2), e12965.
46. Tennant, N. (2017). *Core logic*. Oxford University Press.
47. Urquhart, A. (1972). Semantics for relevant logics. *Journal of Symbolic Logic*, *37*(1), 159–169.

Part I
Philosophical Foundations

A Hierarchy of Relevance Properties

Ethan Brauer

Abstract Being relevant to some topic can be informally understood as making a difference or having something to contribute. Given a sequent $\Delta \Rightarrow \Gamma$, we can say, somewhat schematically, that a component of the sequent is relevant to the sequent when it contributes to the validity of the sequent. Different ways of making precise the idea of *contributing to the validity* and different understandings of the *components* of a sequent lead to a hierarchy of explications of relevance. I identify four key explications, called gaunt validity, perfect validity, relevant validity, and perfectibility. Each is shown to enjoy an interesting variable sharing property. Furthermore, if we begin with a standard sequent calculus for classical logic and introduce some simple constraints on the rules, the result is a fragment of classical logic that proves exactly the gauntly valid sequents.

1 Introduction

In elementary logic and critical thinking, it is common to encounter talk of *relevant* and *irrelevant* premises and conclusions. For instance, fallacies such as an *ad hominem* or an appeal to force are often categorized as fallacies of relevance, which Hurley [10, 125] describes as "shar[ing] the common characteristic that the arguments in which they occur have premises that are logically irrelevant to the conclusion." Similarly, when Lau [14, 107–108] lists four criteria for a good argument, the fourth criterion is that "The premises are all relevant to the conclusion." As a final example, in illustrating how an inconsistent premise set entails everything, Copi et al. [7, 415] write that, "[these] openly inconsistent premises allow us to infer—validly!—an irrelevant and absurd conclusion."

Moreover, when writers on critical thinking and informal logic talk about the (ir)relevance of premises or conclusions, they are not introducing specialized concepts foreign to common sense. The idea that an ad hominem attack is simply

E. Brauer (✉)
University of Oslo, Oslo, Norway

I. Sedlár et al. (eds.), *New Directions in Relevant Logic*, Trends in Logic 63,
https://doi.org/10.1007/978-3-031-69940-5_2

irrelevant to the actual point of an argument is a natural observation to anyone who thinks carefully about reasoning or debating.[1]

Given, then, that there is an intuitive notion of relevance that has a role to play in studying informal logic, we can also ask whether these intuitive ideas about relevance can be made precise and applied within *formal* logic. Is there a robust notion of relevance that can figure in symbolic logic? I believe there is, and in this paper I will present several related ways to formally explicate the concept of relevance and show how they suggest interesting new topics of study, especially in proof theory.

1.1 The Goal

The overall goal of this project is to provide a formally rigorous explication of *relevance* that can be studied in symbolic logic. This gives two desiderata for the success of this project.

First, to make good on the idea that we have explicated the informal notion of relevance, our formal account of relevance should be explicitly linked to a recognizable, intuitive notion of relevance. We started with the observation that there is an intuitive concept of relevance. In the process of explication, we seek to find or develop a formally precise counterpart to that intuitive concept; but we are beholden to the intuitive concept as a measure of success. This is what separates an explication from a stipulative definition.

This is not to say that there must be a *unique* correct explication of relevance. Informal notions are often vague and multifaceted, containing several related but distinct aspects. There may be several plausible intuitive ideas that lie at the core of the conception, so that one has to pick one of these to work with and sharpen; and even once one has picked a single idea as starting point, there may be further choice points as one starts to sharpen that idea, because the idea contains ambiguity, open texture, or other indeterminacy. Indeed, I will suggest that exactly this happens with the concept of relevance. The choice points we face in explicating the concept can lead to several distinct formal definitions which each have some claim to formally explicating the concept of relevance. Thus, I will suggest, one component of studying relevance in logic is a comparative study of these different formal definitions and how they relate to one another.

But even when we have a multiplicity of formal definitions explicating an intuitive concept, it makes sense to ask that each be faithful to the original concept. We cannot necessarily ask that any of them be the unique best way to explicate the informal concept, but we can require that there be a clear line to each of them from a plausible intuitive understanding of the original concept.

[1] Of course, it can be difficult to say exactly when an ad hominem attack amounts to a fallacy and when it actually reveals a bias of the interlocutor that matters for the argument.

The second desideratum for the success of this project is that the explicated notion of relevance has a place in formal logic. That is, the notion should be of clear *logical* interest. If, hypothetically, the only plausible explications of relevance turned on concepts from linguistic pragmatics, then the notion of relevance might be best left to linguists, rather than having something to contribute to the study of formal logic.

In this paper, I will address both of these desiderata, although the emphasis will be more on the second. Beginning from an intuitive concept of relevance (described in the next section), I will argue that there are several plausible and interesting ways of explicating this intuitive concept. This suggests a *comparative study* of relevance, which we can summarize in three broad questions:

1. How do the different explications of relevance relate to one another?
   - For instance, are they arranged in a hierarchy? Is one explication a limiting case of another?
2. Under each of these explications of relevance, what intrinsic properties do relevant arguments exhibit?
   - For instance, do they exhibit variable sharing properties? What about interpolation or disjunction properties?
3. What relations, if any, exist between each of these explications of relevance, and familiar proof systems or fragments thereof?
   - For instance, are there known proof systems that prove only arguments satisfying one of our explications of relevance?

I will provide some answers to each of these questions, but their open-ended nature means that there is still lots to explore. Concerning the first question, I will observe how the different explications of relevance can be arranged in a simple hierarchy. For the second question, I will show that each explication of relevance enjoys a non-trivial variable sharing property. The strongest explication, in particular, enjoys a very demanding variable-sharing property, which has not appeared in the literature before. Finally, for the third question, I will show that if we take a standard sequent calculus for classical logic and impose some simple restrictions on the application of its rules, the result is a system that proves all and only the sequents satisfying the strongest explication of relevance.

While these results are all proved by elementary methods, they are nevertheless interesting and illustrate the potential fruitfulness of studying various explications of relevance along the lines advanced here. Accordingly, part of my goal in this paper is simply to advocate for this comparative study of relevance. In a slogan, rather than studying logics *of* relevance, I want to advocate for studying relevance *within* a logic.

2 Varieties of Relevance Properties

As I suggested above, an important desideratum for a good explication is that it be clearly connected to the intuitive concept being explicated. Let us begin by identifying the intuitive idea we want to explicate.

2.1 The Intuitive Starting Point

In order to explicate relevance, we need a plausible, intuitive statement of what relevance is as our starting point for the work of clarifying, disambiguating, and precisifying. I will take as a starting point the idea that to be relevant is to *make a difference* or *have something to contribute.*

I think this idea of relevance has intrinsic plausibility as an intuitive conception of relevance. Moreover, the same basic conception of relevance appears elsewhere in philosophy. For instance, in theories of confirmation, Bayesians commonly define some evidence E to be irrelevant to some hypothesis H if E *makes no difference* to the probability of H, i.e., if $Pr(H|E) = Pr(H)$.[2] Or, to give this another gloss, in order to be relevant, the evidence E must *contribute something* to determining the probability of H.

Similarly, when it comes to causation, Yablo [31, 321] observes that:

> Almost everyone's first thought about causal relevance is:
>
> (CT) A property P of x is causally relevant to effect y iff y would not have ensued, had x occurred without P.
>
> [...] Had I not seen that water, I would not now be drinking. So, according to [(CT)], my seeing the water was causally relevant. It cannot be said, in most cases, that I would not be drinking had I not earlier been in a rather precise neural condition. So my precise neural condition is not causally relevant.[3]

The presence of P is what *makes a difference* to y occurring, or *contributes* to y's coming about, and that is what makes P relevant to y. The fact that I saw water contributed to, or made a difference to, the fact that I later drank water. By contrast, the *precise* neural state I was in when I perceived the water was not relevant—it made no difference whether I was in that specific neural state or a different neural state consistent with perceiving the water.

Similar ideas can be found elsewhere. For instance, Krämer and Roski [13] offer a theory of difference making grounds as part of an account of (grounding-)explanatory relevance. Recently, Schurz [17] has developed a very general account

---

[2] See [9], who tries to improve on this definition, while maintaining the essential idea. See also [12], who critiques and modifies Gemes' account, but still building on the same essential idea.

[3] Note that Yablo rejects the thesis (CT). But what matters for our purposes here is that it is a common starting point that faithfully captures several intuitive judgments of relevance/irrelevance.

of relevance as difference-making. See also [32] for a general discussion of relevance as difference making, including several other examples of theories of relevance as difference making.

The fact that relevance is understood as *making a difference* or *contributing something* in such a wide range of topics suggests that we are onto a familiar concept with some degree of robustness. This lends a good deal of plausibility to the idea of relevance as *making a difference* or *making a contribution* for the starting point of our formal explication.

To take the first step from this starting point, we can observe that making a difference means making a difference *to* something. In theories of confirmation, we wanted evidence to make a difference to the probability of the hypothesis. In causation, we wanted a genuine cause to contribute to producing the effect. In logic, what feature do we want to make a difference to? An obvious answer is that we want to make a difference to the strength of an argument. Take, for example, the following argument that [14] uses to illustrate irrelevance:

1. Albert Einstein was a physicist.
2. All physicists studied mathematics.
3. Albert Einstein played the violin.

---

4. Therefore, Albert Einstein studied mathematics.

The third premise is clearly irrelevant. But why? Lau's explanation is that: "If we remove this particular premise, it does not affect the strength of the argument at all" [14, 108].

So, we have as our initial starting point the idea that being relevant is making a difference to or making a contribution to *something*; and we add to this the idea that in logic, the something in question is *the strength of the argument*.

Furthermore, we are interested in how this idea applies in *formal* logic, and in formal logic the standard of a *strong* argument gets replaced by the standard of a *valid* argument. Thus, a natural way to understand relevance in formal logic is to say that a component of an argument is relevant to that argument when it contributes something to the validity of the argument. This raises two key questions: which components of an argument are in question here? And what does it take to 'contribute something' to the validity of an argument? These two questions are where we face decision points in our explication. Different answers to these two questions will lead to the different relevance properties studied here. But before we can go further, we need to pin down some technicalities.

2.2 Technical Preliminaries

In the remainder of this section I will introduce a number of precise ways to explicate the idea of relevance as contributing to validity. Because relevance is defined in terms of validity, this presupposes a background logic. I am not engaged in logical reform or advocating any alternative system of logic. For my purposes, I will take classical logic to be the background logic; but the key ideas are schematic, and could be pursued against different background logics. The choice of classical logic partly reflects my personal interests, but it is also interesting because relevance has historically figured in revisionist projects advocating non-classical logics.[4] Focusing on relevance in classical logic is a way to counterbalance that tendency, and, I hope, to help bring relevance into the mainstream of logic. At the same time, I want to emphasize that the general approach to relevance embodied here is not limited to classical logic. A fertile area of research is to develop these ideas against different background logics.[5]

I will focus on propositional logic, because the main interesting questions apply even at this simple level.

To move toward our formal definitions of relevance, informal talk of arguments will be replaced by talk of *sequents*. Because I am focusing on classical logic, and in particular taking for granted a semantic characterization of classical validity, I will regard sequents as pairs of *sets* of sentences, written $\Delta \Rightarrow \Gamma$. Note that any pair of sets of sentences counts as a sequent—but of course they will not all be valid. The sequent $\Delta \Rightarrow \Gamma$ is *valid* when there is no truth assignment that makes every member of Δ true and every member of Γ false.

With these preliminaries in place, the schematic explication of relevance is that a sequent $\Delta \Rightarrow \Gamma$ is relevant when every component of the sequent contributes to the validity of the sequent. To turn this schema into a proper definition, we need to answer two questions. (1) Which *components* of the sequent are in question? (2) What counts as *contributing* to the validity of the sequent?

2.3 Perfect Validity

The first explication of relevance provides these answers to our two questions:

- The components of a sequent that do/do not contribute to its validity are its constituent sentences.
- To count as contributing to the validity of the sequent, each sentence in that sequent must be *essential* for its validity.

---

[4] This is so both in the original Anderson-Belnap tradition [1, 2] and in alternative approaches to relevance such as Tennant's proof-theoretic approach to entailment [20]. Recent work on relevance logic, however, has tended to take a more ecumenical attitude, partly influenced by the pluralist, naturalist, and anti-exceptionalist views in the philosophy of logic advocated by Beall and Restall [3], Shapiro [18], Williamson [30], and others.

[5] Although their work has been developed and motivated independently of mine, Peter Verdée and his collaborators have pursued a similar approach to relevance against non-classical background logics in [28].

This leads us to the formal notion of *perfect validity*. Say that $\Delta' \Rightarrow \Gamma'$ is a subsequent of $\Delta \Rightarrow \Gamma$ when $\Delta' \subseteq \Delta$ and $\Gamma' \subseteq \Gamma$, and if either of these inclusions is proper then it a proper subsequent.

Definition 1 A sequent is *perfectly valid* when it is valid, but no proper subsequent is valid.

For instance, $P, Q \rightarrow R, Q \Rightarrow R$ is not perfectly valid because it has the valid proper subsequent $Q \rightarrow R, Q \Rightarrow R$. And $A, \neg A \Rightarrow B$ is not perfectly valid because it has the valid proper subsequent $A, \neg A \Rightarrow \emptyset$. These and similar examples cover many of the intuitively 'irrelevant' arguments that motivate the remarks of [14] and [7] quoted in the introduction. This provides some assurance that we are on the right track in formally explicating the intuitive concept.

2.4 Redundancy and Relevant Validity

Perfect validity is a reasonable explication of relevance, but it is also a fairly strong notion; and the strength of the notion rules out some arguments that are, in one sense, perfectly legitimate by relevantist lights. For instance,

$$A, B, (A \vee B) \rightarrow C \Rightarrow C$$

is not perfectly valid, because it has the two valid proper subsequents

$$A, (A \vee B) \rightarrow C \Rightarrow C \text{ and } B, (A \vee B) \rightarrow C \Rightarrow C.$$

Nevertheless, there is a clear sense in which every sentence in this sequent contributes to its validity. After all, if A did not contribute to the validity of the argument, then by symmetry neither should B. But if neither contributes anything to the validity of the argument, then we should be able to delete them both and still be left with a valid sequent. Since this is not possible, we must conclude that both A and B do contribute something to the validity of the sequent. In this example, both A and B provide some information that contributes to the validity of the sequent, but what each one contributes is *redundant* with the other.

Somewhat more generally, this scenario can happen not just with single premises being redundant with each other, but also subsequents can be redundant with each other. For instance, consider the sequent:

$$A, A \rightarrow B, A \rightarrow C \Rightarrow B, C.$$

And consider the two subsequents:

$$A \rightarrow B \Rightarrow B \text{ and } A \rightarrow C \Rightarrow C.$$

Neither of these subsequents is essential for the validity of the overall sequent, since either of them could be deleted without rendering the main sequent invalid. But each of these subsequents does contribute something to the validity of the overall sequent. After all, if we deleted them both, we would be left with the invalid sequent $A \Rightarrow \emptyset$.

The second explication of relevance still focuses on sentences as the components that contribute to the validity of a sequent, but it takes a step back from the strength of perfect validity, recognizing that redundant sentences can still contribute to the validity of a sequent, even though—given that they are redundant—they are not *essential* for the validity of the sequent.

Definition 2 In a valid sequent $\Delta \Rightarrow \Gamma$, two subsequents $\Delta_1 \Rightarrow \Gamma_1$ and $\Delta_2 \Rightarrow \Gamma_2$ are *redundant* when $\Delta \setminus \Delta_1 \Rightarrow \Gamma \setminus \Gamma_1$ is valid and $\Delta \setminus \Delta_2 \Rightarrow \Gamma \setminus \Gamma_2$ is valid, but $\Delta \setminus (\Delta_1 \cup \Delta_2) \Rightarrow \Gamma \setminus (\Gamma_1 \cup \Gamma_2)$ is not valid.

Informally, the idea is that two subsequents are redundant when you can delete either one of them and still be left with a valid sequent, but deleting both of them leaves you with an invalid sequent. At least one of them is needed for the sequent to be valid, but given either one, you don't need the other. We can recognize as relevantly acceptable sequents which are either perfectly valid or which fail to be perfectly valid only by including redundant subsequents.

Definition 3 A sequent $\Delta \Rightarrow \Gamma$ is *relevantly valid* when it is valid and for any subsequent $\Delta_1 \Rightarrow \Gamma_1$, if $\Delta \setminus \Delta_1 \Rightarrow \Gamma \setminus \Gamma_1$ is valid, then there is another subsequent $\Delta_2 \Rightarrow \Gamma_2$ which is redundant with $\Delta_1 \Rightarrow \Gamma_1$.

For example, although $A, B \Rightarrow A \vee B$ is not perfectly valid, it is relevantly valid in this new sense. This is because, although you can delete either of the premises A or B and still have a valid sequent, you cannot delete them both. The two subsequents $A \Rightarrow \emptyset$ and $B \Rightarrow \emptyset$ are redundant with each other.

On the other hand, $A, B, \neg B \vee C \Rightarrow C$ is not relevantly valid, because the subsequent $B, \neg B \vee C \Rightarrow C$ is valid, while the subsequent $A \Rightarrow \emptyset$ is not redundant with any other subsequent.

As it turns out, there is an equivalent characterization of relevant validity that does not explicitly mention redundance.

Proposition 4 *A sequent* $\Delta \Rightarrow \Gamma$ *is relevantly valid iff for every* $\phi \in \Delta \cup \Gamma$ *there is a perfectly valid subsequent* $\Delta' \Rightarrow \Gamma'$ *such that* $\phi \in \Delta' \cup \Gamma'$.[6]

This shows that we may alternatively gloss relevant validity as meaning that every sentence in a given sequent is a member of a minimally valid subsequent.

[6] This is not difficult to establish, but a proof may be found in [4].

2.5 Gaunt Validity

Relevant validity is a weakening of perfect validity based on a more inclusive answer to what counts as *contributing* to the validity of a sequent. Gaunt validity, on the other hand, is a strengthening of perfect validity based on a more inclusive conception of which *components* of a sequent can contribute to its validity. For this explication of relevance, we answer the two key questions as follows.

- The components of a sequent that do/do not contribute to its validity are its constituent sentences *and* their internal syntactic structure.
- To count as contributing to the validity of the sequent, each constituent sentence and all of its syntactic structure must be *essential* to its validity.

Consider the sequent $A \wedge B \Rightarrow A \wedge B$. It is perfectly valid, but in order to recognize its validity we did not have to consider the internal structure of the conjunction. Instead, the validity of this sequent is guaranteed by the fact that it is a substitution instance of $C \Rightarrow C$. We might say that the internal syntactic structure of $A \wedge B$ was irrelevant, as revealed by the fact that we could, without loss, focus on the 'coarser' sequent $C \Rightarrow C$.

Definition 5 A sequent $\Delta' \Rightarrow \Gamma'$ is a *coarsening* of a sequent $\Delta \Rightarrow \Gamma$ when there is a uniform substitution σ such that $\sigma(\Delta') = \Delta$ and $\sigma(\Gamma') = \Gamma$.

A coarsening will be considered trivial iff it is obtained by a one-one mapping of atoms to atoms, i.e., if the substitution in question is a mere relettering. Note that even when σ is a mapping of atoms to atoms, a coarsening under σ will still be considered non-trivial if σ is not one-one on the atoms occurring in the sequent.

Definition 6 A sequent $\Delta \Rightarrow \Gamma$ is *gauntly valid* when it is perfectly valid and has no non-trivial coarsening that is perfectly valid.

The decision to count many-one mappings of atoms to atoms as non-trivial is significant. The motivation for this decision is that whether a sequent is valid or not depends on how different atoms are combined into truth-functional sentences and then how those sentences are combined into sequents. It matters not only how atoms are combined into sentences, but also where those atoms are repeated in the sequent. Thus, the repetition of atoms within a sequent is an important syntactic feature that can contribute to the validity of the sequent. For instance, consider the sequent:

$$A \wedge B, A \rightarrow (B \wedge C) \Rightarrow C.$$

The fact that C is repeated from the premise to the conclusion is essential for the validity of the sequent. If these two occurrences of C are coarsened to different atoms, the result

$$A \wedge B, A \rightarrow (B \wedge D) \Rightarrow E,$$

is *not* valid. By contrast, the repetition of B in the premise set is *inessential* for the validity of the sequent, because the coarsening:

$$A \wedge D, A \rightarrow (E \wedge C) \Rightarrow C$$

is still valid. As this example reveals, the notion of gaunt validity is very strong. Here are a few more examples of sequents with valid coarsenings to illustrate the concept.

The sequent	$A \wedge (B \vee C),\ A \rightarrow (C \vee D)\ \Rightarrow\ C \vee D$
coarsens to	$A \wedge P,\ A \rightarrow Q\ \Rightarrow\ Q$
under:	$P \mapsto B \vee C,\ \ Q \mapsto C \vee D$
The sequent	$A \wedge B,\ C \wedge B,\ A \wedge C \rightarrow D \Rightarrow\ D$
coarsens to	$A \wedge E,\ C \wedge F,\ A \wedge C \rightarrow D\ \Rightarrow\ D$
under:	$E \mapsto B,\ \ F \mapsto B$
The sequent	$A \vee B\ \Rightarrow\ B \vee A$
coarsens to	$A \vee C,\ D \vee B\ \Rightarrow\ B \vee A,\ C \vee D$
under:	$C \mapsto B,\ \ D \mapsto A$
The sequent	$A \wedge (B \wedge C)\ \Rightarrow\ (A \wedge B) \wedge C$
coarsens to	$A \wedge D,\ E \wedge (B \wedge F),\ G \wedge (H \wedge C)\ \Rightarrow\ (A \wedge B) \wedge C$
under:	$D \mapsto B \wedge C,\ E \mapsto A,\ F \mapsto C,\ G \mapsto A,\ H \mapsto B$

These last two examples also show how a gauntly valid coarsening of a given sequent can be much longer than the original sequent.

2.6 Perfectibility

Where gaunt validity is a strengthening of perfect validity based on coarsenings, our final explication of relevance will be a weakening of perfect validity based on substitutions. On this view, what it takes to contribute to the validity of a sequent is simply to instantiate a sequent form in which every piece of syntactic structure is essential. It is okay if there is extraneous structure to the sequent, provided it fits within a coarser-grained structure where nothing is extraneous.

Definition 7 A sequent $\Delta \Rightarrow \Gamma$ is *perfectible* iff it is a substitution instance of a gauntly valid sequent, or equivalently, iff it is a substitution instance of a perfectly valid sequent.

(An alternative justification for the perfectible sequents is to focus on deduction. Since proofhood in usual deductive systems is preserved under global uniform substitutions, if we wanted a minimal system of ‘relevance logic’ that captured the perfectly valid or gauntly valid sequents and that was closed under substitutions, we would end up with the perfectible sequents. This is roughly the route that [20] takes.)[7]

2.7 Some Remarks on Sources and Precedents

The definitions of perfect validity and perfectibility are, in the present form, due to Tennant [21], although they are prefigured by conceptually similar ideas in earlier work such as [8, 19, 29], and [15].

Verdée et al. [28] also study a notion of relevance that is essentially perfectibility in a multiset-sequent setting. Their approach is similar to my own in adopting a conservative, proof-theoretic, and logic-relative study of relevance. Of particular note is that, besides considering classical logic as the background, they also study how this approach applies to the Anderson-Belnap logic R. They do not, however, consider properties analogous to gaunt validity or relevant validity in the sense defined above.

The definition of relevant validity comes from [4], although similar ideas had also been applied in areas other than logic; see [32] for a discussion of similar definitions of relevance in areas other than entailment.

The definition of gaunt validity comes from Tennant [23], although the connection to a notion of relevance was not explicitly made in that work. In [5], I explicitly drew the connection between gaunt validity and relevance; somewhat unfortunately, that paper also changed terminology, using *gaunt validity* where Tennant had used the expression *skeletal validity*.[8] Nevertheless, I will stick with my terminology for continuity.

The present approach to relevance owes a clear debt to the work of Tennant. But there are also important differences. Tennant offers perfectibility as *the* correct account of entailment. As a logical monist, he thinks that there is a single correct logic, and that relevance is a necessary condition on true entailments.[9] As I emphasized above, my approach is more ecumenical, being essentially *logic-relative* and taking no stand on the question(s) of logical monism. Thus, to recapitulate the slogan from Sect. 1.1, Tennant argues that his system, called core logic, is a logic *of* relevance, whereas I am interested in studying relevance *in* a logic. This is a more comparative, intra-logical pursuit. Finally, while Tennant provides an account of *entailment* and thinks that relevance is a necessary condition on entailment, his

[7] As an aside, I'll note that if we were to prioritize deduction in our explication of relevance, a natural idea to pursue is that the relevant components of a deduction are the components that are *actually used* in the deduction. This 'real use' criterion of relevance is elegant and has a long history, but is outside the scope of this paper. See, e.g., [1, 18ff] and [16, Ch. VII].

[8] Tennant had used *skeletally valid* as an adjective applied to sequents and *gaunt* as an adjective applied to proofs; but in [5] and [6] I wanted a single term that applied to both proofs and sequents, and adopted 'gaunt' for this purpose.

[9] Specifically, he argues for his system of core logic as the correct logic, see [22, 23, 26].

work does not provide an explication of relevance as such.[10] Of course, this is not to say that Tennant is precluded from developing an account of relevance as such. One contribution of the present work could be seen as providing such an explication.

3 Relations Between the Explications

The relation between gaunt validity, perfect validity, and perfectibility is straightforward. By definition, the class of perfectible sequents is the closure of the perfectly valid sequents under substitution. Every perfectly valid sequent can be obtained from a gauntly valid sequent by substitution. And hence taking the closure of the gauntly valid sequents under substitution again results in the perfectible sequents.

It also follows from Proposition 4 that a sequent $\Delta \Rightarrow \Gamma$ is relevantly valid iff there are perfectly valid sequents $\Delta_1 \Rightarrow \Gamma_1, \ldots, \Delta_n \Rightarrow \Gamma_n$ such that $\Delta = \bigcup_{i \leq n} \Delta_i$ and $\Gamma = \bigcup_{i \leq n} \Gamma_i$. We can summarize this as saying that the relevantly valid sequents are *agglomerations* of perfectly valid sequents. Let $\mathsf{Agg}(S)$ denote the closure of a set of sequents S under agglomeration—that is, combining sequents by taking unions of premise sets and unions of conclusion sets. This suggests the diagram of relevance properties in Fig. 1.

Between the operations of substitution and agglomeration, we may wonder whether it makes a difference which order we apply them in. Is the same class of sequents determined by taking agglomerations of substitution instances as by taking substitution instances of agglomerations? The answer is yes. Slightly misusing a term from algebra, we could say that the diagram in Fig. 1 commutes. To be more exact, however, what we want to say is this.

Proposition 8

1. *For every sequent $\Delta \Rightarrow \Gamma$ that is relevantly valid, there are gauntly valid sequents $\Delta_1 \Rightarrow \Gamma_1, \ldots, \Delta_n \Rightarrow \Gamma_n$ and a substitution σ such that for each i, $\sigma(\Delta_i) \Rightarrow \sigma(\Gamma_i)$ is perfectly valid, and such that $\sigma(\Delta_1), \ldots, \sigma(\Delta_n) \Rightarrow \sigma(\Gamma_1), \ldots, \sigma(\Gamma_n)$ and $\sigma(\Delta_1, \ldots, \Delta_n) \Rightarrow \sigma(\Gamma_1, \ldots, \Gamma_n)$ are both identical to $\Delta \Rightarrow \Gamma$.*

[10] Tennant [25] does sometimes refer to his variable sharing property as an explication of relevance. (See Proposition 10 and footnote 12 below for the variable sharing result in question.) But this is just a clash of terminology, because the variable sharing property does not provide an explication *in the sense of the term that I outlined above*. This variable sharing property has two shortcomings as an explication (in my sense of the term). First, Tennant does not draw a clear line between an informal conception of relevance and the formal definition of his variable sharing property. Second, and more significantly, if the variable sharing property were an adequate explication of relevance in my sense of the term, then exhibiting the variable sharing property should be *sufficient* for exhibiting relevance, which I claim it is not. Specifically, I claim that the premise $D \vee C$ is intuitively irrelevant in the following sequent: $D \vee C, A, (A \vee C) \rightarrow B \Rightarrow B$.

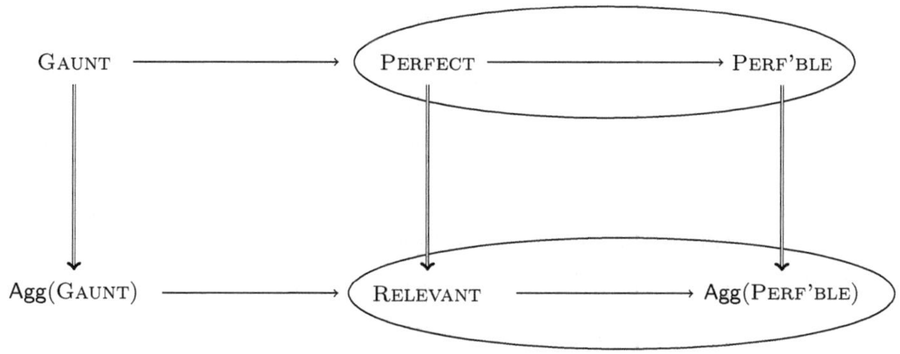

Fig. 1 A hierarchy of relevance properties. The horizontal single arrows represent substitutions, with the vertical double arrows representing agglomerations. Each arrow obviously also indicates a subset relation. Closure of the gauntly valid sequents results in the perfectible sequents, of which the perfectly valid sequents are a subset. Likewise for the bottom row

2. Agg(PERF'BLE) $\subseteq$ sub(RELEVANT). *That is, for every sequent* $\Delta \Rightarrow \Gamma \in$ Agg(PERF'BLE)*, there are perfectly valid sequents* $\Delta_1 \Rightarrow \Gamma_1, \ldots, \Delta_n \Rightarrow \Gamma_n$ *and a substitution* σ *such that* $\sigma(\Delta_1), \ldots, \sigma(\Delta_n) \Rightarrow \sigma(\Gamma_1), \ldots, \sigma(\Gamma_n)$ *and* $\sigma(\Delta_1, \ldots, \Delta_n) \Rightarrow \sigma(\Gamma_1, \ldots, \Gamma_n)$ *are both identical to* $\Delta \Rightarrow \Gamma$.
3. sub(RELEVANT) $\subseteq$ Agg(PERF'BLE). *That is, for every relevantly valid sequent* $\Delta \Rightarrow \Gamma$ *and every substitution* σ*, there are perfectly valid sequents* $\Delta_1 \Rightarrow \Gamma_1, \ldots, \Delta_n \Rightarrow \Gamma_n$ *and substitutions* $\sigma_1, \ldots, \sigma_n$ *such that* $\sigma(\Delta_1, \ldots, \Delta_n) \Rightarrow \sigma(\Gamma_1, \ldots, \Gamma_n)$ *is identical to* $\sigma_1(\Delta_1), \ldots, \sigma_n(\Delta_n) \Rightarrow \sigma_1(\Gamma_1), \ldots, \sigma_n(\Gamma_n)$.[11]

Proof

(1) Given a relevantly valid sequent $\Delta \Rightarrow \Gamma$, we know that there will be some perfectly valid sequents $\Delta_1^p \Rightarrow \Gamma_1^p, \ldots, \Delta_n^p \Rightarrow \Gamma_n^p$ such that $\Delta = \Delta_1^p \cup \ldots \cup \Delta_n^p$ and $\Gamma = \Gamma_1^p \cup \ldots \cup \Gamma_n^p$. Furthermore, we know that each $\Delta_i^p \Rightarrow \Gamma_i^p$ will be a substitution instance of a gauntly valid sequent $\Delta_i \Rightarrow \Gamma_i$. Moreover, we can assume that these Δ_i, Γ_i and Δ_j, Γ_j are atom-disjoint when $i \neq j$. Thus there will be a single substitution σ such that, for each i, $\sigma(\Delta_i) = \Delta_i^p$ and $\sigma(\Gamma_i) = \Gamma_i^p$. Thus,

$$\sigma(\Delta_1) \cup \ldots \cup \sigma(\Delta_n) = \sigma(\Delta_1 \cup \ldots \cup \Delta_n) = \Delta$$

[11] Claim 1 combines two facts. First, we can obtain relevantly valid sequents by starting with gauntly valid sequents, applying a substitution to get a perfectly valid sequents, and then agglomerating them. Second, we can equivalently obtain the relevantly valid sequents by starting with gauntly valid sequents, agglomerating them, and then taking a substitution. It is possible to combine these two facts into one statement because we have an independent characterization of relevant validity. By contrast, the set Agg(PERF'BLE) does not have an independent characterization, hence the need to separate claims (2) and (3) into separate statements.

and

$$\sigma(\Gamma_1) \cup \ldots \cup \sigma(\Gamma_n) = \sigma(\Gamma_1 \cup \ldots \cup \Gamma_n) = \Gamma$$

(2) Let $\Delta_1^b \Rightarrow \Gamma_1^b, \ldots, \Delta_n^b \Rightarrow \Gamma_n^b$ be perfectible sequents that agglomerate into $\Delta \Rightarrow \Gamma$. Find perfectly valid sequents $\Delta_i \Rightarrow \Gamma_i$ of which $\Delta_i^b \Rightarrow \Gamma_i^b$ is an instance; moreover, we can assume that these Δ_i, Γ_i and Δ_j, Γ_j are atom-disjoint when $i \neq j$. Thus there will be a single substitution σ such that, for each i, $\sigma(\Delta_i) = \Delta_i^b$ and $\sigma(\Gamma_i) = \Gamma_i^p$. Thus, as before:

$$\sigma(\Delta_1) \cup \ldots \cup \sigma(\Delta_n) = \sigma(\Delta_1 \cup \ldots \cup \Delta_n) = \Delta$$

and

$$\sigma(\Gamma_1) \cup \ldots \cup \sigma(\Gamma_n) = \sigma(\Gamma_1 \cup \ldots \cup \Gamma_n) = \Gamma$$

(3) If $\Delta \Rightarrow \Gamma$ is relevantly valid, then there are some perfectly valid sequents $\Delta_1 \Rightarrow \Gamma_1, \ldots, \Delta_n \Rightarrow \Gamma_n$ which agglomerate into $\Delta \Rightarrow \Gamma$. Moreover, for any substitution σ, the following obviously hold:

$$\sigma(\Delta_1) \cup \ldots \cup \sigma(\Delta_n) = \sigma(\Delta_1 \cup \ldots \cup \Delta_n)$$

and

$$\sigma(\Gamma_1) \cup \ldots \cup \sigma(\Gamma_n) = \sigma(\Gamma_1 \cup \ldots \cup \Gamma_n)$$

□

4 Variable Sharing Properties

The idea of a variable sharing property first emerged in relevance logic as a formal surrogate for the idea that there should be some meaning in common between the antecedent and consequent of a conditional. On the approach to relevance that I am advocating, variable sharing properties acquire a different significance. In any of the explications introduced above, relevance functions as a sieve, winnowing down the class of valid sequents to a more tightly constrained class. The sieve, however, was primarily semantic, being defined in terms of validity. In this context, variable sharing properties can be viewed as syntactic counterparts of the semantic sieve.

A full-fledged syntactic counterpart of the semantic sieve would be a syntactic or proof-theoretic condition which was both necessary and sufficient for relevance (in one of the explicated senses). In Sect. 5, we will encounter a proof-theoretic characterization of gaunt validity. In this section, however, we will only consider variable-sharing properties as necessary conditions for relevance, giving us syntactic *corollaries* of the semantic sieve.

As we will see, each of our six classes of sequents enjoys some variable sharing property. I begin by recalling some standard definitions, and introducing some new ones.

Definition 9 (Positive/Negative Subformulas; $\bowtie$; $\blacktriangleright\!\!\blacktriangleleft$) I define three notions:

1. Define the *positive* and *negative* occurrences of ϕ inductively:
   - ϕ occurs positively in ϕ;
   - If ϕ occurs positively (negatively) in θ, then ϕ occurs positively (negatively, respectively) in $\theta \vee \psi$, $\psi \vee \theta$, $\theta \wedge \psi$, $\psi \wedge \theta$, and $\psi \rightarrow \theta$;
   - If ϕ occurs positively (negatively) in θ, then ϕ occurs negatively (positively, resp.) in $\neg\theta$ and $\theta \rightarrow \psi$.
2. Further, define $\phi \bowtie \psi$ to mean that some atom occurs positively in ϕ and negatively in ψ, or vice versa. Let $\overline{\bowtie}$ be the ancestral of $\bowtie$.
3. Finally, in the context of a sequent $\Delta \Rightarrow \Gamma$, define $\phi \blacktriangleright\!\!\blacktriangleleft \psi$ to mean that either ϕ and ψ are on the same side of the sequent and $\phi \bowtie \psi$, or ϕ and ψ are on opposite sides of the sequent and $\phi \bowtie \neg\psi$. Let $\overline{\blacktriangleright\!\!\blacktriangleleft}$ be the ancestral of $\blacktriangleright\!\!\blacktriangleleft$.

Proposition 10 *If $\Delta \Rightarrow \Gamma$ is perfectible, then for any two distinct formulas ϕ and ψ in this sequent, $\phi \overline{\blacktriangleright\!\!\blacktriangleleft} \psi$; and if there is only one member ϕ of this sequent, then $\phi \bowtie \phi$. In short, $\Delta \Rightarrow \Gamma$ has no non-trivial $\overline{\blacktriangleright\!\!\blacktriangleleft}$-compartments.*[12]

Proof First, assume that $\Delta \Rightarrow \Gamma$ is perfectly valid, but that there are some non-trivial $\overline{\blacktriangleright\!\!\blacktriangleleft}$-compartments Δ_1, Γ_1 and Δ_2, Γ_2. Since $\Delta_1 \Rightarrow \Gamma_1$ is a proper subsequent of $\Delta \Rightarrow \Gamma$, it will have a countermodel τ_1. Likewise $\Delta_2 \Rightarrow \Gamma_2$ will have a countermodel τ_2. Let τ agree with τ_1 on the atoms in Δ_1, Γ_1 and agree with τ_2 on the atoms in Δ_2, Γ_2. Since Δ_1, Γ_1 is atom-disjoint from Δ_2, Γ_2, we know τ will be well-defined. Moreover, $\tau(\Delta_i) = \tau_i(\Delta_i)$ and $\tau(\Gamma_i) = \tau_i(\Gamma_i)$. So τ will be a countermodel to $\Delta \Rightarrow \Gamma$, a contradiction.

Now, since the variable sharing property is preserved under substitutions, it follows that every perfectible sequent has this property. □

This variable sharing property can be visually pictured by drawing the $\blacktriangleright\!\!\blacktriangleleft$ connections between the formulas in the sequent. The proposition asserts that there is always at least one way of drawing these connections so that there is a path between any two formulas in the sequent. In fact, there will often be more than one way of drawing these connections. For instance, consider the sequent

$$P \wedge Q, Q \rightarrow R, \neg R \vee S, (S \wedge P) \rightarrow T \Rightarrow S \wedge T.$$

[12] This variable sharing property for perfectible sequents comes from [25]. By an induction on proofs, he shows that classical core logic enjoys this property, which entails that the perfectly valid sequents also enjoy this property. The simple semantic argument given here is original.

Then, among other options, we can draw the graph like this:

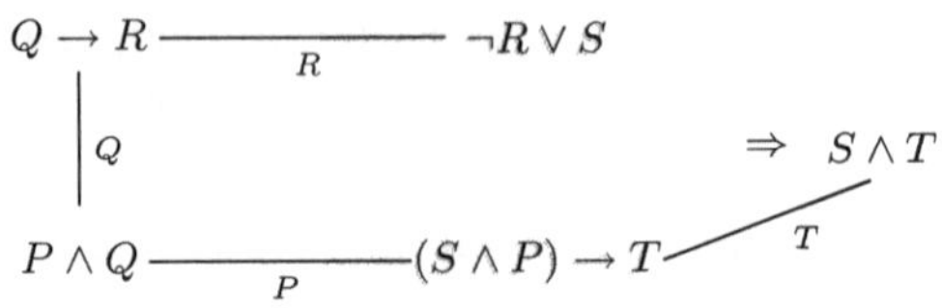

Or we can draw the graph like this:

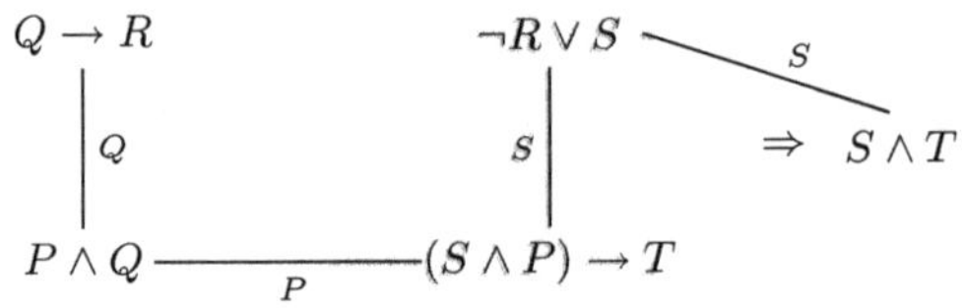

Casting our eye down toward the bottom right corner of our hierarchy, Proposition 10 has a simple corollary concerning agglomerations of perfectible sequents.

Corollary 11 *If* $\Delta \Rightarrow \Gamma \in \mathsf{Agg}(\textsc{Perf'ble})$, *then either (i) there is some atom that occurs with the same parity in* Δ *and* Γ, *or (ii) if* $\Delta \neq \emptyset$, *then there is some atom that occurs both positively and negatively in* Δ, *and if* $\Gamma \neq \emptyset$, *then there is some atom that occurs both positively and negatively in* Γ.

It follows that each of the six notions defined in Sect. 2 is distinct from classical consequence. Since the gauntly valid sequents are also perfectly valid, they are included in Proposition 10. Because gauntness is such a demanding notion, however, the gauntly valid sequents will have a much more tightly constrained syntactic form.

Proposition 12 *If* $\Delta \Rightarrow \Gamma$ *is gauntly valid, then:*

1. *Every atom occurs at most twice in the sequent, and if P does occur twice, then those occurrences witness a* $\bowtie$*-connection.*
2. *The sequent does not contain any cycle* $\phi_1 \bowtie \phi_2 \bowtie \ldots \bowtie \phi_n \bowtie \phi_1$ *in which each* $\bowtie$*-connection is witnessed by a different atom.*
3. *For any two atoms occurring in the sequent, there is at most one formula in which they both occur.*
4. *For any two formulas occurring in the sequent, there is at most one atom that occurs in both formulas.*

Informally, this proposition says that there is a unique graph witnessing the $\overline{\bowtie}$-connectedness of a gauntly valid sequent and that no atom is repeated in the sequent unless with $\bowtie$-connected occurrences. To prove the proposition, I will rely on the following fact, which may in turn be proved by an easy induction on complexity.

Fact 13 If an atom P only occurs positively in a formula ϕ, and $\tau(\phi) = T$, and if $\tau[P \mapsto T]$ assigns T to P and otherwise agrees with τ, then $\tau[P \mapsto T](\phi) = T$. Similarly, if P only occurs negatively in ϕ and $\tau(\phi) = T$, then $\tau[P \mapsto F](\phi) = T$.

Proof of Proposition 12

(1) For simplicity, assume that the instances of an atom P are all in different formulas. This makes the argument clearer, and I will remark on how to avoid this assumption at the end.

Assume that $\Delta \Rightarrow \Gamma$ is gauntly valid, and that some atom P has two occurrences with the same sign in this sequent, say in $\phi_1(P)$ and $\phi_2(P)$. Without loss of generality, assume both of these occurrences are positive, and both occur in Δ. Write $\Delta' := \Delta \setminus \{\phi_1, \phi_2\}$. Let R be some new atom not occurring in $\Delta \Rightarrow \Gamma$. Then $\Delta', \phi_1(P), \phi_2(R) \Rightarrow \Gamma$ has a countermodel. Thus, it has a countermodel where R is true. Now, if P did not have a negative occurrence elsewhere in the sequent, then we could also assume that this has a countermodel where P is true; and hence this would also provide a countermodel to the original sequent $\Delta', \phi_1(P), \phi_2(P) \Rightarrow \Gamma$. This shows that if an atom does not occur with opposite signs, then it can only occur once.

Let $\psi_1, \ldots, \psi_n$ enumerate all the other sentences in Δ in which P occurs. (Without loss of generality, assume P does not occur in Γ.) Write $\Delta'' := \Delta \setminus \{\phi_1, \phi_2, \psi_1, \ldots, \psi_n\}$. Now consider the sequent:

$$\Delta'', \phi_1(P), \phi_2(R), \psi_1(P), \ldots, \psi_n(P), \psi_1(R), \ldots, \psi_n(R) \Rightarrow \Gamma \qquad (\star)$$

If this sequent were valid, then it would have some perfectly valid subsequent; but any perfectly valid subsequent of $(\star)$ would either be a proper subsequent or a proper coarsening of our original sequent $\Delta \Rightarrow \Gamma$, which is impossible.

So there $(\star)$ must have a countermodel τ. Moreover, $\tau(P) \neq \tau(R)$, since otherwise τ would also be a countermodel to $\Delta \Rightarrow \Gamma$. Say that $\tau(P) = F$. Since we know $\tau(\psi_i(P)) = \tau(\psi_i(R)) = T$, it follows that each $\psi_i(P)$ is true whether P is true or false (holding fixed the truth values of the other atoms). Let $\tau' := \tau[P \mapsto T]$. Then $\tau'(\psi_i(P)) = T$; and since P occurs positively in $\phi_1(P)$, $\tau'(\phi_1(P)) = T$. So τ' is a countermodel to the sequent $(\star)$. But since $\tau'(P) = \tau'(R)$, τ' is also a countermodel to the original sequent $\Delta \Rightarrow \Gamma$, which is impossible.

Now to remove the assumption that all the occurrences of P are in different formulas. If there are $n \geq 2$ occurrences of P in Δ, let $\Delta(P, \ldots, P) \Rightarrow \Gamma$ stand for a particular enumeration of members of the sequent with all instances of P displayed. Assume that the first two instances of P are of the same parity, and for the sake of concreteness, assume they are both positive. For the first part of the argument, consider the sequent $\Delta(P, R, \ldots) \Rightarrow \Gamma$. Then argue as above to show that we can find a countermodel to $\Delta \Rightarrow \Gamma$ unless there are further instances of P with negative parity.

For the second part of the argument, in place of (⋆) we consider the sequent:

$$\Delta(P, R, P, \ldots, P), \Delta(P, R, R, \ldots, R) \Rightarrow \Gamma \qquad (\star\star)$$

In other words, first replace the second occurrence of P with R, and then for all $n > 2$ simultaneously, if $\phi(P)$ contains the nth occurrence of P, include both $\phi(P)$ and a copy $\phi(R)$ with P replaced by R. Then from here the argument is the same as above.

(2) Suppose that there were some cycle $\phi_1 \bowtie \ldots \bowtie \phi_n \bowtie \phi_1$ with each $\bowtie$ connection witnessed by a different atom. Let A_1 be the atom witnessing the connection $\phi_n \bowtie \phi_1$ and A_i the atom witnessing the connection $\phi_{i-1} \bowtie \phi_i$. Without loss of generality, suppose that the ϕ_i's are all in the premise set. By (1) we know that A_i will only occur in ϕ_{i-1} and ϕ_i. Let B_i be new atoms not occurring in Δ, Γ. Put $\Delta' := \Delta \setminus \{\phi_1, \ldots, \phi_n\}$, and consider the sequent:

$$\Delta', \phi_1(A_1, A_2), \phi_2(A_2, A_3), \ldots, \phi_n(A_n, B_1), \phi_1(B_1, B_2), \phi_2(B_2, B_3),$$
$$\ldots, \phi_n(B_n, A_1) \Rightarrow \Gamma \qquad (\dagger)$$

As with (⋆) in part (1), this sequent must be invalid. After all, this is a proper coarsening of $\Delta \Rightarrow \Gamma$, so any perfectly valid subsequent would have to be a proper coarsening of a subsequent of $\Delta \Rightarrow \Gamma$, which is impossible. So (†) must have a countermodel τ.

I claim that if τ is any countermodel to †, then for each i, $\tau(A_i) \neq \tau(B_i)$. Argue by induction on i. For the basis step, if $\tau(A_1) = \tau(B_1)$, then $\tau(\phi_n(A_n, B_1)) = \tau(\phi_n(A_n, A_1)) = T$, in which case τ would be a countermodel to the original sequent $\Delta \Rightarrow \Gamma$. For the induction step, suppose $\tau(B_{i+1}) = \tau(A_{i+1})$. Then $\tau(\phi_i(B_i, B_{i+1})) = \tau(\phi_i(A_i, A_{i+1})) = \tau(\phi_i(A_i, B_{i+1})) = T$. By the i.h., $\tau(B_i) \neq \tau(A_i)$. Thus, abusing notation slightly, we have $\tau(\phi_i(T, B_{i+1})) = \tau(\phi_i(F, B_{i+1})) = T$. In other words, ϕ_i is indifferent to the truth value of its first argument place. Suppose that B_i and A_i occur positively in ϕ_{i-1}. Then $\tau[A_i, B_i \mapsto T]$ will assign T to both $\phi_{i-1}(A_{i-1}, A_i)$ and $\phi_{i-1}(B_{i-1}, B_i)$. And we have just observed that $\tau[A_i, B_i \mapsto T]$ will also assign T to $\phi_i(A_i, A_{i+1})$ and $\phi_i(B_i, B_{i+1})$. So $\tau[A_i, B_i \mapsto T]$ will be a countermodel to †, contradicting the i.h. This completes the induction.

Now consider four cases: (i) A_n occurs positively in ϕ_n and $\tau(A_n) = T$, (ii) A_n occurs negatively in ϕ_n and $\tau(A_n) = T$, (iii) A_n occurs positively in ϕ_n and $\tau(A_n) = F$, (iv) A_n occurs negatively in ϕ_n and $\tau(A_n) = F$.

Case (i). Then $\tau(B_n) = F$ and B_n occurs negatively in ϕ_{n-1}. So $\tau(\phi_{n-1}(A_{n-1}, B_n)) = T$, and hence τ is a countermodel to the sequent:

$$\Delta', \phi_1(A_1, A_2), \phi_2(A_2, A_3), \ldots, \phi_{n-1}(A_{n-1}, B_n), \phi_n(B_n, A_1) \Rightarrow \Gamma.$$

Since this is a mere relettering of our original sequent $\Delta \Rightarrow \Gamma$, this is impossible.

Case (ii). Then A_n occurs positively in ϕ_{n-1}, and hence $\tau(\phi_{n-1}(B_{n-1}, A_n)) = T$. So τ provides a countermodel to the sequent:

$$\Delta', \phi_1(B_1, B_2), \phi_2(B_2, B_3), \ldots, \phi_{n-1}(B_{n-1}, A_n), \phi_n(A_n, B_1) \Rightarrow \Gamma.$$

Again, this is a mere relettering of our original sequent, so this is impossible.

Cases (iii) and (iv) are similar.

(3) and (4) are immediate consequences of (1) and (2).

□

This proposition has a simple, but surprising corollary.

Corollary 14 *There is no formula ϕ such that $\emptyset \Rightarrow \phi$ is gauntly valid.*

Proof If $\emptyset \Rightarrow \phi$ were gauntly valid, then we would have $\phi \bowtie \phi$, in contradiction to Proposition 12 (2). □

For instance, $\emptyset \Rightarrow P \vee \neg P$ is not gauntly valid, because it can be coarsened to $\emptyset \Rightarrow P \vee Q, R \vee \neg P$.

5 Gaunt Validity in a Classical Sequent Calculus

In this section I will show how, if we start with a standard sequent calculus for classical logic, imposing some simple restrictions will give us a calculus that proves exactly the gauntly valid sequents. Thus, gaunt validity has a simple proof-theoretic characterization to complement the original semantic characterization in terms of validity.

For simplicity, I will focus only on the logical vocabulary $\neg, \vee, \wedge$, though the inclusion of $\rightarrow$ would be straightforward. As before, I will take the sequents in this calculus to be composed of *sets*, rather than multisets or sequences. The calculus I will study has the following logical rules:

	L-rules	R-rules
$\neg$	$\dfrac{\Delta \Rightarrow \phi, \Gamma}{\Delta, \neg\phi \Rightarrow \Gamma}$	$\dfrac{\Delta, \phi \Rightarrow \Gamma}{\Delta \Rightarrow \neg\phi, \Gamma}$
$\wedge$	$\dfrac{\Delta, \phi \Rightarrow \Gamma}{\Delta, \phi \wedge \psi \Rightarrow \Gamma}$	$\dfrac{\Delta_1 \Rightarrow \phi, \Gamma_1 \qquad \Delta_2 \Rightarrow \psi, \Gamma_2}{\Delta_1, \Delta_2, \Rightarrow \phi \wedge \psi, \Gamma_1, \Gamma_2}$
$\vee$	$\dfrac{\Delta_1, \phi \Rightarrow \Gamma_1 \qquad \Delta_2, \psi \Rightarrow \Gamma_2}{\Delta_1, \Delta_2, \phi \vee \psi \Rightarrow \Gamma_1, \Gamma_2}$	$\dfrac{\Delta \Rightarrow \phi, \Gamma}{\Delta \Rightarrow \phi \vee \psi, \Gamma}$

It should be understood that in $\wedge$-L and $\vee$-R, we may also have $\psi \wedge \phi$ or $\psi \vee \phi$ in place of $\phi \wedge \psi$ and $\phi \vee \psi$. There is one rule for initial sequents:

$$\Delta, \phi \Rightarrow \phi, \Gamma$$

If we want this system to prove only gaunt validities, there are a few obvious restrictions to impose. The first is to restrict the rule of initial sequents so that ϕ is an atom and $\Delta = \Gamma = \emptyset$.

The second is that when $\wedge$-L introduces a new conjunct, it should not have any extraneous logical structure: ψ should be an atom. Likewise when $\vee$-R introduces a new disjunct.

The third restriction concerns $\vee$-L and $\wedge$-R rules. There is a possibility that when we combine the premise sets Δ_1 and Δ_2 or the conclusion sets Γ_1 and Γ_2, this can introduce redundant premises or unnecessary structure. As a simple example, though both upper sequents of the following inference are gauntly valid, the lower sequent clearly is not:

$$\frac{A \wedge B \Rightarrow A \qquad A \wedge B \Rightarrow B}{A \wedge B \Rightarrow A \wedge B}$$

This suggests the restriction that parallel subproofs always be atom-disjoint.

The fourth restriction we need to impose concerns the negation rules, since repeated applications of negation rules can obviously lead to non-gaunt sequents, such as:

$$\frac{\dfrac{A \Rightarrow A}{\Rightarrow \neg A, A}}{\neg A \Rightarrow \neg A}$$

A simple restriction that we can impose is that no single formula be the subject of a $\neg$-L and $\neg$-R rule. In fact, it will suffice to introduce the simpler restriction that no *atom* be the subject of a $\neg$-L and $\neg$-R rule.

These four restrictions are simply the most obvious ones that we might think to impose when we reflect on how the sequent rules might not preserve gaunt validity. It is not too difficult, however, to show that they are also *sufficient* to ensure that the sequent rules preserve gaunt validity.

Proposition 15 *Suppose that $\Delta, \phi \Rightarrow \theta, \Gamma$ is gauntly valid. Then:*

1. *$\Delta, \phi \wedge \psi \Rightarrow \theta, \Gamma$ is gauntly valid iff ψ is a new atom that does not occur elsewhere in the sequent.*
2. *$\Delta, \phi \Rightarrow \theta \vee \psi, \Gamma$ is gauntly valid iff ψ is a new atom that does not occur elsewhere in the sequent.*
3. *$\Delta \Rightarrow \neg\phi, \theta, \Gamma$ is gauntly valid unless ϕ is an atom A, and $\neg A$ occurs somewhere else in Δ, θ, Γ. In that case, if $\Delta^A, \theta^A, \Gamma^A$ denotes the result of replacing any instance of $\neg A$ with A, then $\Delta^A \Rightarrow A, \theta^A, \Gamma^A$ is gauntly valid.*

4. $\Delta, \phi, \neg\theta \Rightarrow \Gamma$ *is gauntly valid unless* θ *is an atom* B*, and* $\neg B$ *occurs somewhere else in* Δ, θ, Γ*. In that case,* $\Delta^B, \phi^B, B \Rightarrow \Gamma^B$ *is gauntly valid.*[13]

Proof

(1) For the 'if' part, suppose for reductio that A is a new atom and that $\Delta, \phi \wedge A \Rightarrow \theta, \Gamma$ has the valid proper coarsening $\Delta', \phi' \wedge A \Rightarrow \theta', \Gamma'$. Then, since A only occurs as indicated, $\Delta', \phi' \Rightarrow \theta', \Gamma'$ would be valid, and this is a proper coarsening of $\Delta, \phi \Rightarrow \theta, \Gamma$, which is impossible.

For the 'only if', suppose A is an atom not occurring in $\Delta, \phi, \theta, \Gamma$. Then $\Delta, \phi \wedge A \Rightarrow \theta, \Gamma$ is a valid coarsening of $\Delta, \phi \wedge \psi \Rightarrow \theta, \Gamma$ under the substitution $\sigma : A \mapsto \psi$. So if $\Delta, \phi \wedge \psi \Rightarrow \theta, \Gamma$ is gauntly valid, then σ must be trivial, which means it must be a 1–1 mapping of atoms; hence ψ must be another atom that does not occur elsewhere in $\Delta, \phi, \theta, \Gamma$.

(2) Similar to (1).

(3) Suppose that $\Delta \Rightarrow \neg\phi, \theta, \Gamma$ has a valid proper coarsening $\Delta' \Rightarrow (\neg\phi)', \theta', \Gamma'$ under the substitution σ. We know that $(\neg\phi)'$ will either be an atom A or of the form $\neg(\phi')$. But if it were of the form $\neg(\phi')$, then we would have $\Delta', \phi' \Rightarrow \theta', \Gamma'$ as a valid proper coarsening of $\Delta, \phi \Rightarrow \theta, \Gamma$, which is impossible. Moreover, if ϕ had any internal logical complexity, then σ could be factored into two non-trivial substitutions $\sigma_1 : A \mapsto \neg A$ and $\sigma_2 : A \mapsto \phi$. Then, since $\Delta' \Rightarrow A, \theta', \Gamma'$ is valid, $\sigma_1(\Delta') \Rightarrow \neg A, \sigma_1(\theta'), \sigma_1(\Gamma')$ will be valid too. Hence $\sigma_1(\Delta'), A \Rightarrow \sigma_1(\theta'), \sigma_1(\Gamma')$ will also be valid; and since σ_2 is non-trivial, this will be a valid non-trivial coarsening of $\Delta, \phi \Rightarrow \theta, \Gamma$. But that is impossible, so ϕ must not have any internal logical complexity, i.e., it is an atom A.

(4) Similar to (3).

□

This proposition provides the basis for showing that any sequent with a proof satisfying our four restrictions will be gauntly valid. As it turns out, the converse claim is also straightforward, using what amounts to a standard proof-search technique.

Theorem 16 $\Delta \Rightarrow \Gamma$ *is gauntly valid iff there is a proof of* $\Delta \Rightarrow \Gamma$ *satisfying the four restrictions that: (1) every initial sequent is of the form* $P \Rightarrow P$*, with* P *atomic, (2) in instances of* $\wedge$*-L and* $\vee$*-R, the newly introduced conjunct/disjunct is a new atom, (3) in instances of* $\wedge$*-R and* $\vee$*-L, the two parallel subproofs are atom-disjoint, and (4) there is no atom that is the subject of both a* $\neg$*-L and* $\neg$*-R rule.*

Proof **Right-to-Left** This follows by an easy induction on the length of proof. The basis case is trivial. For the induction step, when the last rule is $\neg$-L, $\neg$-R, $\wedge$-L, or

[13] Observations 1 and 2 are drawn from [24]. Observations 3 and 4 generalize to classical logic an observation made about intuitionistic logic in [6].

$\vee$-R, the claim follows from Proposition 15. When the last rule is $\wedge$-R, we have an inference:

$$\frac{\Delta_1 \Rightarrow \psi, \Gamma_1 \qquad \Delta_2 \Rightarrow \theta, \Gamma_2}{\Delta_1, \Delta_2 \Rightarrow \psi \wedge \theta, \Gamma_1, \Gamma_2}$$

Suppose that the lower sequent had a valid proper coarsening $\Delta_1', \Delta_2' \Rightarrow (\psi \wedge \theta)', \Gamma_1', \Gamma_2'$. First, observe that the coarsened formula $(\psi \wedge \theta)'$ would have to be of the form $\psi' \wedge \theta'$. Otherwise, since the two upper sequents are atom-disjoint, it would have to be an atom P not occurring elsewhere in the sequent; but this would entail that $\Delta_1, \Delta_2 \Rightarrow \Gamma_1, \Gamma_2$ is valid, in which case each of $\Delta_1 \Rightarrow \Gamma_1$ and $\Delta_2 \Rightarrow \Gamma_2$ would be valid (by Proposition 10, in light of the fact that the two upper sequents are atom-disjoint). Second, given that we have a valid proper coarsening $\Delta_1', \Delta_2' \Rightarrow \psi' \wedge \theta', \Gamma_1', \Gamma_2'$, it follows that $\Delta_1', \Delta_2' \Rightarrow \psi', \theta', \Gamma_1', \Gamma_2'$ is valid; and thus, by the atom-disjointness, $\Delta_1' \Rightarrow \psi', \Gamma_1'$ and $\Delta_2' \Rightarrow \theta', \Gamma_2'$ would be valid. And at least one of these would have to be a proper coarsening of the original sequents, which is impossible.

When the last rule in the proof is $\vee$-L, we argue similarly to the case of $\wedge$-R.

Left-to-Right Argue by induction on the logical complexity of Δ, Γ. In the basis case, if $\Delta \Rightarrow \Gamma$ is gauntly valid, then it must be of the form $P \Rightarrow P$. This has the obvious one line proof.

For the induction step, fix some method of ordering the elements of Δ, Γ, and take the first element ϕ of this ordering that is not atomic. Consider six cases, according to what ϕ's main connective is and whether ϕ is in the antecedent or succedent.

Case 1: $\phi := \neg\psi$ is in the antecedent, so the sequent is $\Delta, \neg\psi \Rightarrow \Gamma$. Then $\Delta \Rightarrow \psi, \Gamma$ must also be gauntly valid. By the i.h., we know there will be a proof Π of $\Delta \Rightarrow \psi, \Gamma$ satisfying our four conditions. Moreover, by Proposition 15, either ψ is not an atom, or, if it is, then there is no other occurrence of $\neg\psi$ in Δ, Γ. Thus, if we append an instance of $\neg$-L to the bottom of Π we will have a proof of $\Delta, \neg\psi \Rightarrow \Gamma$ satisfying our four conditions.

Case 2: $\phi := \neg\psi$ is in the succedent. This is essentially similar to Case 1.

Case 3: $\phi := \psi \wedge \theta$ is in the antecedent, so the sequent is $\Delta, \psi \wedge \theta \Rightarrow \Gamma$. I claim that one of ψ and θ must be an atom not occurring elsewhere in the sequent. If not, then given new atoms P and R, the sequent

$$\Delta, \psi \wedge P, R \wedge \theta \Rightarrow \Gamma$$

would be a valid proper coarsening of our original sequent (and any perfectly valid subsequent would be a proper coarsening and/or proper subsequent of the original sequent). But that is impossible. Thus, ψ, say, must be a new atom P.

So $\Delta, \theta \Rightarrow \Gamma$ must be gauntly valid. By the i.h., we have a proof Π as required of this sequent; and by adding an instance of $\wedge$-L, we get the required proof of $\Delta, P \wedge \theta \Rightarrow \Gamma$.

Case 4: $\phi := \psi \wedge \theta$ is in the succedent, so the sequent is $\Delta \Rightarrow \psi \wedge \theta, \Gamma$. By Corollary 14, we know $\Delta, \Gamma \neq \emptyset$. The variable sharing result Proposition 12 entails that ψ and θ must be atom-disjoint. Moreover, that proposition entails that Δ, Γ can be partitioned into Δ_1, Γ_1 and Δ_2, Γ_2, where Δ_1, Γ_1 $\overline{\bowtie}$-connect with ψ and Δ_2, Γ_2 $\overline{\bowtie}$-connect with θ. We know that, $\Delta_1 \Rightarrow \psi, \Gamma_1$ will be atom-disjoint from $\Delta_2 \Rightarrow \theta, \Gamma_2$, for otherwise there would be a non-trivial $\bowtie$-cycle in our original sequent, in contradiction to Proposition 12.

Now I also claim that $\Delta_1 \Rightarrow \psi, \Gamma_1$ and $\Delta_2 \Rightarrow \theta, \Gamma_2$ are gauntly valid. I will consider only the former sequent, but the argument is identical for the latter. First, observe that there must be a perfectly valid subsequent S_ψ of $\Delta \Rightarrow \psi, \Gamma$ that contains ψ; for otherwise, we would have a perfectly valid subsequent of $\Delta \Rightarrow \Gamma$, which is impossible. Second, because every perfectly valid sequent is $\overline{\bowtie}$-connected, S_ψ must be a subsequent of $\Delta_1 \Rightarrow \psi, \Gamma_1$. Third, if S_ψ were a *proper* subsequent of $\Delta_1 \Rightarrow \psi, \Gamma_1$, then that would give us a *proper* valid subsequent of $\Delta \Rightarrow \psi \wedge \theta, \Gamma$, which is impossible. Thus $\Delta_1 \Rightarrow \psi, \Gamma_1$ is perfectly valid.

Finally, suppose that there were a valid proper coarsening $\Delta_1' \Rightarrow \psi_1', \ldots, \psi_n', \Gamma_1'$, where for some substitution σ we have $\sigma(\Delta_1') = \Delta_1, \sigma(\Gamma_1') = \Gamma_1$, and $\sigma(\psi_1') = \ldots = \sigma(\psi_n') = \psi$. Suppose further that $n > 1$. Then for some m, and some $\delta_1, \ldots, \delta_m \in \Delta_1' \cup \Gamma_1'$,

$$\psi_1' \bowtie \delta_1 \bowtie \ldots \bowtie \delta_m \bowtie \psi_2'.$$

Moreover, δ_1 itself cannot be one of the ψ_i''s; for if it were, then there would be some atom that occurred both positively and negatively in $\psi = \sigma(\psi_1') = \sigma(\delta_1)$, and that contradicts Proposition 12. But now in the context of our original sequent $\Delta_1, \Delta_2 \Rightarrow \psi \wedge \theta, \Gamma_1, \Gamma_2$, we would have

$$\psi \wedge \theta \bowtie \sigma(\delta_1) \bowtie \ldots \bowtie \sigma(\delta_m) \bowtie \psi \wedge \theta.$$

This would be a non-trivial cycle, in contradiction to Proposition 12. So $n = 1$. But now the fact that $\Delta_1' \Rightarrow \psi', \Gamma_1'$ is valid would entail that $\Delta_1', \Delta_2 \Rightarrow \psi' \wedge \theta, \Gamma_1', \Gamma_2$ is a valid proper coarsening of our original sequent, which is impossible.

In sum, we have two gauntly valid, atom-disjoint sequents $\Delta_1 \Rightarrow \psi, \Gamma_1$ and $\Delta_2 \Rightarrow \theta, \Gamma_2$. By the i.h., this gives us two atom-disjoint proofs Π_1 and Π_2 that satisfy our four constraints. So, appending an instance of $\wedge$-R below these two proofs gives us our desired proof of $\Delta \Rightarrow \psi \wedge \theta, \Gamma$.

Case 5: $\phi := \psi \vee \theta$ is in the antecedent. This is essentially similar to Case 4.

Case 6: $\phi := \psi \vee \theta$ is in the succedent. This is essentially similar to Case 3.

□

This theorem is notable for several reasons. First, this theorem takes the form of a soundness and completeness result: it shows how a given proof system suffices to deduce all and only the members of some antecedently given class of arguments. In this respect, Theorem 16 is significant in the way that all soundness and completeness theorems are, showing that a given deductive system is adequate for a certain task. Furthermore, Theorem 16 does not just show that there is *some* deductive system that proves all and only the gauntly valid sequents, it shows how this property is exhibited in a certain neatly circumscribed *fragment* of a standard deductive system for classical logic.

A second reason this theorem is significant is because it entails that every gauntly valid sequent has a proof such that every subproof of that proof is also a proof of a gauntly valid sequent. Proving a gauntly valid sequent will never require us to take a detour through non-gaunt sequents.

A final point of interest from Theorem 16 is that the ultimate proof resulting from this proof search process must be of a tightly constrained form. Suppose we are given a gauntly valid sequent $\Delta \Rightarrow \Gamma$, and we find a proof Π of this sequent by following the proof-search algorithm implicit in Theorem 16. Then, as is easy to verify by inspecting that algorithm, we know that every non-atomic subformula of Δ, Γ will be a critical formula of some inference in Π. Moreover, by induction on the complexity of Δ, Γ, it follows that if there are n connectives in Δ, Γ, then there will be at most $2n + 1$ nodes in Π.[14] Thus, every gauntly valid sequent has a short proof, which means that we can efficiently decide whether a given sequent is gauntly valid or not. This is in stark contrast to the situation with (mere) validities in classical logic, where the question of whether tautologies have short proofs is a major and difficult open problem.[15] Of course, as some of our examples in Sect. 5 illustrated, a gauntly valid sequent can be much longer than its substitution instances. So we cannot expect that the existence of short proofs of gauntly valid sequents will resolve the question of whether all valid sequents have short proofs. It does, however, point to interesting questions for further research, such as whether or when there is an efficient method for finding a gauntly valid coarsening of a given valid sequent.

6 Conclusion

Beginning with the simple observation that it is common to talk about relevance in connection with informal logic, my goal in this paper has been to describe how we can identify a formal and precise counterpart to the informal and vague notion of

[14] And it would be easy to improve on this upper bound by counting the number of $\vee$'s in positive occurrences in Δ or negative occurrences in Γ and the number of $\wedge$'s in negative occurrences in Δ or positive occurrences in Γ.

[15] This problem is actually equivalent to the $NP =_? coNP$ problem. See [11] and [27] for overviews of work on propositional proof complexity and its connections to other areas of logic.

relevance. If relevance means contributing to the validity of an argument, we can make this idea precise by answering two questions. What does it take to contribute to the validity of an argument? And what are the components of an argument that can contribute or fail to contribute to its validity? Different ways of answering those questions led us to four different explications of relevance: relevant validity, perfect validity, gaunt validity, and perfectibility.

Each of these definitions is relative to a background definition of validity. Hence, the interest in these relevance properties is how they can figure in our study of the background logic. I have focused particularly on classical logic. There are broadly three types of questions we can ask about these relevance properties.

First, we can ask how the various explications of relevance relate to each other, for instance whether one is a limiting case of another, and so on. As I observed in Sect. 3, the four explications of relevance that I have discussed here fall into a simple hierarchy based on the two operations of taking substitutions and agglomerations.

Second, we can ask what intrinsic properties the different explications of relevance entail. Do they have variable sharing properties? Do they exhibit structural properties such as interpolation and transitivity? I have addressed some such questions, especially about variable sharing, in Sect. 4.

Third, we can ask about how these properties relate to known deductive systems. When are these relevance properties preserved or not preserved in a deductive system? Are there deductive systems that prove only sequents enjoying one of our relevance properties? In Sect. 5, I answered this latter question by identifying a fragment of a standard sequent calculus that proves all and only the gauntly valid sequents.

Acknowledgments Thanks to an audience at the New Directions in Relevance workshop where I presented this material and to an anonymous referee for comments on the article. This work was partly supported by an Early Career Scheme grant from the Research Grants Council of Hong Kong SAR, China, project number 23606822.

References

1. Anderson, A. R., & Belnap, N. D. (1975). *Entailment: The logic of relevance and entailment, Vol. I*. Princeton: Princeton University Press.
2. Anderson, A. R., Belnap, N. D., & Dunn, J. M. (1992). *Entailment: The logic of relevance and entailment, Vol. II*. Princeton: Princeton University Press.
3. Beall, J., & Restall, G. (2006). *Logical pluralism*. Oxford: Oxford University Press.
4. Brauer, E. (2020). Relevance for the classical logician. *Review of Symbolic Logic*, *13*(2), 436–457.
5. Brauer, E. (2024a). Coarsening natural deduction proofs I: Finding perfect proofs. *Journal of Logic and Computation*. https://doi.org/10.1093/logcom/exad077
6. Brauer, E. (2024b). Coarsening natural deduction proofs II: Finding gaunt proofs. *Journal of Logic and Computation*. https://doi.org/10.1093/logcom/exae003
7. Copi, I. M., Cohen, C., & McMahon, K. (2014). *Introduction to logic* (14th ed.). Pearson Education Limited.
8. Geach, P.T. (1970). Entailment. *Philosophical Review*, *79*(2), 237–239.

9. Gemes, K. (2007). Irrelevance: Strengthening the Bayesian requirements. *Synthese*, *157*(2), 161–166.
10. Hurley, P. J. (2015). *A concise introduction to logic* (12th ed.). Stamford, CT: Cengage Learning.
11. Krajíček, J. (1995). *Bounded arithmetic, propositional logic, and proof complexity*. Cambridge: Cambridge University Press.
12. Krämer, S. (2017). A hyperintensional criterion of irrelevance. *Synthese*, *194*(8), 2917–2930.
13. Krämer, S., & Roski, S. (2017). Difference-making grounds. *Philosophical Studies*, *174*, 1191–1215.
14. Lau, J. Y. F. (2011). *An introduction to critical thinking and creativity*. Hoboken, NJ: Wiley.
15. Lehrer, K. (1973). Relevant deduction and minimally inconsistent sets. *Philosophia*, *3*(3), 153–165.
16. Prawitz, D. (2006). *Natural deduction: A proof-theoretical study*. Orig. published 1965. Mineola, NY: Dover.
17. Schurz, G. (2023). Relevance as difference-making: a generalized theory of relevance and its applications. *Philosophical Studies*. https://doi.org/10.1007/s11098-023-01988-6
18. Shapiro, S. (2014). *Varieties of logic*. Oxford: Oxford University Press.
19. Smiley, T. J. (1959). Entailment and deducibility. *Proceedings of the Aristotelian Society*, *59*, 233–254.
20. Tennant, N. (1979). Entailment and proofs. *Proceedings of the Aristotelian Society*, *79*, 167–189.
21. Tennant, N. (1984). Perfect validity, entailment and paraconsistency. *Studia Logica*, *43*, 179–198.
22. Tennant, N. (1987). *Anti-realism and logic*. Oxford: Clarendon Press.
23. Tennant, N. (1997). *The taming of the true*. Oxford: Oxford University Press.
24. Tennant, N. (2010). *On gaunt logic*. Unpublished MS, originally circa 1990. Version from 2010 consulted.
25. Tennant, N. (2015). The relevance of premises to conclusions of core proofs. *Review of Symbolic Logic*, *8*, 743.
26. Tennant, N. (2017). *Core logic*. New York: Oxford University Press.
27. Urquhart, A. (1995). The complexity of propositional proofs. *Bulletin of Symbolic Logic*, *1*(4), 425–467.
28. Verdée, P., de Bal, I., & Samonek, A. (2019). A non-transitive relevant implication corresponding to classical logic consequence. *Australasian Journal of Logic*, *16*(2), 9–40.
29. von Wright, G. H. (1957). The concept of entailment. In *Logical studies* (pp. 166–191). London: Routledge and Kegan Paul.
30. Williamson, T. (2008). *The philosophy of philosophy*. Malden, MA: Blackwell.
31. Yablo, S. (2003). Causal relevance. *Philosophical Issues*, *13*, 316–328.
32. Yablo, S. (forthcoming). Relevance without minimality. In P. van Elswyck, C. D. Kirk-Giannini, A. Egan, & D. Kindermann (Eds.), *Unstructured content*. Oxford University Press. ISBN: 978-0-198-82355-1.

A Topic-Theoretic Perspective on Variable-Sharing (from the Black Sheep of the Family)

Thomas M. Ferguson

Abstract In the canon of relevant logic—*Entailment*, *Relevant Logics and Their Rivals*, *Directions in Relevant Logic*—space has typically been made for acknowledging *containment logics* in the style of W.T. Parry or R.B. Angell. Although containment logics' satisfaction of the relevant logics' key variable-sharing criterion and are thus relevant, the tenor of such acknowledgement has typically been strained—ranging from grudging to adversarial. This familial estrangement has prevented the relevantist from reaping the fruits offered by her cousin, of course. I would like to suggest in this note that the fruits are not meager—especially in light of current work on the theory of subject-matter—and that the containment logician might have some worthwhile words of advice with respect to variable-sharing.

1 Greetings, Relevantists, from Your Containment Logic Cousins

In Anderson and Belnap's sourcebook [1], space is set aside for systems whose motivations share some common spirit with relevant logics. As Storrs McCall puts it:

> [T]his book has been expanded to include a number of neighbors of **E** in addition to **E** itself. This is not to say that the authors are really very fond of **E**'s neighbors. As they say, they growl a bit about them from time to time. [1, p. 434]

One such cluster of neighbors is the class of Parry-style systems dubbed "conceptivist" by Sylvan, that is, the *containment logics* described by Richard B. Angell in [2] or William Parry in [21], in which validity of an entailment $\varphi \to \psi$ requires (informally) that the "content" of the consequent ψ is "included" within that of the

T. M. Ferguson (✉)
Department of Cognitive Science, Rensselaer Polytechnic Institute, Troy, NY, USA

Institute of Computer Science, Czech Academy of Sciences, Prague, Czech Republic
e-mail: tferguson@gradcenter.cuny.edu

I. Sedlár et al. (eds.), *New Directions in Relevant Logic*, Trends in Logic 63,
https://doi.org/10.1007/978-3-031-69940-5_3

antecedent φ. Formally, such systems are characterized by the so-called *Proscriptive Principle*:

Definition 1 A logic L satisfies the Proscriptive Principle if for every theorem $\varphi \to \psi$, every variable in ψ appears in φ.

Despite a neighborly policy of tolerance—the original *Directions in Relevant Logic* includes contributions from Angell and Parry in [3] and [22], respectively—the relationship between relevant and containment logics was strained from the outset. The space Anderson and Belnap reserve for discussion of Parry's PAI in [1] is somewhat marred for its inclusion in the book's "miscellany" rather than in the spaces devoted to relevant logic proper. The implicit slight is made explicit in the Australasian tradition, with Sylvan describing conceptivist logics as "unimpressive." [25, p. 101] (Further details of Sylvan's ire against containment logics are discussed in [9].)

In fairness, many central figures in the relevant logic were responsible for advancing the study of containment logics—Dunn's and Urquhart's foundational investigations in [7] and [29] lay the groundwork for Fine's semantics introduced in [13]. Even Sylvan eventually acknowledged certain applications for which containment logics were well-suited in [26]. Nevertheless, even in the most ecumenical cases, Parry-style logics—and their legitimacy as authentic members of the relevant family—have been held at one arm's length.

In spite of this, the *variable-sharing criterion* that characterizes relevant logics is satisfied by such logics, as Parry proves in [20]. Indeed, Parry beats the relevantists to variable-sharing by decades as his Proscriptive Principle entails variable-sharing. With their legitimacy unacknowledged, containment systems can with some justification be considered the bastard cousins of relevant logics.

Holding neighbors at too great a distance, however, obscures any neighborly wisdom that they may have to offer. There are lessons for relevant logic that containment logics are uniquely positioned to provide, especially given the recent emphasis in work on *topic* or *subject-matter*.

The prevailing view on relevance is that relevance can be explicated in terms of subject-matter (as we will detail shortly) while the treatments of topic-sensitivity have so far supported a picture closely resembling those of containment logics. For better or for worse, relevant logics' bastard cousins thus enjoy a stronger relationship with current work on theories of topic. Despite its base status as relevant logics' by-blow, the Parry-style account is in a position to translate items from the theory of topic in a way useful to the relevant logician. In other words, if the relevantist seeks after clarity about her guiding principle—that is, *relevance*—it may yet behoove her to deign to receive some advice from these black sheep of the family.

The friendly suggestion offered from the containment perspective is, in a nutshell, this: Containment logics have been responsive to advances in topic-theoretic considerations. Such updated understandings of topic are reflected in refinements to the standard variable inclusion properties of Parry-style systems, which straightforwardly induce corresponding variable-sharing properties. E.g., developments in the theory of topic drive the work on containment logic in the recent [10] and [11]. The logics described in these papers (we will show) exhibit features that are special cases of novel strengthenings of variable-sharing. Thus, our best understanding of topic recommends to us new forms of variable-sharing.

2 Variable-Sharing, Topic, and Relevance

As mentioned, the *variable-sharing property*, introduced and shown to hold of E in [4], is one of the hallmark features of relevant logics:

Definition 2 A logic L enjoys the variable-sharing property if for every L-theorem $\varphi \rightarrow \psi$ there exists an atomic formula common to both φ and ψ.

That a logic enjoys the variable-sharing property is importantly understood as a necessary but not sufficient condition for relevance as it ensures that an entailment's antecedent and consequent share some "common content."

In Anderson and Belnap's words, this common content is guaranteed for the following reasons:

> A formal condition for "common meaning content" becomes almost obvious once we note that commonality of meaning in propositional logic is carried by commonality of propositional variables... If this property fails, then the variables in A and B may be assigned propositional values, in such a way that the resulting propositions have no meaning content in common and are totally irrelevant to each other. [1, p. 33]

Although this notion of *commonality of meaning* is offered as a necessary condition for relevance in the above passage, Anderson and Belnap do not dig further into the matter. Nevertheless, in the intervening years, logicians and philosophers of logic have continued to inquire about what this commonality of meaning amounts to.

2.1 Relevance and Topic

When we look to Anderson and Belnap's "common meaning content," there exist a number of potential interpretations of what precisely this means. The predominant view, one finds, is that the intended sense of *relevance* has something to do with the *overlap of topic* or *subject-matter*. This received view makes relevant logics an especially enticing field as work on the theory of topic—in particular its relationship with logical matters—has exploded over the past decade (e.g., [5, 9, 30]). The reach of this view is quite broad, including e.g. Ed Mares' *Stanford Encyclopedia of Philosophy* article:

> Relevant logicians point out that what is wrong with some of the paradoxes (and fallacies) is that the antecedents and consequents (or premises and conclusions) are on completely different topics... [The variable-sharing property] is a formal principle that relevant logicians apply to force theorems and inferences to "stay on topic." [18]

In other words, relevance can be naturally characterized by shared subject-matter between antecedents and consequents. Likewise, in Lloyd Humberstone's *The Connectives* we find the following:

> The variable-sharing condition on provable implications in FMLA logics, imposed as a necessary condition for a system to qualify as a relevant logic, represents one way of saying what 'relevance' amounts to; it formalizes the idea that for an antecedent to be relevant to its consequent, they must overlap as to subject-matter. [16, p. 336]

One more notable case is found in Neil Tennant's *Core Logic*:

> Moreover, these systems do so while keeping one's reasoning absolutely relevant—that is, while sustaining a proper 'connection of subject matter' between the premises and the conclusion of any proof. [28, p. 2]

I do not want to push much further on this, nor provide a conceptual analysis indicating that relevance really is a matter of topic overlap. My own opinion of the matter is, indeed, rather mercurial. In any case, this is not the goal of this paper. It suffices, rather, to point out that this intuitive understanding of relevance is a sufficiently common illustrative device to be considered one of the prevailing views on the matter. As such, lessons from the theory of topic ought to enjoy some appeal as natural avenues towards clearer understandings of relevance.

2.2 *Variable-Sharing from the Topic-Theoretic Perspective*

Read straightforwardly, Anderson and Belnap's suggestion that "commonality of meaning... is carried by commonality of propositional variables" was doomed from the moment it was printed. As we know well, the vanilla variable-sharing property has received a number of refinements in which stricter conditions are imposed. In each such restriction, commonality of propositional variables alone becomes insufficient to ensure commonality of meaning, or—as common wisdom has it—to ensure topic overlap.

E.g., the *strong variable-sharing property* is a refinement in which the *sign* of an atomic formula factors into whether overlap occurs as regards topic. Rather than shared variables, what is required is that shared variables have a common sign (are both either positively or negatively appearing in a formula), i.e., common content is only guaranteed when two variables are shared with the same sign. Consequently, from the strong relevance perspective, the meaning of an atomic formula is sensitive to sign or, in other words, sign-altering operations (like *negation*) play a nontrivial role in determining a formula's overall subject-matter.

Incidentally, Parry's Proscriptive Principle, too, can be paired with an analogous strong Proscriptive Principle in which the sign of variables plays a role, i.e., every variable in the consequent must appear in the antecedent *with the same sign*. Thus, an acknowledgement of the topic-theoretic influence of sign has precedent in the containment logic tradition itself. A feature of Angell's AC, there has recently been a fair amount of attention on the strong Proscriptive Principle and related matters (see e.g., Hornischer's [15], Randriamahazaka's [24], and Szmuc and Rubin's [27]). Although the strong version is not observed in the system central to this investigation, we will return to consider the strong Proscriptive Principle in the concluding remarks.

Returning to the core relevant logic tradition, the *depth relevance property* described by Ross Brady in [6] requires that the atomic formulae witnessing variable-sharing be nested within the scope of the same number of conditionals. On this picture, common content is only guaranteed when the shared variables have been operated on by conditionals in the same way. Again, to say that two occurrences of the same atomic formula under different depths of nesting within conditionals need not ensure topic overlap is *au fond* a claim that applications of the conditional serve to direct the topic of a subformula. A strong version of depth relevance has been proven to hold of a number of weak relevant logics by Logan in [17].

In [12], it was argued that such well-known refinements can be given a topic-theoretic analysis and that the insufficiency of common variables in each case can be attributed to the transformative power of particular intensional operators. In other words, the distinctions in variable-sharing properties align with the matter of whether or not particular connectives contribute to the overall topic of a complex, that is, whether particular connectives are *topic transparent*.

In the theory of topic, it is generally acknowledged that there exist connectives that exhibit a topic-theoretically transformative function through which the connective itself may influence and shape the topic of a complex sentence. In such cases, the connective may nontrivially contribute to the overall topic of a complex sentence in which it appears. Clearly, such connectives are not topic transparent, i.e., the topic of a sentence in which the connective appears need not coincide with the simple mereological fusion of the topics of its subformulae.

That topic transparency should fail for *intensional* connectives such as relevant implication is a relatively orthodox position. E.g., Franz Berto argues in [5, p. 34] that modal operators can be topic-shifting in this way. In the case of relevant implication, then, we take it as unexceptional that relevant conditionals wield some influence over the ultimate topic of a complex in which they appear. Despite an entrenched commitment to so-called *negation transparency*, the principle that *negations* can wield that same topic-theoretic influence as relevant conditionals has been a mainstay of the relevant enterprise.[1]

Ferguson and Logan [12] notices that a pattern holds between strong relevance and depth relevance. Against the informal assumption that relevance is overlap of topic, the foregoing observation admits a reading as an acknowledgement that intensional negations and conditionals, respectively, maintain a transformational power over the topics of their subformulae. In other words, the canonical revisions to the variable-sharing property each encapsulate a particular regularity governing how a complex's topic is shaped by the negations or conditionals appearing within.

[1] Witnessed by e.g. strong variable-sharing holding of R.

2.3 A Further Wrinkle: Order

If one accepts that the operations of negation and conditionality are topic-theoretically transformative—and that these qualities have downstream effects on appropriate refinements to the variable-sharing property—this is still only a fraction of the task. One still must exhaustively investigate the details and regularities of such influences. Brady's depth relevance of [6] can be seen in such a light in noting that the function is, in a sense, not idempotent, i.e., the number of times the transformation is applied is a crucial element of determining a complex's topic.

We can add to this by noting a further feature common to the topic-theoretic influence of many conditionals: *order*. Simply put, *order matters*. The investigation of [11] takes a cue from Yablo's remarks in [30] in which Yablo asks:

> Why is MAN BITES DOG a better headline than DOG BITES MAN? One thing we can certainly say is that it is on a more interesting topic. A more interesting topic is a different topic. [30]

Insofar as intensional conditionals, too, express relations between antecedent and consequent, Yablo's intuitions apply to the topics of conditionals as well. Thus, [11] offered examples that share the character of Yablo's MAN BITES DOG case:

1 Were John to leap from the bridge, he would break his arm.
2 Were John to break his arm, he would leap from the bridge.

On its face, **[1]** appears to be a more-or-less straightforward truth concerning the physiological consequences of falling from a great height. In contrast, **[2]** is a relatively shocking statement about John. Like Yablo's example, **[2]** is clearly about a more *interesting* topic than **[1]** and, by similar reasoning, the two must be about *different* topics.

3 Conceptivist Logics

The class of conceptivist logics refines the relevant intuition by demanding that validity requires not only *overlap* but *inclusion* of the topic of the consequent within that of the antecedent. The prototypical such system is due to William Parry, who introduced a logic of *analytic implication* in [20], inspired by a thesis that if validity is to be *analytic* the subject-matter of the consequent cannot *exceed* that of the antecedent.

Again, it is important to note that Parry's PAI satisfies the variable-sharing property and is thus in a strong sense a *relevant logic*. It may be part of a side of the family that is less polished than the systems favored by Anderson and Belnap or Sylvan and Meyer, but the claim to legitimacy made by PAI and subsystems is strong.

3.1 Parry's Analytic Implication

Parry's logic is motivated by a desire to give a formal account of entailment as drawing the consequent from the antecedent through a process of *analysis*. As discussed in [9], a guiding intuition is that a necessary condition for the *analyticity* of an entailment relation is that the consequent of a valid entailment should not *include subject-matter* (or *concepts*) *not present* in the antecedent. Parry's PAI went without a semantic analysis until Fine's [13] provided a model theory based on the following structures:

Definition 3 A PAI Fine model is a tuple $\langle W, R, \mathcal{T}, \oplus, v, t\rangle$ with:

- $\langle W, R\rangle$ is an S4 Kripke frame
- For each $w \in W$, $\langle \mathcal{T}_w, \oplus_w\rangle$ is a join semilattice
- v is a valuation from atomic formulae to W
- For each $w \in W$, t_w is a function mapping atomic formulae to $\mathcal{T}_w$
- If wRw' and $t_w(p) \leq_w t_w(q)$ for atoms p, q then $t_{w'}(p) \leq_{w'} t_{w'}(q)$

The topic assignment function and truth conditions are extended recursively:

Definition 4 The topic assignment function t_w is extended through the language:

- $t_w(\neg\varphi) = t_w(\varphi)$
- $t_w(\varphi \star \psi) = t_w(\varphi) \oplus_w t_w(\psi)$ for binary connectives $\star$

Definition 5 Truth conditions are defined so that:

- $w \Vdash p$ if $w \in v(p)$
- $w \Vdash \neg\varphi$ if $w \nVdash \varphi$
- $w \Vdash \varphi \wedge \psi$ if $w \Vdash \varphi$ and $w \Vdash \psi$
- $w \Vdash \varphi \rightarrow \psi$ if $\begin{cases} \text{for all } w' \text{ such that } wRw', \text{ if } w' \Vdash \varphi \text{ then } w' \Vdash \psi \\ t_w(\psi) \leq_w t_w(\varphi) \end{cases}$

Note that the fusion $\oplus_w$ is intended to capture a mereological fusion of two topics or subject-matters, e.g., where the topic $t_w(p)$ may be **Cats** and the topic $t_w(q)$ may be **Dogs**, the fusion $t_w(p) \oplus_w t_w(q)$ would be the complex topic of **Cats-And-Dogs**.

These definitions provide what is needed to define validity:

Definition 6 $\Gamma \vDash_{\mathsf{PAI}} \varphi$ if for every PAI model and every world w, if $w \Vdash \gamma$ for each $\gamma \in \Gamma$ then $w \Vdash \varphi$

So far, there is no refinement to variable-sharing properties that seems to arise from PAI; the Proscriptive Principle implies vanilla variable-sharing and nothing more. In order to discover refined versions of the Proscriptive Principle, we must survey more recent subsystems of PAI.

3.2 Conditional-Agnostic Analytic Implication

Recent modifications of the Parry-style systems have been introduced in which new developments in the theory of topic are applied to PAI in order to generate more refined systems of containment logic. These systems clearly should—although have not thus far been proven to—impose stricter requirements on containment than Parry's favored condition. *These* refinements promise to have corresponding variable-sharing consequences that might be reimported into the mainstream relevant tradition.

In particular, we focus on the system CA/PAI of *conditional-agnostic analytic implication* introduced in [11]. The design of this system was motivated by an acknowledgement both that conditionals can transform topic and that the regularities of such transformations can vary wildly depending on the type of conditional whose topic is being determined. CA/PAI thus characterizes a Parry-style system making minimal assumptions about the topic assigned to any given conditional in the language. Model theoretically, CA/PAI models can be defined by modifying Fine's models in the following way:

Definition 7 A CA/PAI Fine model is a tuple $\langle W, R, \mathcal{T}, \oplus, \multimap, v, t, h\rangle$ differing from Definition 3 by the addition of:

- For each $w \in W$, $\multimap_w$ is a binary function from $\mathcal{T}_w \times \mathcal{T}_w \to \mathcal{T}_w$
- For all w, w' such that wRw', $h_{w,w'} : \mathcal{T}_w \to \mathcal{T}_{w'}$ is a homomorphism such that:
  - for atoms p, $h_{w,w'}(t_w(p)) = t_{w'}(p)$
  - $h_{w,w'}(a \oplus_w b) = h_{w,w'}(a) \oplus_{w'} h_{w,w'}(b)$
  - $h_{w,w'}(a \multimap_w b) = h_{w,w'}(a) \multimap_{w'} h_{w,w'}(b)$

While the homomorphism $h_{w,w'}$ do not play a role in what follows, their presence is required to ensure preservation of topic inclusion at accessible worlds.

Definition 8 The topic assignment function t_w is extended through the language:

- $t_w(\neg\varphi) = t_w(\varphi)$
- $t_w(\varphi \star \psi) = t_w(\varphi) \oplus_w t_w(\psi)$ for $\star$ extensional
- $t_w(\varphi \to \psi) = t_w(\varphi) \multimap_w t_w(\psi)$

Definitions of truth at a world and validity can be straightforwardly borrowed from the foregoing subsection.

Informally, we can approach the $\multimap$ function as encapsulating the *agnosticism* of CA/PAI, allowing the model nearly complete freedom to assign a topic to a conditional. Despite this freedom, the device is nevertheless strong enough to ensure some interesting variations of the Proscriptive Principle hold, as we will see in the next section.

To explore further, we will need to avail ourselves of some facts about the proof theory, the provision of which demands some discussion of notation. First, we follow Parry's notation in which $f(\varphi)$ is any formula in which φ appears and treat

$f(\psi)$ as the result of replacing one or more instances of φ with ψ in $f(\varphi)$. We have two further notational devices to consider. First, consider the following definition:

Definition 9 The *armature* of a sentence φ is defined as follows:

- $\mathcal{L}_{\mathbf{A}}(p) = \{p\}$
- $\mathcal{L}_{\mathbf{A}}(\neg\varphi) = \mathcal{L}_{\mathbf{A}}(\varphi)$
- $\mathcal{L}_{\mathbf{A}}(\varphi \star \psi) = \mathcal{L}_{\mathbf{A}}(\varphi) \cup \mathcal{L}_{\mathbf{A}}(\psi)$ for $\star$ extensional
- $\mathcal{L}_{\mathbf{A}}(\varphi \rightarrow \psi) = \{\langle \mathcal{L}_{\mathbf{A}}(\varphi), \mathcal{L}_{\mathbf{A}}(\psi)\rangle\}$

Then the formula $g(\varphi)$ stands in for any formula such that $\mathcal{L}_{\mathbf{A}}(\varphi) \subseteq \mathcal{L}_{\mathbf{A}}(g(\varphi))$. Finally, the notation $\mathbf{t}_\varphi$ represents the formula $\varphi \supset \varphi$.

In any case, we are now prepared to offer an axiomatization of $\mathsf{CA/PAI}$:

Definition 10 The logic of *conditional agnostic analytic implication* $\mathsf{CA/PAI}$ is determined by the following axioms:

$[A1]$	$(\varphi \wedge \psi) \rightarrow (\psi \wedge \varphi)$
$[A2]$	$\varphi \rightarrow (\varphi \wedge \varphi)$
$[A3]$	$\varphi \rightarrow \neg\neg\varphi$
$[A4]$	$\neg\neg\varphi \rightarrow \varphi$
$[A5]$	$(\varphi \wedge (\psi \vee \xi)) \rightarrow ((\varphi \wedge \psi) \vee (\varphi \wedge \xi))$
$[A6]$	$(\varphi \vee (\psi \wedge \neg\psi)) \rightarrow \varphi$
$[A7^\dagger]$	$((\varphi \rightarrow \psi) \wedge (\psi \rightarrow \xi)) \supset (\varphi \rightarrow \xi)$
$[A8^\dagger]$	$(\varphi \rightarrow (\psi \wedge \xi)) \supset (\varphi \rightarrow \psi)$
$[A9^\dagger]$	$((\varphi \rightarrow \xi) \wedge (\psi \rightarrow \zeta)) \supset ((\varphi \wedge \psi) \rightarrow (\xi \wedge \zeta))$
$[A10^\dagger]$	$((\varphi \rightarrow \xi) \wedge (\psi \rightarrow \zeta)) \supset ((\varphi \vee \psi) \rightarrow (\xi \vee \zeta))$
$[A11^\dagger]$	$(\varphi \rightarrow \psi) \supset (\varphi \supset \psi)$
$[A12^\dagger]$	$((\varphi \leftrightarrow \psi) \wedge f(\varphi)) \supset f(\psi)$
$[A13^\dagger]$	$g(\varphi) \rightarrow (\varphi \supset \varphi)$
$[A14^\dagger]$	$((\neg\varphi \rightarrow \varphi) \wedge (\varphi \rightarrow \psi)) \supset (\neg\psi \rightarrow \psi)$
$[D1]$	$(\mathbf{t}_\varphi \leftrightarrow \mathbf{t}_\psi) \supset (\mathbf{t}_{f(\varphi)} \rightarrow \mathbf{t}_{f(\psi)})$
$[D2]$	$(\varphi \rightarrow \psi) \supset (\neg(\varphi \rightarrow \psi) \rightarrow (\varphi \rightarrow \psi))$

and rules:

$[MP1]$	$\varphi, \varphi \supset \psi \Rightarrow \psi$
$[MP2]$	$\varphi, \varphi \rightarrow \psi \Rightarrow \psi$
$[ADJ]$	$\varphi, \psi \Rightarrow \varphi \wedge \psi$
$[MOD]$	$\varnothing \Rightarrow \neg\varphi \rightarrow \varphi$ if φ is an axiom

Provability can be defined in the usual sense, that is, the existence of a Hilbert-style proof.

So far, so good; we have rehearsed the details of a containment logic whose understanding of topic inclusion is stricter than that required by Parry. What remains is to extract from lessons from the system CA/PAI that can be brought to bear on questions in mainstream relevance logics.

4 Words of Advice for Relevantists

So what, precisely, are these promised words of advice for relevantists? To answer, we take a cue from the freedom of the $\multimap$ function of CA/PAI models. If considerations of topic overlap are taken seriously—and the transformative nature of intensional conditionals is acknowledged (as I think it should be)—then relevance seems to demand that the influence such connectives wield over subject-matter means Anderson and Belnap's assumption that "commonality of meaning in propositional logic is carried by commonality of propositional variables" was far too hasty. If we remain agnostic about the particulars about this influence—as encoded in CA/PAI—to ensure commonality of meaning, we need to track the genealogy—including order—of the applications of the conditional.

4.1 Intensional Genealogy and Variable-Sharing

We will now make this more precise. In order to formalize this intuition of genealogy, let us define a class of objects recording a formula's genealogy. We use decorated formulae (e.g., $\dot{p}$ or $\ddot{\xi}$) to indicate a particular occurrence of a formula and the notation $\varphi[\dot{\psi}]$ to emphasize that $\dot{\psi}$ is a focal occurrence of ψ occurring in φ. $\varphi[\dot{\psi}] \wedge \xi$, for example, indicates that the focal occurrence of ψ is in the first conjunct.

Definition 11 An *intensional genealogy* of an occurrence $\dot{\psi}$ of a formula ψ in a formula φ is a string defined recursively:

- If $\varphi = \psi$ then $\iota(\varphi[\dot{\psi}]) = \langle\rangle$
- $\iota(\neg\varphi[\dot{\psi}]) = \iota(\varphi[\dot{\psi}])$
- $\iota(\varphi[\dot{\psi}] \star \xi) = \iota(\xi \star \varphi[\dot{\psi}]) = \iota(\varphi[\dot{\psi}])$ for $\star \in \{\wedge, \vee\}$
- $\iota(\varphi[\dot{\psi}] \to \xi) = \iota(\varphi[\dot{\psi}]) ^\frown \mathsf{L}$
- $\iota(\xi \to \varphi[\dot{\psi}]) = \iota(\varphi[\dot{\psi}]) ^\frown \mathsf{R}$

Essentially, $\iota(\varphi[\psi])$ provides a sort of *provenance* of ψ. This provenance provides a record of the history of applications of the intensional conditional to the subformula—including the matter of whether ψ was included as antecedent or consequent—from which one may recover a detailed account of the influences over the overall subject-matter of φ exerted by intensional conditionals.

For example, consider a formula $((p \wedge q) \to \neg(r \to s)) \to t$ with unique variables to eliminate ambiguity. Then:

- $\iota((((p \wedge q) \to \neg(r \to s)) \to t)[\dot{p}]) = \text{LL}$
- $\iota((((p \wedge q) \to \neg(r \to s)) \to t)[\dot{q}]) = \text{LL}$
- $\iota((((p \wedge q) \to \neg(r \to s)) \to t)[\dot{r}]) = \text{LRL}$
- $\iota((((p \wedge q) \to \neg(r \to s)) \to t)[\dot{s}]) = \text{RRL}$
- $\iota((((p \wedge q) \to \neg(r \to s)) \to t)[\dot{t}]) = \text{R}$

This allows the introduction of a novel variable-sharing property:

Definition 12 Two formulae φ and ψ have *shared genealogy* if there exists an atomic formula appearing in φ and ψ with the same intensional genealogy, i.e., an atom p with occurrences $\dot{p}$ and $\ddot{p}$ such that $\iota(\varphi[\dot{p}]) = \iota(\psi[\ddot{p}])$.

Definition 13 A logic has the *genealogical variable-sharing property* if for all theorems of the form $\varphi \to \psi$, φ and ψ have shared genealogy.

This notion of variable-sharing, like depth relevance, acknowledges that the interaction between conditionals and atoms affects the matter of whether there is shared topic. To better understand the precise details, we will turn to CA/PAI once more and the associated containment properties that it exhibits.

Given the foregoing definitions, we have enough material to provide an appropriate refinement of Parry's Proscriptive Principle:

Definition 14 A formula ψ will be *genealogically included* in a formula φ if every occurrence of an atomic formula p appearing in ψ occurs in φ with the same genealogy.

Definition 15 We say that a logic satisfies the *genealogical Proscriptive Principle* if for all theorems of the form $\varphi \to \psi$, ψ is genealogically included in φ.

Before proceeding, let us explicitly tie together genealogical inclusion with the formalism of armatures. First, note the following lemma:

Lemma 16 *For any ψ and p, the following are equivalent:*

- *there is an instance $\dot{p}$ such that $\iota(\psi[\dot{p}]) = \langle\rangle$*
- *$p \in \mathcal{L}_{\mathbf{A}}(\psi)$.*

Proof By induction on complexity of formulae. In case $p = \psi$ then this is immediate by definition as $\iota(p[\dot{p}]) = \langle\rangle$ and $\mathcal{L}_{\mathbf{A}}(p) = \{p\}$.

- If $\psi = \neg\varphi$, then $\iota(\neg\varphi[\dot{p}]) = \iota(\varphi[\dot{p}])$ which, by induction hypothesis, is $\langle\rangle$ precisely when $p \in \mathcal{L}_{\mathbf{A}}(\varphi)$. But this latter set is identical to $\mathcal{L}_{\mathbf{A}}(\neg\varphi)$.
- If $\psi = \varphi \star \xi$ for $\star$ extensional, then $\iota((\varphi \star \xi)[\dot{p}]) = \langle\rangle$ if and only if either $\iota(\varphi[\dot{p}]) = \langle\rangle$ or $\iota(\xi[\dot{p}]) = \langle\rangle$. This disjunction is by induction hypothesis equivalent to the disjunction that either $p \in \mathcal{L}_{\mathbf{A}}(\varphi)$ or $p \in \mathcal{L}_{\mathbf{A}}(\xi)$, which is just to say that $p \in \mathcal{L}_{\mathbf{A}}(\varphi) \cup \mathcal{L}_{\mathbf{A}}(\xi)$, i.e., $p \in \mathcal{L}_{\mathbf{A}}(\varphi \star \xi)$.
- If $\psi = \varphi \to \xi$ then both $\iota(\psi[\dot{p}]) \neq \langle\rangle$ and $\mathcal{L}_{\mathbf{A}}(\psi)$ is a singleton set whose element is an ordered pair (and consequently $p \notin \mathcal{L}_{\mathbf{A}}(\psi)$).

□

Followed by a definition and a fact concerning the definition:

Definition 17 Call a formula ψ an *extensionally immediate subformula* of φ if ψ appears within the scope of no instances of $\to$ and call the set of all extensionally immediate subformulae of φ $\mathtt{Ext}(\varphi)$.

Lemma 18 *If* $\psi \in \mathtt{Ext}(\varphi)$ *then* $\iota(\psi[\dot{p}]) = \iota(\varphi[\dot{p}])$.

Proof Simple induction on complexity of φ, noting it holds trivially when $\varphi = \psi$ and that the property is preserved through the application of extensional connectives to subformulae. □

This allows us to prove a crucial fact:

Lemma 19 *If* $\mathcal{L}_{\mathtt{A}}(\psi) \subseteq \mathcal{L}_{\mathtt{A}}(\varphi)$ *then* ψ *is genealogically included in* φ.

Proof We prove that if $\mathcal{L}_{\mathtt{A}}(\psi) \subseteq \mathcal{L}_{\mathtt{A}}(\varphi)$ and $\iota(\psi[\dot{p}]) = \sigma$ then there is a $\ddot{p}$ such that $\iota(\varphi[\ddot{p}]) = \sigma$, proceeding by induction on length of σ. The basis step is the case in which $\sigma = \langle\rangle$. By Lemma 16, $p \in \mathcal{L}_{\mathtt{A}}(\psi)$. By inclusion, $p \in \mathcal{L}_{\mathtt{A}}(\varphi)$ and by Lemma 16, $\iota(\varphi[\ddot{p}]) = \langle\rangle$. (Note that ψ and φ are arbitrary here.) As induction hypothesis, assume that the property holds for all τ shorter than σ. Only two cases need be considered:

- If $\sigma = \tau \frown \mathrm{L}$, then there is a subformula $\xi \to \theta \in \mathtt{Ext}(\psi)$ such that $\iota(\xi[\dot{p}]) = \tau$ and $\langle \mathcal{L}_{\mathtt{A}}(\xi), \mathcal{L}_{\mathtt{A}}(\theta)\rangle \in \mathcal{L}_{\mathtt{A}}(\psi)$ (and thus $\langle \mathcal{L}_{\mathtt{A}}(\xi), \mathcal{L}_{\mathtt{A}}(\theta)\rangle \in \mathcal{L}_{\mathtt{A}}(\varphi)$). This element is introduced into $\mathcal{L}_{\mathtt{A}}(\varphi)$ in virtue of the particular syntactic structure of φ, whence there exists a $\xi' \to \theta' \in \mathtt{Ext}(\varphi)$ such that $\mathcal{L}_{\mathtt{A}}(\xi) = \mathcal{L}_{\mathtt{A}}(\xi')$. As $\iota(\xi[\dot{p}]) = \tau$ and $\mathcal{L}_{\mathtt{A}}(\xi) \subseteq \mathcal{L}_{\mathtt{A}}(\xi')$, the induction hypothesis means that there is a $\ddot{p}$ such that $\iota(\xi'[\ddot{p}]) = \tau$, whence by definition, $\iota((\xi' \to \theta')[\ddot{p}]) = \tau \frown \mathrm{L}$. As $\xi' \to \theta'$ is in $\mathtt{Ext}(\varphi)$, by Lemma 18, we infer that $\iota(\varphi[\ddot{p}]) = \tau \frown \mathrm{L}$.
- If $\sigma = \tau \frown \mathrm{R}$, then symmetric reasoning establishes that the property holds.

□

4.2 A Genealogically Relevant Logic

We first make amends for an inexcusable omission in [11]. Although, as a conceptivist logic, it would have been natural to investigate the refinements to Parry's Proscriptive Principle that hold of CA/PAI, such an investigation was—for whatever reason—neglected.

Our interest is not, of course, merely the correction of the gap in coverage of [11]. Showing that CA/PAI observes this genealogical property has consequences for our conclusions concerning variable-sharing; just as vanilla variable-sharing is guaranteed by the Proscriptive Principle, so, too, will genealogical variable-sharing follow from the genealogical Proscriptive Principle. Correcting the gap, then, has the side-effect of providing a witness that genealogical variable-sharing holds of *some* relevant logics—even if the witness's own genealogy places it among the black sheep of the family tree.

First, let us fix some further definitions:

Definition 20 Let $\preccurlyeq$ be an complete ordering of the language. The *canonical* $\mathcal{L}_{\mathbf{A}}$ *projection* of an armature is defined recursively as follows:

- $\pi(p) = p$
- $\pi(X) = \bigwedge_{\preccurlyeq} \pi[X]$
- $\pi(\langle X, Y\rangle) = \pi(X) \rightarrow \pi(Y)$

where $\pi[X]$ is the image of X under π and $\bigwedge_{\preccurlyeq}$ is conjunction ordered by $\preccurlyeq$.[2]

In the sequel, we assume $\preccurlyeq$ to be fixed. Now, by the above definition, each element of $\mathcal{L}_{\mathbf{A}}$ may be associated with a unique formula (given the ordering $\preccurlyeq$). As each formula is associated with a unique object in $\mathcal{L}_{\mathbf{A}}$, this suggests a way to get a sort of *canonical proxy* for an object $\mathcal{L}_{\mathbf{A}}(\varphi)$:

Definition 21 Let $\preccurlyeq$ be a total ordering of the language. The *canonical proxy* of a formula φ—$\boldsymbol{\lambda}(\varphi)$—is defined so that $\boldsymbol{\lambda}(\varphi) = \pi(\mathcal{L}_{\mathbf{A}}(\varphi))$.

We will now prove some facts concerning canonical proxies, relying on the following lemmas, both of which immediately follow from observations made in [11]:

Lemma 22 $\vdash_{\mathsf{CA/PAI}} \mathbf{t}_{\neg\varphi} \leftrightarrow \mathbf{t}_{\varphi}$.

Lemma 23 $\vdash_{\mathsf{CA/PAI}} \mathbf{t}_{\varphi\wedge\psi} \leftrightarrow \mathbf{t}_{\varphi\vee\psi}$.

Now, we can prove appropriate facts about the $\boldsymbol{\lambda}$ function, notably, how it relates to conjunction, the conditional, and the $\mathbf{t}$ device.

Lemma 24 $\vdash_{\mathsf{CA/PAI}} (\boldsymbol{\lambda}(\varphi) \wedge \boldsymbol{\lambda}(\psi)) \leftrightarrow \boldsymbol{\lambda}(\varphi \wedge \psi)$.

Proof Per the definition, $\boldsymbol{\lambda}(\varphi) \wedge \boldsymbol{\lambda}(\psi)$ is $(\bigwedge_{\preccurlyeq} \pi[\mathcal{L}_{\mathbf{A}}(\varphi))] \wedge (\bigwedge_{\preccurlyeq} \pi[\mathcal{L}_{\mathbf{A}}(\psi)])$, i.e., a conjunction whose conjuncts are precisely those (with some reorderings and redundancies possible) as $\bigwedge_{\preccurlyeq} \pi[\mathcal{L}_{\mathbf{A}}(\varphi \wedge \psi)]$. This latter conjunction is, of course, simply $\boldsymbol{\lambda}(\varphi \wedge \psi)$. Some applications of $[A1]$ and $[A2]$ will thus ensure equivalence. □

Lemma 25 $\vdash_{\mathsf{CA/PAI}} (\boldsymbol{\lambda}(\varphi) \rightarrow \boldsymbol{\lambda}(\psi)) \leftrightarrow \boldsymbol{\lambda}(\varphi \rightarrow \psi)$.

Proof Both $\boldsymbol{\lambda}(\varphi) \rightarrow \boldsymbol{\lambda}(\psi)$ and $\boldsymbol{\lambda}(\varphi \rightarrow \psi)$ are defined as $\bigwedge_{\preccurlyeq} \pi[\mathcal{L}_{\mathbf{A}}(\varphi)] \rightarrow \bigwedge_{\preccurlyeq} \pi[\mathcal{L}_{\mathbf{A}}(\psi)]$, so both sides of the biconditional are identical. □

Lemma 26 $\vdash_{\mathsf{CA/PAI}} \mathbf{t}_{\varphi} \leftrightarrow \mathbf{t}_{\boldsymbol{\lambda}(\varphi)}$.

Proof As basis step in which φ is an atomic formula, $\varphi = \boldsymbol{\lambda}(\varphi)$. Assume as induction hypothesis that this holds for all subformulae of φ.

- If $\varphi = \neg\psi$, then we have equivalence of $\mathbf{t}_{\neg\psi}$ and $\mathbf{t}_{\psi}$ (by Lemma 22) and $\mathbf{t}_{\boldsymbol{\lambda}(\psi)}$ (by induction hypothesis) and, finally, $\mathbf{t}_{\boldsymbol{\lambda}(\neg\psi)}$ (by definition of $\boldsymbol{\lambda}$).

[2] Note that π is defined for a broader domain than just armatures.

- If $\varphi = \psi \wedge \xi$ then by Lemma 24, $\mathbf{t}_{\psi \wedge \xi}$ is equivalent to $\mathbf{t}_{\psi} \wedge \mathbf{t}_{\xi}$. This is equivalent to $\mathbf{t}_{\boldsymbol{\lambda}(\psi)} \wedge \mathbf{t}_{\boldsymbol{\lambda}(\xi)}$ (by applications of $[A9^{\dagger}]$ and induction hypothesis) which by Lemma 24 is equivalent to $\mathbf{t}_{\boldsymbol{\lambda}(\psi \wedge \xi)}$. The case in which $\varphi = \psi \vee \xi$ follows from this fact in concert with Lemma 23.
- If $\varphi = \psi \rightarrow \xi$, then because we have as induction hypothesis the provability of $\mathbf{t}_{\psi} \leftrightarrow \mathbf{t}_{\boldsymbol{\lambda}(\psi)}$ and $\mathbf{t}_{\xi} \leftrightarrow \mathbf{t}_{\boldsymbol{\lambda}(\xi)}$, appeals to instances of $[D1]$ (e.g. $(\mathbf{t}_{\psi} \leftrightarrow \mathbf{t}_{\boldsymbol{\lambda}(\psi)}) \supset (\mathbf{t}_{\psi \rightarrow \xi} \rightarrow \mathbf{t}_{\boldsymbol{\lambda}(\psi) \rightarrow \xi})$) ultimately yield the validity of $\mathbf{t}_{\psi \rightarrow \xi} \leftrightarrow \mathbf{t}_{\boldsymbol{\lambda}(\psi) \rightarrow \boldsymbol{\lambda}(\xi)}$. But by Lemma 25, the latter is equivalent to $\mathbf{t}_{\boldsymbol{\lambda}(\psi \rightarrow \xi)}$.

□

Now, we can start tying these observations back to the armatures:

Lemma 27 $\mathcal{L}_{\mathbf{A}}(\varphi) = \mathcal{L}_{\mathbf{A}}(\psi)$ *iff* $\vdash_{\mathsf{CA/PAI}} \mathbf{t}_{\varphi} \leftrightarrow \mathbf{t}_{\psi}$.

Proof By Lemma 26, we can work with $\mathbf{t}_{\boldsymbol{\lambda}(\varphi)} \leftrightarrow \mathbf{t}_{\boldsymbol{\lambda}(\psi)}$ without loss of generality, so we proceed to prove this by induction on complexity of $\boldsymbol{\lambda}(\varphi)$ and $\boldsymbol{\lambda}(\psi)$. In the basis step, in which $\boldsymbol{\lambda}(\varphi)$ and $\boldsymbol{\lambda}(\psi)$ are conjunctions of atomic formulae, $\mathcal{L}_{\mathbf{A}}(\varphi) = \mathcal{L}_{\mathbf{A}}(\psi)$ is clearly equivalent to the validity of $\mathbf{t}_{\boldsymbol{\lambda}(\varphi)} \leftrightarrow \mathbf{t}_{\boldsymbol{\lambda}(\psi)}$. Assume that this equivalence holds for all subformulae of $\boldsymbol{\lambda}(\varphi)$ and $\boldsymbol{\lambda}(\psi)$ and that $\mathcal{L}_{\mathbf{A}}(\varphi) \neq \mathcal{L}_{\mathbf{A}}(\psi)$. One of two cases emerges:

- If there is an atomic formula in the complement—say a $p \in \mathcal{L}_{\mathbf{A}}(\varphi) \setminus \mathcal{L}_{\mathbf{A}}(\psi)$—then we can find a model in which $t_w(p) \gneqq_w \bigoplus_{\xi \in \mathcal{L}_{\mathbf{A}}(\psi)} t_w(\xi)$. This will witness the invalidity of $\mathbf{t}_{\varphi} \leftrightarrow \mathbf{t}_{\psi}$.
- In the remaining case, either $\boldsymbol{\lambda}(\varphi)$ or $\boldsymbol{\lambda}(\psi)$ includes a conditional $\xi \rightarrow \zeta$ among its conjuncts not found in the other. Suppose without loss of generality that φ includes this conjunct. Then we can pick a model in which $(t_w(\xi) \multimap_w t_w(\zeta)) \gneqq_w \bigoplus_{\xi \in \mathcal{L}_{\mathbf{A}}(\psi)} t_w(\xi)$, whence $t_w(\mathbf{t}_{\varphi}) \neq t_w(\mathbf{t}_{\psi})$ and invalidity of $\mathbf{t}_{\varphi} \leftrightarrow \mathbf{t}_{\psi}$.

Each case shows by contraposition that the validity of $\mathbf{t}_{\varphi} \leftrightarrow \mathbf{t}_{\psi}$ entails that $\mathcal{L}_{\mathbf{A}}(\varphi) = \mathcal{L}_{\mathbf{A}}(\psi)$. For the other direction, note that $\mathcal{L}_{\mathbf{A}}(\varphi) = \mathcal{L}_{\mathbf{A}}(\psi)$ then $\mathbf{t}_{\boldsymbol{\lambda}(\varphi)}$ and $\mathbf{t}_{\boldsymbol{\lambda}(\psi)}$ are the same formula. By Lemma 26, then, $\mathbf{t}_{\varphi}$ will be equivalent to $\mathbf{t}_{\psi}$. □

In the concluding moments, we can appeal to an old trick of Angell's (which Anderson and Belnap attribute to Hintikka's [14]) asserting the interdefinition of inclusion and equivalence via the following, which holds in CA/PAI:

Lemma 28 $\vdash_{\mathsf{CA/PAI}} \mathbf{t}_{\varphi} \rightarrow \mathbf{t}_{\psi}$ *iff* $\vdash_{\mathsf{CA/PAI}} \mathbf{t}_{\varphi \wedge \psi} \leftrightarrow \mathbf{t}_{\varphi}$.

Finally, this wins us the fundamental lemma:

Lemma 29 $\vdash_{\mathsf{CA/PAI}} \varphi \rightarrow \psi$ *only if* $\mathcal{L}_{\mathbf{A}}(\psi) \subseteq \mathcal{L}_{\mathbf{A}}(\varphi)$.

Proof By Theorem 2 of [11], $\vdash_{\mathsf{CA/PAI}} \varphi \rightarrow \psi$ holds only if $\vdash_{\mathsf{CA/PAI}} \mathbf{t}_{\varphi} \rightarrow \mathbf{t}_{\psi}$ holds. By Lemmas 27 and 28, then, this requires that $\mathcal{L}_{\mathbf{A}}(\varphi) = \mathcal{L}_{\mathbf{A}}(\varphi \wedge \psi)$. By definition, of course, $\mathcal{L}_{\mathbf{A}}(\varphi \wedge \psi) = \mathcal{L}_{\mathbf{A}}(\varphi) \cup \mathcal{L}_{\mathbf{A}}(\psi)$, so this requirement is equivalent to $\mathcal{L}_{\mathbf{A}}(\psi) \subseteq \mathcal{L}_{\mathbf{A}}(\varphi)$. □

Together, Lemmas 19 and 29 allow us to establish the main theorem concerning the genealogical Proscriptive Principle:

Theorem 30 CA/PAI *observes the genealogical Proscriptive Principle.*

My sense is that Theorem 30 itself is—or *should* be—an interesting result for relevant logics. Of course, we promised results interesting to *mainstream* relevant logics as well, requiring that we bring the above theorem back to bear on variable-sharing proper.

First, a lemma relating shared genealogy to the armatures that fueled the proof of the genealogical Proscriptive Principle:

Lemma 31 *If* φ *and* ψ *do not have shared genealogy, then* $\mathcal{L}_{\mathbf{A}}(\varphi) \cap \mathcal{L}_{\mathbf{A}}(\psi) = \varnothing$.

Proof Trivial by induction on complexity of formulae. □

This allows us to immediately infer the next theorem:

Theorem 32 CA/PAI *has the genealogical variable-sharing property.*

Thus, we have shown that a more restrictive notion of variable-sharing is available, is well-motivated by trends in the theory of topic, and that there exist relevant logics that enjoy the property.

5 Concluding Remarks

The details of [12] introduce a strong variable-sharing property of *strong cn-relevance* that bears some resemblance insofar as variable-sharing compares a sort of provenance of each appearance of an atomic formula. In this case, the provenance records the sequence of negations and conditionals within which an atomic formula is nested without regard for whether it appears in the antecedent or consequent position. Two items from this work motivate future directions in which the above remarks can be taken.

First, it is shown in [12] that the weak relevant logic B enjoys strong cn-relevance. Given cn-relevance's ability to track both sign and depth of conditionals, it seems likely that the following conjecture may hold:

Conjecture 33 B has the genealogical variable-sharing property.

Considering the not-unreasonable desideratum to produce a mainstream relevant logic witnessing genealogical variable-sharing, B is a good place to start.

Second, strong cn-relevance also takes into account the nestings of negations. Considering that e.g., R enjoys the strong variable-sharing property, the relevant tradition carries an innate allergy to the thesis that a sentence's topic must be identified with that of its negation.

This is an assumption that is relatively widespread. In e.g. the work of David Nelson (in [19]) and Vit Punčochář (in [23, pp. 118–120]) we find a great deal of reason to discount the thesis that $\varphi \rightarrow \psi$ and $\neg(\varphi \rightarrow \psi)$ share meaning. On Nelson's account, e.g., the former is intensional while the latter is extensional, meaning no more than simply $\varphi \wedge \neg\psi$; given the analysis in [11], this alone is cause for skepticism. It is implicit in other formalisms as well, such as the computational treatment of relevance logics in [8].

A modified definition of genealogy in which a substring N is added to record applications of negation would clearly induce a strong notion of genealogical relevance and wed both genealogical and strong versions of the Proscriptive Principle. Whether such a variable-sharing property holds of any relevant logics—whether mainstream or black sheep—is a further open question.

Acknowledgments I appreciate the helpful remarks of an anonymous referee and the suggestions of the participants of the *New Directions in Relevant Logic* event. This paper was written with the support of the MetaMuSo project (Czech Science Foundation project GA22-01137S).

References

1. Anderson, A. R., & Belnap, Jr., N. D. (1975). *Entailment: The logic of relevance and necessity* (Vol. I). Princeton: Princeton University Press.
2. Angell, R. B. (1977). Three systems of first degree entailment. *Journal of Symbolic Logic*, *42*(1), 147.
3. Angell, R. B. (1989). Deducibility, entailment and analytic containment. In J. Norman & R. Sylvan (Eds.), *Directions in relevant logic, Reason and argument* (pp. 119–143). Boston, MA: Kluwer Academic Publishers.
4. Belnap, Jr., N. D. (1960). Entailment and relevance. *The Journal of Symbolic Logic*, *25*(2), 144–146.
5. Berto, F. (2022). *Topics of thought*. Oxford: Oxford University Press.
6. Brady, R. T. (1984). Depth relevance of some paraconsistent logics. *Studia Logica*, *43*(1–2), 63–73.
7. Dunn, J. M. (1972). A modification of Parry's analytic implication. *Notre Dame Journal of Formal Logic*, *13*(2), 195–205.
8. Ferguson, T. M. (2016). Faulty Belnap computers and subsystems of FDE. *Journal of Logic and Computation*, *26*(5), 1617–1636.
9. Ferguson, T. M. (2017). *Meaning and proscription in formal logic*. Cham: Springer.
10. Ferguson, T. M. (2021). The subject-matter of intensional conditionals. In *Proceedings of the Twelfth Smirnov Readings in Logic* (pp. 59–63). Moscow: Russian Society of History and Philosophy of Science Publishing House.
11. Ferguson, T. M. (2023). Subject-matter and intensional operators I: Conditional-agnostic analytic implication. *Philosophical Studies*, *180*(7), 1849–1879.
12. Ferguson, T. M., & Logan, S. A. (2023, to appear). Topic transparency and variable sharing in weak relevant logics. *Erkenntnis*.
13. Fine, K. (1986). Analytic implication. *Notre Dame Journal of Formal Logic*, *27*(2), 169–179.
14. Hintikka, J. (1959). Existential presuppositions and existential commitments. *Journal of Philosophy*, *56*, 125–137.
15. Hornischer, L. (2020). Logics of synonymy. *Journal of Philosophical Logic*, *49*(4), 767–805.
16. Humberstone, L. (2011). *The connectives*. Cambridge, MA: MIT Press.
17. Logan, S. A. (2021). Strong depth relevance. *Australasian Journal of Logic*, *18*(6), 645–656.
18. Mares, E. (2022). Relevance logic. In E. N. Zalta & U. Nodelman (Eds.), *The Stanford encyclopedia of philosophy* (Fall 2022 edition).
19. Nelson, D. (1949). Constructible falsity. *Journal of Symbolic Logic*, *14*(1), 16–26.
20. Parry, W. T. (1933). Ein Axiomensystem für eine neue Art von Implikation (analytische Implikation). In *Ergebnisse eines mathematischen Kolloquiums* (Vol. 4, pp. 5–6).

21. Parry, W. T. (1968). The logic of C. I. Lewis. In P. A. Schilpp (Ed.), *The philosophy of C. I. Lewis, The Library of Living Philosophers* (pp. 115–154). La Salle, IL: Open Court.
22. Parry, W. T. (1989). Analytic implication: Its history, justification, and varieties. In J. Norman & R. Sylvan (Eds.), *Directions in relevant logic, Reason and argument* (pp. 101–118). Boston: Kluwer Academic Publishers.
23. Punčochář, V. (2019). *Paradoxy klasické logiky*. Prague: Filosofia.
24. Randriamahazaka, T. (2022). A note on the signed occurrences of propositional variables. *Australasian Journal of Logic*, *19*(1), 59–72.
25. Routley, R., Plumwood, V., Meyer, R. K. & Brady, R. (1982). *Relevant logics and their rivals* (Vol. 1). Atascadero, CA: Ridgeview Publishing.
26. Sylvan, R. (1988). Relevant containment logics and certain frame problems of AI. *Logique et Analyse*, *31*, 11–24.
27. Szmuc, D., & Rubin, M. (2022). Pure refined variable inclusion logic. *Australasian Journal of Logic*, *19*(5), 147–166.
28. Tennant, N. (2017). *Core logic*. Oxford: Oxford University Press.
29. Urquhart, A. (1973). A semantical theory of analytic implication. *Journal of Philosophical Logic*, *2*(2), 212–219.
30. Yablo, S. (2014). *Aboutness*. Princeton: Princeton University Press.

Withered Relevance

Evaluating the Anderson-Belnap Account of Relevant Logics

Tore Fjetland Øgaard

we will shake you off and fly once more

Meyer – Maximum Leader

Abstract The two "relevance" criteria set out by Anderson and Belnap are discussed. It is argued that the motivation backing the variable sharing property is far weaker than it is commonly made out to be, and that the use-criterion does not distinguish between relevant logics such as **E** and **R** and "irrelevant" logics such as **S4**, intuitionistic and classical logic. In short, then, the paper argues that Anderson and Belnap's two criteria of relevance are both motivationally unsound, and do not accomplish what they are commonly taken to accomplish.

1 Introduction

Inspired by Ackermann's work, Anderson and Belnap—*A&B* from now on—set out to

> give a formal analysis of the notion of logical implication, variously referred to also as "entailment," or "the converse of deducibility" (Moore 1920), expressed in such logical locutions as "if . . . , then —," "implies," "entails," etc., and answering to such conclusion-signalling logical phrases as "therefore," "it follows that," "hence," "consequently," and the like. [4, p. 21]

I would very much like to thank the audience at the conference *New Directions in Relevant Logic* as well as the referees for helpful comments and suggestion. An earlier version of this paper appeared in my PhD dissertation *A Study in Relevant Logics* [42]. I would very much like to thank my PhD supervisor, Ole Thommessen Hjortland, as well as the opponents, Greg Restall and Shawn Standefer, for engaging conversations.

T. F. Øgaard (✉)
Department of Philosophy, University of Bergen, Bergen, Norway
e-mail: Tore.Ogaard@uib.no

I. Sedlár et al. (eds.), *New Directions in Relevant Logic*, Trends in Logic 63,
https://doi.org/10.1007/978-3-031-69940-5_4

A&B thought that both classical and intuitionistic logic, as well as the modal logic **S4** conceived of as a logic of strict implication, all validated so-called *implicational paradoxes*. Such paradoxes, according to A&B, are claims of validity where the premises are *irrelevant*, that is fail to be *relevance-related*, to the conclusion somehow. Their project, simply put, was to restore "relevance" onto validity, insisting that for "more than two millenia logicians have taught that a necessary condition for the validity of an inference from A to B is that A be relevant to B" [4, p. 33]. Just how, however, to make sense of A being relevant to B, is a difficult task. A&B set up two criteria for validity which they claimed to do so. The first pertains to *meaning*: roughly, that the content of the conclusion should somehow be connected to that of the premises. The other pertains to, to borrow Read [48, p. 114]'s term, *derivational utility*: that the premises must somehow be *used* if the conclusion is to follow *from* them. This paper discusses these two "relevance" criteria for validity.

Requiring premises to be meaning-connected to the conclusion in a valid entailment is an idea with a pedigree starting with the very earliest records of philosophical discussion of logic. How to cash it out, however, varies considerably. A&B did so by requiring the now well-known formal property of variable sharing: if $A \to B$ is to be logically true, then A and B must share a propositional variable. It is argued, however, that the formulation of this property badly reflects the motivation backing it. It is shown that the availability of a deduction theorem akin to A&B's "Entailment theorem" allows one to "lift" the variable sharing property to its rightful possessor, namely the consequence relation itself. Even if thus rectified, however, the property is incredibly weak. I argue that not only is sharing a propositional variable unimpressive, but A&B's reason for claiming that this is a necessary criterion of validity is also far from convincing: Their argument in favor of the variable sharing property can, in short, be stated as follows:[1]

(1) If A entails B, then A is relevant to B.

(2) If A is relevant to B, then A is meaning-related to B.

(3) If A is meaning-related to B, then A and B share a propositional variable.

∴ If A entails B, then A shares a propositional variable with B.

It is argued, then, that A&B's reasons for regarding (3) true, only seem to tell in favor of thinking that its converse,

(3′) If A and B share a propositional variable, then A is meaning-related to B,

is true. The conclusion drawn, then, is that since the above argument with (3) replaced by (3′) is obviously not valid, A&B have failed to substantiate that their argument is sound. A&B's account of entailment, however, is such that for a large

[1] Note, however, that A&B never stated their argument like this.

class of logics **L** including not only relevant logics such as **E** and **R**, but also classical and intuitionistic logic, and **S4**, is it the case that A entails B according to **L** if and only if $A \to B$ is a logical truth of **L**. Consequently, it follows that A&B have failed to properly argue in favor of the variable sharing property itself.

The main claim with regards to the *use*-criterion is that the formal property thought to account for a notion of use is downright useless. I argue that the motivational support for the criterion has been wrongly conceived of, but even when this is cleared up, the notion of use turns out to be rather hollow. It will be shown that by and large every logic can be outfitted with the following-*from* proof-technique which defines A&B's notion of consequence, and that A&B's so-called "Entailment theorem" holds for every such logic. What is distinctive of A&B's notion of consequence is that it yields no logical theorems. It is shown, however, that for certain logics—logics such as A&B's three rival logics: **S4**, intuitionistic, and classical logic—the standard Hilbertian consequence relation is simply an extension by definition in the case of the empty set of premises of A&B's notion of consequence. The conclusion drawn is that A&B's use-condition for logical consequence should not be regarded as an umpire of logics as A&B thought.

A&B argued that it is extensional conjunction which interprets premise-aggregation. Some relevant logicians have argued that this is wrong claiming rather that premises are to be viewed as *intensionally* aggregated. The accompanying notion of premise-use for such a view is that *every* premise must be used. It is shown, however, that the same considerations apply to this notion of use as to that of A&B.

The notion of relevance has been a uniting one even though it is in fact quite hard to get a clear sense from the writings of relevant logicians of what the phenomenon is supposed to be. This, then, is in rather stark contrast to what A&B claim, namely that

> one of the principal merits of [Ackermann's] system of *strengen Implikation* is that it, and its neighbors, give us for the first time a mathematically satisfactory way of grasping the elusive notion of *relevance* [5, p. xxi]

Even though, then, Anderson and Belnap do focus on *relevance*, claiming that the meaning-criterion amounts to a semantical account of relevance, and that the use-criterion amounts to a proof-theoretic one [cf. 5, p. 33], I have, especially in the light of the lightness of the use-criterion, come to think of the notion of relevance as simply a red herring.[2] Both so for getting a clear idea of what A&B and other relevant logicians were and are after, but also when trying to evaluate the philosophical foundation of logics called 'relevant' or 'relevance' logics. The view expressed in this paper is that what A&B were primarily after was a theory of *entailment* which they tried to unearth using an unclear notion of relevance. Throughout this paper I will therefore focus on the motivational ideas pertaining to

[2] My critique, then, differs radically from for instance that found in Copeland [18] where the primary critique is on Anderson and Belnap's failure to give a coherent account of relevance with which to motivate **E**.

logical consequence as well as the properties set forth as the formal counterpart of these ideas. Although I will on occasion refer to other writers on relevant logics, my aim is primarily to critically discuss and assess some of the ideas of relevant logics as express by A&B. There are many different philosophies of relevant logics—too many to even survey in a paper like this. If, and if so to what extent, my critique of A&B also pertains to other philosophical motivations for relevant logics will therefore not be discussed.[3]

The paper is structured as follows: Sect. 2 defines the concept of a Hilbert proof as well as introducing the "minimal" logic **ML** and defines the relevant logics **E** and **R**, classical and intuitionistic logic, as well as the modal logic **S4**. Section 3 discusses the notion of meaning-relatedness as a requirement on logical consequence. Section 4 gives a lengthy discussion of the use-criterion in light of both A&B's motivational ideas, as well as previous critiques of it, before Sect. 5 summarizes.

This paper is, it is hoped, readable without too much prior knowledge of relevant logics. Some of the more technical results pertaining to the *use*-account of relevance referred to in this paper are to be found in chapter 12 of this volume [43].

2 The Hilbert Consequence Relation and Some Logics Defined

Definition 1 (The Hilbert Consequence Relation) A HILBERT PROOF of a formula A from a set of formulas Γ in the logic $\mathbf{L}$ is defined to be a finite list $A_1, \ldots, A_n$ such that $A_n = A$ and every $A_{i \leq n}$ is either a member of Γ, a logical axiom of $\mathbf{L}$, or there is a set $\Delta \subseteq \{A_j \mid j < i\}$ such that $\Delta \Vdash A_i$ is an instance of a rule of $\mathbf{L}$. The existential claim that there is such a proof is written $\Gamma \vdash_{\mathbf{L}}^{H} A$ and expressed as "there exists a Hilbert proof (H-proof) of A from Γ in the logic $\mathbf{L}$." If $\varnothing \vdash_{\mathbf{L}}^{H} A$, then A is said to be a logical theorem of $\mathbf{L}$.

[3] Despite this, however, I would like to at least *mention* two different approaches: the *information* based approach and the metaphysical *state-of-affairs* based approach. Both of these take as their starting point the Routley-Meyer semantics and motivate the logical laws using intuitions and theories of information/state-of-affairs. With regards to the information approach, at least Slaney [55], Restall [49], and Mares [28, 29] ought to be mentioned. With regards to the metaphysical approach, see Jago [25]. A&B's approach, it should be mentioned, antedates the discovery of the Routley-Meyer semantics. Two other approaches ought also to be mentioned, namely the non-transitive one of Neil Tennant, and the more intensional one of Arnon Avron. See Tennant [58] and Avron [7, 8] for these two approaches.

Definition 2 The following list of axioms and rules defines the logic **ML**:

(A1) $A \to A$

(A2) $A \wedge B \to A$

(A3) $A \wedge B \to B$

(R1) $\{A, A \to B\} \Vdash B$

(R2) $\{A, B\} \Vdash A \wedge B$

(R3) $\{A \to B, A \to C\} \Vdash A \to B \wedge C$

(R4) $\{A \to B, B \to C\} \Vdash A \to C$

Definition 3 (E, R, S4, IL & CL) **E** consists of (A1)–(A3) and (R1)–(R2), together with (A4)–(A15) below. **R** is obtained from **E** by replacing (A15) by (A16). Classical logic **CL** is obtained from either **E** or **R** by adding (A17). Intuitionistic logic **IL** is obtained from **E** or **R** by adding (A17) and replacing (A14) by (A18). **S4** is obtained from **E** by adding (A19).[4]

(A4) $(A \to B) \wedge (A \to C) \to (A \to B \wedge C)$

(A5) $A \to A \vee B$

(A6) $B \to A \vee B$

(A7) $(A \to C) \wedge (B \to C) \to (A \vee B \to C)$

(A8) $A \wedge (B \vee C) \to (A \wedge B) \vee (A \wedge C)$

(A9) $(A \to B) \to (C \to A) \to (C \to B)$

(A10) $(A \to B) \to (B \to C) \to (A \to C)$

(A11) $(A \to (A \to B) \to (A \to B)$

(A12) $(A \to {\sim}A) \to {\sim}A$

(A13) $(A \to {\sim}B) \to (B \to {\sim}A)$

(A14) ${\sim}{\sim}A \to A$

(A15) $(((A \to A) \wedge (B \to B)) \to C) \to C$

(A16) $A \to ((A \to B) \to B)$

(A17) $A \to (B \to A)$

(A18) $A \to ({\sim}A \to B)$

(A19) $A \to (B \to B)$

[4] The necessity operator of **S4** can be defined as it is in **E**: $\Box A =_{df} (A \to A) \to A$. See Meyer [35] for a host of results pertaining to the relationship between **E** and **S4**.

3 The Idea of Meaning-Relatedness

We have already seen that A&B claimed that the idea of relevance being a necessary ingredient of logical consequence runs through millennia of logical research. At least the thought that logical consequence requires that the relata be *meaning-related*, seems to be quite common idea throughout the ages. For current purposes it will have to do to note some of the more recent history of the idea.[5]

Although various authors had express uneasiness of the "irrelevant" features of classical, intuitionistic and modal logics, it was arguably Ackermann's essay, *Begründung einer Strengen Implikation* [1], which initiated A&B's search for a new theory of logical consequence, or 'entailment' as they rather preferred. Ackermann's logic is simply classical logic with an added "rigorous implication" connective with regards to which he writes that

> [t]he rigorous implication, expressed as $A \to B$, expresses that there exists a logical connection between A and B; that the content of B is part of the content of A, or how now best to put it. [1, p. 113][6]

Like so many before him, Ackermann didn't make the notion of content precise, nor how parthood thereof is to be understood. This is also the case with the three authors quoted by Belnap [11]. First of all, Nelson writes that "[e]ntailment, not being defined in terms of truth-values, is a necessary connexion between meanings" [39, p. 445]. Secondly, Duncan-Jones suggested that entailment be the "relation which only holds between two propositions, p and q, when q arises out of the meaning of p" [19, p. 71]. The last author cited by Belnap suggests that a proposition A implies a proposition B if and only if "the intensional meaning of B is identical with a part of the intensional meaning of A" [10, p. 397].

All these four authors seem to agree that meaning-relatedness is essential to logical consequence. Beyond this, however, it is hard to get a clear idea of what kind of relation they had in mind. The story of logic in the twentieth century is a story of an increasingly formal notion of logic. Thus, the goal of the research program of relevant logics initiated by A&B was from the very onset viewed as giving a *formal* account of both the theory of logical consequence, but also of the motivating features undergirding it. Whereas previous writers relied on a more intuitive notion of meaning connection, Belnap stressed that

> [i]n regard to formal theories proposed as explications of the notion of entailment, it becomes important to have at our disposal some *formal* account of the demand for relevance as between antecedent and consequent of an entailment. [11, p. 144]

---

[5] See Bobzien [13] and Dutilh Novaes [22] for interesting entries on the, respectively, ancient and medieval history of the idea.

[6] My own translation.

Although Belnap fails to mention Ackermann, we have already noted that he did quote Nelson, Duncan-Jones and Baylis. These writers, Belnap notes,

> agree that in a formal theory of entailment, we would wish to reject $A \to B$ if A and B have nothing to do with one another, if, that is, A and B are totally disparate in meaning. The problem is to restate this condition in formal terms.
>
> Confining our attention to propositional logic, a partial solution becomes almost obvious once we note that in propositional logic, commonality of meaning is carried by identity of propositional variables. Thus, for A to be relevant to B in the required sense, a necessary condition is that A and B have some propositional variable in common. [11, p. 144]

Belnap then states that **E** "satisfies this condition" and goes on to prove that **E** has what is now known as the *variable sharing property*:

Definition 4 A logic **L** has the VARIABLE SHARING PROPERTY (VSP) just in case for every formulas A and B, $\varnothing \vdash_{\mathbf{L}} A \to B$ only if A and B share a propositional variable.

In addition to showing that **E** satisfies (VSP), Belnap noted that his proof also covers Ackermann's logic Π'.[7] It was later shown that the algebraic model used by Belnap also validates the stronger logic **R** and so this logic satisfies (VSP) as well.[8]

Note that relevance is in Belnap [11] regarded as a necessary property of entailment. We shall later see that A&B seem to have thought the concept of relevance to be such as to also *suffice* for entailment. Note, then, that it is not relevance, but primarily the concept of *entailment* that A&B are after.

Belnap seems to regard A as entailing B just in case $A \to B$ is logically true. The next section goes on to argue that the variable sharing property needs to be reformulated so as to make it apply directly to the relation of logical consequence, but let's put this aside for now. The question, then, is whether the variable sharing property legitimately does formalize the view that meaning-relatedness is a necessary condition for logical consequence.

The argument in favor of this is the variable sharing property's alleged connection to a notion of meaning-connection. With regards to this, note that both Ackermann, Beylis, Duncan-Jones and Nelson seem to want to explicate the concept of entailment using notions of meaning-connection. At least the three first authors seem to view this notion as a rather strong one—they write of meaning-*containment* and propositions arising out of the meanings of others. Belnap's view of meaning-connection seems to be much weaker as any old *connection* seem to be all that is required: Surely, any propositions $p \to p$ and p have *something* to do with each other, yet there need not be a "necessary connexion" between their meanings (Nelson), nor be the case that the meaning of p should "arise" out of the meaning of $p \to p$ (Duncan-Jones), nor that the intensional meaning of p be part of that of $p \to p$ (Baylis), nor that content of p be part of the content of $p \to p$ (Ackermann).

---

[7] Π' can be identified as **E** with the γ-rule $\{A, {\sim}A \vee B\} \Vdash B$ added as a primitive rule.

[8] Relevant logics are sometimes formulated so as to include propositional constants. See Øgaard [41, § 6] for a discussion of the variable sharing property for such logics.

$p \rightarrow p$ cannot entail p, yet although the ideas appealed to by Ackermann, Baylis, Duncan-Jones and Nelson seem to rule in accordance with this, neither Belnap's intuitive idea, nor its restated formal counterpart, do: Belnap's aim seems simply to be of a much less substantial notion of meaning-connection than the other four authors were aiming at. Even though Belnap does state that his notion of meaning-connection is only a necessary condition for entailment, it bares notice that it is at best a rather weak such condition. As Meyer put it, "Sharing a single variable is not, after all, a very impressive guarantee that two propositions are deeply entwined logically; there may yet be 20 million unshared variables" [37, pp. 610f]. On the other hand, Craig's interpolation theorem holds for classical logic which yields that if A entails C according to this logic, then there is a B made up of all and only the common variables of A and C such that A entails B which in turn entails C—unless, that is, $\sim A$ or C are logical theorems. One would have thought that such a theorem would hold unrestrictedly for relevant logics, but it turns out that not even the restricted version holds for a large class of such [cf. 59].

The real question, however, is not with its strength, but whether the variable sharing property ought to be viewed as a necessary criterion of logical consequence to begin with. Now Belnap does take it to rule out A entailing any B "totally disparate in meaning" [11, p. 144] and notes that "commonality of meaning is carried by identity of propositional variables" (ibid.). Even though this may be the case, what is needed for legitimizing the variable sharing property is that *only* identity of propositional variables carries commonality of meaning (in the propositional case). If this is the case, then any two formulas A and B which do not share a propositional variable will be "totally disparate in meaning." If it is possible, however, for A and B to not share any variables but still be meaning-connected, then the variable-sharing property should not be taken as a necessary condition for *validity*. In short, then, sharing a propositional variable can be acknowledge as a sufficient condition for a "meaning-connection." This, however, seems to be all that Belnap's "note" establishes. What is needed, however, is an argument to the effect that it is *necessary*, that *only* propositional variables can instigate commonality of meaning. This is a substantial thesis about meanings which is far from being "almost obvious."

Note, furthermore, that A&B must have been content with Belnap's original note as they repeat it in both Anderson and Belnap [4, p. 34] and Anderson and Belnap [5, p. 33], and thus simply restate that sharing a propositional variable is an "almost obvious" way of ensuring what is

> suggested by the consideration that informal discussion of implication or entailment have frequently demanded "relevance" of A to B as a necessary condition for the truth of $A \rightarrow B$, where relevance is now construed as involving some "meaning content" common to both A and B. [5, pp. 32f]

What is needed is an argument to the effect that if A and B do not share a propositional variable, then they are "totally irrelevant to each other." It is not my purpose to argue against the view that commonality of meaning is (in the propositional case) *only* carried by commonality of propositional variables, only that the view *needs* arguing for as it is far from being "almost obvious." I've not found any such argument in any of A&B's writings.

I should note that there are philosophical interpretations of relevant logics which try to work out not only a notion of meaning-connection, but a more encompassing notion of meaning-*containment* akin to what we saw both Ackermann and Beylis seemed to appeal to. Amongst relevant logicians it is first and foremost Ross Brady who have tried to developed such a theory. Note, then, that even Brady simply repeats A&B's view of the variable sharing property:

> A way of ensuring that there is a real connection between the content of the antecedent and that of the consequent of an inference of sentential logic is to require that the logic satisfies the [variable sharing property]. A variable in common between A and B would ensure that some content is in common between the antecedent A and the consequent B of the inference A → B. [...]. It is clear that [...] an inference A → B with no variables in common [...] will enable A and B to be totally irrelevant to each other and thus have no content connection whatever. [16, pp. 4f]

As with the above critique of Belnap's idea: although one need not deny that *a way* of ensuring connectivity between antecedent and consequent of a logically true conditional is to enforce variable sharing. A substantial argument is needed, however, to make it clear that an inference without such sharing would amount to total irrelevance.

This paper does not provide reasons for thinking that no such argument can be made. I should like, however, to offer two avenues on which one could build an argument to the effect that a valid argument can have premises relevant to the conclusion even in the absence of variable sharing. The first is simply to insist that *relevance* is not a very substantial notion after all: if A entails B, then A is relevant to B. This is the view express by for instance Meyer:

> what is the relation linking premiss to conclusion, in virtue of which we have a valid argument? It is scarcely anything but the relation of logical entailment itself. The relevance *requisite* for inference is precisely that *supplied* when one proposition stands to another in the entailment *relation*. [...] If A does entail B, then A *must* have been relevant to B, in the wanted logical sense. Defenders of strict implication (Prior, for example) have made just this point. Since, according to them, A & -A entails every B, they conclude that A & -A is (logically) relevant to absolutely everything. [37, pp. 609f]

The "point" that Meyer attributes to Prior is probably that which Prior expressed in his *Formal Logic* when writing with regards to Nelson [39]'s claim that entailment

> must consist in a definite inner connexion between antecedent and consequent, and cannot be present through some peculiarity (necessity, impossibility, truth, or falsehood) attaching to either proposition alone. [47, p. 196]

Prior, however goes on to parenthetically note that "[f]or some reason he does not consider that necessary and impossible propositions as such *have* a definite inner connexion with all propositions whatever" [47, p. 196].

The other, and more substantial, avenue I should like to mention is David Lewis' theory of *aboutness* which was designed precisely to answer the relevantists demand for meaning-connectivity between premise and conclusion of a valid argument.

Note, furthermore, that Lewis' theory in fact yields that "whenever a premise implies a conclusion, in the ordinary sense of truth-preservation, then also the premise is relevant to the conclusion" [27, p. 161]. One may have reservations with regards to Lewis's theory, and, of course, the relevantist may deny that Lewis' notion of relevance really does ensure the correct kind of connectivity. However, the burden of proof is surely on the relevantist to show that no theory of aboutness or of meaning-connectivity, meaning-containment, relevance, or related notions, will ever be sufficient for ensuring that no premise ever entail a variable-wise unrelated conclusion, save one that measures up to the variable sharing property.

The variable sharing property was put forth as a formal property to ensure that the theory of entailment, that is logical consequence, not validate entailments between propositions "totally disparate in meaning." We have found, however, that although sharing a propositional variable may be acknowledged to be a sufficient condition for a meaning-connection between formulas, an argument to the effect that it is also *necessary* is wanting. In the absence of such an argument it seems that the variable sharing property cannot be claimed as backed by the more intuitive idea that entailment involves a notion of meaning-connectedness. The burden of proof upon the relevantist is even more taxing since, as we saw, there are *alternative* theories of meaning-connectivity—Lewis' theory of aboutness was mentioned—which do seem to yield meaning-connectedness but without requiring variable sharing. The conclusion, then, is that A&B's formal meaning-criterion of entailment is, as it stands, in need of proper arguing.

We saw above that Ackermann intended his "rigorous implication" connective to be read as expressing a notion of logical consequence. Anderson and Belnap likewise regarded the $\rightarrow$ of their logic **E** as expressing the relation of entailment itself. Although the variable sharing property relates to object-language conditionals, its intent is that it should regulate the relation of logical consequence itself. The next section argues this and introduces also A&B's special concept of derivation which will be used in Sect. 4.2 where their other formal criterion of entailment is discussed.

4 On the *Use*-Criterion

This section looks into A&B's second main motivational idea for relevance being relevant to logical consequence, namely that there is a special notion of premise-*use* which relevant logics can account for, but which "irrelevant" logics cannot. Although, as we shall see in Sect. 4.4, other authors have been critical towards this idea, it is also an idea which has survived untarnished within major contemporary strains of the philosophy of relevant logic. The most ardent support for the criterion is found in the writings of Edwin Mares who advertises the topic of his book as follows: "The defining feature of relevant logic is that it forces the premises of an argument to be really used ('relevant') in deriving its conclusion" [31, p. i].

This section is divided into six subsections. The first subsection gets clear on the formal aspects of premise use in the case of A&B's use-modified Hilbert calculus;

the second on A&B motivation for the notion of premise use as such. The third subsection gives a brief account of the notion of premise use in relation to the A&B's Fitch calculi which makes Sect. 4.4's account of some of the previous critiques of the use criterion easier to grasp. The fifth subsection goes into a related notion of *use* for which *every* premise must be used in arriving at the conclusion, before the last subsection summarizes.

4.1 On How to Properly State the One Formal Relevance Criterion

Belnap's formulation of the variable sharing property does not properly reflects its true intent, namely that B should not be regarded as a *logical consequence* of A unless A and B are meaning-related. It is the theory of logical consequence which is the theory under scrutiny, not merely—which Belnap's formulation can be read as—which conditionals of a certain sort are logically true. Which consequence relation, then, is the correct one?

As lamented by Avron [7], it is not always clear which consequence relation is in fact intended. If one looks at relevant logics in action, so to speak—to instances where relevant logics are applied to axiomatic theories, then the dominant consequence relation is arguably the Hilbertian one.[9] The philosophically pressing issue, however, is whether this is the correct account of consequence. To borrow a thought from Priest, note that $\{A\} \vdash^H B \to B$ although $\varnothing \nvdash^H A \to (B \to B)$ for every logic normally called *relevant*. Thus, taking $\vdash^H$ to be the relation of entailment

> divorces the conditional from entailment [...]. This is not necessarily an objection; but it does make the consequence relation look a bit odd from a relevant perspective, and it raises the question of what, exactly, the conditional means. [46, p. 60]

I should only like to add that not only does it raise the question of what the conditional means, but it also undermines why even the variable sharing property ought to hold for it. This property was introduced to guard against the paradoxes of implication which the material, strict and intuitionist conditional, read as conditionals *expressing* logical implication or entailment, were thought to be marred with. It is under the interpretation that these conditionals express entailment,

---

[9] The literature on relevant arithmetic, for instance, is a noteworthy example. See Dunn [20], Routley [51, § A.9], Slaney et al. [56], Meyer and Restall [38], and Restall [50] for five different (in terms of arithmetical and logical axioms and rules) approaches to relevant arithmetics where it is the Hilbert consequence relation, be it proof-theoretically or semantically cloaked, which is the relation the arithmetical theories answer to. McKubre-Jordens and Weber McKubre-Jordens and Weber [34] explicitly acknowledges the use of the Hilbert consequence relation in investigating relevant analysis, whereas the naïve set theories investigated in Routley [51, § A.5–8], Brady [16], Priest [45], Weber [60, 61, 62, 63] and Istre and McKubre-Jordens [24] are all closed under Hilbert consequence.

however, that instances such as $(A \wedge {\sim}A) \supset B$ and $A \supset (B \supset B)$ were viewed as paradoxical.[10] Without this particular reading of $\supset$, however, these formulas are relevantly unproblematic as they are in fact logical theorems of both **E** and **R**. Since $\rightarrow$ does not express $\vdash^H$—at least not in the familiar sense that would require that a deduction theorem on the form

$$\{A_1, \ldots, A_n\} \vdash^H_{\mathbf{E}} B \Longleftrightarrow \varnothing \vdash^H_{\mathbf{E}} A_1 \wedge \ldots \wedge A_n \rightarrow B$$

to hold—it seems strange to demand that the variable sharing property must hold for $\rightarrow$ when doing so evidently does not suffice for a corresponding property holding for the relation assumed to be the relation of entailment, namely $\vdash^H$. $\vdash^H$ fails to satisfy what is arguably the hallmark of relevance, namely variable sharing, and so should not be regarded as the intended consequence relation for any relevant logic.[11]

A&B thought of **E**'s connective $\rightarrow$ as expressing *entailment*—that is *logical consequence*—and preferred reading $A \rightarrow B$ as *that A entails that B* [cf. 5, A5]. They never thought of $\rightarrow$ as expressing $\vdash^H$, however, a relation they rather disapprovingly called the "Official" position for which

> terms like "deduction," "from," "premiss," "proof," "hypothesis," "valid," "demonstration," "consequence," and "argument" have virtually no connection with the same terms as understood outside the Official context—even by the Officer himself, when he takes off his uniform [...]. [5, p. 258]

A&B offered a different way of defining a consequence relation over a set of axioms and rules than the one that was used in defining $\vdash^H$, and it is rather this relation—what they called the relation of *entailment* in the case of **E**—which $\rightarrow$ is thought of as expressing. Their favorite formal presentation of this relation utilizes a Fitch-style natural deduction calculus. However, they also provided a Hilbert calculus for it, which proves easier to compare to the "Official" notion of logical consequence, and so I will stick to it in the following. I will state a version of A&B's notion of entailment rigorously in a while. The rough idea, however, is easily explainable as a simple proof-cleaning strategy: Their trick is to introduce an assumption-tracking device into the Hilbertian definition of a proof. They demand that every assumption be marked by a star and that the conclusion be marked by a star as a result of passing forward the stars from the assumptions using the rules. Some Hilbert proofs will typically not make the cut with this tracking devise enforced: the list A, B, for instance, is a perfectly OK Hilbert proof of any axiom B from the assumption A. Since no rules are applied, however, no star-rule can be

[10] In fact, A&B found any reading of $\supset$ as even a kind of *conditional* offensive. Øgaard [41] argues that this is unwarranted even by their own standards.

[11] Avron's account of relevant logics is one where the Hilbert consequence relation is explicitly acknowledged as the intended one. On his account it still is the case that $\{A\} \vdash^H B \rightarrow B$ even though there is a sense in which the conditional "reflects" the consequence relation—for his "strong relevance logics" it is the case that $\Delta \cup \{A\} \vdash^H B$ if and only if either $\Delta \vdash^H B$ or $\Delta \vdash^H A \rightarrow B$. See Avron [8] for his definition of a relevant logic, and Øgaard [44] for a critical discussion of it.

invoked so as to make B starred as a consequence of A being starred. Such a proof, therefore, will not be an acceptable one by A&B's standards for entailment.

A&B's notion of entailment only applies to logics with no more primitive rules than modus ponens and adjunction [cf. 5, pp. 277f] and furthermore only truly applies to certain rather strong logics amongst such. I will get back to this later. It's generalization is as follows:

Definition 5 (Conjunctive Entailment) A C-PROOF of a formula A from a set of formulas Γ in a logic **L** extending **ML**, is a Hilbert-proof $A_1, \ldots, A_n$ of A from Γ such that it is possible to mark the A_i's with # according to the following rules:

(1) If $A_i \in \Gamma$, then A_i is marked.[12]
(2) R1 is not used on premises A_j and $A_j \to A_i$ where $A_j \to A_i$ is marked.
(3) If A_i is obtained from A_j and $A_j \to A_i$ using R1, then A_i is marked if A_j are marked.
(4) R2 is only used on premises which are either both marked or both unmarked.
(5) If A_i is obtained from A_j and A_k using R2 and both of A_j and A_k are marked, then A_i is marked.
(6) Other rules than R1 and R2 are only used on unmarked formulas.
(7) No other formulas are marked.
(8) As a consequence of (1–7), A_n is marked.

The existential claim that there is such a proof is written $\Gamma \vdash_{\mathbf{L}}^{C} A$ and expressed as "A is C-derivable from Γ in the logic **L**" or as "Γ C-entails A in the logic **L**."

The 'C' is here simply short for 'conjunction' since it is the extensional conjunction $\wedge$ which interprets premise combination. For comparison, note that A&B used *entails* where I have used *C-entails*, and that they used a star as a marking sign instead of the #-mark used here. A&B's notion of entailment is obtained by deleting marking rule (6), and by replacing (2) and (3) by the rule that the conclusion of an application of modus ponens is stared if at least one of the premises are stared. It is easily shown, however, that C-entailment and A&B's notion of entailment are equivalent for logics such as **E** and **R**.

I have already noted that the standard deduction theorem fails for **E** with regards to $\vdash^{H}$. Using $\vdash^{C}$ allows for a type of deduction theorem, however. A&B stated in Anderson and Belnap [5, p. 278] what they call the "Entailment theorem," which reads:

> ENTAILMENT THEOREM. If there is a proof in **E** that $A_1, \ldots, A_n$ entail(s) B, then $A_1 \& \ldots \& A_n \to B$ is provable in **E**.

[12] There is, as one of the referee's stressed, a complication here: A_i could both be a member of Γ, but also derivable as a logical theorem. The intent is that the Hilbert-proof $A_1, \ldots, A_n$ comes with a given justification, and if, then, the justification given for A_i is that $A_i \in \Gamma$, then A_i is to be marked. For a more precise account, see Øgaard [43].

This theorem, using A&B's notion of entailment, can be shown to hold not only for **E**, but also for **R** and even *classical logic*.[13] Using C-entailment, however, the theorem generalizes quite radically. As a consequence, we get the following:[14]

Theorem 1 (C-Entailment Theorem) *For any logic **L** extending **ML**,*

$$\{A_1, \ldots, A_n\} \vdash^C_{\mathbf{L}} B \Longleftrightarrow \varnothing \vdash^H_{\mathbf{L}} A_1 \wedge \ldots \wedge A_n \rightarrow B$$

We have seen that taking $\vdash^H$ as the explication of the notion of entailment undercuts the motivation for the variable sharing property. A&B thought that $\vdash^C$ was a better formal account of entailment. Since the variable sharing property was introduced to ensure that some sense of commonality of meaning is preserved by entailment and thereby guard against the paradoxes of implication, it seems prudent to reformulate the variable sharing property so as to apply directly to consequence relations:

Definition 6 A set-formula[15] consequence relation $\vdash$ for a logic **L** has the CONSEQUENTIAL VARIABLE SHARING PROPERTY just in case for every set of formulas $\Theta \cup \{A\}$, if $\Theta \vdash_{\mathbf{L}} A$, then A shares a propositional variable with at least one of the formulas in Θ.

What the C-entailment theorem allows, then, is to formally connect the variable sharing property to entailment:

Corollary 1 $\vdash^C_{\mathbf{L}}$ *satisfies the consequential variable sharing property for any logic **L** extending **ML** if and only if **L** satisfies (VSP).*

To better reflect the motivation stated for the variable sharing property, then, it ought to be reformulated as a consequential variant akin to the above consequential variable sharing property. As argued in the previous section, however, an argument as to *why* commonality of meaning cannot hold without involving variable sharing is needed in order to properly motivate also such a consequential version of the variable sharing property.

The next section will look closer at the *use*-criterion for entailment, that the premises somehow must be *used* if they are to entail the conclusion. As we shall see, having a clear conception of the proof-technique used in C-proofs will make this criterion much easier to get a clear idea of. To that end, it is befitting also

---

[13] See Øgaard [40, thm. 9] for details.

[14] The proof of the theorem, as well as a comparison of the current #-rules with those set forth by Anderson and Belnap as well as the generalization of these found in Brady [15] are found in Øgaard [43].

[15] This can obviously be generalized to also cover more general consequence relations beyond those merely taking sets of formulas as premise aggregates and delivering formulas as conclusions. For present purposes, however, the given definition will suffice.

to note the conditions under which the consequence relations $\vdash^C$ and $\vdash^H$ become indistinguishable:

Theorem 2 *For $\Delta \neq \varnothing$ and **L** extending **ML** which is such that*

$$(i)\ \varnothing \vdash^H_{\mathbf{L}} A \Longrightarrow \varnothing \vdash^H_{\mathbf{L}} B \to A$$

$$(ii)\ \{A, A \to B\} \vdash^C_{\mathbf{L}} B$$

but which does not have any primitive *rules other than modus ponens and adjunction, then*

$$\Delta \vdash^H_{\mathbf{L}} A \Longleftrightarrow \Delta \vdash^C_{\mathbf{L}} A$$

Proof *The right-to-left part is trivial since every C-proof is a H-proof.*

For the other direction, assume that $\Delta \vdash^H_{\mathbf{L}} A$ and let $B_1, \ldots, B_n$ be a proof of this. The following proof is to the effect that $\Delta \vdash^C_{\mathbf{L}} B_i$ for every $i \leq n$.

*Base case: if $B_i \in \Delta$, then it holds rather trivially that $\Delta \vdash^C_{\mathbf{L}} B_i$. Else if B_i is an axiom of **L**, then it follows from* (i) *that $\varnothing \vdash^H_{\mathbf{L}} \delta \to B_i$ for some $\delta \in \Delta$. From Thm. 1 it then follows that $\Delta \vdash^C_{\mathbf{L}} B_i$ since structural weakening obviously holds for $\vdash^C$.*

Induction case: assume for inductive hypothesis that both $\Delta \vdash^C_{\mathbf{L}} B_j$ and $\Delta \vdash^C_{\mathbf{L}} B_k$, and let $C^{(\#)}_{j_1}, \ldots, C^{(\#)}_{j_m}, B^\#_j$ and $C^{(\#)}_{k_1}, \ldots, C^{(\#)}_{k_m}, B^\#_k$ be C-proofs of this. If B_i is obtained from B_j and B_k using adjunction then since

$$C^{(\#)}_{j_1}, \ldots, C^{(\#)}_{j_m}, B^\#_j, C^{(\#)}_{k_1}, \ldots, C^{(\#)}_{k_m}, B^\#_k, (B_j \wedge B_k)^\#$$

is a C-proof it follows that $\Delta \vdash^C_{\mathbf{L}} B_i$. If B_i is obtained using modus ponens from B_j and $B_k = B_j \to B_i$, then it follows using the assumption (ii) *the C-proof*

$$C^{(\#)}_{j_1}, \ldots, C^{(\#)}_{j_m}, B^\#_j, C^{(\#)}_{k_1}, \ldots, C^{(\#)}_{k_m}, B^\#_k$$

can be expanded to one of B_i from Δ, and thus that $\Delta \vdash^C_{\mathbf{L}} B_i$. □

Note that the conditions of the above theorem are satisfied by both intuitionistic and classical logic as well as **S4**. Thus, $\vdash^C$ and $\vdash^H$ are identical consequence relations for these logics when it comes to non-empty premise sets. Since, however, $\varnothing \nvdash^C_{\mathbf{L}} A$ for every formula A and every logic **L**, it follows that $\vdash^H_{\mathbf{L}}$ is simply an extension by definition of $\vdash^C_{\mathbf{L}}$ for these three logics, in the sense that for all sets of formula $\Gamma \cup \{A\}$,

$$\Gamma \vdash^H_{\mathbf{L}} A \Leftrightarrow \Gamma \vdash^C_{\mathbf{L}} A \text{ OR } \Delta \vdash^C_{\mathbf{L}} A \text{ for every set of formulas } \Delta \neq \varnothing.$$

There are two possible take-home lessons from this. Either that the sense in which $\vdash^C$ differs from $\vdash^H$ for these logics is a rather uninteresting one. At least for logics for which the two relations collapse for non-empty sets of premises, this conclusion seems to be the correct one. Such logics, we have seen, validate the weakening rule as an admissible rule, and so validates implicational "paradoxes" such as $A \to (B \to B)$. The relevantist, however, who sticks by the intuition that such sentences express truly paradoxical, i.e. false, claims of entailment and rather insists upon upholding the variable sharing property, should be relieved that there *is* a notion of premise use hard-coded into $\vdash^C$. It must be admitted that it turned out to be a rather meek notion of use—certainly weaker than A&B led on—but, after all, a non-trivial one: $\vdash^C$ does require premises to be *used* in order that any conclusion should follow; in short, $\vdash^C$ ensure that nothing follows from nothing. It also, it should be noted, ensure that the variable sharing property is properly lifted to apply to where it ought to apply, namely to the intended account of entailment.

4.2 Anderson and Belnap on Premise Use and Derivational Utility

One of the "prejudices" the Entailment theorem reflects is, according to Anderson and Belnap [5, p. 278], that the conclusion ought to follow *from* the premises. The thought is that if the conclusion follows without *using* the premises, then it doesn't really follow *from* them. Or, as I will argue, at least this is how A&B *should* have pitched their idea. However, the idea is, or so this section argues, a hollow one even if pitched perfectly.

We have already seen that A&B derided $\vdash^H$ as a consequence relation since it fails to account for this *from*-ness aspect of logical consequence. Alonzo Church had already appealed to a notion of premise use in his proof of a deduction theorem for the implicational fragment of **R**, although he didn't put the notion to any work beyond this [cf. 17, p. 25]. A&B's very first presentation of **E** is to be found in Anderson and Belnap [3] which is an abstract of a twenty-minute presentation given at the twenty-third annual meeting of the Association of Symbolic Logic. The notion of premise use is not mentioned therein, but what came to be the explication of the notion is. In it, A&B state that their

> principal interest in E (as opposed to Π' and Π'') is that for E we have the following: *Entailment theorem.* $\vdash \mathbf{A} \to \mathbf{B}$ if and only if there is a proof that **A** entails **B**. [3, p. 458]

The notion of entailment here is defined using the star-rules applied to a Hilbert derivation akin to the #-rules of C-proofs. These rules are "designed to explicate what we might mean by saying that **B** 'really depends on' **A** (Ackermann's *logischer Zusammenhang*)" [3, p. 458]. Naturally, no proof of the Entailment theorem is provided since the paper is only an abstract. What is most interesting, however, is their claim that it does not hold for Ackermann's logics Π' and Π''. The easiest way to define Π' is to view it as **E** with $\{A, \sim A \vee B\} \Vdash B$ added as a primitive rule.[16] As such it has more rules than A&B's notion of entailment is defined for, and because of this it rather trivially fails to satisfy the Entailment theorem. It seems, however, that A&B took their notion of entailment and the coupled Entailment theorem to be a more substantial differentiating property. Just how these two elements—the *entailment*-consequence relation and the Entailment theorem—go together to make the requisite differentiating property is at times hard to get a clear idea of. As we shall see, however, this bundle became tied to a notion of premise use and eventually came to be regarded as a necessary and sufficient condition for relevance.

The roots of the Entailment theorem is to be found in Anderson [2]. The difference is that Anderson therein used a Fitch system with a technique akin to the star-rules to facilitate the same kind of dependence tracking. Just as the star-rules in Anderson and Belnap [3], Anderson views the Fitch system as a way to explicate "the notion of 'depends on the logical content of' " [2, p. 203] which he equates with Ackermann's *logischer Zusammenhang*. That paper is the first presentation of a Fitch system for **E** and its main result is that this proof calculus is theorem-wise identical to the Hilbert calculus for **E**. I will get back to the Fitch calculus, but for now notice that Anderson [2] does not actually define a consequence relation. What Anderson does is to claim that the fact that the Fitch calculus for **E** proves all and only the logical theorems provable using the Hilbert calculus shows "that $\mathfrak{B}$ depends on the logical content of $\mathfrak{A}$ if and only if $\mathfrak{A} \to \mathfrak{B}$ is provable in E" [2, p. 210]. No further analysis of what depending on the logical content of means is offered, however. What is at least tempting, then, is to read into this a *version* of Entailment theorem, namely, that B is, let's say, *F-entailed* by A in the Fitch calculus for **E** if and only if $A \to B$ is a logical theorem of **E**.

Anderson's paper is first and foremost a technical paper and the notion of premise use is only mentioned in connection with the Fitch system having a restricted $\to$-introduction rule so as to "keep track of all the hypotheses which are actually *used* as premisses" [2, pp. 203f].[17] Belnap's account of Anderson's paper, however, highlights the *use*-interpretation to a larger extent claiming that Anderson's account "explicates the notion 'depends on the logical content of' in terms of the actual use of a premiss" [11, p. 144]. Anderson and Belnap [4, p. 34] goes even further and ties both *relevance* and *use* to the restriction on the $\to$-introduction rule as they claim that it is restricted so as to be applied "when A is relevant to B in the sense that A is *used* in arriving at B."

[16] Π'' is a further extension of Π' using the false propositional constant **f**. See Anderson et al. [6, §§ 44–46] for details on Ackermann's logics.

[17] The $\to$-introduction rule in a natural deduction calculus is usually stated using the notion of a *subproof*, where, then, from a subproof with assumption A and derived formula B the $\to$-intro rule allows one to close off the subproof and infer $A \to B$. Anderson's indices allow one to do without subproofs, and so his $\to$-introduction rule is simply that from a proof of B_a on the hypothesis A_k to infer $(A \to B)_{a-\{k\}}$, provided k occurs in a [cf. 2, p. 204]. k will, then, occur in a just in case A_k has been properly *used* in deriving B_a.

Now there is nothing *technically* wrong with Anderson's result, just as the Entailment theorem itself is technically correct. However, that the Fitch calculus in itself, or even in conjunction with the Entailment theorem, somehow shows that Ackermann's notion of logical connection can be explicated in terms of use, is less than clear. Nor is the issue really cleared up in Anderson and Belnap [4] wherein both the use-criterion and the meaning-criterion of relevance make their first joint appearance. They only discuss the implicational fragment of certain logics, but state that

> In arguing that E_I satisfies a principle of relevance, we venture, somewhat gingerly, on new ground. We offer two conditions, the first as necessary and sufficient, the second as necessary only. [4, p. 46]

They go on to repeat Belnap [11]'s informal discussion of the meaning-criterion and then state that "we propose as a necessary (but not sufficient) condition for the relevance of A to B in the pure theory of entailment, that A and B must share a variable" [4, p. 48]. Notice first of all that their proposal is strictly speaking not the variable sharing property, that the theory of entailment be such that if $A \rightarrow B$ is logically true, A and B should share a propositional variable. What they seem to be after is rather an explication of the notion of *relevance itself*, and more specifically of how to understand the phrase "A is relevant to B." This is also confirmed in their note with regards to the first criterion:

> The subscripting technique as applied in W_I[18] and E_I may be construed as a formal analysis of the intuitive idea that for A to be relevant to B it must be possible to use A in a deduction of B from A. [4, p. 46]

However, in showing that the implicational fragment of **E** satisfies the second merely necessary condition for relevance, A&B simply prove that the variable sharing property holds [cf. 4, pp. 48f]. It seems, then, that they use "A is relevant to B" more or less interchangeable with "$A \rightarrow B$ is a logical theorem." This suspicion is confirmed also in their note with regards to the first, allegedly necessary and sufficient, condition, which

> has to do with entailment in its guise as the converse of deducibility, and in this sense is a purely syntactical completeness theorem: A is *relevant* to B, when A $\rightarrow$ B, just if there exists a proof (satisfying certain conditions) of B *from* the hypothesis A. [4, pp. 49]

The received view, then, seems legitimate in that it attributes to A&B the view that the variable sharing property is a necessary relevance property, whereas the Entailment theorem is a jointly necessary and sufficient relevance property. What is more urgent, however, is that it allows the reformulation of the quote from Anderson and Belnap [4, p. 46] above to:

> The subscripting technique as applied in W_I and E_I may be construed as a formal analysis of the intuitive idea that for ['that A entails that B' to be logically true] it must be possible to use A in a deduction of B from A.

[18] The implicational fragment of **R**.

What is of primary interest here, however, is the notion of use which is not so much connected to the Entailment theorem as such, but rather to the subscripting technique in the Fitch case and the marking-rules in the case of $\vdash^C$. Notice first of all that the modality here—it must be *possible* to use A in a deduction of B from A—seems to be just plainly wrong. Now A&B reject that the correct modality is necessity rather than possibility.[19] They note that even though Rosser's assumption of mere consistency is weaker than that of ω-consistency, but "surely no one would hold that ω-consistency was *irrelevant* to Gödel's result" [4, p. 46]. They end their discussion of the first criterion with the following quote from Smiley [57, p. 249]:

> ...inferences may be justified on more grounds than one, and the present theory requires not that there should be *no* cogent way of reaching the conclusion *without* using all the premisses, but only that there should be *some* cogent way of reaching it *with* all the premisses used. [4, p. 48]

They qualify Smiley only in specifying that *all* premises need not be used, only the conjunction of *some* of them: "What is required is that there be *some* cogent way of reaching the conclusion with *some* of the (conjoined) premisses used" (ibid.). But this still seems wrong. A&B's worry seems to be that conclusions typically have multiple possible premises from which they follow. Say, then, that B follows from both A_1 and A_2 so that it be acceptable to claim both that A_1 entails B and that A_2 entails B. Even though, then, it is possible to use either A_1 or A_2 in order to obtain B, that does not imply that it is possible to use A_1 to show that B follows from A_2. The case where "the antecedent is used in the proof of the consequent, though it need not be" [4, p. 46] seems to be precisely the case with the standard Hilbert notion of proof where any logical axiom follows from any premises, although the premise need not be used although it quite often can: the list $B, A \rightarrow A$, for instance, is perfectly fine Hilbert proof of $A \rightarrow A$ from B, in any logic in which $A \rightarrow A$ is an axiom. If the logic also has $B \rightarrow (A \rightarrow A)$ as an axiom, then one can expand the list to $B, B \rightarrow (A \rightarrow A), A \rightarrow A$ in which $A \rightarrow A$ is *derived* from the previous two formulas using modus ponens. What, rather, the subscripting and the marking techniques show, is that there is a notion of following from where a conjunction of some of the premises *must* be used to reach *any* conclusion: if B does follow *from* A in the sense that $\{A\} \vdash^C B$, then A *must* be used in any admissible proof of B *from* A. That B also follows from some other C, is simply beside the point. My final update of the above quote, then, is

> The marking-rules of C-entailment may be construed as a formal analysis of the intuitive idea that for A to truly entail B, every deduction of B *from* A must use A.

The question, then, is whether this can be regarded as relevant for relevance. Since C-entailment applies to any extension of **ML**, and thus to **E** and **R** as well as to classical logic, intuitionistic logic and **S4** alike, explicating the notion of premise use in this way will not yield an umpire of logics which was, after all, what A&B

[19] That the modality is that of possibility and not necessity is also claimed by Meyer [cf. 37, p. 612].

were after: according to them, classical logic, intuitionistic logic and **S4** all validate implicational paradoxes, whereas **E** and **R** do not. What differentiates classical logic from **E** is the *axiom* $A \rightarrow (B \rightarrow A)$, not the notion of following from encoded in the *entailment*-consequence relation, nor in the appended Entailment theorem: even A&B's *original* star-rules apply equally to these three "irrelevant" logics as it does apply to **E** and yield in all cases the Entailment theorem. In short, then, the notion of premise-use explicated using C-entailment does nothing to tell either for or against which axioms ought to make up the list of the true logic of entailment.

4.3 Premises and the Fitch Calculus

While Anderson and Belnap's star-modified Hilbert calculus quite evidently takes premises to be conjuncted together into a conditional antecedent, this is not evident in the case of their Fitch calculus. This section gives a brief review of how premises are dealt with in this latter calculus as well as to give a brief account of how the notion of premise use was thought materialized within it.

A Fitch calculus is a certain sort of natural deduction calculus. As was mentioned in the previous section, Anderson—one of Fitch's doctoral students—developed such a calculus for **E**. Anderson's stated aim with regards to this calculus was to alter Fitch [23]'s notion of a subproof

> in such a way as to keep track of all the hypotheses which are actually *used* as premisses in the application of rules leading to a proposition $\mathfrak{A}$, and then to restrict the introduction rule for implication [...] in such a way that $\mathfrak{A} \rightarrow \mathfrak{B}$ will be said to follow from a proof of $\mathfrak{B}$ on the hypothesis $\mathfrak{A}$, only if $\mathfrak{A}$ is used in arriving at $\mathfrak{B}$. [2, pp. 203f]

Not only do A&B seem to have retained this idea—that a certain restriction on conditional introduction for a natural deduction system would yield a substantive notion of premise use over what a standard Hilbert calculus affords—throughout their joint work on relevant logics, but it seems also to have captured the fancy of several other relevant logicians. For instance, in the entry on relevant logics in the Stanford Encyclopedia of Philosophy, Mares writes with regards to the Fitch calculus for **R** that "when an implication is proven relevantly, the assumption of the antecedent must really be used to prove the conclusion" [30, § 3]. Mares even claims that such a logic even captures a whole *theory* of relevance which "can be understood in terms of how the corresponding natural deduction system records the real use of premises" [30, § 3].

Anderson, as noted in the previous section, did not explicitly define what logical consequence is supposed to be. Even though I think it is beyond doubt that Anderson and Belnap regarded what I've called *C-entailment* as the proper analysis of logical consequence, this is not always clear. This is especially confusing when digging into the details of the Fitch calculus. The reason for this is the way in which premises were handled in Anderson's original Fitch calculus for **E** and which was retained in his joint work with Belnap and which seems to have been adopted by most later writers as well.

The easiest type of natural deduction proof from assumptions Δ using Fitch's original calculus can be displayed graphically as follows:

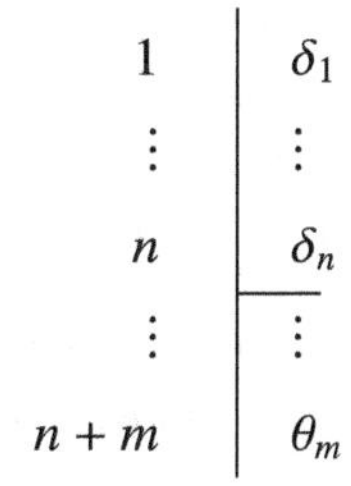

All the $\delta_{i\leq n}$ are elements of Δ and the last formula θ_m is the conclusion of the proof. Formulas that occur below the horizontal bar are either derived premises of subproofs, or derived from earlier formulas or subproofs. The number of vertical lines to a formula's left is the subproof-complexity of the formula. The conclusion of any proof must have one and only one vertical line to its left—subproof complexity 0—whereas the other formulas occurring below the horizontal bar may occur in nested subproofs of any complexity. Such subproofs "look" like an ordinary proof, with the only exception that they have one and only one premise whereas the list of δ_i's can be of any finite length including 0. Without going further into details, let's call θ_m FH-DERIVABLE from Δ just in case there is a Fitch proof as depicted above.

Although strictly speaking unnecessary, Anderson's Fitch calculus for **E** in fact disallows "ordinary" premises to occur in any proof. Any such, however, may occur in a subproof, and so one may start a proof with the same δ_i's as in Fitch's original calculus, but listed as follows:

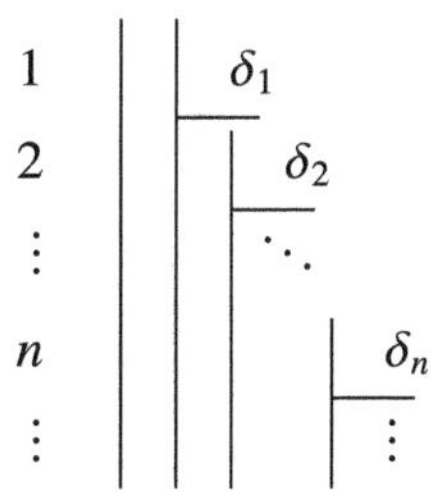

Here each such subproof should be thought of as a separate context of relevance—each of the δ_i's are assigned their separate "relevance index" which the rules of the calculus then carry forward in specific ways depending on the logic in question.

One way to define B following from Δ given Anderson's Fitch calculus for **E** is, then, as the existence of a proof object with an initial segment as displayed above, and with B as the conclusion of the subproof which has δ_n as its premise. This effectively yields that B follows from Δ just in case $\delta_1 \rightarrow (\delta_2 \rightarrow \ldots (\delta_n \rightarrow B) \ldots)$

is a logical truth, where the δ_i's are any formulas from Δ. I'll get back to this idea of defining logical consequence later. It is not, however, how Anderson and Belnap thought of logical consequence: For instance $A \wedge B$, according to A&B, is a logical consequence of $\{A, B\}$ [cf. 5, § 21.2.1], but for no logic which satisfies the variable sharing property and has (A2) and (A3) as logical axioms is it possible to pick δ_i's in $\{A, B\}$ such that $\delta_1 \rightarrow (\delta_2 \rightarrow \ldots (\delta_n \rightarrow A \wedge B) \ldots)$ is a logical truth.[20]

One of the "prejudices" that the Entailment theorem reflects, according to A&B, is that it is the *conjunction* of the premises which entails the conclusion [cf. 5, p. 278]. An equivalent definition of C-entailment, then—the consequence relation which Anderson and Belnap argued to be the correct account of entailment—using this calculus is obtained by defining $\Delta \vdash^{F_c} B$ just in case there exists a conjunction $\delta_1 \wedge \cdots \wedge \delta_n$ of the formulas from Δ and an Fitch-Anderson proof object on the form

$$\begin{array}{c||l}
1 & \delta_1 \wedge \cdots \wedge \delta_n \\ \hline
\vdots & \vdots \\
m & B
\end{array}$$

I should emphasize that A&B never, as far as I know, defined logical consequence this way. One could, of course, liberalize the calculus somewhat so as to allow subproofs with multiple premises. The calculus, furthermore, works as expected also when "ordinary" premises are permitted. Thus, one can define Hilbert-consequence in the standard way in which Fitch defined it, and therefore also logical theoremhood. In fact, Anderson's initial result is that his Fitch-type calculus proves all and only the logical theorems obtainable using a Hilbert calculus. Interestingly, they suppress the initial vertical line, and so instead of the leftmost Fitch-type proof that $A \rightarrow A$ is a logical theorem, they write it as displayed on the righthand side instead.

1	$\vert\ \vert\ A$	hypothesis	1	$\vert\ A$	hypothesis
2	$\vert\ \vert\ A$	1, repetition	2	$\vert\ A$	1, repetition
3	$\vert\ A \rightarrow A$	1–2, $\rightarrow$Intro	3	$A \rightarrow A$	1–2, $\rightarrow$Intro

Although A&B did not define C-entailment using their Fitch calculus, they did, however, give a rigorous definition of theoremhood in the case of the implication-fragment of certain logics which accords with the standard one [cf. 5, § 8.1].

[20] To see why, simply let A and B be logical theorems with no propositional variable sin common. Then $\delta_1 \rightarrow (\delta_2 \rightarrow \ldots (\delta_n \rightarrow A \wedge B) \ldots)$ would yield $\delta_n \rightarrow A \wedge B$, and therefore both $\delta_n \rightarrow A$ and $\delta_n \rightarrow B$ where one of these, then, would have violated the variable sharing property.

The upside to defining C-entailment using the Fitch calculus is that it displays the essence of the Entailment theorem quite nicely, namely as mere instances of the introduction and elimination rule for the conditional:

1	$\quad\vert\ \delta_1 \wedge \cdots \wedge \delta_n$	hypothesis
$\vdots$	$\quad\vert\ \vdots$	
m	$\quad\vert\ B$	
$m+1$	$\delta_1 \wedge \cdots \wedge \delta_n \to B$	1–m, →Intro

$\vdots$	$\vdots$	
n	$\delta_1 \wedge \cdots \wedge \delta_n \to B$	
$n+1$	$\quad\vert\ \delta_1 \wedge \cdots \wedge \delta_n$	hypothesis
$n+2$	$\quad\vert\ B$	$n, n+1$, →Elim

Therefore, Δ entails B just in case there is a logically true conditional which has B as its consequent and a conjunction of some of the Δ's as its antecedent.

The notion of premise use which the Fitch calculus was thought to harbor relates to the rule of conditional introduction. Consider the following derivation of the "positive paradox":

1	$\quad\vert\ A$	hypothesis
2	$\quad\vert\quad\vert\ B$	hypothesis
3	$\quad\vert\quad\vert\ A$	1, Reit
4	$\quad\vert\ B \to A$	2–3, →Intro
5	$A \to (B \to A)$	1–4, →Intro

What, according to A&B, is wrong with this proof is that B is not used in any meaningful way in deriving A in line 3, and so the conditional introduction on line 4 is illegitimate. Anderson and Belnap dispensed with the need for explicit subproofs and rather supplied every formula with a "relevance index"—intuitively the set of formulas on which it depends in the proof under consideration. This index is then carries forward, with varying restrictions depending on the logic in question, from premise to conclusion using any of the standard natural deduction rules just as the #-mark is carried forward in the C-proofs. The purpose of any one index is to track the usage of the assumption of a subproof throughout that subproof. In the case of conditional introduction, then—the only rule which ends subproofs [cf. 6, § R.3]—the premise set is reduced: If a subproof with premise A where A has relevance index $\{k\}$, and ending with a formula B with relevance index α, then one is allowed to end the subproof in the usual way with conclusion $A \to B$ which then has $\alpha \setminus \{k\}$ as its relevance index. The catch is that this is only permitted provided $k \in \alpha$—intuitively the condition that A has been *used* in obtaining B.

It is important to stress that the notion of use thought materialized in this Fitch-type calculus is credited solely to this use-restriction on conditional introduction, and not to any feature of the other rules or combinations of such. For instance, in commenting on the natural deduction proof

1	$\quad B$	hypothesis
2	$\quad A \to A$	axiom
3	$B \to (A \to A)$	1–2, $\to$Intro

A&B write[21]

> In this example we indeed proved $A \to A$, but it is crashingly obvious that we did not prove it *from* the hypothesis B: the defect lies in the definition, which fails to take seriously the word "from" in "proof *from* hypothesis." And this fact suggests a solution to the problem: we should devise a technique for keeping track of the steps used, and then allow application of the introduction rule only when A is relevant to B in the sense that A is *used* in arriving at B. [4, p. 34]

This view of the restriction of the conditional introduction rule is also adhered to by more recent writers on relevant logics. Mares, for instance, writes that the "proviso that k occur in α is essential. It ensures that, in this case, that A is really used in the derivation of B" [32, p. 611]. Mares furthermore states with regards to the same proviso that it "helps to eliminate non-sequitures" such as the argument from "My dog barks at rubbish collectors" to "Either it is raining in Bolivia right now or it is not" [31, p. 7].

This section has given an informal presentation of Anderson and Belnap's Fitch calculus with the aim of explaining how premises are handled within it as well as how the notion of premise-use was explicated. We've seen that C-entailment has a rather contrived definition using the Fitch calculus, namely as holding between a set of formulas Δ and B just in case a Fitch-proof on the form

1	$\quad\quad \delta_1 \wedge \cdots \wedge \delta_n$	hypothesis
$\vdots$	$\quad\quad \vdots$	
m	$\quad\quad B$	

exists, where the δ_i's are at least some of the members of Δ.

[21] The same passage, but with an added digression, is also found in Anderson and Belnap [5, p. 18].

The next section looks into previous critiques of the use-criterion base upon both the use-modified Hilbert calculus as well as the Fitch calculus. In the process it will become evident that the use-judgement with regards to C-proofs apply equally to proofs using the Fitch calculus.

4.4 *Previous Critiques of the* Use-*Criterion*

The view of A&B's notion of premise use expressed here is not entirely new. The first critical point in this direction was voiced by Bennett in his review of Anderson and Belnap [4] in which he writes that

> The view that $A \rightarrow A$ follows from any B "fails to take seriously the word 'from'" (p. 34). This is in the anti-scientific spirit of: "The view that the set of primes has the same cardinality as the set of rationals fails to take 'same' seriously." Furthermore, the stipulation that B is deduced from A only if "A is used in arriving at B" (p. 34) by itself achieves nothing: Lewis has shown how to use B in proving $A \vee {\sim}A$. The reason why E_I lacks $B \rightarrow (A \rightarrow A)$ is that it is over-all weak, not that it obeys any obvious, re-applicable rules about relevance or serious from-ness. [12]

At least in hindsight one may think it reasonable that Bennett's comments should have caused A&B to worry. It is, however, worth pointing out that Bennett is not overly clear on why he thinks the implication-fragment of **E** fails to prove $B \rightarrow (A \rightarrow A)$—being *over-all weak* is not, after all, a precise term. As far as I have been able to verify, Bennett's review is mentioned once by A&B. They write, after having repeated the line from Anderson and Belnap [4] that the subscripting technique used in the Fitch calculus for the implication-fragment of **R** and **E** "may be construed as a formal analysis of the intuitive idea that for A to be relevant to B it must be possible to *use* A in a deduction of B from A" [5, p. 30], the following with regards to Bennett's critique:

> the distinctions drawn and the techniques elaborated thus far do *not* "wither in the sun outside the tiny area covered by $\mathbf{E}_{\rightarrow}$," to quote an elegant conceit of Bennett 1965. Quite the contrary: [...]. The bulk of this book will be devoted to a comprehensive and rigorous mathematical proof of this fact. [5, p. 41]

A&B's reaction, then, seems to have been to buckle down and expand upon their account, rather than to retract or alter it.

Bennett's review was also mentioned in Meyer [36]. Meyer took Bennett's view to be that the use criterion is "motivationally unsound, or that it foreshadows collapse back to K in the wider context of truth functions" [36, p. 56]. Meyer noted that, what might have been Bennett's view all along, that "the classicist might say (though none have been *quite* so nasty), one can use A to get $B \vee \bar{B}$; plug A in as antecedent of the logical truth $A \rightarrow B \vee \bar{B}$ [36, p. 57]. Even though Meyer does claim that "Eventually, even the worthy old use criterion has to go" [36, p. 56], it is not at all evident that he himself took heed of this. I will get back to this in connection with a slightly different notion of use for which *all* the premises must be used, rather than merely some, as we have seen was A&B's view.

A similar point to that of Bennett's was also made in Copeland [18], but there with regards to the Fitch-system. Copeland noted that the Fitch calculus for **E** can be augmented by a propositional constant to yield classical logic without altering "the fact that $A \rightarrow B$ is asserted on the basis of a proof of B on hypothesis A, only if A is used in arriving at B" [18, p. 330].[22] Copeland, however, claims his point valid only given that "all the truth functional connective [are] in the picture," but hastens to add that "the use account does suffice to motivate a system of relevant entailment in the restricted case where $\rightarrow$ is the only logical symbol" [18, p. 330]. This, however, seems rather doubtful as the same kind of rule-extension which Copeland envisioned can be used to obtain the positive paradox. To see this, note that one of the rules that Copeland added was that Λ is inferable from A and $\sim A$ retaining the union of relevance indices of A and $\sim A$. Using this together with the use-restricted conditional introduction rule one can then easily derive $A \rightarrow (\sim A \rightarrow \Lambda)$.[23] However, by modifying the reiteration rule so that A is inferable from A and B retaining the union of the relevance indices of these, one can, then, derive $A \rightarrow (B \rightarrow A)$ using the use-restricted conditional introduction rule:

1	$A_{\{1\}}$	hypothesis
2	$B_{\{2\}}$	hypothesis
3	$A_{\{1,2\}}$	1, 2, Reit*
4	$B \rightarrow A_{\{1\}}$	2–3, →Intro
5	$A \rightarrow (B \rightarrow A)_{\varnothing}$	1–4, →Intro

Copeland's claim with regards to his Λ-rules, that

> [n]onetheless the rules do not block the tracking of hypotheses through the proof, and their presence in the system does not alter the fact that $A \rightarrow B$ is asserted on the basis of a proof of B on hypothesis A, only if A is used in arriving at B. [18, p. 330]

seems equally true of the modified reiteration rule used to derive the positive paradox above. His claim that the "use account of relevance, then, is no account of

[22] Copeland's rules for his false propositional constant Λ translates in the case of **E** to the following four axioms: $(A \rightarrow \Lambda) \rightarrow \sim A$, $A \wedge \sim A \rightarrow \Lambda$, $\sim(A \rightarrow B) \rightarrow ((A \rightarrow B) \rightarrow \Lambda)$ and $\Lambda \rightarrow A$. The addition yields an *extension* of classical logic—what Copeland [18, p. 330] calls a "classical entailment system." If added to **R**, however, the addition simply yields classical logic.

[23] Copeland's rule is easily seen to yield $A \wedge \sim A \rightarrow \Lambda$ for every formula A regardless of which logic is at play. It also yields $B \rightarrow (\sim B \rightarrow \Lambda)$, but since the reiteration rule is restricted to $\rightarrow$-formulas in the case of **E**, only provided that B is such a $\rightarrow$-formula. No such restriction holds in the case of **R**, however, and so Copeland's rule yields the unrestricted schema and therefore full classical logic when added to **R**.

relevance at all" [18, p. 330] should not be made contingent, as Copeland insists, on the presence of the "truth functional connectives," including, then, the propositional constant Λ. That the positive paradox is thus derivable also undermines Copeland's claim that it "is to its non classical treatment of negation that E owes its paradox freedom" [18, p. 330]. Just as we've seen that what differentiates classical logic from logics such as **E** and **R** in the context of the #-version of the Hilbert calculus was their *axioms*, and not the notion of following from encoded in the *entailment*-consequence relation, so also with the Fitch calculus: it is the index-rules tied to the connective rules and which formulas with which indices that are permitted to be reiterated into a subproof which truly determine which formulas end up as logical theorems, not the use-restricted rule of conditional introduction. Of course, one may argue that there is something wrong with the modified reiteration rule, or with rules for Copeland's propositional constant, but the fundamental point here is that the use-account of relevance is solely tied to the use-restricted conditional introduction rule, not to a specific set of structural and logical rules.[24]

Although it is listed in the extended bibliography of Anderson et al. [6], Copeland's result was, as far as I have seen, not mentioned in any of Anderson and Belnap's writings.[25] It was, however, mentioned in Routley [52], wherein it is taken to show that the "use account" of relevance is "only of limited value, since accounts of the same type can reflect a wide range of logics including even irrelevant logics, in particular [...] classical logic, that end point of irrelevance" [52, p. 146]. The same point, albeit without mentioning Copeland, is also made in Routley et al. [53, p. 258], but therein the discussion is abruptly ended with a claim that A&B's Entailment theorem is after all desirable, yet rather discriminating [cf. 53, p. 258]. No further hints as to how it is discriminating—-and as we have seen, even a weak logic such as **ML** satisfies A&B's Entailment theorem—or why it is desirable is given, nor any proof that the theorem actually holds "for a very large class of relevant, and also irrelevant, logics" [53, p. 258].

It is also worth noticing the rather puzzling fact that Anderson and Belnap [5, § 27.2] did provide a Fitch calculus with the same use restriction on the conditional introduction rule for the logic **RM**, a logic which in the same book is acknowledged

[24] Another way to obtain classical logic is by adding either the axiom called *Summation*, or the axiom called *Factor*—$(A \to B) \to (A \vee C \to B \vee C)$ and $(A \to B) \to (A \wedge C \to B \wedge C)$, respectively—to **R**. An Andersonian Fitch calculus for **R** augmented by both these principles was provided in Brady [14] and shown to be theorem-wise identical to the corresponding Hilbert calculus. Dunn Dunn [21] subsequently showed that **R** augmented by Factor is in fact classical logic whereupon Brady somewhat strangely noted that his Fitch calculus for this logic "should be ignored as it can be shown that it is classical logic" [15, p. 568]. The case of Summation is quite instructive as it shows that A&B's rule for disjunction elimination is restricted so as not to permit one to infer C from $A \vee B$ together with subproofs of C from hypothetical assumptions A and B.

[25] The same goes for Bennett's review. Furthermore, neither of these two critiques were mentioned in Brady [14] in which a range of relevant logics (as well as what was in Brady [15, p. 568] acknowledge to be classical logic!) were outfitted with Fitch-calculi, nor in Read [48] or Mares [31]—two monographs on relevant logics which both place an important emphasis on the notion of premise use.

to have $\sim(A \rightarrow A) \rightarrow (B \rightarrow B)$ as a logical theorem [cf. 5, § 29.5]. A&B viewed **RM**'s conditional to be rather similar to the material conditional in classical logic—as *no* kind of conditional—due to $(A \rightarrow B) \vee (B \rightarrow A)$ being a logical truth of **RM**. It is strange, then, that they should not have realized that their use restriction is not sufficient to even stay clear of the "implicational paradoxes," or at least to offer some kind of reflection on their Fitch calculus for **RM** in relation to the notion of premise use which the calculus, as we have seen, was designed to necessitate.[26]

This subsection has shown forth some of the critiques of A&B's notion of premise use as a guarantee that the premises are relevant to the conclusion. Although the current one is in line with both Bennett's and that of Copeland, their critiques are rather unclear. Bennett, we saw, simply claimed that the reason why relevant logics such as **E** fails to prove "implicational paradoxes" is that it is "over-all weak." Copeland is the clearest account, but his claim that the use-criterion does motivate an account of relevant entailment if it only be restricted to the pure $\rightarrow$-fragment, does call into question the depth of his critique. Meyer seems at times to voice the same type of critique as Copeland, but as we'll see in the next subsection, also explicitly adhered to the use-criterion as a defining feature of relevant consequence.

A&B's concept of entailment is a notion of logical consequence for which the extensional conjunction is the natural object-language interpretation of premise-aggregation. Some relevant logicians have argued that this is wrong, however, claiming that entailment is even more intensional than A&B made it out to be. The next section looks swiftly at this notion with regards to the use-criterion finding it to suffer the same fate also in such an intensional setting.

4.5 Using All *the Premises*

A&B listed two "prejudices" that the Entailment theorem supposedly reflects. We have already looked at the *from*-prejudice. It is now time to look at the other one which relates to whether it is *all* of the premises which entail a given conclusion, or if *some* in some cases may suffice.

A&B insisted that *extensional conjunction* is the correct premise combinator: "it is the conjunction of the premisses that entails the conclusion" [5, p. 278]. If, then, some set of premises C-entails some B, then any superset of that set does as well. A&B's notion of use, therefore, is one for which only *some* of the premises of a

[26] As stressed by the referee, one reason for this might have been that A&B viewed the use-restriction as only guaranteeing that the $\rightarrow$-fragment of any logic will be free of implicational paradoxes. In fact, A&B showed that the negation-implicational fragment of **RM** does in fact satisfy the variable sharing property [cf. 5, § 14.7]. Although not an impossible interpretation, I should like to stress that A&B never, at least to my knowledge, made such limitation on the significance of the use criterion. Indeed, it is with regards to the whole of **E** that the use-criterion was first formulated and then tied to the restriction on the introduction rule for implication, as we saw earlier in the quote from Anderson [2, pp. 203f].

given premise-set need to be used. They did acknowledge that there is a stronger notion of use for which *every* premise has to be used. However, they argued against such notion of entailment, claiming that

> If in fact a proof fails to use all the apparatus in the hypothesis, the argument is faulted on grounds of inelegance rather than logical incorrectness—and it is only the latter problem which is of overriding importance for **E**. [5, p. 279]

Building on work by Church and Moh,[27] Kron [26] showed that a stronger notion of entailment (for **E**) can be defined so as to yield that $A_1, \ldots, A_n$ entail(s) B if and only if $A_1, \ldots, A_{n-1}$ entail(s) $A_n \rightarrow B$. As a corollary to this we get that $A_1, \ldots, A_n$ entail(s) B if and only if $A_1 \rightarrow (\ldots \rightarrow (A_n \rightarrow B))$ is a logical truth of **E**. Kron's result covers not only **E**, but also **T** and **R** as well. Brady [15] subsequently generalized Kron's result so as to cover a wide range of logics including the weak contractionless logic **B**. Let's call this notion of entailment *I-entailment*—short for *iterated* entailment.[28] I-entailment is defined in much the same way as C-entailment is, namely as a restriction on a standard Hilbert deduction system. But whereas C-proofs mark all premises with the same mark, an I-proof demands that every premise be marked with distinct marks and that *every* such mark be carried forward to the conclusion using the rules. It is this feature which ensures that every premise must be used. These marks, then, are typically sets of indices, where premise n is assigned the singleton $\{n\}$ as its index. Rules are, just as in the C-entailment case, restricted: If we take the rather simple case of **R** it is

---

[27] See Church [17] and Shaw-Kwei [54, pp. 63f].

[28] I-entailment is in fact a substructural consequence relation for many logics. In the case of logics such as **R**, for instance, A follows from one instance of A, but not from two, corresponding, then, to the fact that $A \rightarrow A$ is, but $A \rightarrow (A \rightarrow A)$ is not a logical theorem of **R**. Going into details here would take us too far afield, but let me note that there is a related consequence relation which, like C-entailment, is fully structural in the case of A&B's logics **T**, **E** and **R**, but which as far as I know has not been given much attention in the literature beyond Avron's rejection of it. Avron [9] suggests that one consequence relation for **R** could be defined as follows:

$$\Delta \vdash^i_{\mathbf{R}} A \Longleftrightarrow [\exists \delta_{i \leq n} \in \Delta : \varnothing \vdash^H_{\mathbf{R}} \delta_1 \rightarrow (\delta_2 \rightarrow (\ldots(\delta_n \rightarrow A)\ldots))].$$

Avron rejects $\vdash^i_{\mathbf{R}}$ as a plausible candidate for logical consequence, however, due to, amongst other features, the failure of $\{A, B\} \vdash^i_{\mathbf{R}} A \wedge B$ which Avron finds high counterintuitive in light of the extensional nature of $\wedge$ has in **R** [cf. 9, p. 58].

A third option which would be more in line with the requirement of using *all* the premises than what is encoded into Avron's suggestion $\vdash^i$, would be as follows: For any finite set Θ,

$$\Theta \vdash^m_{\mathbf{R}} A \Longleftrightarrow \varnothing \vdash^H_{\mathbf{R}} \theta_1 \rightarrow (\theta_2 \rightarrow (\ldots(\theta_n \rightarrow A)\ldots))]$$

where every formula in Θ appears at least once as one of the $\theta_{i \leq n}$. As with $\vdash^i$, one can choose to set $\varnothing \nvdash^m_{\mathbf{L}} A$ for every A, or to rather set $\varnothing \vdash^m_{\mathbf{L}} A \Leftrightarrow \varnothing \vdash^H_{\mathbf{L}} A$. Such a notion of consequence is a set-formula relation just like C-entailment. However, it too is substructural for certain logics since structural weakening must fail in order to uphold the variable sharing property: $\{p \rightarrow p\} \vdash^m_{\mathbf{L}} p \rightarrow p$, for every propositional variable p, but if q is distinct from p, then $\{p \rightarrow p, q \rightarrow q\} \nvdash^m_{\mathbf{L}} p \rightarrow p$ if **L** satisfies the variable sharing property.

demanded that adjunction only be applied to premises with the same index set and yields a conclusion with the same index set as its premises. The conclusion of modus ponens is assigned the union of the index sets of the premises. We can display this graphically as follows:

$$\{A^{\iota_1}, B^{\iota_1}\} \Rightarrow (A \wedge B)^{\iota_1} \qquad \{A^{\iota_1}, (A \rightarrow B)^{\iota_2}\} \Rightarrow B^{\iota_1 \cup \iota_2}$$

Some relevant logicians have argued that it is I-entailment which is the correct formal account of the use-criterion and that it is the Church-Moh-Kron version of the Entailment theorem which gives the correct "statement of the Deduction Equivalence" [48, p. 50].[29] Regardless of the merits of such a view, however, note that it is rather evident that simply having a notion of I-entailment defined for a logic for which a corresponding I-entailment theorem holds cannot, just as in the C-entailment case, in itself be regarded as a differentiating property between logics which respects that *every* premise be *used* and those that do not. Just as the C-entailment theorem does require certain features of the logic, it might be the case that certain logics do not have a notion of I-entailment for which a corresponding I-entailment theorem holds. Note, however, that this is not the case for either classical or intuitionistic logic, nor for **S4**. In fact, by taking the described notion of I-entailment for **R** one obtains a notion of entailment which yields an I-entailment theorem also for these three "irrelevant" logics. The simple reason why this is so, is that both Kron and Brady prove their I-entailment theorems using quite standard induction proofs. The key to take note of, however, is that their success do not depend on axioms being *absent*, only that certain axioms be present.[30] Just as these logics are indistinguishable with regards to the notion of use formalized using C-entailment, the three logics which A&B thought of as the primary rivals of **E** are also indistinguishable from **E** with regards to the notion of use formalized using I-entailment.

Even though the case against the use criterion being a substantial one in the case of C-entailment extends also to I-entailment, some authors have held on to the latter claiming that it does yield a substantial notion of use. Meyer, as we have seen, was quite dismissive of A&B's use-criterion claiming that it "motivates whatever system one wants to motivate, within very broad limits," [36, p. 56]. The quote, however, continues

> depending on one's antecedent understanding of what constitutes a deduction; and that crucial to the latter is an understanding of how premisses can be combined and broken up in the course of a deduction. I.e., as it's now put, to an understanding of fusion. [36, pp. 56f]

[29] Besides Read's book just quoted from, this view is also adhered to in Mares and Meyer [33, § 3] and Mares [31, ch. 9].

[30] Nor do they depend on the presence of the double negation elimination axiom (A14). See Øgaard [43] for details.

It should be noted, then, that fusion does have a precise reading, according to Meyer:

> It's not, of course, that '.' doesn't have an English reading. It does; it's read 'and'. But it's read 'and' as in the sentence, "*A* and *B*, together, relevantly, and without superfluous redundancy, entail *C*." [36, p. 57]

What Meyer seems to have thought wrong with A&B's use criterion was that they mixed extensional and intensional elements. Mares and Meyer in fact explicitly require that it be "*possible* to use all the premises in a relevant deduction" [33, p. 284] and explicates this using the use-restricted conditional introduction rule together with an ultimate insult to A&B in claiming that "[relevant logicians] deny that the premises in a relevantly valid argument are conjoined to one another with standard conjunction" [33, p. 285]. We have already seen that *possible* is the wrong modality, but it is not only this error that Mares and Meyer borrow from A&B, but also the whole use-idea. That premises must be "fused" together is beside the point; what matters is the appeal to a notion of premise use to differentiate the relevant from the non-relevant. Even though, then, Meyer did claim that A&B's use-criterion "must go," as we've seen he put it, it is difficult to square his account with a total rejection of A&B's use-idea.

We saw in Sect. 4.1 that the Hilbert consequence for **S4**, intuitionistic, and classical logic, can be viewed as a mere extension by definition of C-entailment. This, then, is also the case with I-entailment for these logics.[31] The three notions of logical consequence diverge more radically for other logics, but for present purposes it is beside the point to go into the details of just *how* they diverge, for which logics this is the case, or whether there are good reasons for preferring one consequence relation over others. What matter here is that the notion of *use* thought to be explicated using either C- or I-entailment is not a notion which can be an umpire of logics as relevant logicians have thought.

4.6 *Summary of the* Use-*Account of Relevance*

A&B appealed to a notion of premise use in order to motivate a formal criterion of relevance. Throughout A&B's writings it is difficult to get a clear sense of what they in fact mean by *relevance*. Although it makes for a catchy phrase, what is philosophically important is what use it is put to. I've tried to show that by rather focusing on the *criteria* of relevance—the meaning-criterion and the use-criterion—one can simply circumvent the problem of what A&B took relevance to be as what A&B were after was primarily criteria of *logical consequence*. Their main agenda was to argue for a new theory of logical consequence, one which could account for both the meaning-criterion—that the premises must be meaning-connected to the conclusion—and the use-criterion—that the premises must be used in obtaining the

[31] Again, see Øgaard [43] for details.

conclusion in order for it to truly follow *from* them. This section has shown that the use-criterion is rather useless: it fails to distinguish between the logics A&B thought it to and in particular fails to distinguish between their favored logics **E** and **R** and "irrelevant" logics such as **S4**, classical and intuitionistic logic. A&B's notion of logical consequence is different from the standard Hilbertian one also for these latter three logics. As we saw, however, this is so in a rather uninteresting way as the Hilbertian consequence relation is obtainable through a simple extension by definition of A&B's notion of entailment in the case of the empty set of premises.

In light of this rather negative judgement, let me stress that the current critique is directed at A&B's notion of premise use. As such it does not rule out there being a well-motivated additional proof-theoretic criterion which can be coupled with A&B's *use*-idea so as to ensure that premises only be used in a relevantly acceptable way. The aim of this paper is merely to evaluate the Anderson-Belnap account of relevance in the hope that coming new directions of relevant research will do a better job at motivating and specifying the technical and philosophical assumptions and aims than what has been the case so far.

Being *rather useless* need not entail being without merit, and the sober judgement of the use-criterion ought not to be that it is without significance. Rather, its true value lies in the fact that it raises the variable sharing property from merely applying to logically true conditionals to applying directly to *logical consequence*, which, after all, was the intent behind the variable sharing property to begin with.

5 Summary

Anderson and Belnap thought that the standard theory of logical consequence was wrong and sought a new one free of the "paradoxes" they thought both classical and intuitionistic logic, as well as the modal logic **S4**, to be infested with. Their judgment was that these logics all fail to take account of the fact that "a necessary condition for the validity of an inference from A to B is that A be relevant to B" [4, p. 33]. To give content to this relevance-idea, Anderson and Belnap appealed to two features: the first was that the premises must be meaning-connected to the conclusion. This was then formalized as the well-known variable sharing property. I've argued that the formal property needs to be reformulated so as to apply directly to the consequence relation itself—and not only derivatively from the fact that the conditional thought of as expressing that relation satisfies the variable sharing property—if it is to connect up with the motivation stated for it. It was shown that this can be done in most cases through a generalization of Anderson and Belnap's "Entailment theorem." It was argued, however, that even if this is conceded, the property is still motivationally wanting: what is needed is not only the acknowledgement that variable sharing *suffices* for meaning-connectedness, but also that it is *necessary*. Establishing this, it was argued, requires much more than what Belnap's "obvious note" that "commonality of meaning is carried by identity of propositional variables" [11, p. 144] establishes.

Anderson and Belnap also thought that the standard Hilbertian notion of logical consequence was itself wrong: "the defect lies in the definition, which fails to take seriously the word 'from' in 'proof *from* hypothesis' " [4, p. 34]. The thought was that if the conclusion follows without *using* the premises, then it doesn't really follow *from* them. I've argued that Anderson and Belnap stated the motivation incorrectly as they only seem to require that it be *possible* to use the premises in deriving the conclusion. Even if, however, this is rectified so that the premises *must* be used, I've shown that the formal property which was thought of as ensuring that premises be used, also holds for by and large every logic. Even if one requires that *every* premise be used, and not only that *some* as Anderson and Belnap were content to, this notion of use cannot be used to rule out "irrelevant" logics such as **S4**, intuitionistic or classical logic.

In short, then, this paper has argued that the use-criterion cannot as it stands be regarded as an umpire of logics and that the idea of meaning-connectedness as a necessary condition of logical consequence does not so easily motivate the variable sharing property as commonly thought. The Anderson-Belnap approach to entailment, then, or so this paper has tried to show, is in dire need of a new account of what relevance amounts to.

This paper has evaluated one aspect of the Anderson-Belnap program of trying to restore relevance onto logical consequence. It has done so using primarily Anderson and Belnap's theory by itself, so to speak, without, that is, going too much into the reasons *why* they thought a notion of relevance was lacking from the "standard" account of logical consequence. Anderson and Belnap judged that entailment involved both meaning-relatedness as well as a robust notion of *following from*. This paper has shown that the formal properties Anderson and Belnap came up with do not capture their motivational ideas accurately. Whether or not other properties can, or whether Anderson and Belnap's motivation for the properties can be updated somehow, are questions for another occasion.

References

1. Ackermann, W. (1956). Begründung einer strengen Implikation. *Journal of Symbolic Logic*, *21*(2), 113–128. https://doi.org/10.2307/2268750
2. Anderson, A. R. (1960). Completeness theorems for the systems E of entailment and EQ of entailment with quantifiication. *Zeitschrift für mathematische Logik und Grundlagen der Mathematik*, *6*, 201–216. https://doi.org/10.1002/malq.19600060709
3. Anderson, A. R., & Belnap, N. D. (1958). A modification of Ackermann's "rigorous implication". *Journal of Symbolic Logic*, *23*(4), 457–458. https://doi.org/10.2307/2964046
4. Anderson, A. R., & Belnap, N. D. (1962). The pure calculus of entailment. *Journal of Symbolic Logic*, *27*(1), 19–52. https://doi.org/10.2307/2963676
5. Anderson, A. R., & Belnap, N. D. (1975). *Entailment: The logic of relevance and necessity* (Vol. 1). Princeton: Princeton University Press.
6. Anderson, A. R., Belnap, N. D., & Dunn, M. J. (1992). *Entailment: The logic of relevance and necessity* (Vol. 2). Princeton: Princeton University Press. https://doi.org/10.1515/9781400887071

7. Avron, A. (1992). Whither relevance logic? *Journal of Philosophical Logic*, *21*(3), 243–281. https://doi.org/10.1007/BF00260930
8. Avron, A. (2014). What is relevance logic? *Annals of Pure and Applied Logic*, *165*(1), 26–48. https://doi.org/10.1016/j.apal.2013.07.004
9. Avron, A. (2022). On formal criteria for relevance. In K. Bimbó (Ed.), *Relevance logics and other tools for reasoning: Essays in honor of J. Michael Dunn* (pp. 54–64). London: College Publications.
10. Baylis, C. A. (1931). Implication and subsumption. *The Monist*, *41*(3), 392–399. https://doi.org/10.5840/monist193141325
11. Belnap, N. D. (1960). Entailment and relevance. *Journal of Symbolic Logic*, *25*(2), 144–146. https://doi.org/10.2307/2964210
12. Bennett, J. (1965). Review of *Entailment and deducibility* by T. J. Smiley, and *The pure calculus of entailment* by Alan Ross Anderson and Nuel D. Belnap, Jr. *The Journal of Symbolic Logic*, *30*(2), 240–241. https://doi.org/10.2307/2270141
13. Bobzien, S. (2019). Dialectical school. In E. N. Zalta (Ed.), *The Stanford encyclopedia of philosophy* (spring 2019 edition). Metaphysics Research Lab, Stanford University. https://plato.stanford.edu/archives/spr2019/entries/dialectical-school/
14. Brady, R. T. (1984). Natural deduction systems for some quantified relevant logics. *Logique et Analyse*, *27*(108), 355–377. http://www.jstor.org/stable/44084099
15. Brady, R. T. (1993). Rules in relevant logic—II: Formula representation. *Studia Logica*, *52*(4), 565–585. https://doi.org/10.1007/BF01053260
16. Brady, R. T. (2006). *Universal logic*. Stanford: CSLI Publication.
17. Church, A. (1951). The weak theory of implication. In A. Menne, A. Wilhelmy, & H. Angsil (Eds.), *Kontroliertes Denken: Untersuchungen zum Logikkalkül und zur Logik der Einzelwissenschaften* (pp. 22–37). Munich: Kommissions-Verlag Karl Alber.
18. Copeland, B. J. (1980). The trouble Anderson and Belnap have with relevance. *Philosophical Studies*, *37*(4), 325–334. https://doi.org/10.2307/4319379
19. Duncan-Jones, A. (1935). Is strict implication the same as entailment? *Analysis*, *2*(5), 70–78. https://doi.org/10.1093/analys/2.5.70
20. Dunn, J. M. (1979). Relevant Robinson's arithmetic. *Studia Logica*, *38*(4), 407–418. https://doi.org/10.1007/BF00370478
21. Dunn, J. M. (1987). Relevant predication 1: The formal theory. *Journal of Philosophical Logic*, *16*(4), 347–381. https://doi.org/10.1007/BF00431183
22. Dutilh Novaes, C. (2020). Medieval theories of consequence. In E. N. Zalta (Ed.), *The Stanford encyclopedia of philosophy* (Fall 2020 edition). Metaphysics Research Lab, Stanford University. https://plato.stanford.edu/archives/fall2020/entries/consequence-medieval/
23. Fitch, F. B. (1952). *Symbolic logic: An introduction*. New York: The Rondald Press Company.
24. Istre, E., & McKubre-Jordens, M. (2019). The difficulties in using weak relevant logics for naive set theory. In C. Başkent & T. M. Ferguson (Eds.), *Graham Priest on Dialetheism and Paraconsistency* (pp. 365–381). Cham: Springer International Publishing. https://doi.org/10.1007/978-3-030-25365-3_17
25. Jago, M. (2020). Truthmaker semantics for relevant logic. *Journal of Philosophical Logic*, *49* (4), 681–702. https://doi.org/10.1007/s10992-019-09533-9
26. Kron, A. (1973). Deduction theorems for relevant logics. *Mathematical Logic Quarterly*, *19* (3–6), 85–92. https://doi.org/10.1002/malq.19730190306
27. Lewis, D. (1988). Relevant implication. *Theoria*, *54*(3), 161–174. https://doi.org/10.1111/j.1755-2567.1988.tb00716.x
28. Mares, E. (1997). Relevant logic and the theory of information. *Synthese*, *109*(3), 345–360. https://doi.org/10.1007/BF00413865
29. Mares, E. (2010). The nature of information: a relevant approach. *Synthese*, *175*(1), 111–132. https://doi.org/10.1007/s11229-010-9737-z
30. Mares, E. (2020). Relevance logic. In E. N. Zalta (Ed.), *The Stanford encyclopedia of philosophy* (Winter 2020 edition). Metaphysics Research Lab, Stanford University. https://plato.stanford.edu/archives/win2020/entries/logic-relevance/

31. Mares, E. D. (2004). *Relevant logic: A philosophical interpretation.* Cambridge: Cambridge University Press. https://doi.org/10.1017/CBO9780511520006
32. Mares, E. D. (2006). Relevance logic. In *A companion to philosophical logic* (pp. 609–627). John Wiley & Sons, Ltd. https://doi.org/10.1002/9780470996751.ch39
33. Mares, E. D., & Meyer, R. K. (2001). Relevant logics. In *The Blackwell guide to philosophical logic* (pp. 280–308). Malden: Blackwell Publishers Ltd. https://doi.org/10.1002/9781405164801.ch13
34. McKubre-Jordens, M., & Weber, Z. (2012). Real analysis in paraconsistent logic. *Journal of Philosophical Logic, 41*(5), 901–922. https://doi.org/10.1007/s10992-011-9210-6
35. Meyer, R. K. (1970). **E** and **S4**. *Notre Dame Journal of Formal Logic, 11*(2), 181–199. https://doi.org/10.1305/ndjfl/1093893935
36. Meyer, R. K. (1974). New axiomatics for relevant logics, I. *Journal of Philosophical Logic, 3* (1), 53–86. ISSN 1573-0433. https://doi.org/10.1007/BF00652071
37. Meyer, R. K. (1985). A farewell to entailment. In G. Dorn & P. Weingartner (Eds.), *Foundations of logic and linguistics: Problems and their solutions* (pp. 577–636). Boston, MA: Springer US. https://doi.org/10.1007/978-1-4899-0548-2_25
38. Meyer, R. K., & Restall, G. (1999). "*Strenge*" arithmetics. *Logique & Analyse, 42*(167/168), 205–220. https://www.jstor.org/stable/44084657
39. Nelson, E. J. (1930). Intensional relations. *Mind, 39*(156), 440–453. https://doi.org/10.1093/mind/XXXIX.156.440
40. Øgaard, T. F. (2021). Non-Boolean classical relevant logics I. *Synthese, 198*, 6993–7024. https://doi.org/10.1007/s11229-019-02507-z
41. Øgaard, T. F. (2021). Non-Boolean classical relevant logics II: Classicality through truth-constants. *Synthese, 199*, 6169–6201. https://doi.org/10.1007/s11229-021-03065-z
42. Øgaard, T. F. (2023). *A study in relevant logics*. PhD thesis, University Bergen.
43. Øgaard, T. F. (2024). Entailment generalized. In S. Standefer, I. Sedlár, & A. Tedder (Eds.), *New directions in relevant logics*, chapter 12. Springer.
44. Øgaard, T. F. (forthcoming). Suppressing irrelevance: Notes on Arnon Avron's view of relevance. *Logique et Analyse*.
45. Priest, G. (2011). Paraconsistent set theory. In D. DeVidi, M. Hallett, & P. Clarke (Eds.), *Logic, mathematics, philosophy: Vintage enthusiasms, The Western Ontario series in philosophy of science* (Vol. 75, pp. 153–169). Dordrecht: Springer. https://doi.org/10.1007/978-94-007-0214-1_8
46. Priest, G. (2015). Fusion and confusion. *Topoi, 34*(1), 55–61. https://doi.org/10.1007/s11245-013-9175-x
47. Prior, A. N. (1962). *Formal logic* (2nd ed.). Oxford: Oxford University Press.
48. Read, S. (1988). *Relevant logic: A philosophical examination of inference*. Oxford: Basil Blackwell.
49. Restall, G. (1996). Information flow and relevant logics. In J. Seligman & D. Westerståhl (Eds.), *Logic, language and computation* (pp. 463–477). Stanford: CSLI Publications.
50. Restall, G. (2010). Models for substructural arithmetics. *Australasian Journal of Logic, 8*, 82–99. https://doi.org/10.26686/ajl.v8i0.1814
51. Routley, R. (1980). *Exploring Meinong's jungle and beyond.* Departmental Monograph, Philosophy Department, RSSS, Australian National University, vol. 3. Canberra: RSSS, Australian National University, Canberra. https://hdl.handle.net/11375/14805
52. Routley, R. (1981). How disjunctive syllogism can be seen as a fallacy of relevance. *Bulletin of the Section of Logic, 10*(4), 144–149. http://www.filozof.uni.lodz.pl/bulletin/pdf/10_4_1.pdf
53. Routley, R., Meyer, R. K., Plumwood, V., & Brady, R. T. (1982). *Relevant logics and their rivals* (Vol. 1). Atascadero, CA: Ridgeview Publishing Company.
54. Shaw-Kwei, M. (1950). The deduction theorems and two new logical systems. *Methodos, 2* (5), 56–75.
55. Slaney, J. (1990). A general logic. *Australasian Journal of Philosophy, 68*(1), 74–88. https://doi.org/10.1080/00048409012340183

56. Slaney, J. K. Meyer, R. K., & Restall, G. (1996). Linear arithmetic desecsed. *Logique & Analyse*, *39*(155–156), 379–387. https://www.jstor.org/stable/44084577
57. Smiley, T. J. (1958). Entailment and deducibility. *Proceedings of the Aristotelian Society*, *59*, 233–254. https://doi.org/10.2307/454461
58. Tennant, N. (2005). Relevance in reasoning. In S. Shapiro (Ed.), *The Oxford handbook of philosophy of mathematics and logic* (pp. 696–726). Philosophia Verlag. https://doi.org/10.1093/0195148770.003.0023
59. Urquhart, A. (1993). Failure of interpolation in relevant logics. *Journal of Philosophical Logic*, *22*(5), 449–479. https://doi.org/10.2307/30226507
60. Weber, Z. (2010). Extensionality and restriction in naive set theory. *Studia Logica*, *94*(1), 87–104. https://doi.org/10.1007/s11225-010-9225-y
61. Weber, Z. (2010). Transfinite numbers in paraconsistent set theory. *The Review of Symbolic Logic*, *3*(1), 71–92. https://doi.org/10.1017/S1755020309990281
62. Weber, Z. (2012). Transfinite cardinals in paraconsistent set theory. *The Review of Symbolic Logic*, *5*(2), 269–293. https://doi.org/10.1017/S1755020312000019
63. Weber, Z. (2013). Notes on inconsistent set theory. In K. Tanaka, F. Berto, E. Mares, & F. Paoli (Eds.), *Paraconsistency: Logic and applications* (pp. 315–328). Dordrecht: Springer Netherlands. https://doi.org/10.1007/978-94-007-4438-7_17

Variable-Sharing as Relevance

Shawn Standefer

Abstract A challenge for relevant logicians is to delimit their area of study. I propose and explore the definition of a relevant logic as a logic satisfying a variable-sharing property and closed under detachment and adjunction. This definition is, I argue, a good definition that captures many familiar logics and raises interesting new questions concerning relevant logics.

As is familiar to readers of *Entailment* or *Relevant Logics and Their Rivals*, the motivations for relevant logics have a strong intuitive pull.[1] The philosophical picture put forward by Anderson and Belnap [1], for example, is compelling and has led to many fruitful developments. With some practice, one can develop a feel for what sorts of axioms or rules lead to violations of relevance in standard relevant logics. These sorts of intuitions only go so far, as some principles that lead to violations of relevance in stronger logics are compatible with it in weaker logics. There is a large number of relevant logics, but there is not much discussion of precise characterizations of the class of relevant logics.[2]

It is well known that the standard relevant logics avoid C. I. Lewis's paradoxes of implication, such as $A \to (B \to B)$ and $A \to (B \to A)$, but avoiding the paradoxes does not provide an adequate characterization for two reasons. First, it is a negative characterization that is also open-ended, insofar as the list of paradoxes of implication is open-ended. Second, it is not clearly based on a formal principle. Lewis's paradoxes of implication present some counter-intuitive features of the

[1] For an overview of different motivations for relevant logics, see [58].

[2] Avron [3] expresses this concern and provides a characterization, but it is less inclusive than the one adopted here.

S. Standefer (✉)
Department of Philosophy, National Taiwan University, Taipei, Taiwan
e-mail: standefer@ntu.edu.tw

I. Sedlár et al. (eds.), *New Directions in Relevant Logic*, Trends in Logic 63,
https://doi.org/10.1007/978-3-031-69940-5_5

classical material implication, but it is not clear what formal principles should be extracted from the intuitions motivating their rejection.

The lack of a general, abstract characterization of the class of relevant logics is unfortunate, because investigations of relevant logics typically proceed either by focusing on a particular logic or by presenting a more or less arbitrary selection of axioms and rules combination of which determine the logics under consideration. Further, while the motivations mentioned above are often compelling, it is not clear that they succeed in isolating only the proponents' preferred logics.[3] Different logicians have defended different views about relevant logics and these views may not lead to the same groups of logics. There is, then, a need to reconsider foundational aspects of relevant logics, one aspect of which is a principled delimitation of the area of study.

In contrast to the situation with relevant logics, the (classical) modal logician has a ready answer to the analogous question "what is a modal logic?" The modal logician can say that a (normal) modal logic is any logic extending classical logic with the (K) axiom, $\Box(A \rightarrow B) \rightarrow (\Box A \rightarrow \Box B)$, and closed under the rule (Nec), $A \Rightarrow \Box A$, to be understood as saying that if A is a logical truth, then so is $\Box A$.[4] Similarly, the paraconsistent logician can delimit their area of study as those consequence relations where $A, {\sim}A \not\vdash B$, for some formulas A and B and a negation operator '$\sim$'. These abstract characterizations allow the modal or paraconsistent logician to take a broader view of their areas, shifting the focus from the more or less standard logics, such as K or S4, to more general classes of logics. So, it seems like an abstract characterization of relevant logics would be useful.

The question is how to formulate an abstract characterization of relevant logics. There are prominent groups of relevant logics, such as the preferred logics of Anderson and Belnap, but those are not the only relevant logics. One could try to give a model-based characterization, such as using the ternary relational models of Sylvan and Meyer, but the models for relevant logics can also be used to adequately model logics containing $A \rightarrow (B \rightarrow B)$ and $A \rightarrow (B \rightarrow A)$. Additionally, it would be good to have a model-independent characterization, and this points us to the basis of the proposal.

A key feature of relevant logics, one of the few that is generally agreed upon, is that they satisfy Belnap's *variable-sharing criterion*:

Definition 1 (Variable-Sharing Criterion) A logic L *satisfies the variable-sharing criterion* iff whenever $A \rightarrow B$ is a logical truth, then A and B share a propositional variable.

[3] As an example, see [30] or [35, 37, 39], for example.

[4] This may not be adequate as a definition once other intensional operators are permitted, such as the actuality operator. If one has contingent logical truths with actuality, as defended by [34], the rule of Necessitation can fail in what is intuitively a modal logic. I thank Ben Blumson for pointing this out.

It is often said that the variable-sharing criterion is a necessary but not sufficient condition on being a relevant logic. Indeed, as Anderson and Belnap say,

> A formal condition for "common meaning content" becomes almost obvious once we note that commonality of meaning in propositional logic is carried by commonality of propositional variables. So we propose as a necessary, but by no means sufficient, condition for the relevance of A to B in the pure calculus of entailment, that A and B must share a variable.[5]

This view is fairly standard, as illustrated by Mares, who says,

> The variable sharing principle is only a necessary condition that a logic must have to count as a relevance logic. It is not sufficient. Moreover, this principle does not give us a criterion that eliminates all of the paradoxes and fallacies. Some remain paradoxical or fallacious even though they satisfy variable sharing.[6]

Further examples are not hard to find. The variable-sharing criterion presents a *minimal* requirement of formal relevance for implicational logical truths. There are natural things one can add to it and ways to strengthen it, some of which I will return to in the final section of this paper, but the basic variable-sharing criterion provides a succinct specification of a minimal level of formal relevance. While it is commonplace to say that variable-sharing is not sufficient for being a relevant logic, it is worth considering how things look if we take it to be sufficient, with another minimal condition, seeing what follows from that and what requires more. The proposal, then, is to say that relevant logics are those that satisfy the variable-sharing criterion, along with two other conditions. I will explore this proposal, prove some initial results concerning this class of logics, and then discuss how more familiar families of logics fit into this class. In Sect. 1, I will present the relevant background and the proposal for a definition of the class of relevant logics. In Sect. 2, I explore the proposal, identify some consequences of the definition, and identify further conditions one might use to identify subclasses of logics. In Sect. 3, I relate the proposal to some extant views about relevant logics and respond to some potential criticisms, closing with a brief discussion of other forms of variable-sharing that one can consider.

1 The Proposal

Let the language under consideration be made from countably many propositional variables, $p, q, r, \ldots$, and the connectives $\{\wedge, \vee, \sim, \rightarrow\}$, with $A \leftrightarrow B$ defined, as usual, as $(A \rightarrow B) \wedge (B \rightarrow A)$.[7] To facilitate comparisons, we will take classical

[5] Anderson and Belnap [1, 33].

[6] Mares [26].

[7] A comment on the language is in order. The Ackermann and Church truth constants, t and $\top$, are excluded from the language. The primary reason is that for formulas that contain them, the variable-sharing criterion does not apply. See [65] or [36] for discussion of variable-sharing in the

logic to be formulated in this vocabulary and to include $(A \to B) \leftrightarrow ({\sim}A \vee B)$ as a theorem. The language can be pared back, but the key thing is that there is an implication connective, $\to$, in the language. The other connectives are optional. We will, for this paper, focus just on logical truths, as relevant logics are most often studied under that guise. Indeed, the logical truths seem to be the key to distinguishing relevant logics.[8] We will assume some familiarity with relevant logics, but only the better known ones, such as R, T, and B.[9]

We will treat logics as sets of formulas closed under uniform substitution.[10] A substitution is a function σ from propositional variables to formulas that is extended to the a function σ^+ on the whole language such that

- $\sigma^+(p) = \sigma(p)$, for p a propositional variable,
- $\sigma^+({\sim}A) = {\sim}\sigma^+(A)$, and
- $\sigma^+(A \star B) = \sigma^+(A) \star \sigma^+(B)$, for $\star \in \{\wedge, \vee, \to\}$.

With that definition in hand, we can define logics.

Definition 2 (Logics) A set of formulas X is a *logic* iff for any substitution σ and formula A, if $A \in X$, then $\sigma^+(A) \in X$

A logic L is a *sublogic* of another logic M iff L $\subseteq$ M.

Throughout the paper, I will use formula schemes to present various principles, where the instances of the scheme are obtained by replacing the displayed letters with arbitrary formulas. When presenting an instance of a scheme, I will use a specific formula displaying the propositional variables. A logic *includes* a formula scheme iff it includes every instance of that scheme. A logic *excludes* a scheme iff there is an instance of the scheme it does not contain.

Before I get to the main definition, I will present two preliminary definitions of classes of logics.

Definition 3 (Proto-Relevant Logics) A logic L is a *proto-relevant logic* iff L satisfies the variable-sharing criterion.

The class of proto-relevant logics captures an important feature of relevant logics, but below we will argue that a little bit more should be added. For the next definition, it will be useful to introduce the notation for rules: A logic L is closed under a rule, $X \Rightarrow A$, iff for all substitutions σ, if $\sigma^+(B) \in$ L, for all $B \in X$, then $\sigma^+(A) \in$ L.

presence of truth constants. The formulation of a general definition of relevant logics that includes the truth constants, or at least the Ackermann constant, is left as an open problem.

We have omitted fusion from the language, but we can add it, making some minor changes below.

[8] Avron [2, 3] looks at consequence relations and contains an alternative, less inclusive, characterization of relevant logics.

[9] For axiomatic presentations of these logics, see [10]. For a general overview of relevant logics, see [5, 18, 43], or [26].

[10] This is the framework FMLA of [23, ch. 1.2].

Definition 4 (DA-Logics) A logic L is a *DA-logic* iff both

(1) it is closed under detachment, also known as *modus ponens*, $A, A \to B \Rightarrow B$, and
(2) it is closed under adjunction, $A, B \Rightarrow A \wedge B$.

If one wants to consider only the $\to$-fragment of logics, then condition (2) can be dropped. DA-logics include many familiar logics, such as classical logic, intuitionistic logic, and R. As should be clear from the examples, some DA-logics are not proto-relevant logics.

The two classes of logics, proto-relevant logics and DA-logics, are brought together in our main definition.

Definition 5 (Relevant Logics) A logic L is a *relevant logic* iff L is a proto-relevant logic and is a DA-logic.

I claim that this definition provides a good delimitation of the area of relevant logics, and I will explore and defend this definition in the remaining sections of this paper. I will begin by commenting on the parts of Definition 5.

First, being a proto-relevant logic immediately guarantees that every relevant logic will satisfy Belnap's variable-sharing criterion. It immediately rules out logics that include some of the implicational paradoxes, such as $A \to (B \to B)$. These are, on their face, incompatible with being a relevant logic.[11]

The proto-relevant logics include many logics that are non-transitive, in the sense that $A \to B$ and $B \to C$ are logical truths but $A \to C$ is not.[12] The implicational form of the steps of C. I. Lewis's argument for explosion will do.

- $(A \wedge {\sim}A) \to ({\sim}A \wedge (A \vee B))$
- $({\sim}A \wedge (A \vee B)) \to B$
- $(A \wedge {\sim}A) \to B$

The first two are cleared by the variable-sharing criterion but not the third. While non-transitive logics are not unheard of, they are also typically excluded from the family of relevant logics.[13] As Anderson and Belnap [1, 154] say,

> And what this shows is that connection of meaning, though necessary, is not a sufficient condition for entailment, since the latter relation is transitive. Any criterion according to which entailment is non-transitive, is *ipso facto* wrong. It seems in fact incredible that anyone should admit that B follows from A, and that C follows from B, but feel that some further argument was required to establish that A entails C. What better evidence for $A \to C$ could one want?

---

[11] Even the well known relevant logics, such as R and B, contain some instances of the scheme $A \to (B \to B)$, e.g. $(p \to p) \to (p \to p)$. The important thing is that they exclude the scheme by omitting some instances, such as $p \to (q \to q)$. I would like to thank Lloyd Humberstone for highlighting this point.

[12] Szmuc [61].

[13] See [45, 56], or [64], for examples of non-transitive logics.

The proto-relevant logics include many logics that are not transitive, and so, arguably, are not logics of *entailment*. Even the logics taken to be central relevant logics include some that are not logics of entailment, such as R. So, there does not seem to be an essential conflict between being a relevant logic and being non-transitive. This is a point of agreement with the criticisms of [15, 331], which doubts that non-transitive logics should be excluded from consideration as relevant logics.

Requiring closure under detachment seems like a reasonable condition for a connective being an *implication* connective. Together with satisfying variable-sharing, it excludes many of the remaining paradoxes of implication, in particular $A \to (B \to A)$.

Theorem 6 *No relevant logic contains* $A \to (B \to A)$.

Proof To see this, let C be $p \to (q \to p)$ and let D be $r \to (s \to r)$. Suppose that $A \to (B \to A)$ is a logical truth. Then $C \to (D \to C)$ and C will be logical truths, as they are instances of the scheme and so $D \to C$ will be as well, but that violates variable-sharing. □

Another paradox that is excluded is $A \to ({\sim}A \to B)$. Let C be $p \to ({\sim}p \to q)$ and let D be $r \to ({\sim}r \to s)$. Then $C \to ({\sim}C \to D)$ and C are both in the logic, so by detachment, ${\sim}C \to D$ is as well, but ${\sim}C$ and D do not share a variable. Indeed, a reason not to claim that the proto-relevant logics should be the target class of logics for the relevant logician is that the logic consisting of every instance of $A \to (B \to A)$ satisfies variable-sharing, while violating the condition of closure under detachment.

Being a DA-logic ensures that all relevant logics are theories.

Definition 7 (Theory) A *theory* is a set X of formulas such that (i) if $A \to B \in X$ and $A \in X$, then $B \in X$ and (ii) if $A \in X$ and $B \in X$, then $A \wedge B \in X$.

Theories are typically defined with respect to a particular background logic under whose implications the theory is supposed to be closed, but since all the sets of formulas we are considering are supposed to be logics, the definition requires closure under their own implications. Theories are important in the development of models for relevant logics, and being a theory is a natural constraint on being a logic. Nonetheless, it is not clear that being a theory is an essential feature of being a relevant logic, so one could adopt a broader definition of relevant logics as those proto-relevant logics closed under detachment but not necessarily closed under adjunction.

2 Exploring the Proposal

Let us explore the proposed definition of relevant logics, Definition 5, in order to help justify the claim that this is a good definition. At the outset, I noted that modal logicians have a snappy definition of a normal modal logic. That definition sets a baseline, minimal condition. There is a minimal (normal) modal logic, namely the logic K. Any extension of K is a modal logic. On Definition 5, there is a unique, minimal relevant logic, namely the empty logic, which will appear again later. Unlike the situation with modal logics, there are extensions of the minimal logic that are not relevant logics. With the relevant logics, the interest is more on the *upper bounds* for relevant logics. The two parts of the definition do different jobs. Being a DA-logic forces certain formulas to be in the logic, given that others are. Being a proto-relevant logic, by contrast, does not require the inclusion of anything, but rather states that certain formulas cannot be in the logic.

Given that the upper bounds of the class of relevant logics is the point of interest, the natural question is whether there is a unique, strongest relevant logic. There is not, but before getting to this theorem and its proof, I need to briefly explain a connection between matrices and logics. A matrix is a non-empty set of values, V, a set of operations on V under which V is closed, and a non-empty set $D \subseteq V$, which is set of designated values.[14] An assignment on a matrix assigns values from V to atoms and it is extended to the whole language using the set of operations. Matrices are often used to show that a logic has the variable-sharing property. For example, Belnap's matrix M_0 and Meyer's crystal lattice can be used to demonstrate variable-sharing.[15] A matrix can be used to define a logic as the set of formulas that take designated values on all assignments based on that matrix. We will call the logics of the crystal lattice and M_0, respectively, CL and M_0. Swirydowicz [59] has shown that there are two maximal extensions of R that satisfy variable-sharing. These are, in fact, CL and M_0, the logics of the crystal lattice and M_0. With that in mind, we turn to the theorem.

Theorem 8 *There is not a strongest relevant logic, ordered by the sublogic relation.*

Proof CL and M_0 both satisfy conditions on being a DA-logic. Their union, when closed under detachment and adjunction, does not satisfy the variable-sharing criterion. □

A natural question that arises at this point is how to characterize the maximal relevant logics. As shown by Robles and Méndez [46, 47], there is a general sufficient condition on a matrix that can be used to show variable-sharing. The key feature, for our purposes, is that there are two disjoint sets of values, X and Y such that each is closed under the operations of the matrix and for $x \in X$ and $y \in Y$, $x \to y$ is not a designated value. There are some matrices that can be used to show variable-sharing that do not yield maximal relevant logics. As an example, the product matrix of M_0 with the crystal lattice yields a relevant logic, which is to say a DA-logic that satisfies variable-sharing, but is not maximal.

Before proceeding to further investigate the proposed definition of relevant logic, I want to pause to consider an alternative proposed demarcation that is sharp, albeit

[14] See [17, ch. 7] for more on matrices.

[15] See, respectively, [1, 252ff.] and [51, §3.6].

less snappy. Since R is often viewed as the strongest standard relevant logic, one might be tempted to define *relevant logics* as logics that are sublogics, or rather sublogics that are also DA-logics, of CL and M_0. While this class would not have a greatest logic, it would have exactly two maximal logics. It has, however, a major problem: It would exclude some logics that have a strong claim to being in the relevant logic family. One example is TM, the logic T with the addition of the mingle axiom, $A \to (A \to A)$. As is well known, RM, the result of adding mingle to R, violates variable-sharing, as $\sim(A \to A) \to (B \to B)$ is a logical truth of RM. TM, however, maintains variable-sharing.[16] TM is not, however, a sublogic of R or its two maximal extensions with variable-sharing. The proposed definition in terms of CL and M_0 is, then, inadequate, so I will return to the main proposal.

A virtue of the proposal of Definition 5 is that it is *presentation-independent*. It does not involve presenting a proof system. Nor does it rely on having a class of models or frames. Rather, it isolates some properties of a logic, and uses those to characterize the broad area. This approach to logic, while perhaps not the most common, is in fact well-known and useful. It lets us ask general and abstract questions about logics and relations between logics, such as whether a given logic is identical to the intersection of two others, as Anderson and Belnap asked about their logic E in relation to R and S4.[17] We can ask about operations under which classes of logics are closed.

Lemma 9 *The class of DA-logics is closed under intersection.*

Proof Let L and K be DA-logics. Suppose A and $A \to B$ are in both. By condition (1) of Definition 4, both are closed under detachment, so $L \cap K$ contains B. Next, suppose that A and B are in both. By condition (2) of Definition 4, then, $A \wedge B \in L \cap K$. □

Lemma 10 *Let* L *be a proto-relevant logic. Let* K *be a logic. Then,* $L \cap K$ *satisfies variable-sharing, which is to say that it is a proto-relevant logic.*

Proof Suppose L be a proto-relevant logic. Let A and B be arbitrary formulas such that $A \to B \in L \cap K$. It follows that $A \to B \in L$, so $A \to B$ share a propositional variable. Since A and B were arbitrary, it follows that $L \cap K$ satisfies variable-sharing. □

The following theorem is then a corollary of the previous two lemmas.

Theorem 11 *The class of relevant logics is closed under intersection.*

Proof Let L and K be relevant logics. By Lemma 9, $L \cap K$ is a DA-logic. By Lemma 10, it is also a proto-relevant logic. □

We can also ask about operations under which classes of logics are not closed.

---

[16] Méndez et al. [28].

[17] For the interested reader, see [1, §8.10].

Lemma 12 *The class of DA-logics is not closed under union.*

Proof Let L be the least DA-logic containing all substitution instances of excluded middle, $A \vee \sim A$. Let K be the least DA-logic containing $A \to A$. $\mathsf{L} \cup \mathsf{K}$ is not a DA-logic. To see this, note that $(p \vee \sim p) \wedge (q \to q) \notin \mathsf{L} \cup \mathsf{K}$. This conjunction is neither in L, as L contains no formulas whose main connective is $\to$, nor in K, as K contains no formulas whose main connective is disjunction. □

Rather than simple union, the DA-logics are closed under the operation $\overline{\cup}$, where $\mathsf{L}\overline{\cup}\mathsf{K}$ is defined as the least DA-logic containing $\mathsf{L}\cup\mathsf{K}$. The DA-logics form a lattice with $\cap$ as meet and $\overline{\cup}$ as join.

In contrast to the DA-logics, the proto-relevant logics are closed under $\cup$.

Lemma 13 *The proto-relevant logics are closed under union.*

Proof Let L and K be proto-relevant logics. Let $A \to B \in \mathsf{L} \cup \mathsf{K}$. Then either $A \to B \in \mathsf{L}$ or $A \to B \in \mathsf{K}$. In both cases, A and B share a propositional variable. □

This lemma, combined with Lemma 10, tells us that the proto-relevant logics form a lattice with $\cap$ and $\cup$ as meet and join, respectively. The proto-relevant logics, however, are not closed under the $\overline{\cup}$ operation that acts as join for the lattice of DA-logics.

Lemma 14 *The proto-relevant logics are not closed under* $\overline{\cup}$.

Proof The logic $\mathsf{CL}\,\overline{\cup}\,\mathsf{M}_0$ violates variable-sharing. □

Many examples provide alternative proofs of the preceding lemma, such as $\mathsf{TM}\overline{\cup}\mathsf{R}$. Lemma 14 and Theorem 11 have the following corollary, given that the empty logic is a relevant logic.

Corollary 15 *The class of relevant logics is not a sublattice of the lattice of DA-logics, with* $\cap$ *as meet and* $\overline{\cup}$ *as join.*

Corollary 16 *The class of relevant logics is a meet semi-lattice with a bottom element, where* $\cap$ *is the meet and* $\emptyset$ *is the bottom element.*

While the class of proto-relevant logics is closed downwards under the sublogic relation, the class of DA-logics, and so the class of relevant logics, is not. The reason is that closure under adjunction, as well as closure under detachment, need not be preserved by sublogics. The appropriate ordering for considering relevant logics is, rather, the ordering defined by the meet operation of the semi-lattice, $\mathsf{L} \leq \mathsf{K}$ iff $\mathsf{L} \cap \mathsf{K} = \mathsf{L}$. Lemma 14 has the following as a corollary.

Corollary 17 *There is no greatest relevant logic according to the semi-lattice ordering,* $\leq$.

Next, I will comment upon some logics that are, on this definition, relevant logics. First, all the standard relevant logics in the literature, roughly those contained between R and B, are relevant logics, in the sense of Definition 5. They are all theories and they all satisfy variable-sharing.

Second, lattice R, linear logic, and Brady's logic MC count, even though they fail to extend B, seeing as they lack the distribution axiom, $(A \wedge (B \vee C)) \rightarrow ((A \wedge B) \vee (A \wedge C))$.[18] Since they fail to extend B, the standard ternary relational models cannot model them adequately. Despite this, these logics have many affinities with the more familiar relevant logics, and they are naturally included with them.

Third, many logics of analytic containment, studied by Parry and others, will be relevant.[19] As with the non-distributive logics, the analytic containment logics do not extend B. Nonetheless, they have many affinities with relevant logics, and they are often presented with similar ends.

Fourth, the set of classical tautologies in the vocabulary $\{\wedge, \vee, \sim\}$, closed under substitutions, is a relevant logic, as it contains no implications as theorems.[20] This is a degenerate case of the definition, much like the empty logic, as it contains no implications. In fact, any logic lacking implications will qualify as a relevant logic under Definition 5. While there may be much about these logics to dissatisfied with, the dissatisfaction would have to be based on something apart from considerations of relevance, as expressed by the implication.[21] For example, one might object to such logics on the basis of having bad principles for the truth-functional connectives, but this is not a matter solely of relevance. In practice, one typically wants to consider a relevant logic that contains at least some implications, so these edge cases do not seem to tell against the definition.

It is worth pausing for a moment to consider the much-loved system, FDE, as it is not a logic in our sense.[22] As defined by [1], the axiomatic form of FDE is a set of formulas $A \rightarrow B$, where neither A nor B contain any implications. This set of formulas is not a logic, in our sense, as it isn't closed under substitutions. We can obtain a logic from FDE by taking the set of formulas and closing it under substitutions. Let us call this system FDE$^+$. FDE$^+$ contains theorems, such as $(p \wedge (p \rightarrow q)) \rightarrow p$, which FDE does not, however it treats the embedded implications the same as atoms. FDE$^+$ is a logic, and in fact it is a sublogic of E and B, since FDE is contained in B and B is closed under substitutions.[23]

Some salient subclasses of relevant logics are easy to specify. This is parallel to the situation in modal logics. As noted above, the modal logician has an easy

---

[18] See [44, 63], and [12], respectively. Note that by 'linear logic', I here mean MALL, multiplicative-additive linear logic. The exponentials are not included. The reason is that the exponentials bring with them violations of variable-sharing.

[19] See [19, 7–8]. See also [60] and [20].

[20] To be more precise, one takes the set of formulas in the vocabulary $\{\wedge, \vee, \sim\}$ that are classical tautologies and closes that under substitutions from the vocabulary $\{\wedge, \vee, \sim, \rightarrow\}$, so that the result is a logic in the full vocabulary.

[21] I would like to thank an anonymous referee for pressing this issue.

[22] See [40] for a nice overview of this system.

[23] I would like to thank an anonymous referee for raising questions about the status of FDE.

answer to the question "What is a modal logic?"[24] A simple answer is that a modal logic is a logic containing all classical tautologies, the (K) axiom, and closed under detachment and the rule (Nec). This identifies a large class of logics, but the necessity operators of those logics may, in some cases, be thought to be too strong. The modal logician can broaden the definition, following [53], and others. The definition just given is for a *normal* modal logic. A *monotonic* modal logic is a logic containing all classical tautologies, closed under detachment and the rule (Mono), $A \to B \Rightarrow \Box A \to \Box B$. A *congruential* modal logic is a logic containing all classical tautologies, closed under detachment and the rule (Cong), $A \leftrightarrow B \Rightarrow \Box A \leftrightarrow \Box B$. This gives the modal logician clear areas of study, at least when considering singulary necessity operators. Further refinements of these classes can be obtained by placing some conditions on the classes.

Much as the modal logician can broaden the characterization of a modal logic to permit weaker and weaker modal logics, the relevant logician can add some conditions to constrain the upper bounds of their areas of interest. For example, one way to avoid the degenerate logic of all the classical tautologies in the vocabulary $\{\wedge, \vee, \sim\}$ is to require that *some* implication $A \to B$ be in the logic. Relevant logics containing an implicational logical truth we can call the *implicational* logics.[25]

We might be interested in the *lattice relevant logics*, those that contain the following logical truths.

(L1) $A \wedge B \to A$, $A \wedge B \to B$
(L2) $((A \to B) \wedge (A \to C)) \to (A \to (B \wedge C))$
(L3) $A \to A \vee B$, $B \to A \vee B$
(L4) $((B \to A) \wedge (C \to A)) \to ((B \vee C) \to A)$

These are the logics in which conjunction and disjunction behave in the more or less familiar lattice-connective fashion. The *distributive relevant logics* are the lattice relevant logics that contain

(L5) $(A \wedge (B \vee C)) \to ((A \wedge B) \vee (A \wedge C))$.

While lattice R, linear logic, and MC are all lattice relevant logics, they fail to be distributive. Most of the standard relevant logics fall into the class of distributive relevant logics, but distribution seems like an optional extra.

Above I noted that the class of relevant logics permits there to be non-transitive relevant logics in the sense that a logic could contain $A \to B$ and $B \to C$ but lack $A \to C$. The implication of such logics does not express a sufficiency relation or entailment, whereas expressing entailment was central to the project of Anderson and Belnap. To that end, it is natural to consider the class of *transitive logics*, those closed under the rule $A \to B, B \to C \Rightarrow A \to C$. Alternatively, one might consider

---

[24] Segerberg [54] notes that this question presents some difficulties, but we will set those aside for the time being.

[25] The implicational logics would include the logic S of [27], modifying the definition of DA-logic in the absence of conjunction.

the class of *affixing logics*, those closed under the rules $A \to B \Rightarrow (B \to C) \to (A \to C)$ and $A \to B \Rightarrow (C \to A) \to (C \to B)$.

Let us say that the γ-*relevant logics* are those that are closed under the rule γ, also known as disjunctive syllogism, $A, {\sim}A \vee B \Rightarrow B$. The question of whether a logic is a member of this class has played an important role in the development of the technical apparatus of relevant logics. Many of the standard relevant logics are γ-relevant logics, the verification of which fact was much celebrated.[26]

Anderson and Belnap were interested in logics that contained all the tautologies of classical logic in the vocabulary $\{\wedge, \vee, {\sim}\}$. As we have set things up, their logics are sublogics of classical logic, although this does not seem to be essential to the study of relevant logics. One might want to consider relevant logics that are *contra-classical*, in the sense that they have logical truths that classical logic lacks. One example is (the $\to$-fragment of) Abelian logic, which contains $((A \to B) \to B) \to A$ as a distinguishing principle.[27] Another example is the family of connexive logics, which contain principles such as ${\sim}(A \to {\sim}A)$ and ${\sim}({\sim}A \to A)$.[28] There are relevant logics that contain the connexive principles, although those logics cannot be as strong as R.[29] Including some contra-classical logics in the family of relevant logics seems like another virtue of the proposal.

One further thing that speaks in favor of the utility of a presentation-independent definition of relevant logic is that it combines neatly with additional criteria used for extensions of the language. For example, at the outset we provided a characterization of normal modal logics for classical logic. We can specify that a normal modal relevant logic is a relevant logic that contains the (K) axiom, is closed under the rule (Nec), and also contains the $(\Box\wedge)$ axiom, $(\Box A \wedge \Box B) \to \Box(A \wedge B)$. A more appropriate definition for the relevant logic context is that a modal relevant logic is a relevant logic closed under the rule (Mono), $A \to B \Rightarrow \Box A \to \Box B$, and containing the $(\Box\wedge)$ axiom. It is, I think, a virtue of the proposal that a specification of relevant modal logics can be provided using it, in a manner similar to that of normal modal logics in the case of classical logic.

The definition excludes many logics, as it should. Classical logic (with '$\to$' in the vocabulary), intuitionistic logic,[30] and RM all fail to be relevant logics.[31] Further, the addition of Boolean negation to some familiar relevant logics result in a non-relevant logic.[32]

---

[26] Meyer and Dunn [31].

[27] See Meyer and Slaney [29], Butchart and Rogerson [14], and Paoli et al. [42].

[28] Omori and Wansing [41], Francez [22].

[29] Routley et al. [51, 343].

[30] Since the logical truths of Tennant's [62] core logic are those of intuitionistic logic, that will fail to be a relevant logic on this definition.

[31] It is worth noting that the addition of the mingle axiom, $A \to (A \to A)$, to the implicational fragment of R, results in an implicational logic, $\mathsf{RM0}$, that has variable-sharing. See [1, 148] or [23, 333,363] for details and discussion. I would like to thank Lloyd Humberstone for pointing this out to me.

[32] See [32, 33]. I'd like to thank Tore Fjeltand Øgaard for this suggestion.

Smiley [56] defines an entailment relation via substitutions. We can adapt this definition for a kind of implication connective, by saying $A \to B$ is valid iff for some substitution σ and formulas $C, D, \sigma^{+}(C \to D) = A \to B$, where $C \to D$ is a classical tautology but neither $\sim C$ nor D is a classical tautology. Smiley presents this as an alternative account of entailment to E. Smiley's logic, however, fails to be a relevant logic, on the present criterion. The reason is that it fails to be closed under detachment. To see this, note that $(r \to r) \to (s \to (r \to r))$ is a substitution of $p \to (q \to p)$ and neither $\sim p$ nor $q \to p$ is a tautology. Further, $r \to r$ is a substitution of $p \to p$, and neither $\sim p$ nor p is a tautology. To see that $s \to (r \to r)$ is not valid on Smiley's criterion, we consider the options for substitution: p, $p \to q$, $p \to (q \to q)$, and $p \to (q \to r)$. Of these, only the third is a tautology, but its consequent is itself a tautology.[33]

As noted in this section, Definition 5 includes many logics that it should and it excludes many as well. It includes many relevant logics beyond the usual suspects. In the next section, I will discuss this feature, along with other consequences of the definition and some objections.

3 Discussion

I will begin with a comment on the proposal and its relation to extant work on relevant logics. The proposed definition is not an attempt to provide a formal criterion that captures Anderson and Belnap's philosophical motivations for relevant logics. Nor is it meant to capture the philosophical views of other relevant logicians, such as Sylvan, Meyer, or Brady. The proposal, instead, offers a formal criterion that includes all the standard relevant logics, as well as many of their close neighbors, and that uses only a concept that is widely agreed to be central to the study of relevant logics, namely variable-sharing. The philosophical reasons behind the importance of variable sharing, whether they rely on use, suppression-avoidance, meaning containment, or something else, may vary, but relevant logicians, by and large, have taken variable-sharing to be a feature of relevant logics. The suggestion is to take variable-sharing, along with being a DA-logic, as sufficient as well as necessary for being a relevant logic, with additional conditions imposed to reach one's preferred logic or class of logics.

It is plausible that the sentiment that variable-sharing is only necessary, not sufficient, for pinning down the relevant logics remains when closure under detachment and adjunction is added. One reason for this sentiment is, I think, that variable-sharing is not sufficient for pinning down R and its major sublogics,

[33] An alternative way of adapting Smiley's definition for an implication connective is to focus on the implications between formulas in the $\{\sim, \wedge, \vee\}$ vocabulary and substitution instances of those. Adding to this all the substitution instances of classical tautologies in the vocabulary $\{\sim, \wedge, \vee\}$ results in a relevant logic, as defined above.

For further discussion of these logics defined by substitution in this manner, see [16].

roughly the standardly studied relevant logics. When Anderson and Belnap were developing relevant logics, they had identified E and R, along with some of their neighbors. These logics are, in many ways, special: They are all transitive and they have important connections to combinatory logic and the structural rules of sequent systems.[34] These logics can be adequately modeled using classes of Routley-Meyer ternary relational frames.[35] Variable-sharing is not sufficient to isolate these logics.

The class of relevant logics, in the sense of Definition 5, includes logics stronger than R and some incomparable with R. From our perspective, standing on the shoulders of giants, we can see that the class of relevant logics naturally extends beyond and around R. Yet, one might ask how the current definition relates to what Anderson and Belnap were doing. One approach to this question is to provide an additional condition to isolate a subclass of logics of R. It seems like something like this is operative in the Use Criterion of Anderson and Belnap, which highlights R, E, and some other logics as salient relevant logics. If this is right, then it is natural to view those logics as combining two features, relevance as variable-sharing and use, specified in a proof-theoretic way.[36] Similarly, the logics motivated by the sufficiency view of [50] or the meaning containment view of [11] are special in many ways, but they can plausibly be isolated by imposing additional constraints on being a relevant logic.[37]

It is a perhaps surprising fact about the proposal that none of the standard relevant logics, namely those discussed in depth in any of the surveys of the area, really stand out as distinctive. For example, neither B nor R is natural stopping points according to the definition.[38] They may feature prominently in certain subclasses of logics, e.g. those that are complete with respect to a class of Routley-Meyer frames, but there are many subclasses of logics in which they do not mark a minimal or maximal logic. From an abstract point of view, they are not particularly special. This is not to say that they are not special. Indeed, the attention that they, and their neighbors, have received is because they have many qualities, formal and philosophical, that mark them off as distinctive.

The fact that the proposal does not highlight or privilege the standard relevant logics should be viewed as a virtue. The reason is that it underlines a sense in which the net of relevant logics has been cast too narrowly. This shift of focus will, I think, help to isolate the aspects of relevant logics that are philosophically important, as proponents of a particular logic or family of logics will have the burden of articulating the additional philosophical features and their formal counterparts

[34] For more on combinatory logic, see [6, 7]. For more on structural rules, see [44] and [8], among others.

[35] See [51, ch. 4].

[36] That the Use Criterion and variable-sharing were distinct, separable aspects of relevant logics was raised, as a criticism, by [15].

[37] Philosophically, it isn't correct to view either of their views as combining relevance with another element, sufficiency or meaning containment, as they see relevance as arising from the other element. Formally, there is not a problem with proceeding in this way.

[38] Avron [3] highlights R as distinctive of relevant logics.

that exclude the undesired relevant logics. This process of making explicit intuitive criteria may help to clarify the philosophical positions that support various relevant logics, such as R and B.[39]

The next topic to discuss concerns principles that are, or are not, ruled out by the proposal. On that score, it is worth emphasizing an important point, namely that whether a particular axiom can be in a relevant logic is something that cannot, in general, be read off the axiom. An example is the mingle axiom, $A \to (A \to A)$. Its antecedent and consequent share a variable. Adding it to R results in a violation of variable-sharing. On the other hand, adding it to T allows one to stay within the class of relevant logics. The axiom form of γ, $(A \wedge ({\sim}A \vee B)) \to B$, has a variable shared between its antecedent and consequent, but it cannot be added to any transitive, distributive relevant logic.[40] Even the weakening axiom, $A \to (B \to A)$, shares a variable between antecedent and consequent while leading to immediate failures of variable-sharing.

A natural question is whether the proposal picks out only logics that exclude all the standard paradoxes of implication. As we saw above, two of the positive Lewis paradoxes, $A \to (B \to A)$ and $A \to (B \to B)$, are both ruled out. There are some paradoxes involving other connectives that will be excluded from the class of relevant logics, just in virtue of their form, such as $(A \wedge {\sim}A) \to B$ and $A \to (B \vee {\sim}B)$. It makes no difference what principles govern $\vee$, $\wedge$, and $\sim$ in these cases, as those violate variable-sharing on their face.

Other putative paradoxes will depend on the principles of the logic. Two examples of such are

- Dummett's axiom, $(A \to B) \vee (B \to A)$, and
- $A \vee (A \to B)$.

Robles and Méndez [47] have provided a matrix whose logic satisfies variable-sharing while also containing Dummett's axiom. In a similar vein, the logic CL contains the second formula, $A \vee (A \to B)$. While neither of these principles is, I think, particularly *compelling* from a relevant logical point of view, their unattractiveness is somewhat lessened by keeping in mind that logics containing them will, typically, not be *prime*, where a logic is prime iff whenever it contains a disjunction, it contains at least one disjunct. Typically when either of these axioms is in a logic, the logic will fail to be prime, and so containing one does not entail that either of its disjuncts is valid. Further, neither of the particular logics containing the above axioms is closed under γ, so the consequences the disjunctive principles above are more restricted than one might expect.[41] The logic containing

[39] Brandom [13].

[40] See [37, 38] for discussion of γ, in rule and axiom form, in the context of relevant logics.

[41] As a witness to the failure of γ in the Robles and Méndez's logic with Dummett's axiom, ${\sim}((p \to p) \to {\sim}(q \to q))$ and $((p \to p) \to {\sim}(q \to q)) \vee ({\sim}(q \to q) \to (p \to p))$ are theorems, but ${\sim}(q \to q) \to (p \to p)$ is not, on pain of violating variable-sharing. As a witness to the failure of γ in the logic of the crystal lattice, ${\sim}{\sim}(p \to p)$ and ${\sim}(p \to p) \vee ({\sim}(p \to p) \to q)$

Dummett's axiom has fairly strong principles, and the second formula above is valid in an extension of R. So, worries about their potential for undermining variable-sharing should be somewhat tempered by the fact that they can be combined with comparatively strong logical principles without resulting in violations of variable-sharing. One might have philosophical reasons for wanting to exclude the above paradoxes from consideration, but they will go beyond the minimal sense of relevance of Definition 5.

Another example of an axiom that can be added to some, but not all, relevant logics is $\sim A \to (A \to B)$ In these logics, one may not have any logical truths whose main connective is a negation. One such example is the least DA-logic containing $\sim A \to (A \to B)$. In such a logic, every implicational logical truth has a variable shared between antecedent and consequent. If in the logic, negation obeys the plausible principle, $A \to \sim\sim A$, or even $A \to \sim^k A$, where $k \geq 1$ and '$\sim^k$' is a k-long sequence of '$\sim$'s, then $\sim A \to (A \to B)$ will be excluded. As proof, let us suppose $A \to \sim^k A$ is in the logic, $k = m + 1$, and let C be $\sim p \to (p \to q)$. Then C is a theorem, as is $\sim\sim^m C \to (\sim^m C \to B)$. From the assumptions, $\sim\sim^m C$ is a theorem, so then $\sim^m C \to B$ is as well. As B can be any formula, we can suppose that it contains only the atom r, so we have a violation of variable sharing.

Reflecting on two of the negative Lewis paradoxes, namely $A \to (\sim A \to B)$ and $\sim A \to (A \to B)$, we can pull out a general lesson about variable-sharing and singulary connectives.

Theorem 18 *Let # be a singulary connective. Then no relevant logic can include the theorem* $A \to (\#A \to B)$. *Further, no relevant logic can have as theorems, for all formulas* A *and* B *and formula contexts* C, *both* $A \to \#C(A)$, *and* $\#A \to (A \to B)$.

Proof For the first part, let D be $p \to (\#p \to q)$. Then, D is a theorem and so is $D \to (\#D \to r)$. Thus, $\#D \to r$ is a theorem as well, which is a violation of variable sharing. For the second, let A be $\#p \to (p \to q)$ and let B contain no atoms in $C(A)$. It follows from the assumptions that $\#C(A)$ is a theorem. As $\#C(A) \to (C(A) \to B)$ is a theorem, it then follows by detachment that $C(A) \to B$ is a theorem, which is a violation of variable sharing. □

Although the negative Lewis paradoxes feature negation, their relevance-violating features do not depend on features specific to negation at all. Rather, they provide a lesson about connectives obeying certain principles being compatible with relevance.[42]

Definition 5 casts a wide net, and some may worry that the net is cast *too* widely. As noted, some of the paradoxes of implication are excluded from the class of relevant logics at the outset and some are excluded from subclasses that contain

are both theorems, but $\sim(p \to p) \to q$ is not, on pain of violating variable-sharing. I would like to thank Tore Fjetland Øgaard for raising this issue with me.

[42] See [57] for a general discussion of relevant connectives.

plausible principles governing certain connectives. Some relevant logics, on our definition, can contain some odd or surprising principles, such as the following.[43]

(i) $(A \to B) \to (B \to A)$
(ii) $\sim(A \to A) \to (A \to B)$
(iii) $(A \to B) \to (B \wedge C)$
(iv) $(A \vee \sim A) \to A$

One obtains a relevant logic by taking exactly one of (i)–(iv) as the sole axiom and closing under detachment and adjunction. Those four logics, however, have very few principles governing their connectives, and so their logical behavior may diverge from the usual understanding of the connectives. It should be no surprise that in extremely weak logics, some odd looking axioms are compatible with variable-sharing. When some plausible principles are added, variable-sharing may be violated. For example, axiom (i) and (iii) lead to violations of variable sharing in the presence of the axiom (L1) and transitivity rules, axiom (iv) immediately yields violations of variable-sharing in any logic with excluded middle, and axiom (ii) violates variable sharing in logics with permutation, $(A \to (B \to C)) \to (B \to (A \to C))$.[44] For principles that are compatible with variable-sharing and many plausible principles, it is up to their critics to formulate the reasons, formal and philosophical, that they should be excluded.

I will close with a short discussion of variable-sharing. The form of variable-sharing used in Definition 3 can be called *simple variable-sharing*. There are stronger forms of variable-sharing that one might consider for isolating classes of logics.

One of the stronger, more refined forms of variable-sharing is known as *strong variable-sharing*.[45] For strong variable-sharing, each formula is assigned a polarity, positive or negative, based on its place a formula.[46] In $A \to B$, A is negative and B is positive. Being in consequent position, being a conjunct, and being a disjunct all maintain polarity, whereas being in antecedent position or being negated changes polarity. Strong variable-sharing says that if $A \to B$ is a logical truth, then A and B share a variable with the same polarity. Belnap's original proof of variable-sharing actually demonstrated strong variable-sharing. Strong variable-sharing would be sufficient to rule out $\sim A \to (A \to B)$, which as we saw above was not ruled out by simple variable-sharing.

Another stronger form of variable-sharing is Avron's strong variable-sharing property, which I will call *basic variable-sharing*, as the property of the previous

---

[43] I thank an anonymous referee for raising this issue and supplying some example axioms to consider.

[44] Judgments of plausibility vary, and some have found excluded middle and permutation to be implausibly strong. See [55], for example.

[45] See [1], where the variable-sharing theorem establishes strong variable-sharing. Similarly, [46, 47] also discuss strong variable-sharing.

[46] The terminology 'consequent part' and 'antecedent part', or 'cp' and 'ap', is commonly used in this context.

paragraph is well established with the name 'strong variable-sharing'.[47] The basic variable-sharing property says that if (i) $\vdash_L (A \wedge B) \rightarrow C$ and (ii) A and $B \rightarrow C$ do not share a propositional variable, then $\vdash_L B \rightarrow C$. Avron shows that, under fairly minimal conditions, having the basic variable-sharing property entails having the simple variable-sharing property. This is a compelling principle, but it requires conjunction as well as implication to state, so we have not adopted it for the main definition of variable-sharing.

Another form of variable-sharing, proposed by Brady [9], is known as *depth relevance*.[48] The *depth* of an occurrence of a propositional variable in a formula is, roughly, the number of implications it is under in the formulas parse tree. For example, in $p \rightarrow (p \rightarrow q)$, the leftmost p has depth 1, whereas the rightmost p has depth 2. The depth relevance criterion requires that if $A \rightarrow B$ is a logical truth, then A and B share a variable at the same depth. Clearly $\sim A \rightarrow (A \rightarrow B)$ is ruled out by the depth relevance criterion.

Finally, strong variable-sharing and depth relevance can be combined into the *strong depth relevance criterion*, as studied by Logan [24, 25]. The strong depth relevance criterion says that if $A \rightarrow B$ is a logical truth, then A and B share a propositional variable at the same depth with the same polarity.

These stronger variable-sharing criteria isolate what seem to be interesting and natural classes of relevant logics.[49] They capture stronger senses of connection than is required by (simple) variable-sharing. For purposes of identifying the area of relevant logics, taking the broader definition as provided in Sect. 1, seems the most natural. The stronger criteria all require additional specification that goes beyond the core of the simple connection idea found in the variable-sharing criterion.

To conclude I will take stock of what has been done. I have motivated and presented a definition of the class of relevant logics as those logics (1) all of whose implicational logical truths share a variable between antecedent and consequent, and (2) that are closed under detachment and adjunction. Some notable consequences of this definition is that there is no greatest relevant logic and many of Lewis's paradoxes of implication are excluded from relevant logics. All of the standard relevant logics and many of their close neighbors are included in the class of relevant logics, in the sense of Definition 5. I noted that we can see the work of Anderson and Belnap as isolating a subclass of relevant logics, implicitly using some additional conditions for being a relevant logic, and I responded to the objection that the definition is too inclusive. Finally, I noted some alternative forms of variable-

[47] Avron [4] presents his basic relevance property for consequence relations, and he proposes an implicational form, what is here being called the 'basic variable-sharing property'. Under fairly minimal conditions, the consequence relation and implicational properties are equivalent.

[48] See also [48, 49] and [52].

[49] There are further variable-sharing properties not discussed here. Some examples are offered by the analytic containment logics, for which see [19, 20]. Another example is *cn-relevance*, described by [21]. There are further properties one might want to consider as well. One example is the "no loose pieces" principle discussed by [47]. Another example is the Ackermann property, discussed by [1, 243].

sharing that could be used to identify interesting subclasses of logics. As noted, the definition of relevant logics given by Definition 5 generalizes neatly to include relevant modal logics, and, as demonstrated by Theorem 18, it can be useful in proving limitative results about what sorts of connectives can or cannot be included in relevant logics. Further investigation along these lines will be pursued in future work, and it will, I hope, prove useful in clarifying the foundations of relevant logics.

Acknowledgments I would like to thank Greg Restall, Rohan French, Ed Mares, Tore Fjetland Øgaard, Ren-June Wang, Ben Blumson, Shay Logan, and Lloyd Humberstone for discussion and feedback on the ideas of this paper. This work was supported by research grant 111-2410-H-002-006-MY3 from the National Science and Technology Council of Taiwan.

References

1. Anderson, A. R., & Belnap, N. D. (1975). *Entailment: The logic of relevance and necessity, Vol. I*. Princeton University Press.
2. Avron, A. (1992). Whither relevance logic? *Journal of Philosophical Logic*, *21*(3), 243–281.
3. Avron, A. (2014). What is relevance logic? *Annals of Pure and Applied Logic*, *165*(1), 26–48.
4. Avron, A. (2022). On formal criteria for relevance. In K. Bimbó (Ed.), *Relevance logics and other tools for reasoning: Essays in honor of J. Michael Dunn* (pp. 54–64). College Publications.
5. Bimbó, K. (2007). Relevance logics. In D. Jacquette (Ed.), *Philosophy of logic*, volume 5 of *Handbook of the philosophy of science* (pp. 723–789). Elsevier.
6. Bimbó, K. (2009). Combinatory logic. In E. N. Zalta (Ed.), *Stanford encyclopedia of philosophy*. The Metaphysics Research Lab, Center for the Study of Language and Information, Stanford University.
7. Bimbó, K. (2011). *Combinatory logic: Pure, applied and typed*. Taylor & Francis.
8. Bimbó, K. (2014). *Proof theory: Sequent calculi and related formalisms*. CRC Press.
9. Brady, R. T. (1984a). Depth relevance of some paraconsistent logics. *Studia Logica*, *43*(1–2), 63–73.
10. Brady, R. T. (1984b). Natural deduction systems for some quantified relevant logics. *Logique Et Analyse*, *27*(8), 355–377.
11. Brady, R. T. (2006). *Universal logic*. CSLI Publications.
12. Brady, R. T., & Meinander, A. (2013). Distribution in the logic of meaning containment and in quantum mechanics. In F. Berto, E. Mares, K. Tanaka, & F. Paoli (Eds.), *Paraconsistency: Logic and applications* (pp. 223–255). Springer.
13. Brandom, R. (1994). *Making it explicit: Reasoning, representing, and discursive commitment*. Cambridge, MA: Harvard University Press.
14. Butchart, S., & Rogerson, S. (2014). On the algebraizability of the implicational fragment of Abelian logic. *Studia Logica*, *102*(5), 981–1001.
15. Copeland, B. J. (1980). The trouble Anderson and Belnap have with relevance. *Philosophical Studies*, *37*(4), 325–334.
16. Dunn, J. M. (1980). A sieve for entailments. *Journal of Philosophical Logic*, *9*(1), 41–57.
17. Dunn, J. M., & Hardegree, G. M. (2001). *Algebraic methods in philosophical logic*. Oxford University Press.
18. Dunn, J. M., & Restall, G. (2002). Relevance logic. In D. M. Gabbay & F. Guenthner (Eds.), *Handbook of philosophical logic* (Vol. 6, 2nd ed., pp. 1–136). Kluwer.
19. Ferguson, T. M. (2017). *Meaning and proscription in formal logic: Variations on the propositional logic of William T. Parry*. Springer Verlag.

20. Ferguson, T. M. (2024). A topic-theoretic perspective on variable-sharing (from the black sheep of the family). In I. Sedlár, S. Standefer, & A. Tedder (Eds.), *New directions in relevant logic*, chapter 2. Springer.
21. Ferguson, T. M., & Logan, S. A. (2024, Forthcoming). Topic transparency and variable sharing in weak relevant logics. *Erkenntnis*, 1–28. https://link.springer.com/article/10.1007/s10670-023-00748-6.
22. Francez, N. (2019). Relevant connexive logic. *Logic and Logical Philosophy*, *28*(3), 409–425.
23. Humberstone, L. (2011). *The connectives*. MIT Press.
24. Logan, S. A. (2021). Strong depth relevance. *The Australasian Journal of Logic*, *18*(6), 645–656.
25. Logan, S. A. (2022). Depth relevance and hyperformalism. *Journal of Philosophical Logic*, *51*(4), 721–737.
26. Mares, E. (2022). Relevance logic. In E. N. Zalta & U. Nodelman (Eds.), *The Stanford encyclopedia of philosophy* (Fall 2022 edition). Metaphysics Research Lab, Stanford University.
27. Martin, E. P., & Meyer, R. K. (1982). Solution to the P-W problem. *Journal of Symbolic Logic*, *47*(4), 869–887.
28. Méndez, J. M., Robles, G., & Salto, F. (2012). Ticket entailment plus the mingle axiom has the variable-sharing property. *Logic Journal of the IGPL*, *20*(1), 355–364.
29. Meyer, R., & Slaney, J. (1989). Abelian logic (from A to Z). In G. Priest, R. Routley, & J. Norman (Eds.), *Paraconsistent logic: Essays on the inconsistent* (pp. 245–268). Munich: Philosophia.
30. Meyer, R. K. (1985). A farewell to entailment. In G. Dorn & P. Weingartner (Eds.), *Foundations of logic and linguistics, problems and their solutions* (pp. 577– 636). Plenum Press.
31. Meyer, R. K., & Dunn, J. M. (1969). E, R, and γ. *Journal of Symbolic Logic*, *34*(3), 460–474.
32. Meyer, R. K., & Routley, R. (1973). Classical relevant logics I. *Studia Logica*, *32*, 51–66.
33. Meyer, R. K., & Routley, R. (1974). Classical relevant logics II. *Studia Logica*, *33*(2), 183–194.
34. Nelson, M., & Zalta, E. N. (2012). A defense of contingent logical truths. *Philosophical Studies*, *157*(1), 153–162.
35. Øgaard, T. F. (2020). Farewell to suppression-freedom. *Logica Universalis*, *14*(3), 297–330.
36. Øgaard, T. F. (2021a). Boolean negation and non-conservativity II: The variable-sharing property. *Logic Journal of the IGPL*, *29*(3), 363–369.
37. Øgaard, T. F. (2021b). Non-Boolean classical relevant logics I. *Synthese*, *198*, 6993–7024.
38. Øgaard, T. F. (2021c). Non-Boolean classical relevant logics II: Classicality through truth-constants. *Synthese*, *199*, 6169–6201.
39. Øgaard, T. F. (2023). *A study in relevant logics*. PhD thesis, University of Bergen.
40. Omori, H., & Wansing, H. (2017). 40 years of FDE: An introductory overview. *Studia Logica*, *105*(6), 1021–1049.
41. Omori, H., & Wansing, H. (2019). Connexive logics. an overview and current trends. *Logic and Logical Philosophy*, *28*(3), 371–387.
42. Paoli, F., Spinks, M., & Veroff, R. (2008). Abelian logic and the logics of pointed lattice-ordered varieties. *Logica Universalis*, *2*(2), 209–233.
43. Read, S. (1988). *Relevant logic: A philosophical examination of inference*. Blackwell.
44. Restall, G. (2000). *An introduction to substructural logics*. Routledge.
45. Ripley, D. (2013). Revising up: Strengthening classical logic in the face of paradox. *Philosophers' Imprint*, 13.
46. Robles, G., & Méndez, J. M. (2011). A class of simpler logical matrices for the variable-sharing property. *Logic and Logical Philosophy*, *20*(3), 241–249.
47. Robles, G., & Méndez, J. M. (2012). A general characterization of the variable-sharing property by means of logical matrices. *Notre Dame Journal of Formal Logic*, *53*(2), 223–244.
48. Robles, G., & Méndez, J. M. (2014a). Curry's paradox, generalized modus ponens axiom and depth relevance. *Studia Logica*, *102*(1), 185–217.
49. Robles, G., & Méndez, J. M. (2014b). Generalizing the depth relevance condition: Deep relevant logics not included in R-mingle. *Notre Dame Journal of Formal Logic*, *55*(1), 107–127.

50. Routley, R. (2019). *Ultralogic as universal?: The Sylvan jungle - Volume 4*. Cham: Springer Verlag. This includes a republication of the Appendix of *Exploring Meinong's Jungle and Beyond* (1980), pp. 893–962.
51. Routley, R., Plumwood, V., Meyer, R. K., & Brady, R. T. (1982). *Relevant logics and their rivals* (Vol. 1). Ridgeview.
52. Salto, F., Robles, G., & Méndez, J. M. (2018). Curry's paradox, generalized contraction rule and depth relevance. In K. Boudouris (Ed.), *Proceedings XXIII World Congress Philosophy* (pp. 35–39). Charlottesville: Philosophy Documentation Center.
53. Segerberg, K. (1971). *An essay in classical modal logic*. Filosofiska Föreningen Och Filosofiska Institutionen Vid Uppsala Universitet.
54. Segerberg, K. (1982). *Classical propositional operators: An exercise in the foundations of logic*. Oxford University Press.
55. Slaney, J. K. (1989). RWX is not curry paraconsistent. In G. Priest, R. Sylvan, & J. Norman (Eds.), *Paraconsistent logic: Essays on the inconsistent* (pp. 472–480). Philosophia Verlag.
56. Smiley, T. J. (1959). Entailment and deducibility. *Proceedings of the Aristotelian Society, 59*, 233–254.
57. Standefer, S. (2022). What is a relevant connective? *Journal of Philosophical Logic, 51*(4), 919–950.
58. Standefer, S. (2024). Routes to relevance: Philosophies of relevant logics. *Philosophy Compass, 19*(2), e12965.
59. Swirydowicz, K. (1999). There exist exactly two maximal strictly relevant extensions of the relevant logic R. *Journal of Symbolic Logic, 64*(3), 1125–1154.
60. Szmuc, D., & Rubin, M. (2022). Pure refined variable inclusion logics. *The Australasian Journal of Logic, 19*(5), 147–166.
61. Szmuc, D. E. (2021). The (greatest) fragment of classical logic that respects the variable-sharing principle (in the FMLA-FMLA framework). *Bulletin of the Section of Logic, 50*(4), 421–453.
62. Tennant, N. (2017). *Core logic*. Oxford University Press.
63. Thistlewaite, P. B., Mcrobbie, M. A., & Meyer, R. K. (1988). *Automated theorem-proving in non-classical logics*. Pitman Publishing.
64. Weir, A. (2013). A robust non-transitive logic. *Topoi, 34*(1), 99–107.
65. Yang, E. (2013). R and relevance principle revisited. *Journal of Philosophical Logic, 42*(5), 767–782.

Part II
Model Theory

Algorithmic Correspondence for Relevance Logics. II. Inductive Formulae in Flat Languages for Relevance Logics

Willem Conradie and Valentin Goranko

Dedicated to the memory of our teacher and collaborator Dimiter Vakarelov

Abstract This paper is a follow-up to our recent work, where we apply and extend the theory and methods of algorithmic correspondence theory previously developed for modal logics to the language $\mathcal{L}_R$ of relevance logics with respect to their standard Routley-Meyer relational semantics. In the above mentioned precursor of the present work we develop a non-deterministic algorithmic procedure PEARL for computing first-order equivalents in terms of that semantics and proving canonicity of formulae of a language $\mathcal{L}_R^+$ extending $\mathcal{L}_R$ with all residuals and adjoints of the connectives in $\mathcal{L}_R$. We identified a large syntactically defined class of *inductive formulae* in $\mathcal{L}_R$, and showed that PEARL succeeds for every so defined inductive formula.

In this work we present an alternative approach to defining inductive formulae for relevance logics, originally developed for polyadic modal logics in earlier works by Goranko and Vakarelov, based on a rewriting of the language $\mathcal{L}_R$ that allows composing logical connectives into composite 'box' and 'diamond' terms that enable a 'flat' representation of formulae of $\mathcal{L}_R$, thus providing a technically simplified definition of the class of inductive formulae. We show that, modulo translation, the two approaches define the same class of inductive formulae, and explain how PEARL can be modified to work on 'flat' inductive formulae.

W. Conradie (✉)
School of Mathematics, University of the Witwatersrand, Johannesburg, South Africa

National Institute for Theoretical and Computational Sciences, Stellenbosch, South Africa
e-mail: willem.conradie@wits.ac.za

V. Goranko
Department of Philosophy, Stockholm University, Stockholm, Sweden

(Visiting Professor) School of Mathematics, University of the Witwatersrand, Johannesburg, South Africa
e-mail: valentin.goranko@philosophy.su.se

I. Sedlár et al. (eds.), *New Directions in Relevant Logic*, Trends in Logic 63,
https://doi.org/10.1007/978-3-031-69940-5_6

1 Introduction

This paper is a follow-up to [4], where we apply and extend the theory and methods of algorithmic correspondence theory previously developed for modal logics to the language $\mathcal{L}_R$ of relevance logics with respect to their standard Routley-Meyer (RM) relational semantics. In [4][1] we developed a non-deterministic algorithmic procedure PEARL for computing first-order RM-frame equivalents of formulae of a language $\mathcal{L}_R^+$ extending $\mathcal{L}_R$ with all residuals and adjoints of the connectives in $\mathcal{L}_R$. PEARL is an adaptation of the previously developed algorithmic procedures SQEMA [6] (for normal modal logics) and ALBA [7] (for distributive and non-distributive modal logics). We then identified a large syntactically defined class of *inductive formulae* in $\mathcal{L}_R$, and show that PEARL succeeds for every so defined inductive formula. Subsequently, in [5] we show that all $\mathcal{L}_R$ formulae on which PEARL succeeds are canonical and thus provide complete axiomatizations with respect to the Routley-Meyer semantics. Consequently, every inductive formulae in $\mathcal{L}_R$ is canonical.

Taking these results into account, the idea of identifying the class of inductive formulae is quite relevant (no pun intended) for relevance logics, because nearly all important axioms of relevance logics identified in the literature turn out to be inductive (see in [4] a long list of over 35 axioms for important relevance logics which have been identified there as inductive), whereas only some of them are of the more restricted syntactic type of the natural analogues of Sahlqvist formulae for relevance logics, as defined in [14] and [1]. Furthermore, proving directly canonical completeness of relevance logics with respect to the Routley-Meyer semantics is technically more complicated than in the case of modal logics. Thus, we argue that algorithmic correspondence theory for relevance logics is not only of theoretical value but also of essential practical use for relevance logicians, and that was the main motivation for the present work.

Contributions of This Paper

In this work we present an alternative approach to defining inductive formulae for relevance logics, originally developed for polyadic modal logics in [11] and [12], based on re-defining the language $\mathcal{L}_R$ in a way that allows composing logical connectives into composite 'box' and 'diamond' terms that enable a 'flat' representation of formulae of $\mathcal{L}_R$, thus providing a technically simplified definition of the class of inductive formulae. We show that the two approaches define the same class of inductive formulae, and explain how the algorithm PEARL can be modified to work on 'flat' inductive formulae.

These definitions and results have several technical, conceptual and practical benefits. The definition of inductive formulae presented in [4] and [5] is given

[1] We refer the reader to [4] for an informal overview of algorithmic correspondence theory, and for all technical details providing background for the present paper and not included in the preliminaries here.

in terms of a list of conditions on the signed generation trees of formulae or in equalities or formulae, and follows the pattern established in [10] and refined, generalized and developed in the literature on unified correspondence [3, 7, 8]. This style of definition is advantageous for logics where classical negation is unavailable. In settings where classical negation is available, however, correspondence-theoretic definitions can be greatly simplified through assumptions on the polarities of variable and connective occurrences which there can be made without loss of generality.

For all their technical usefulness, generation tree style definitions can be somewhat intricate. These definitions are moreover not 'constructive', in the sense that, unlike in the classical cases [2, 12], they do not provide (at least explicitly) a procedure for building inductive and Sahlqvist formulae for simpler constituent parts.

The present paper responds to these considerations, precisely by developing such a 'constructive definition' in the setting of relevance logic. This is facilitated by the ability, offered by flat languages, to compose connectives into compound modalities which greatly helps to make the structure of inductive and Sahlqvist formulae clear. Not only does this reconnect subsequent research with the original definition of inductive formulae [12] but it also provides the conceptual and technical ideas needed to reformulate in constructive (and flat) terms the very general definitions of inductive and Sahlqvist formulae for arbitrary logics captured by normal lattice expansions [8].

The Structure of the Paper

In the preliminary Sect. 2 we summarise the basics of the syntax and Routley-Meyer relational semantics of relevance logic and of an extension of its language needed for correspondence-theoretic computations. We recall the standard first-order translation of these languages and the definitions of inductive and Sahlqvist relevance formulae from [4]. Section 3 introduces flat polyadic language for relevance logics together with their semantics and a translation to the usual language of relevance logic. Section 4 introduces the definitions of regular and inductive flat relevance formulae. These definitions are illustrated with examples and a suitable equivalence (modulo translation) with the inductive formulae from [4] is established. In Sect. 5 we review the algorithm PEARL and show how it may be applied to inductive flat formulae by pre-translating these formulae. We define new rewrite rules for the compound connectives of flat languages which, when added to the calculus used by PEARL, makes it capable of successfully handling inductive flat relevance formulae directly. We conclude in Sect. 6 with some remarks on future lines of research.

2 Preliminaries

We will assume the reader is familiar with the basic syntax and relational semantics of modal and relevance logics, general references for which are e.g. [2] (for modal logics) and [9, 13] (for relevance logics). In the interest of keeping the paper self contained, we will briefly recall some essential details and give a few additional definitions.

2.1 Syntax and Routley-Meyer Relational Semantics for Relevance Logics

Hereafter we consider the language of propositional relevance logics $\mathcal{L}_R$ over a fixed set of propositional variables VAR containing the classical connectives $\wedge$, $\vee$, plus the relevant connectives **fusion** $\circ$, **(relevant) negation** $\sim$, **(relevant) implication** $\rightarrow$, and the special constant **(relevant) truth t**. For correspondence theoretic computations we will also require an extension of this language $\mathcal{L}_R^+$ obtained by adding connectives which are related, as residuals or adjoints, to some of the connectives in $\mathcal{L}_R$. In particular, we add the left adjoint $\sim^\flat$ and the right adjoint $\sim^\sharp$ of $\sim$, the Heyting implication $\Rightarrow$ (as right residual of $\wedge$), the co-Heyting implication $\prec$ as the left residual of $\vee$ and the operation $\hookrightarrow$ as the residual of $\circ$ in the second coordinate and of $\rightarrow$ in the first coordinate. $\mathcal{L}_R^+$ also includes two (countably infinite) sets, $\mathsf{NOM} = \{\mathbf{j}_0, \mathbf{j}_1, \mathbf{j}_2, \ldots\}$ and $\mathsf{CNOM} = \{\mathbf{m}_0, \mathbf{m}_1, \mathbf{m}_2, \ldots\}$, of special variables, respectively called **nominals** and **co-nominals**. Informally, we will denote nominals by $\mathbf{i}, \mathbf{j}, \mathbf{k}$, possibly with indices, while co-nominals will be denoted by $\mathbf{m}, \mathbf{n}$, possibly with indices. We will use capital Roman letters $A, B, C, D, \ldots$ as meta variables for the formulae of $\mathcal{L}_R$ and $\mathcal{L}_R^+$, which are defined by the following grammar:

$$A = p \mid \mathbf{i} \mid \mathbf{m} \mid \top \mid \bot \mid \mathbf{t} \mid \sim A \mid (A \wedge A) \mid (A \vee A) \mid (A \circ A) \mid (A \rightarrow A) \mid$$
$$\sim^\flat A \mid \sim^\sharp A \mid (A \prec A) \mid (A \Rightarrow A) \mid (A \hookrightarrow A)$$

where $p \in \mathsf{VAR}, \mathbf{i} \in \mathsf{NOM}$ and $\mathbf{m} \in \mathsf{CNOM}$. We denote $\mathsf{ATOMS} := \mathsf{VAR} \cup \mathsf{NOM} \cup \mathsf{CNOM}$. The elements of ATOMS will be called **atoms**. The formulae of $\mathcal{L}_R$ are obtained by omitting the clauses for $\mathbf{i}$, $\mathbf{m}$, $\top$, $\bot$, $\sim^\flat$, $\sim^\sharp$, $\prec$, $\Rightarrow$ and $\hookrightarrow$.
A **relevance frame** is a tuple $\mathcal{F} = \langle W, O, R, ^* \rangle$, where:

- W is a non-empty set of states (possible worlds);
- $O \subseteq W$ is the subset of **normal** states;
- $R \subseteq W^3$ is a **relevant accessibility relation**;
- $^*: W \rightarrow W$ is a function, called the **Routley star**, used to provide semantics for $\sim$.

The following binary relation $\preceq$ is defined in every relevance frame:

$$u \preceq v \text{ iff } \exists o(o \in O \wedge Rouv)$$

A **Routley-Meyer frame** (for short, **RM-frame**) is a relevance frame satisfying the following conditions for all $u, v, w, x, y, z \in W$:

1. $x \preceq x$
2. If $x \preceq y$ and $Ryuv$ then $Rxuv$.
3. If $x \preceq y$ and $Ruyv$ then $Ruxv$.
4. If $x \preceq y$ and $Ruvx$ then $Ruvy$.
5. If $x \preceq y$ then $y^* \preceq x^*$.
6. O is upward closed w.r.t. $\preceq$, i.e. if $o \in O$ and $o \preceq o'$ then $o' \in O$.

These properties ensure that $\preceq$ is reflexive and transitive, hence a preorder, and that the semantics of the logical connectives has the monotonicity properties stated further.

A **Routley-Meyer model** (for short, **RM-model**) is a tuple $\mathcal{M} = \langle W, O, R,^*, V\rangle$, where $\langle W, O, R,^* \rangle$ is a Routley-Meyer frame and $V : \mathsf{VAR} \to \mathcal{P}(W)$ is a mapping, called a **relevant valuation**, assigning to every atomic proposition $p \in \mathsf{VAR}$ a set $V(p)$ of states *upward closed* w.r.t. $\preceq$.

Truth of a formula A **in a RM-model** $\mathcal{M} = \langle W, O, R,^*, V\rangle$ at a state $u \in W$, denoted $\mathcal{M}, u \Vdash A$, is defined as follows:

- $\mathcal{M}, u \Vdash p$ iff $u \in V(p)$;
- $\mathcal{M}, u \Vdash \mathbf{t}$ iff there is $o \in O$ such that $o \preceq u$; equivalently, iff $u \in O$;
- $\mathcal{M}, u \Vdash {\sim}A$ iff $\mathcal{M}, u^* \nVdash A$;
- $\mathcal{M}, u \Vdash A \wedge B$ iff $\mathcal{M}, u \Vdash A$ and $\mathcal{M}, u \Vdash B$;
- $\mathcal{M}, u \Vdash A \vee B$ iff $\mathcal{M}, u \Vdash A$ or $\mathcal{M}, u \Vdash B$;
- $\mathcal{M}, u \Vdash A \to B$ iff for every v, w such that $Ruvw$, if $\mathcal{M}, v \Vdash A$ then $\mathcal{M}, w \Vdash B$.
- $\mathcal{M}, u \Vdash A \circ B$ iff there exist v, w such that $Rvwu$, $\mathcal{M}, v \Vdash A$ and $\mathcal{M}, w \Vdash B$.

The additional connectives of $\mathcal{L}_R^+$ are interpreted in the same Routley-Meyer models as $\mathcal{L}_R$, except that the notion of valuation need to be adjusted so that instead of $V : \mathsf{VAR} \to \mathcal{P}(W)$, we have $V : \mathsf{ATOMS} \to \mathcal{P}(W)$ and V maps nominals to principal up-sets and co-nominals to complements of principal down-sets, i.e., for all $\mathbf{i} \in \mathsf{NOM}$ and all $\mathbf{m} \in \mathsf{CNOM}$ we have $V(\mathbf{i}) = {\uparrow}w$ for some $w \in W$ and $V(\mathbf{m}) = ({\downarrow}v)^c$ for some $v \in W$. The semantics of the additional connectives of $\mathcal{L}_R^+$ are given as follows:

- $\mathcal{M}, w \Vdash \mathbf{i}$ iff $w \in V(\mathbf{i})$
- $\mathcal{M}, w \Vdash \mathbf{m}$ iff $w \in V(\mathbf{m})$
- $\mathcal{M}, w \Vdash \top$
- $\mathcal{M}, w \nVdash \bot$
- $\mathcal{M}, w \Vdash {\sim}^\flat A$ iff there is a v such that $v^* = w$ and $\mathcal{M}, v \nVdash A$
- $\mathcal{M}, w \Vdash {\sim}^\sharp A$ iff for all v such that $v^* = w$, it is the case that $\mathcal{M}, v \nVdash A$.
- $\mathcal{M}, w \Vdash A \prec B$ iff there exists v such that $v \preceq w$, $\mathcal{M}, v \Vdash A$ and $\mathcal{M}, v \nVdash B$

- $\mathcal{M}, w \Vdash A \Rightarrow B$ iff for all $v \succeq w$, if $\mathcal{M}, v \Vdash A$ then $\mathcal{M}, v \Vdash B$
- $\mathcal{M}, w \Vdash A \hookrightarrow B$ iff for all $v, u \in W$, if $Rvwu$ and $\mathcal{M}, v \Vdash A$ then $\mathcal{M}, u \Vdash B$

For every RM-model $\mathcal{M}$ and $\mathcal{L}_R^+$ formula A we define the **extension of** A **in** $\mathcal{M}$ as

$$[\![A]\!]_{\mathcal{M}} := \{u \in \mathcal{M} \mid \mathcal{M}, u \Vdash A\}$$

A $\mathcal{L}_R^+$ formula A is declared:

- **true in an RM-model** $\mathcal{M}$, denoted by $\mathcal{M} \Vdash A$, if $O \subseteq [\![A]\!]_{\mathcal{M}}$, i.e., $\mathcal{M}, o \Vdash A$ for every $o \in O$.
- **valid in an RM-frame** $\mathcal{F}$, denoted by $\mathcal{F} \Vdash A$, iff it is true in every RM-model over that frame.
- **RM-valid**, denoted by $\Vdash A$, iff it is true in every RM-model.

An important property of this semantics is *Monotonicity*: for every RM-model $\mathcal{M}$ and formula A, the set $[\![A]\!]_{\mathcal{M}}$ is $\preceq$*-upward closed*.

A formula A of $\mathcal{L}_R$ not containing variables will be called a **constant formula**. Clearly, the truth of a constant formula in a RM-model does not depend on the valuation, i.e., the extension $[\![A]\!]_{\mathcal{M}}$ is the same for every RM-model $\mathcal{M}$ based on a RM-frame $\mathcal{F}$, so we will identify it with validity in $\mathcal{F}$ and denote it by $[\![A]\!]_{\mathcal{F}}$. Then, $\mathcal{F}, u \Vdash A$ iff $O \subseteq [\![A]\!]_{\mathcal{F}}$.

Formulae A and B from $\mathcal{L}_R$ are: **semantically equivalent**, hereafter denoted $A \equiv B$, iff they are true at the same states in every RM-model; **RM-model-equivalent**, if they are true at the same RM-models; **RM-frame-equivalent**, if they are valid in the same RM-frames. Hereafter, 'equivalent formulae of $\mathcal{L}_R$' will mean 'semantically equivalent formulae', unless otherwise specified.

Clearly, Routley-Meyer frames are first-order structures for the first-order language with unary predicate symbol O, unary function symbol $*$, ternary relation symbol R, and individual variables $x_1, x_2, x_3, \ldots$, informally denoted x, x', x'' etc. We will call this language FO_R. Moreover, the semantics of relevance logic can be transparently expressed in FO_R and every relevance formula is then equivalently translated into a formula in FO_R by the following **standard translation** $ST : \mathcal{L}_R \to \mathrm{FO}_R$, parametric in a first-order individual variable:

$$
\begin{aligned}
ST_x(p) &= P(x)\\
ST_x(\mathbf{t}) &= O(x)\\
ST_x(\sim A) &= \exists x'(x' = x^* \wedge \neg ST_{x'}(A))\\
ST_x(A \wedge B) &= ST_x(A) \wedge ST_x(B)\\
ST_x(A \vee B) &= ST_x(A) \vee ST_x(B)\\
ST_x(A \circ B) &= \exists x'x''(Rx'x''x \wedge ST_{x'}(A) \wedge ST_{x''}(B))\\
ST_x(A \to B) &= \forall x'x''(Rxx'x'' \wedge ST_{x'}(A) \to ST_{x''}(B))
\end{aligned}
$$

$$ST_x(\mathbf{j}_i) = x \succeq y_i$$

$$ST_x(\mathbf{m}_i) = \neg(x \preceq z_i)$$

$$ST_x(\top) = x = x$$

$$ST_x(\bot) = \neg(x = x)$$

$$ST_x(\sim^\flat A) = \exists x'((x')^* = x \wedge \neg ST_{x'}(A))$$

$$ST_x(\sim^\sharp A) = \forall x'((x')^* = x \rightarrow \neg ST_{x'}(A))$$

$$ST_x(A \prec B) = \exists x'(x' \preceq x \wedge ST_{x'}(A) \wedge \neg ST_{x'}(B))$$

$$ST_x(A \Rightarrow B) = \forall x'(x' \succeq x \wedge ST_{x'}(A) \rightarrow ST_{x'}(B))$$

$$ST_x(A \hookrightarrow B) = \forall x' \forall x''(Rx'xx'' \wedge ST_{x'}(A) \rightarrow ST_{x''}(B))$$

where x' and x'' are fresh individual variables and $x \preceq x'$ is shorthand for $\exists x''(O(x'') \wedge R(x''xx'))$.

It is routine to check that for every Routley-Meyer model $\mathcal{M}$, state w in $\mathcal{M}$ and $\mathcal{L}_R^+$-formula A, it holds that $\mathcal{M}, w \Vdash A$ iff $\mathcal{M} \models ST_x(A)[x := w]$, where $[x := w]$ indicates that the free variable x in $ST_x(A)$ is interpreted as w.

For the purpose of the algorithm PEARL (see Sect. 5) we will combine formulae of $\mathcal{L}_R^+$ into inequalities and quasi-inequalities, as follows: an **inequality** is an expression of the form $A \leq B$ for $A, B \in \mathcal{L}_R^+$, while a **quasi-inequality** is an expression $A_1 \leq B_1, \ldots, A_n \leq B_n \implies A \leq B$ where $A_1, \ldots, A_n, B_1, \ldots, B_n, A, B \in \mathcal{L}_R^+$. The semantics of these expressions are as expected: an inequality $A \leq B$ is true in a Routley-Meyer model $\mathcal{M}$, denoted

$$\mathcal{M} \Vdash A \leq B,$$

iff $[\![A]\!]_{\mathcal{M}} \subseteq [\![B]\!]_{\mathcal{M}}$, while a quasi-inequality $A_1 \leq B_1, \ldots, A_n \leq B_n \implies A \leq B$ is true in $\mathcal{M}$, denoted

$$\mathcal{M} \Vdash A_1 \leq B_1, \ldots, A_n \leq B_n \implies A \leq B$$

iff $[\![A_i]\!]_{\mathcal{M}} \not\subseteq [\![B_i]\!]_{\mathcal{M}}$ for some $1 \leq i \leq n$ or $[\![A]\!]_{\mathcal{M}} \subseteq [\![B]\!]_{\mathcal{M}}$. Now, the notions of validity of inequalities and quasi-inequalities in a RM-frame is defined in the expected way, as validity in all RM-models on that frame.

The formulae in $\mathcal{L}_R^+$ will be treated as a special type of inequality, viz. a formula A will be identified with the inclusion $\mathbf{t} \leq A$. Clearly, this identification complies with the semantics of formulae and inclusions.

2.2 *Sahlqvist and Inductive Formulae in $\mathcal{L}_R$*

In this section we recall the notions of Sahlqvist and Inductive formulae for $\mathcal{L}_R$ as developed in [4] and [5]. These definitions can be seen as specializations of the general purpose definition given in [8] for modal logics algebraically captured by classes of normal lattice-expansions, which are based purely on the order-theoretic properties of the interpretations of the connectives. We refer readers who are interested in this level of generality and in the algebraic and order-theoretic analysis of the Sahlqvist and inductive classes to that paper.

The classes of Sahlqvist and inductive formulae are defined in terms of their *signed generation trees* (Sect. 2.2.1) and the existence of suitable partitions of certain branches in the latter, following e.g. [8]. However, by exploiting the special features of the syntax of $\mathcal{L}_R$, it is possible to give much simpler definitions which are more convenient for practically identifying inductive and Sahlqvist $\mathcal{L}_R$ formulae, but less suitable for generalization or for use in proofs—for this simpler (but conceptually more limited) definition we refer the reader to [4, Section 4.4.1].

2.2.1 Signed Generation Trees, Polarity Types and Critical Branches

We now define the notion of a signed-generation tree of a formula. The definitions of Sahlqvist and inductive formulae are conveniently given in terms of these trees. This style of definition first appears in the definition of the Sahlqvist formulae for distributive modal logic [10].

Definition 1 (Signed Generation Tree) The **positive** (resp. **negative**) *generation tree* of any $\mathcal{L}_R^+$-formula A is defined by labelling the root node of the generation tree of A with the sign $+$ (resp. $-$), and then propagating the labelling on each remaining node as follows:

- For any node labelled with $\vee$, $\wedge$ or $\circ$, assign the same sign to its children.
- If a node is labelled with $\sim$, $\sim^\sharp$ or $\sim^\flat$ assign the opposite sign to its child.
- If a node is labelled with $\rightarrow$, $\Rightarrow$ or $\hookrightarrow$ assign the opposite sign to its left child and the same sign to its right child.
- If a node is labelled with $\prec$ assign the same sign to its left child and the opposite sign to its right child.

Nodes in signed generation trees are **positive** (resp. **negative**) if they are signed $+$ (resp. $-$).

An occurrence of a variable p in a formula A is **positive** (**negative**) if the corresponding leaf node in the positive signed generation tree of A is positive (negative). A is **positive (negative) in** p if all leaf nodes labelled with p in the positive signed generation tree of A are positive (negative). An **inequality** $A \leq B$ **is**

positive (negative) in p if A is negative (positive) in p while B is positive (negative) in p.[2]

Definition 2 A **polarity-type** over a set of propositional variables S is a map $\epsilon : S \to \{+, -\}$. For every polarity-type ϵ, we denote its **opposite polarity-type** by ϵ^∂, that is, $\epsilon^\partial(p) = +$ iff $\epsilon(p) = -$, for every $p \in S$. We will sometime talk of a **polarity-type over a formula** A when we mean a polarity type over the set $var(A)$ of propositional variables occurring in A.

A positive (negative) occurrence of variable p in a formula A **agrees with a polarity type** ϵ **over** A if $\epsilon(p) = +$ ($\epsilon(p) = -$), otherwise it **disagrees with** ϵ.

For any formula $A(p_1, \ldots, p_n)$, any polarity-type ϵ over $\{p_1, \ldots p_n\}$, and any $1 \leq i \leq n$, an ϵ**-critical node** in a signed generation tree of A is a leaf node labelled with $\epsilon(p_i)p_i$. An ϵ**-critical branch** in the tree is a branch with an ϵ-critical leaf node. As we will see later, the variable occurrences corresponding to ϵ-critical nodes will be of special importance in the algorithm.

Hereafter we will use sg to denote $+$ or $-$. For every formula $A = A(p_1, \ldots p_n)$ and every polarity-type ϵ, we say that the signed formula $\mathsf{sg}\, A$ **agrees with** ϵ, and write $\epsilon(\mathsf{sg}\, A)$, if every leaf in the signed generation tree of $\mathsf{sg}\, A$ is ϵ-critical, i.e., signed in agreement with ϵ. Given an occurrence of a subformula ψ of a formula A, we will write $\mathsf{sg}'\psi \prec \mathsf{sg}\, A$ to indicate that the sign of that occurrence of ψ in the signed generation tree $\mathsf{sg}\, A$ is sg'. We will also write $\epsilon(\psi) \prec \mathsf{sg}\, A$ (resp. $\epsilon^\partial(\psi) \prec \mathsf{sg}\, A$) to indicate that the signed subtree of the given occurrence of the subformula ψ of the signed tree $\mathsf{sg}\, A$, agrees with ϵ (resp. with ϵ^∂).

2.2.2 Sahlqvist and Inductive Formulae

To define Sahlqvist and inductive formulae, we classify nodes in signed generation trees according to Table 1. Note that this differs from the classification given in [4] and [5] in that the signs have been reversed. We have chosen to do this since, in the present paper, we focus on formulae rather than inequalities of formulae. The classification used in [4, 5] is aimed at inequalities and would require us to consider the negative generation tree of a formula, which is arguably less intuitive than taking the positive tree. This change also facilitates an easier comparison with the flat inductive formulae, which we undertake in Sect. 4.2. The terminology is indicative of the PEARL rules (see Sect. 5) that will be needed to process nodes when calculating frame correspondents. For a more extensive discussion, we refer the reader to [4].

[2] Note that in other papers (e.g. [4, 8]) we typically consider the positive generation tree $+A$ for the left-hand side and the negative one $-B$ for the right-hand side. Although initially counter-intuitive, this convention is well motivated and goes back to [10]. However, in the present paper, as we focus primarily on formulae rather than inequalities, we depart from this convention, as will be pointed out in the relevant places further.

Table 1 Classification of nodes

	Inner nodes	
Approximation nodes	Adjunction nodes	Residuation nodes
$-$ $\vee$ $\wedge$ $\circ$ $\sim$	$-$ $\wedge$ $\sim$	$-$ $\vee$ $\rightarrow$
$+$ $\wedge$ $\vee$ $\rightarrow$ $\sim$	$+$ $\vee$ $\sim$	$+$ $\wedge$ $\circ$

Definition 3 A branch in a signed generation tree $\mathsf{sg}\,A$, is called a **good branch** if it can be represented as a concatenation of two paths P_1 and P_2, any of which may possibly be of length 0, such that P_1 is a path starting from the leaf of the branch and consisting (apart from variable nodes) only of inner-nodes, and P_2 consists (apart from variable nodes) only of approximation-nodes. A good branch is **excellent** if it is a concatenation of paths P_1 and P_2 as above (each possibly empty), where only adjunction-nodes (and no residuation-nodes) can occur on P_1.

Remark 4 Note that, since all inner nodes that are not residuation nodes are also approximation nodes, we can define an excellent branch equivalently as one that contains no inner nodes. This is particular to the signature of $\mathcal{L}_R$ and may not be the case for other signatures, e.g. in the signature of modal relevance logic.

Definition 5 Given an order type ϵ and a formula $A = A(p_1, \ldots p_n)$ of $\mathcal{L}_R$, the signed generation tree $\mathsf{sg}\,A$ of A is **ϵ-Sahlqvist** if every ϵ-critical branch is excellent. A is **Sahlqvist** if $+A$ is ϵ-Sahlqvist for some ϵ.

See [4] for several examples showing that some important axioms are of Sahlqvist type. We now introduce the more general classes of *inductive formulae and inclusions for* $\mathcal{L}_R$:

Definition 6 (Inductive Formulae and Inclusions) For any polarity-type ϵ and any strict partial order Ω on $p_1, \ldots p_n$, the signed generation tree $\mathsf{sg}\,A$ of a formula $A = A(p_1, \ldots p_n)$ is **(Ω, ϵ)-inductive** if

1. every ϵ-critical branch is good (cf. Definition 3);
2. for every residuation-node occurring in a critical branch ending in leaf $\epsilon(p_i)p_i$ and labelled with a subformula of the form $C \odot B$ or $B \odot C$ where $\odot$ is a signed residuation-connective, the following hold:
   (a) the leaf $\epsilon(p_i)p_i$ of the critical branch occurs in the subtree corresponding to the subformula B,
   (b) $\epsilon^\partial(C) \prec \mathsf{sg}\,A$, i.e., the signed subtree corresponding to subformula C has no critical leaves, and
   (c) $p_k <_\Omega p_i$ for every p_k occurring in C.

   Thus, clauses (b) and (c) above say that every leaf in the signed subtree corresponding to the subformula C is non-critical (i.e. of the form $\epsilon^\partial(p_j)p_j$) with $p_j <_\Omega p_i$. (Note that we write $p_j <_\Omega p_i$ for $\Omega(p_j, p_i)$.)

We will refer to Ω as the **dependency order** on the variables. A **formula** A **is inductive** if $+A$ is (Ω, ϵ)-inductive for some Ω and ϵ.

See [4] for examples of some well-known axioms of relevance logic that are not Sahlqvist but are inductive. As shown in [4], modulo renaming of variables, every Sahlqvist formula as defined in either of [14] and [1] is an inductive formula.

3 Flat Polyadic Language $\mathcal{L}_R^f$ for Relevance Logics

We have already considered one extension of $\mathcal{L}_R$, namely from $\mathcal{L}_R^+$. In this section, we will extend the language of $\mathcal{L}_R$ in a different way, by allowing re-writing of formulae of $\mathcal{L}_R$ into a shape that is more convenient to re-define the class of inductive formulae, as well as possibly define other, larger classes of FO-definable formulae of $\mathcal{L}_R$ on which suitable extensions of PEARL should succeed.

The idea in a nutshell is to enable composing nested connectives of the same type (respectively, of box or diamond type) into *composite terms*, treated as new polyadic connectives, and then to extend the set of formulae by using these newly defined connectives, too. That will produce an extension $\mathcal{L}_R^f$ of $\mathcal{L}_R$. That extension will be only 'syntactic sugar' but it will enable us to define the class of inductive formulae by means of their global structure when suitably re-written in the richer language, rather than locally ('branch-wise'), in terms of the patterns of occurrence of each variable, as in Sect. 2.2.2.

In this section, we are partly following [11] and [12].

3.1 Flat Polyadic Language $\mathcal{L}_R^f$: Syntax

The **flat polyadic language** $\mathcal{L}_R^f$ builds on the vocabulary of $\mathcal{L}_R$, but treats the logical connectives $\sim$, $\wedge$, $\vee$, $\circ$, $\rightarrow$, as well as the identity connective ι added for technical purposes, as terms which can be composed into more complex terms, thus defining composite polyadic modal operators respectively of Box and Diamond types (cf. discussion on these in Sect. 2.2.2), which will be added to the basic connectives in the definition of formulae of $\mathcal{L}_R^f$.

First, we are going to define polyadic relevance terms and the polarities of their arguments, using the following notation. Given an n-ary term α, by $\epsilon^1(\alpha), \ldots, \epsilon^n(\alpha)$ we denote the polarities ($+$ or $-$) of the respective arguments of α. We will also be using a vector notation: $\epsilon(\alpha) = (\epsilon^1(\alpha), \ldots, \epsilon^n(\alpha))$ and will say that **the term** α **is positive** (respectively, **negative**) if all arguments of α have a positive (respectively, negative) polarity, i.e. if $\epsilon(\alpha) = (+, \ldots, +)$ (respectively, $\epsilon(\alpha) = (-, \ldots, -)$).

Definition 7 The set of **pure box-terms (pure B-terms)** BT, the set of **pure diamond-terms, (pure D-terms)** DT, their **arity function** ρ, and the **polarities** of their arguments, are defined by a simultaneous mutual induction given below.

(BD-CT) Every constant formula A of $\mathcal{L}_R$ is a pure B-term and a pure D-term with $\rho(A) = 0$.

(BD-IT) The identity connective ι is a pure B-term and a pure D-term with $\rho(\iota) = 1$ and $\epsilon(\iota) = (+)$.

(BT1) The connective $\vee$ is a pure B-term with $\rho(\vee) = 2$ and $\epsilon(\vee) = (+, +)$.

(BT2) The connective $\rightarrow$ is a pure B-term with $\rho(\rightarrow) = 2$ and $\epsilon(\rightarrow) = (-, +)$.

(BT3) If δ is a pure D-term with $\rho(\delta) = n$ then $\sim \delta$ is a pure B-term with $\rho(\sim \delta) = n$ and $\epsilon(\sim \delta) = \overline{\epsilon(\delta)}$, where $\overline{\epsilon(\delta)}$ is the vector with flipped polarities of $\epsilon(\delta)$,
i.e. $\overline{\epsilon^i(\delta)} = -$ iff $\epsilon^i(\delta) = +$, for each $i = 1, \dots, n$.

(BT4) Let β be a pure B-term with $\rho(\beta) = n$ and $\beta_1, \dots, \beta_n$ be pure B-terms or D-terms, such that for every $i = 1, \dots, n$, if $\epsilon^i(\beta) = +$ then β_i is a pure B-term and if $\epsilon^i(\beta) = -$ then β_i is a pure D-term. Then $\beta(\beta_1, \dots, \beta_n)$ is a pure B-term with $\rho(\beta(\beta_1, \dots, \beta_n)) = \rho(\beta_1) + \dots + \rho(\beta_n)$.
In particular, this holds if β is a positive pure B-term and $\beta_1, \dots, \beta_n$ are pure B-terms.
The polarity vector of $\beta(\beta_1, \dots, \beta_n)$ is defined as follows. Let k be the index of the j-th argument of the subterm β_i in the term $\beta(\beta_1, \dots, \beta_n)$, i.e., $k = \rho(\beta_1) + \dots + \rho(\beta_{i-1}) + j$. Then $\epsilon^k(\beta(\beta_1, \dots, \beta_n)) = +$ iff $\epsilon^i(\beta) = \epsilon^j(\beta_i)$, that is, the polarity is positive in the composite term iff it is either positive in the subterm β_i which is in a positive position in the term β, or it is negative in the subterm β_i which is in a negative position in β.

(DT1) The connective $\wedge$ is a pure D-term with $\rho(\wedge) = 2$ and $\epsilon(\wedge) = (+, +)$.

(DT2) The connective $\circ$ is a pure D-term with $\rho(\circ) = 2$ and $\epsilon(\circ) = (+, +)$.

(DT3) If β is a pure B-term with $\rho(\beta) = n$ then $\sim \beta$ is a pure D-term with $\rho(\sim \beta) = n$ and $\epsilon(\sim \beta)$ is defined just like $\epsilon(\sim \beta)$ in (BT3).

(DT4) Let δ be a pure D-term with $\rho(\delta) = n$ and $\delta_1, \dots, \delta_n$ be pure B-terms or D-terms, such that for every $i = 1, \dots, n$, if $\epsilon^i(\delta) = +$ then δ_i is a pure D-term and if $\epsilon^i(\delta) = -$ then δ_i is a pure B-term. Then $\delta(\delta_1, \dots, \delta_n)$ is a pure D-term with $\rho(\delta(\delta_1, \dots, \delta_n)) = \rho(\delta_1) + \dots + \rho(\delta_n)$.
In particular, this holds if δ is a positive pure D-term and $\delta_1, \dots, \delta_n$ are pure D-terms.
The polarity vector $\epsilon(\delta(\delta_1, \dots, \delta_n))$ is defined just like in (BT4).

For every pure B-term β with $\rho(\beta) = n > 0$ and $k \leq n$ and $\epsilon^k(\beta) = +$ ($\epsilon^k(\beta) = -$) we define the D-term (B-term) β^{-k}, called the **k-inverse of** β. The polarity vector $\epsilon(\beta^{-k})$ is such that $\epsilon^k(\beta^{-k}) = \epsilon^k(\beta)$ and for $j \neq k$ it is the case that $\epsilon^j(\beta^{-k}) = \overline{\epsilon^j(\beta)}$ if $\epsilon^k(\beta) = +$ and $\epsilon^j(\beta^{-k}) = \epsilon^j(\beta)$ if $\epsilon^k(\beta) = -$.

Likewise every pure D-term δ with $\rho(\delta) = n > 0$ and $k \leq n$ and $\epsilon^k(\delta) = +$ ($\epsilon^k(\delta) = -$) we define the B-term (D-term) δ^{-k}, called the **k-inverse of δ**. The polarity vector $\epsilon(\delta^{-k})$ is such that $\epsilon^k(\delta^{-k}) = \epsilon^k(\delta)$ and for $j \neq k$ it is the case that $\epsilon^j(\delta^{-k}) = \overline{\epsilon^j(\delta)}$ if $\epsilon^k(\delta) = +$ and $\epsilon^j(\delta^{-k}) = \epsilon^j(\delta)$ if $\epsilon^k(\delta) = -$.

Pure B-terms and D-terms will be called collectively **BD-terms**.

A simple example of a non-pure B-term is $\rightarrow (\iota, \circ)$ and a simple non-pure D-term is $\circ(\iota, \rightarrow)$.

Definition 8 The set $\mathsf{For}(\mathcal{L}_R^f)$ of **formulae of $\mathcal{L}_R^f$**, the subset of **flat formulae** $\mathsf{FlFor}(\mathcal{L}_R^f)$ **of $\mathcal{L}_R^f$**, their **leading terms** and **main components**, are defined inductively, as follows:

- (F) Every formula of $\mathcal{L}_R$ is also a formula of $\mathcal{L}_R^f$. It has a leading term ι and is its only main component.
- (BF) If $A_1, \ldots, A_n$ for $n > 0$ are formulae of $\mathcal{L}_R^f$ and α is a pure n-ary B-term with $\rho(\alpha) = n$, then $A = [\alpha](A_1, \ldots, A_n)$ is a B-formula of $\mathcal{L}_R^f$, with **leading term** α and **main components** $A_1, \ldots, A_n$.
- (DF) If $A_1, \ldots, A_n$ for $n > 0$ are formulae of $\mathcal{L}_R^f$ and α is a pure D-term with $\rho(\alpha) = n$, then $A = \langle\alpha\rangle(A_1, \ldots, A_n)$ is a D-formula of $\mathcal{L}_R^f$, with **leading term** α and **main components** $A_1, \ldots, A_n$.

In the clauses above, when $A_1, \ldots, A_n$ are propositional variables $p_1, \ldots, p_n$, the formula $[\alpha](p_1, \ldots, p_n)$ is called a **flat B-formula** and $\langle\alpha\rangle(p_1, \ldots, p_n)$ is called a flat **D-formula**.

The polarities of the arguments $A_1, \ldots, A_n$ in the formula A in each clause above are defined as the respective polarities of the BD-term α. In case of flat formulae, these polarities define the **positive** and **negative occurrences of variables** in the formula.

In particular, every constant formula A of $\mathcal{L}_R$ is a flat B-formula and a flat D-formula, with leading term A itself. Likewise, every propositional variable is a flat B-formula and a flat D-formula, with leading term ι.

As we will see further, BD-terms define flat formulae with semantics that corresponds to the standard semantics in $\mathcal{L}_R$.

Note that constant formulae of $\mathcal{L}_R$ are regarded as both 0-ary terms and formulae. This ambiguity of the syntax should not cause confusion if properly handled, and we have adopted it for the sake of technical convenience.

Many formulae of $\mathcal{L}_R$ can be re-written as flat formulae, even if they do not look flat at all. For instance, the formula

$$((\sim (\mathbf{t} \rightarrow \sim p) \rightarrow \sim ((\sim \mathbf{t}) \circ p)) \vee (p \rightarrow \sim \mathbf{t}))$$

can be re-written as the flat B-formula $[\beta]p$, where β is the pure unary B-term

$$\vee(\rightarrow(\sim\rightarrow(\mathbf{t},\sim\iota),\sim\circ(\sim\mathbf{t},\iota)),\rightarrow(\iota,\sim\mathbf{t}))$$

3.2 Semantics of $\mathcal{L}_R^f$

The **semantics** of $\mathcal{L}_R^f$ is a straightforward combination of the standard Routley-Meyer semantics for $\mathcal{L}_R$ given in Sect. 2 and the semantics of PDL-type polymodal languages (cf. [11] and [12]), after taking into account the fact that the logical connectives are now treated as terms.

First, let us define a general composition of relations. Let $R_\alpha, R_{\beta_1}, \ldots, R_{\beta_n}$ be relations in the set W of respective arities $n+1, b_1+1, \ldots, b_n+1$. Then we define the relation

$$R_\alpha(R_{\beta_1}, \ldots, R_{\beta_n}) := \big\{(w, w_{11}, \ldots, w_{1b_1}, \ldots, w_{n1}, \ldots, w_{nb_n}) \in W^{b_1+\cdots+b_n+1} \mid$$

$$\exists u_1 \ldots \exists u_n (R_\alpha w u_1 \ldots u_n \wedge \bigwedge_{i=1}^{n} R_{\beta_i} u_i w_{i1} \ldots w_{ib_i})\big\}.$$

We also define the operation π of **cyclic permutation** on relations which moves the last argument of the relation to the first position, i.e. $\pi(R)vu_1 \ldots u_n$ holds iff $Ru_1 \ldots u_n v$.

Definition 9 Given a RM-frame $\mathcal{F} = \langle W, O, R, *\rangle$, for every pure term α with $\rho(\alpha) = n$ we define a $(n+1)$-ary relation R_α recursively on the combined inductive definition of pure B-terms and D-terms.

(RBD-CT) For every constant formula A of $\mathcal{L}_R$, $R_A = [\![A]\!]_{\mathcal{F}}$.
(RBD-IT) $R_\iota = \{(w, w) \mid w \in W\}$,
(RBT1) $R_\vee = \{(w, w, w) \mid w \in W\}$.
(RBT2) $R_\rightarrow = R$.
(RBT3) If δ is a D-term with $\rho(\delta) = n$ then $R_{\sim\delta} := R_*(R_\delta)$, where $R_* = \{(u, u^*) \mid u \in W\}$.
(RBT4) If β is a B-term with $\rho(\beta) = n$ and $\beta_1, \ldots, \beta_n$ are respective B-terms or D-terms, then $R_{\beta(\beta_1,\ldots,\beta_n)} := R_\beta(R_{\beta_1}, \ldots, R_{\beta_n})$.
(RDT1) $R_\wedge = \{(w, w, w) \mid w \in W\}$.
(RDT2) $R_\circ = \pi(R)$.
(RDT3) If β is a B-term with $\rho(\beta) = n$ then $R_{\sim\beta} := R_*(R_\beta)$.
(RDT4) If δ is a D-term with $\rho(\delta) = n$ and $\delta_1, \ldots, \delta_n$ are B-terms or D-terms, then $R_{\delta(\delta_1,\ldots,\delta_n)} := R_\delta(R_{\delta_1}, \ldots, R_{\delta_n})$.

(RBD-Inv) If β is a B-term with $\rho(\beta) = n > 0$ and $k \leq n$ we define $R_{\beta^{-k}}$ as obtained from R_β by transposing the 0-th and the k-th arguments. Formally,

$$R_{\beta^{-k}} = \{(u_0, u_1, \ldots, u_{k-1}, u_k, u_{k+1}, \ldots, u_n) \mid$$
$$(u_k, u_1, \ldots, u_{k-1}, u_0, u_{k+1}, \ldots, u_n) \in R_\beta\}$$

For any D-term δ with $\rho(\beta) = n > 0$ and $k \leq n$, the relation $R_{\delta^{-k}}$ is defined in terms of R_δ in the same way.

Definition 10 **Truth of a formula** $A \in \mathsf{For}(\mathcal{L}_\mathsf{R}^\mathsf{f})$ **at a state** u **in a Routley-Meyer model** $\mathcal{M} = \langle W, O, R, ^*, V\rangle$, denoted $\mathcal{M}, u \models^f A$, is defined as follows:

- $\mathcal{M}, u \models^f A$ iff $\mathcal{M}, u \models A$ for any formula A in $\mathcal{L}_R$.
- $M, w \models^f [\alpha](A_1, \ldots, A_n)$, where α is a pure n-ary B-term, iff for all $u_1, \ldots, u_n \in W$ such that $R_\alpha w u_1 \ldots u_n$, it holds that
  - $M, u_i \models^f A_i$ for some $i \in \{1, \ldots, n\}$ such that $\epsilon(\alpha) = +$,
    or
  - $M, u_i \not\models^f A_i$ for some $i \in \{1, \ldots, n\}$ such that $\epsilon(\alpha) = -$,
- $M, w \models^f \langle\alpha\rangle(A_1, \ldots, A_n)$, where α is a pure n-ary D-term, iff there exist $u_1, \ldots, u_n \in W$ such that $R_\alpha w u_1 \ldots u_n$ and:
- $M, u_i \models^f A_i$ for all $i \in \{1, \ldots, n\}$ such that $\epsilon(\alpha) = +$,
  and
- $M, u_i \not\models^f A_i$ for all $i \in \{1, \ldots, n\}$ such that $\epsilon(\alpha) = -$.

For every RM-model $\mathcal{M}$ and a formula $A \in \mathsf{For}(\mathcal{L}_\mathsf{R}^\mathsf{f})$ we define the **f-extension of** A **in** $\mathcal{M}$ as

$$[\![A]\!]^f_{\mathcal{M}} := \{u \in \mathcal{M} \mid \mathcal{M}, u \models^f A\}$$

The notion of satisfaction of $\mathcal{L}_R^f$ formulae is naturally extended to inequalities of $\mathcal{L}_R^f$ formulae, viz.

$$\mathcal{M} \models^f A \leq B \text{ iff } [\![A]\!]^f_{\mathcal{M}} \subseteq [\![B]\!]^f_{\mathcal{M}}.$$

The next lemma now follows directly from the definitions:

Lemma 11 *For any Routley-Meyer model* $\mathcal{M}$*, any* n*-ary pure* B*-term* β *and* n*-ary pure* D*-term* δ *it holds that:*

1. $\mathcal{M} \models^f A \leq [\beta](B_1, \ldots, B_k, \ldots, B_n)$ *iff* $\mathcal{M} \models^f \langle\beta^{-k}\rangle(B_1, \ldots, A, \ldots, B_n) \leq B_k$, *for* $\epsilon^k(\beta) = +$.
2. $\mathcal{M} \models^f A \leq [\beta](B_1, \ldots, B_k, \ldots, B_n)$ *iff* $\mathcal{M} \models^f B_k \leq [\beta^{-k}](B_1, \ldots, A, \ldots, B_n)$, *for* $\epsilon^k(\beta) = -$.

3. $\mathcal{M} \models^f \langle\delta\rangle(B_1, \ldots, B_k, \ldots, B_n) \leq A$ *iff* $\mathcal{M} \models^f B_k \leq [\delta^{-k}](B_1, \ldots, A, \ldots, B_n)$, *for* $\epsilon^k(\delta) = +$.
4. $\mathcal{M} \models^f \langle\delta\rangle(B_1, \ldots, B_k, \ldots, B_n) \leq A$ *iff* $\mathcal{M} \models^f \langle\delta^{-k}\rangle(B_1, \ldots, A, \ldots, B_n) \leq B_k$, *for* $\epsilon^k(\delta) = -$.

3.3 *Translation from $\mathcal{L}_R^f$ to $\mathcal{L}_R$*

The language $\mathcal{L}_R^f$ is expressively equivalent to the sublanguage $\mathcal{L}_R$, as every formula of $\mathcal{L}_R^f$ can be translated (essentially, rewritten) into a semantically equivalent (as we show further) formula of $\mathcal{L}_R$. We define the following straightforward translation $\tau(\cdot)$ of the formulae of $\mathcal{L}_R^f$ into $\mathcal{L}_R$, by induction on flat formulae and terms, as follows.

- $\tau(A) = A$ for any formula $A \in \mathcal{L}_R$,
- $\tau([\iota]A) = \tau(A)$,
- $\tau([\vee](A_1, A_2)) = \tau(A_1) \vee \tau(A_2)$,
- $\tau([\rightarrow](A_1, A_2)) = \tau(A_1) \rightarrow \tau(A_2)$,
- $\tau([\sim \beta](A_1, \ldots, A_n)) = \sim \tau(\langle\beta\rangle(A_1, \ldots, A_n))$, for any n-ary D-term β.
- $\tau([\beta(\beta_1, \ldots, \beta_n)](A_1^1, \ldots, A_{m_1}^1, \ldots, A_1^n, \ldots, A_{m_n}^n)) =$
  $\tau([\beta](\tau([\beta_1](A_1^1, \ldots, A_{m_1}^1)), \ldots, \tau([\beta_n](A_1^n, \ldots, A_{m_n}^n))))$,
  for any B-term β and respective to the polarities of the arguments BD-terms $\beta_1, \ldots, \beta_n$ with $\rho(\beta) = n$, $\rho(\beta_1) = m_1, \ldots, \rho(\beta_n) = m_n$.
- $\tau(\langle\iota\rangle A) = \tau(A)$,
- $\tau(\langle\iota\rangle(A_1 \wedge A_2)) = \tau(A_1) \wedge \tau(A_2)$,
- $\tau(\langle\iota\rangle(A_1 \circ A_2)) = \tau(A_1) \circ \tau(A_2)$,
- $\tau(\langle\sim \beta\rangle(A_1, \ldots, A_n)) = \sim \tau([\beta](A_1, \ldots, A_n))$, for any n-ary B-term β.
- $\tau(\langle\delta(\delta_1, \ldots, \delta_n)\rangle(A_1^1, \ldots, A_{m_1}^1, \ldots, A_1^n, \ldots, A_{m_n}^n)) =$
  $\tau(\langle\delta\rangle(\tau(\langle\delta_1\rangle(A_1^1, \ldots, A_{m_1}^1)), \ldots, \tau(\langle\delta_n\rangle(A_1^n, \ldots, A_{m_n}^n))))$,
  for any D-term δ and respective to the polarities of the arguments BD-terms $\delta_1, \ldots, \delta_n$ with $\rho(\delta) = n$, $\rho(\delta_1) = m_1, \ldots, \rho(\delta_n) = m_n$.

For example:
$\tau(\langle\circ(\sim \iota, \sim \rightarrow)\rangle(p, q, r)) =$
$\tau(\langle\circ\rangle(\tau(\langle\sim \iota\rangle(p)), \tau(\langle\sim \rightarrow\rangle(q, r)))) =$
$\tau(\langle\sim \iota\rangle(p)) \circ \tau(\langle\sim \rightarrow\rangle(q, r)) =$
$\sim \tau([\iota](p)) \circ \sim \tau([\rightarrow](q, r)) =$
$\sim p \circ \sim (q \rightarrow r)$.

Proposition 12 *For every formula A of $\mathcal{L}_R^f$ and any RM-model $\mathcal{M}$ and a state u in it:*

$$\mathcal{M}, u \Vdash^f A \text{ iff } \mathcal{M}, u \Vdash \tau(A)$$

Consequently, $[\![A]\!]^f_{\mathcal{M}} = [\![\tau(A)]\!]_{\mathcal{M}}$.

Proof An almost routine induction on the definition of $\Vdash^f$ in $\mathcal{L}_R^f$, using Definition 9. The only not completely trivial cases are those of $A = [\rightarrow](A_1, A_2)$ and $A = [\circ](A_1, A_2)$,

For the former case, note that $\epsilon(\rightarrow) = (-, +)$ and for any $v, w \in W$, "if $\mathcal{M}, v \Vdash \tau(A_1)$ then $\mathcal{M}, v \Vdash \tau(A_2)$" is equivalent to "$\mathcal{M}, v \nVdash \tau(A_1)$ or $\mathcal{M}, v \Vdash \tau(A_2)$". For the latter case, it suffices to note that for any $u, v, w \in W$, $\pi(R)uvw$ is equivalent (by definition) to $Rvwu$. ■

The translation $\tau(\cdot)$ can be reversed non-deterministically to a translation from $\mathcal{L}_R$ to $\mathcal{L}_R^f$. E.g., the $\mathcal{L}_R$-formula $\sim p \circ \sim (q \rightarrow r)$ can be translated back to $\langle\circ\rangle\big([\sim\iota](p), \langle\sim\rightarrow\rangle(q, r)\big)$. Such inverse translation is not unique, because constant formulae and the connectives ι and $\sim$ are regarded as both pure B-terms and D-terms and because the nested connectives can be composed in different orders. For instance, the formula above can also be inverse-translated to $\langle\circ\rangle\big(\langle\sim\iota\rangle(p), \langle\sim\rightarrow\rangle(q, r)\big)$, as well as to the flat D-formula $\langle\circ(\sim\iota, \sim\rightarrow)\rangle\big(p, q, r\big)$, as the earlier example shows. Still, it is easy to see that every formula in $\mathcal{L}_R$ is the translation of at least one (not necessarily flat) formula in $\mathcal{L}_R^f$ and, according to the proposition above, all pre-images under $\tau(\cdot)$ are semantically equivalent.

The important feature of the language $\mathcal{L}_R^f$ is the flat and uniform representation of formulae, from which their semantic conditions can be computed in a way that will enable us to define a very general class of FO-definable formulae of $\mathcal{L}_R$.

Another effect of the mutual definition of BD-terms and formulae of $\mathcal{L}_R^f$ is that it allows construction of relevance BD-terms that are parameterized with constant formulae. Further we will extend that idea by also treating arguments which are variables as parameters. For instance, the formula $(q \rightarrow (\sim\mathbf{t} \rightarrow p)) \vee (\sim\mathbf{t} \rightarrow \mathbf{t})$ can be translated to $[\beta(q)]p$ where $\beta(q)$ is the unary parameterised B-term $\vee(\rightarrow(q, \rightarrow(\sim\mathbf{t}, \iota)), \rightarrow(\sim\mathbf{t}, \mathbf{t}))$ obtained by treating the constant sub-formulae as terms and the variable q as a parameter imported into the leading term.

4 Regular and Inductive Formulae in $\mathcal{L}_R^f$

4.1 Regular Flat Polyadic Relevant Formulae

Definition 13 A **box-formula** is a $\mathcal{L}_R^f$-formula of one of the following two types:

- $B = [\beta](C_1, \ldots, C_m)$ where β is a pure m-ary B-term, for $m \geq 1$, and $C_1, \ldots, C_m$ are formulae in $\mathsf{For}(\mathcal{L}_\mathsf{R}^\mathsf{f})$. A formula of this type will be called a **headless box**.
- $B = [\beta](p, C_1, \ldots, C_m)$ where β is a pure $(m+1)$-ary B-term, for $m \geq 0$, $C_1, \ldots, C_m$ are formulae in $\mathsf{For}(\mathcal{L}_\mathsf{R}^\mathsf{f})$, and p is a variable. A formula of this type will be called a **headed box**, and the variable p is called the **head** of the formula. The head need not be the first argument of a headed box, but to simplify the notation we will usually put it in first position.

Headed and headless diamond-formulae are defined likewise, by replacing B-terms with D-terms and $[\beta]$ with $\langle\beta\rangle$ in the definitions above.

Here are more examples of box-formulae and diamond-formulae:

- $[\rightarrow](p, q)$ can be seen as a box with head p or as a box with head q.
- $[\rightarrow (\iota, \sim)](p, [\rightarrow](\sim p, q)) \equiv p \rightarrow \sim(\sim p \rightarrow q)$ is a box with head p.
- $\langle\wedge(\iota, \circ(\iota, \iota))\rangle(p, q, [\rightarrow](p, \sim q)) \equiv p \wedge (q \circ (p \rightarrow \sim q))$ is a diamond with head p and a diamond with head q.
- $[\vee](\sim p, \sim \langle\circ\rangle(\sim q, [\rightarrow](p, \sim q)))$, is a headed box-formula and is equivalent to the flat headed box formula $[\vee(\sim\iota, \sim\circ(\sim\iota, \rightarrow (\iota, \sim\iota)))](p, q, p, q)$.

Definition 14 A **regular formula in** $\mathsf{For}(\mathcal{L}_\mathsf{R}^\mathsf{f})$ is any box-formula $A = [\beta](B_1, \ldots, B_n)$ where

1. B_i is a diamond-formula whenever $\epsilon^i(\beta) = +$, and
2. B_i is a box-formula whenever $\epsilon^i(\beta) = -$.

Example 15 Here are some examples of regular formulae, where $[\iota]$ is often omitted, thus writing just p instead of $[\iota]p$:

1. $A_1 = [\rightarrow (\wedge, \rightarrow)]([\rightarrow](p, q), [\rightarrow](q, r), p, r)$, translated as $((p \rightarrow q) \wedge (q \rightarrow r)) \rightarrow (p \rightarrow r)$, is a regular formula, because the first 3 arguments have negative polarities in the B-term $\rightarrow (\wedge, \rightarrow)$ and are (headed) boxes, whereas the last argument has positive polarity and is (vacuously) a diamond-formula (with head p). This formula is Axiom B2 from [13], see also [4, Appendix 2].
2. Likewise, $A_2 = [\rightarrow (\wedge, \rightarrow)]([\rightarrow](p, r), [\rightarrow](q, r), [\vee](p, q), r)$, translated as $((p \rightarrow r) \wedge (q \rightarrow r)) \rightarrow ((p \vee q) \rightarrow r)$, is a regular formula, because the first 3 arguments have negative polarities in the B-term $\rightarrow (\wedge, \rightarrow)$ and are (headed) boxes, whereas the last argument has positive polarity and is the diamond $\langle\iota\rangle p$. This is Axiom A7 from [13], see also [4, Appendix 2].

3. $A_3 = [\to (\iota, \iota)]([\to](p, p \circ q), \langle\circ\rangle(p, q))$, translated as $(p \to (p \circ q)) \to (p \circ q)$, is regular, as $\to (\iota, \iota)$ is a pure B-term with $\epsilon^1(\to (\iota, \iota)) = -$ and $\epsilon^2(\to (\iota, \iota)) = +$, since $[\to](p, p \circ q)$ is a box-formula and $\langle\circ\rangle(p, q)$ is a diamond-formula.
4. $A_4 = [\to (\iota, \iota)]([\to (\sim, \iota)](p, \sim p \circ q), \langle\circ\rangle(p, q))$, translated as $(\sim p \to (\sim p \circ q)) \to (p \circ q)$ is also regular.
5. $A_5 = [\to (\iota, \iota)]([\to (\iota, \iota)](p \to p, q), q) \equiv ((p \to p) \to q) \to q$ is regular, too.
6. $A_6 = [\to](\langle\wedge\rangle(p, \sim p), q) \equiv (p \wedge \sim p) \to q$ is not a regular formula, but is equivalent to the regular flat formula $[\to (\wedge, \iota)](p, [\sim\iota]p, q)$.
7. Similarly, $A_7 = [\to](p, [\vee](r, \langle\circ\rangle(r, q))) \equiv p \to (r \vee (r \circ q))$ is not regular but is equivalent to the regular flat formula $[\to (\iota, \vee)](p, r, \langle\circ\rangle(r, q))$.

Definition 16 An occurrence of a variable in a regular formula A is **essential in** A if it is a head of a main component of the formula, otherwise it is **inessential in** A.

A variable occurring in a regular formula A is **essential in** A if it has at least one essential occurrence in it, otherwise it is **inessential in** A.

A regular formula A is **lean** if every variable occurring in it is essential in A.

For example, the variables p and r are essential in $[\to (\iota, \sim\iota)](p, [\vee](r, \sim\langle\circ\rangle(r, q)))$, while q is inessential there.

4.2 *Inductive Flat Polyadic Relevant Formulae*

Definition 17 Given a regular formula $A = [\alpha](H_1, \ldots, H_n, N_1, \ldots, N_k)$, where $H_1, \ldots, H_n$ are the headed boxes, the **dependency digraph of** A is a digraph $G = \langle \mathbf{V}_A, \prec_A \rangle$ where $\mathbf{V}_A = \{p_1, \ldots, p_n\}$ is the set of heads in A, and $p_i \prec_A p_j$ iff p_i occurs as an inessential variable in a formula from $\{H_1, \ldots, H_n\}$ with a head p_j. A digraph is called **acyclic** if it does not contain oriented cycles along $\prec_A$.

Definition 18 **An inductive flat (polyadic) relevance formula** is any regular formula A in $\mathsf{For}(\mathcal{L}_{\mathsf{R}}^{\mathsf{f}})$ with an acyclic dependency digraph such that for every variable p occurring in A:

1. no essential occurrence of p and inessential occurrence of p have the same polarity in A,
2. all essential occurrences of p have the same polarity in A, and
3. all inessential occurrences of p have the same polarity in A.

The class of inductive formulae in $\mathcal{L}_R^f$ will be denoted by $\mathbf{IF}(\mathcal{L}_R^f)$.

Example 19 Here are some examples and non-examples of inductive relevance formulae in $\mathsf{For}(\mathcal{L}_{\mathsf{R}}^{\mathsf{f}})$. As before, we often omit $[\iota]$, e.g., just writing p instead of $[\iota]p$.

1. Recall the regular formula $A_1 = [\to (\wedge, \to)]([\to](p, q), [\to](q, r), p, r)$ from Example 15, translated as $((p \to q) \wedge (q \to r)) \to (p \to r)$ (Axiom B2 from

[13]). Taking the arguments of $[\rightarrow(\wedge,\rightarrow)]$ from left to right, and declaring q as the head in $[\rightarrow](p,q)$, r as the head in $[\rightarrow](q,r)$, p as the head of p, and q headless, we find that the negative occurrences of p, q and r are essential while the positive ones are inessential, so the polarity conditions of Definition 18 are met. Moreover dependency digraph becomes $p \prec q \prec r$, which is acyclic, hence the formula is inductive.

2. Likewise, the regular formula $A_2 = [\rightarrow\ (\wedge,\rightarrow)]([\rightarrow](p,r),[\rightarrow](q,r),[\vee](p,q),r)$ from Example 15, translated as $((p\rightarrow r)\wedge(q\rightarrow r))\rightarrow((p\vee q)\rightarrow r)$ (Axiom A7 from [13]), is inductive. Indeed, let us identify the heads of the arguments to be the underlined variable occurrences as follows: $[\rightarrow\ (\wedge,\rightarrow)]([\rightarrow](\underline{p},r),[\rightarrow](\underline{q},r),[\vee](p,q),\underline{r})$. Then we see that the positive occurrences of p, q and r are essential while their negative occurrences are inessential. Moreover, the arcs of the resulting dependency digraph are $r \prec p$ and $r \prec q$, so it is acyclic.
3. Consider the regular formula $A_3 = [\rightarrow(\iota,\iota)]([\rightarrow](p, p\circ q),\langle\circ\rangle(p,q))$ from Example 15, translated as $(p\ \rightarrow\ (p\circ q))\ \rightarrow\ (p\circ q)$. It is not inductive. Indeed, to try and satisfy the polarity conditions of Definition 18, we would have to declare the first occurrence of p (which is positive in A) the head of $[\rightarrow](p, p\circ q)$, so p must then also be essential in $\langle\circ\rangle(p,q)$ as its occurrence there is also positive in A and hence both the negative and positive occurrences of q would have to be inessential (as a box or diamond formula is allowed at most one head), violating the requirements of the definition.
4. Consider now the regular formula $A_4 = [\rightarrow(\iota,\iota)]([\rightarrow(\sim,\iota)](p,\sim p\circ q),\langle\circ\rangle(p,q))$ from Example 15, translated as $(\sim p\rightarrow(\sim p\circ q))\rightarrow(p\circ q)$. Taking p to be the head in $[\rightarrow\ (\sim,\iota)](p,\sim p\circ q)$ and q the head in $\langle\circ\rangle(p,q)$ we would have that all essential occurrences of p in A_4 are negative and all inessential ones are positive, whereas all essential occurrences of q are positive and all inessential ones are negative, thus satisfying the polarity conditions of Definition 18. Moreover, this is the only possible choice of heads that would satisfy these conditions. However, note that the dependency digraph would then be cyclic, and thus A_4 is not inductive.
5. The formula A_5 in Example 15, which is an axiom in [13], is not inductive, either, because of the unavoidable loop $p \prec p$ in its dependency graph. Notably, as shown in [4], this formula is nevertheless first-order definable, though the algorithm PEARL, as defined there, does not succeed on it.

We now compare the inductive flat formulae in $\mathsf{For}(\mathcal{L}^{\mathsf{f}}_{\mathsf{R}})$ (Definition 18) with the inductive formulae in $\mathcal{L}_R$ (Definition 6). The following two theorems establish the fact that, modulo translation, these two classes coincide.

Theorem 20 *If A is an inductive flat formula in $\mathsf{For}(\mathcal{L}^{\mathsf{f}}_{\mathsf{R}})$, then $\tau(A)$ is an inductive formula in $\mathcal{L}_R$.*

Proof Suppose $A = [\alpha](H_1,\ldots,H_n,N_1,\ldots,N_k)$ is an inductive flat formula in $\mathsf{For}(\mathcal{L}^{\mathsf{f}}_{\mathsf{R}})$, where $H_1,\ldots,H_n$ are its headed main components and $N_1,\ldots,N_k$ are its headless main components. We will argue that the positive generation tree of $\tau(A)$

is (Ω, ϵ_0)-inductive where Ω is the dependency order on the variables (acyclic by the assumption on A) and $\epsilon_0(p) = +$ if the essential occurrences of p are positive in A, respectively $\epsilon_0(p) = -$ if the essential occurrences of p are negative in A. Again, by the assumption that A is inductive, ϵ_0 so specified is well-defined.

Let $q_1, \ldots, q_n$ and $r_1, \ldots, r_k$ be propositional variables occurring A. A routine induction on α shows that all nodes in the positive signed generation tree of $\tau(A(q_1, \ldots, q_n, r_1, \ldots, r_k))$ are approximation nodes.

Next, consider any of the headed main components, say H_i. If $\epsilon^i(\alpha) = +$ then H_i is of the form $\langle\delta\rangle(p_i, C_1, \ldots, C_m)$ where δ is a D-term. Again a straightforward induction on δ shows that in the positive signed generation tree of $\tau(\langle\delta\rangle(p_i, C_1, \ldots, C_m))$ all nodes on the branch ending in the occurrence corresponding to the head p_i are inner nodes. Moreover, any other signed variable node in the positive signed generation tree of $\tau(\langle\delta\rangle(p_i, C_1, \ldots, C_m))$ is less than the head according to Ω and is of the opposite polarity specified by ϵ. Similar considerations prove that the same situation obtains in the case of the negative generation tree when $\epsilon^i(\alpha) = -$ and H_i is of the form $[\beta](p_i, C_1, \ldots, C_m)$ for a B-term β.

From these observations it should be clear that, in the positive generation tree of $\tau(A)$, every ϵ_0-critical occurrence of a variable p_i corresponds to the head of one of the headed main components, either $\langle\delta\rangle(p_i, C_1, \ldots, C_m)$ or $[\beta](p_i, C_1, \ldots, C_m)$, and that they are leaves on good branches. Moreover, whenever such an occurrence is in the scope of an inner residuation node (i.e., $-\vee$, $- \rightarrow$, $+\wedge$ or $+\circ$), that residuation node arises from the translation of the main component and hence all other signed variable nodes in its scope (other than the critical one) are not ϵ_0-critical and strictly less that p_i according to Ω. The claim now follows. ■

Theorem 21 *Every inductive formula in $\mathcal{L}_R$ can be rewritten as a semantically equivalent conjunction of flat inductive formulae in* $\mathsf{For}(\mathcal{L}_{\mathsf{R}}^{\mathsf{f}})$.

Proof Suppose A is an (Ω, ϵ_0)-inductive $\mathcal{L}_R$ formula. We begin by observing that, by applying the distributive properties of the connectives, A can be rewritten as a conjunction $A_1 \ldots A_m$ of (Ω, ϵ_0)-inductive $\mathcal{L}_R$ formulae where:

(1) any negative (respectively, positive) occurrence of $\vee$ with an ϵ_0-critical variable in its scope is in a subformula which is the right-hand argument of a negative (respectively, positive) occurrence of $\rightarrow$,
(2) every positive (respectively, negative) occurrence of $\wedge$ with an ϵ_0-critical variable in its scope is in a subformula which is either the left-hand argument of a negative (respectively, positive) occurrence of $\rightarrow$, or an argument of a positive (negative) occurrence of $\circ$.

Equivalently, on every ϵ_0-critical branch in the positive generation tree of each A_i, there are no occurrences of $+\wedge$ or $-\vee$ on the segment of the branch consisting of approximation nodes and there are no occurrences of $-\wedge$ or $+\vee$ on the segment of the branch consisting of inner nodes. We will call formulae satisfying the conditions imposed on all components A_i **definite (Ω, ϵ_0)-inductive $\mathcal{L}_R$ formulae**.

We may thus assume, without loss of generality, that A is, in fact, a definite (Ω, ϵ_0)-inductive $\mathcal{L}_R$ formula. Under this assumption, we note that A can be written in the form $A = A'(H_1, \ldots, H_n, N_1, \ldots, N_k)$ where A' is of the form $A'(q_1, \ldots, q_n, r_1, \ldots, r_k)$ with $q_1, \ldots, q_n$ and $r_1, \ldots, r_k$ placeholder variables not occurring in A and:

1. the positive generation tree of A' consists entirely of approximation nodes (excluding $+\wedge$ and $-\vee$),
2. if A' is positive (respectively, negative) in its ith argument $(1 \leq i \leq n)$ then the positive (respectively, negative) generation tree of H_i contains exactly one ϵ_0-critical leaf, say p_i, with the path from the root to that leaf consisting of inner nodes excluding $-\wedge$ and $+\vee$, and with all other leaves either signed constants or signed variables that are strictly less than p_i according to Ω, and
3. if A' is positive (respectively, negative) in its ith argument $(n + 1 \leq i \leq n + k)$ then the positive (respectively, negative) generation tree of H_i contains no ϵ_0-critical leaves.

A straightforward induction on A' establishes that $A'(q_1, \ldots, q_n, r_1, \ldots, r_k)$ can be written as $[\alpha](q_1, \ldots, q_n, r_1, \ldots, r_k)$ for some pure B-term α, i.e., such that $\tau([\alpha](q_1, \ldots, q_n, r_1, \ldots, r_k)) = A'(q_1, \ldots, q_n, r_1, \ldots, r_k)$.

Next, we show that each H_i can be written as either a headed Box-formula $[\beta](p_i, C_1, \ldots, C_\ell)$, if A' is negative in the ith argument, or a headed Diamond-formula $\langle\delta\rangle(p_i, C_1, \ldots, C_\ell)$, if A' is positive in the ith argument. We proceed by induction to prove a slightly stronger claim, namely that each subformula B of H_i containing p_i can be written either as a headed Box-formula $[\beta](p_i, C_1, \ldots, C_\ell)$, if the occurrence of B is negative in A, or as a headed Diamond-formula $\langle\delta\rangle(p_i, C_1, \ldots, C_\ell)$, if the occurrence of B is positive in A. In the base case, B is just a variable p_i which can be written in the appropriate form either as $\langle\iota\rangle p_i$ or $[\iota]p_i$. Next consider the induction step when B is of the form $B' \to B''$ and the occurrence B is negative A. This splits into two cases, depending on whether the occurrence of p_i comes from B' or B''. If the occurrence of p_i comes from B' then, by induction hypothesis we can write B' as $\langle\delta'\rangle(p_i, C_1, \ldots, C_\ell)$ for some D-term δ' and hence B can be written as $[\to (\delta', \iota)](p_i, C_1, \ldots, C_\ell, B'')$. Otherwise, if the occurrence of p_i comes from B'', then, by induction hypothesis, we can write B'' as $[\beta'](p_i, C_1, \ldots, C_\ell)$ for some B-term β' and hence B can be written as $[\to (\iota, \beta')](B', p_i, C_1, \ldots, C_\ell)$. The other inductive cases, namely those for B of the forms $B' \vee B''$ or $\sim B'$ with the occurrence of B negative in A, and those for B of the forms $B' \circ B''$, $B' \wedge B''$ or $\sim B'$ with the occurrence of B positive in A', are similar.

Finally, putting these claims together and modulo permutation of arguments, we can write A as

$$[\alpha]([\beta_1](p_1, C_{1_1}, \ldots, C_{\ell_{11}}), \ldots, [\beta_m](p_m, C_{1_m}, \ldots, C_{\ell_{mm}}), \langle\beta_{m+1}\rangle(p_{m+1},$$

$$C_{1_{m+1}}, \ldots, C_{\ell_{m+1_{m+1}}}), \ldots, \langle\beta_n\rangle(p_n, C_{1_n}, \ldots, C_{\ell_{nn}}), N_1, \ldots, N_k)$$

which is a flat inductive formula in $\mathsf{For}(\mathcal{L}^{\mathsf{f}}_{\mathsf{R}})$. Note that the $p_1, \ldots, p_n$ need not be distinct. ∎

Corollary 22 *Every Sahlqvist formula in $\mathcal{L}_R$ can be rewritten as a semantically equivalent conjunction of flat inductive formulae in* $\mathsf{For}(\mathcal{L}^{\mathsf{f}}_{\mathsf{R}})$ *which are of the form* $[\alpha](p_1, \ldots, p_n, N_1, \ldots, N_k)$ *with* $N_1, \ldots, N_k$ *the headless main components.*

Proof It is sufficient to inspect the induction argument in the proof of Theorem 21 used to show that in $A = A'(H_1, \ldots, H_n, N_1, \ldots, N_k)$ the formula H_i can be written as a headed box formula $[\beta](p_i, C_1, \ldots, C_\ell)$ if A' is negative in the ith argument or as headed Diamond-formula $\langle\delta\rangle(p_i, C_1, \ldots, C_\ell)$ if A' is positive in the ith argument. Here we note that, since A is Sahlqvist, a branch in its positive generation tree leading to a critical occurrences of p_i does not contain any inner nodes (see Remark 4) and thus $\beta = \iota$. Hence $[\beta](p_i, C_1, \ldots, C_\ell)$ becomes simply $[\iota]p_i$ and $\langle\delta\rangle(p_i, C_1, \ldots, C_\ell)$ becomes simply $\langle\iota\rangle p_i$. Lastly, the ι's can be absorbed into α in the main box of the constructed inductive flat formula, yielding a formula of the desired form. ∎

5 The Algorithm PEARL for Flat Inductive Formulae, Correspondence and Canonicity

In this section we briefly review the first-order correspondence algorithm for relevance logic PEARL which was introduced and studied in [4] and [5]. We then show how it can be adapted to find first-order frame correspondents and simultaneously prove the canonicity of $\mathcal{L}_R^f$ formulae.

5.1 PEARL *for* $\mathcal{L}_R$

PEARL is an algorithm in the style of SQEMA [6] and ALBA [7, 8] and is based on a calculus of rewrite rules, which we will now recall. The algorithm accepts (inequalities of) $\mathcal{L}_R^+$ formulae as input and, if it succeeds, it produces a first-order formula in the language of RM-frames that is valid on an RM-frame if and only if the input inequalities are valid on this RM-frame.

5.1.1 PEARL Rules

The rules manipulate **quasi-inequalities**[3] of $\mathcal{L}_R^+$ formulae, i.e., expressions of the form $D_1 \le E_1, \ldots, D_n \le E_n \implies D \le E$, with $D, E, D_i, E_i \in \mathcal{L}_R^+$.

Any formula $D \in \mathcal{L}_R^+$ can be treated as the inequality $\mathbf{t} \le D$, which is a quasi-inequality with no assumptions. The inequalities not affected by the application of the rule are regarded as a **context**, which will be denoted by Γ. Given a set

[3] In [4] these are treated set-theoretically and are called there 'quasi-inclusions'.

of inequalities Γ, we say that Γ is positive (respectively, negative) in an atom a whenever each member of Γ is positive (respectively, negative) in a. We will write $\Gamma(A/p)$ for the set of inequalities obtained by uniformly substituting A for atom p in each member of Γ.

All rules below that are indicated by a double line are invertible.

Monotone Variable Elimination Rules

$$\frac{\Gamma(p) \implies C(p) \leq B(p)}{\Gamma(\bot/p) \implies C(\bot/p) \leq B(\bot/p)}\,(\bot) \qquad \frac{\Delta(p) \implies B(p) \leq C(p)}{\Delta(\top/p) \implies B(\top/p) \leq C(\top/p)}\,(\top)$$

where $B(p)$ and $\Delta(p)$ are positive in p, while $C(p)$ and $\Gamma(p)$ are negative in p.

First Approximation Rule

$$\frac{\Gamma \implies D \leq E}{\mathbf{j} \leq D,\ E \leq \mathbf{m},\ \Gamma \implies \mathbf{j} \leq \mathbf{m}}$$

where $\mathbf{j}$ is a nominal and $\mathbf{m}$ is a co-nominal both not occurring in the premise.

Approximation Rules

$$\frac{F \rightarrow D \leq \mathbf{m},\ \Gamma \implies A \leq B}{\mathbf{j} \leq F,\ \mathbf{j} \rightarrow D \leq \mathbf{m},\ \Gamma \implies A \leq B}\,(\rightarrow\text{Appr-L}) \qquad \frac{F \rightarrow D \leq \mathbf{m},\ \Gamma \implies A \leq B}{D \leq \mathbf{n},\ F \rightarrow \mathbf{n} \leq \mathbf{m},\ \Gamma \implies A \leq B}\,(\rightarrow\text{Appr-R})$$

$$\frac{\mathbf{i} \leq F \circ D,\ \Gamma \implies A \leq B}{\mathbf{j} \leq F,\ \mathbf{i} \leq \mathbf{j} \circ D,\ \Gamma \implies A \leq B}\,(\circ\text{Appr-L}) \qquad \frac{\mathbf{i} \leq F \circ D,\ \Gamma \implies A \leq B}{\mathbf{j} \leq D,\ \mathbf{i} \leq F \circ \mathbf{j},\ \Gamma \implies A \leq B}\,(\circ\text{Appr-R})$$

$$\frac{{\sim} D \leq \mathbf{m},\ \Gamma \implies A \leq B}{D \leq \mathbf{n},\ {\sim}\mathbf{n} \leq \mathbf{m},\ \Gamma \implies A \leq B}\,({\sim}\text{Appr-L}) \qquad \frac{\mathbf{i} \leq {\sim} D,\ \Gamma \implies A \leq B}{\mathbf{j} \leq D,\ \mathbf{i} \leq {\sim}\mathbf{j},\ \Gamma \implies A \leq B}\,({\sim}\text{Appr-R})$$

where $\mathbf{j}$ a nominal and $\mathbf{n}$ is a co-nominal both not appearing in the premises.

Residuation Rules

$$\frac{D \leq F \vee E,\ \Gamma \implies A \leq B}{D \mathbin{-\!\!<} F \leq E,\ \Gamma \implies A \leq B}\,(\vee\text{Res}) \qquad \frac{D \wedge F \leq E,\ \Gamma \implies A \leq B}{D \leq F \Rightarrow E,\ \Gamma \implies A \leq B}\,(\wedge\text{Res})$$

$$\frac{D \leq F \rightarrow E,\ \Gamma \implies A \leq B}{D \circ F \leq E,\ \Gamma \implies A \leq B}\,(\rightarrow\text{Res}) \qquad \frac{E \leq D \hookrightarrow F,\ \Gamma \implies A \leq B}{D \circ E \leq F,\ \Gamma \implies A \leq B}\,(\hookrightarrow\text{Res})$$

Adjunction Rules

$$\frac{D \vee F \leq E,\ \Gamma \ \Longrightarrow\ A \leq B}{D \leq E,\ F \leq E,\ \Gamma \ \Longrightarrow\ A \leq B}\,(\vee\mathrm{Adj}) \qquad \frac{E \leq D \wedge F,\ \Gamma \ \Longrightarrow\ A \leq B}{E \leq D,\ E \leq F,\ \Gamma \ \Longrightarrow\ A \leq B}\,(\wedge\mathrm{Adj})$$

$$\frac{{\sim}D \leq E,\ \Gamma \ \Longrightarrow\ A \leq B}{{\sim}^{\flat} E \leq D,\ \Gamma \ \Longrightarrow\ A \leq B}\,({\sim}\mathrm{Adj\text{-}L}) \qquad \frac{D \leq {\sim}E,\ \Gamma \ \Longrightarrow\ A \leq B}{E \leq {\sim}^{\sharp} D,\ \Gamma \ \Longrightarrow\ A \leq B}\,({\sim}\mathrm{Adj\text{-}R})$$

Not to clutter the procedure with extra rules, we allow commuting the arguments of $\wedge$ and $\vee$ whenever needed before applying the rules ($\wedge$Adj) and ($\vee$Adj) above. These rules are applied exhaustively in the downward direction, and produce the same results regardless of how an expression is parenthesized.

Ackermann-Rules

The Right Ackermann-rule (RAR) and Left Ackermann-rule (LAR) are stated as follows:

$$\frac{A \leq p,\ \Gamma(p) \ \Longrightarrow\ C(p) \leq B(p)}{\Gamma(A/p) \ \Longrightarrow\ C(A/p) \leq B(A/p)}\,(\mathrm{RAR}) \qquad \frac{p \leq A,\ \Delta(p) \ \Longrightarrow\ B(p) \leq C(p)}{\Delta(A/p) \ \Longrightarrow\ B(A/p) \leq C(A/p)}\,(\mathrm{LAR})$$

They are subject to the following conditions:

- p does not occur in A,
- B is positive in p,
- C is negative in p,
- Γ is negative in p,
- Δ is positive in p,

Note that the rules ($\bot$) and ($\top$) are, in fact, special cases of the Ackermann-rules (RAR) and (LAR), respectively.

Simplification Rules

$$\frac{\Gamma,\ \mathbf{i} \leq D \ \Longrightarrow\ \mathbf{i} \leq E}{\Gamma \ \Longrightarrow\ D \leq E}\,(\mathrm{Simpl\text{-}Left}) \qquad \frac{\Gamma,\ E \leq \mathbf{m} \ \Longrightarrow\ D \leq \mathbf{m}}{\Gamma \ \Longrightarrow\ D \leq E}\,(\mathrm{Simpl\text{-}Right})$$

In the rule (Simpl-Left) the nominal **i** must not occur in D, or E, or any inequality in Γ. Likewise, in the rule (Simpl-Right) the co-nominal **m** must not occur in D, or E, or any inequality in Γ. These rules are usually applied in the post-processing, to eliminate nominals and co-nominals introduced by the approximation rules.

5.1.2 Description of PEARL

PEARL is a non-deterministic algorithmic procedure, the purpose of which is to eliminate propositional variables from inclusions, while maintaining frame validity. It consists of 3 main phases:

I. Pre-processing The algorithm starts with an input formula $E \in \mathcal{L}_R$ and preprocesses it by applying distribution to surface conjunctions (and negative occurrences of disjunction). The algorithm proceed separately on each main conjunct produced in the process, which is justified as the frame correspondent of a conjunction is the conjunction of the correspondents of the conjuncts. If a main conjunct is positive (respectively, negative) in a given variable, that variable may be eliminated by uniform substitution with $\perp$ (respectively, $\top$) while maintaining equivalence on frames. Each resulting main conjunct A is then written equivalently as the quasi-inequality $\varnothing \implies \mathbf{t} \leq A$. To this we apply the **First approximation rule** (FA) to produce the quasi-inequality $\mathbf{j} \leq \mathbf{t},\ A \leq \mathbf{m} \implies \mathbf{j} \leq \mathbf{m}$. Now the inequality in the consequent contains no propositional variables and thus all steps after this point are aimed only at eliminating propositional variables from the two inequalities in the antecedent.

II. Main (Elimination) Phase In this phase the resulting system of quasi-inclusions is transformed by alternatingly applying the **approximation, residuation and adjunction rules** to bring the quasi-inequality into a shape to which an Ackermann rule is applicable and then applying the **Ackermann-rules** (RAR) and (LAR) to eliminate propositional variables. Eventually, the algorithm either succeeds to eliminate all variables or it reaches a stage where there are still variables but no further applications of the Ackermann-rules can be enabled. Then the algorithm fails.

III. Post-processing This phase applies if/when the algorithm succeeds to eliminate all variables, thus ending with **pure quasi-inequalities**, containing only nominals and co-nominals, but no variables. The purpose of the post-processing is to produce a first-order condition equivalent to the input formula. Each pure quasi-inequality produced in the elimination phase is post-processed separately to produce a corresponding FO condition, and all these are then taken conjunctively to produce the corresponding FO condition of the input formula. For further details and examples of the post-processing, see [4] and [5].

5.1.3 PEARL's Completeness for Inductive Formulas

Theorem 23 ([4]) PEARL *succeeds on all inductive formulas.*

Theorem 24 ([4]) *Every inductive $\mathcal{L}_R$–formula, has an effectively computable first-order frame correspondent on Routley-Meyer frames.*

5.1.4 Computing Correspondents of $\mathcal{L}_R^f$ Formulae with PEARL

One may apply PEARL to an $\mathcal{L}_R^f$ formula A by simply feeding in it the translation $\tau(A)$. For several examples of the execution of PEARL, we refer the reader to [4] where it was proved that PEARL successfully computes first-order correspondents on RM-frames for all inductive $\mathcal{L}_R$ formulae, and to [5], where a Python implementation of PEARL was presented and it was also shown that a successful PEARL run guarantees the canonicity of the input formula. We therefore have the following theorem.

Theorem 25 *All flat inductive $\mathcal{L}_R^f$ formulae have effectively computable first-order correspondents on RM-frames and are canonical.*

5.2 *Extending* PEARL *to* $\mathcal{L}_R^f$

While PEARL can be used to search for first-order frame correspondents for $\mathcal{L}_R^f$ formulae by first applying the translation τ to translate such formulae into $\mathcal{L}_R$ and then running PEARL, it can also be streamlined by adding rules to handle directly the modalities $\langle\alpha\rangle$ and $[\beta]$ for any pure D and B terms α and β, respectively, as follows.

Approximation Rules for Compound Modalities

$$\frac{[\beta](C_1,\ldots,C_i,\ldots C_m)\leq \mathbf{m},\ \Gamma\ \Longrightarrow\ A\leq B}{C_i\leq \mathbf{n},\ [\beta](C_1,\ldots,\mathbf{n},\ldots C_m)\leq \mathbf{m},\ \Gamma\ \Longrightarrow\ A\leq B}\ \text{(Box-Appr-Pos)}$$

$$\frac{[\beta](C_1,\ldots,C_j,\ldots C_m)\leq \mathbf{m},\ \Gamma\ \Longrightarrow\ A\leq B}{\mathbf{j}\leq C_j,\ [\beta](C_1,\ldots,\mathbf{j},\ldots C_m)\leq \mathbf{m},\ \Gamma\ \Longrightarrow\ A\leq B}\ \text{(Box-Appr-Neg)}$$

where $\epsilon^i(\beta)=+$ and $\epsilon^j(\beta)=-$ and **n** and **j** are a fresh co-nominal and nominal, respectively.

$$\frac{\mathbf{i}\leq\langle\delta\rangle(C_1,\ldots,C_i,\ldots C_m),\ \Gamma\ \Longrightarrow\ A\leq B}{\mathbf{j}\leq C_i,\mathbf{i}\leq\langle\delta\rangle(C_1,\ldots,\mathbf{j},\ldots C_m),\ \Gamma\ \Longrightarrow\ A\leq B}\ \text{(Diamond-Appr-Pos)}$$

$$\frac{\mathbf{i}\leq\langle\delta\rangle(C_1,\ldots,C_j,\ldots C_m),\ \Gamma\ \Longrightarrow\ A\leq B}{C_j\leq\mathbf{n},\mathbf{i}\leq\langle\delta\rangle(C_1,\ldots,\mathbf{n},\ldots C_m),\ \Gamma\ \Longrightarrow\ A\leq B}\ \text{(Diamond-Appr-Neg)}$$

where $\epsilon^i(\delta)=+$ and $\epsilon^j(\delta)=-$ and **n** and **j** are a fresh co-nominal and nominal, respectively.

Residuation Rules for Compound Modalities

$$\frac{D \leq [\beta](C_1, \ldots, C_i, \ldots, C_n),\ \Gamma \implies A \leq B}{\langle \beta^{-i} \rangle (C_1, \ldots, D, \ldots, C_n) \leq C_i\ \Gamma \implies A \leq B}\ ([\beta]\text{Pos-Res})$$

where $\epsilon^i(\beta) = +$.

$$\frac{D \leq [\beta](C_1, \ldots, C_j, \ldots, C_n),\ \Gamma \implies A \leq B}{C_j \leq [\beta^{-j}](C_1, \ldots, D, \ldots, C_n)\ \Gamma \implies A \leq B}\ ([\beta]\text{Neg-Res})$$

where $\epsilon^j(\beta) = -$.

$$\frac{\langle \delta \rangle (C_1, \ldots, C_i, \ldots, C_n) \leq D,\ \Gamma \implies A \leq B}{C_i \leq [\delta^{-i}](C_1, \ldots, D, \ldots, C_n),\ \Gamma \implies A \leq B}\ (\langle \delta \rangle \text{Pos-Res})$$

where $\epsilon^i(\delta) = +$.

$$\frac{\langle \delta \rangle (C_1, \ldots, C_j, \ldots, C_n) \leq D,\ \Gamma \implies A \leq B}{\langle \delta^{-j} \rangle (C_1, \ldots, D, \ldots, C_n) \leq C_j\ \Gamma \implies A \leq B}\ (\langle \delta \rangle \text{Neg-Res})$$

where $\epsilon^j(\delta) = -$.

The soundness of these rules follows from Lemma 11. We note that, modulo partial rewriting into $\mathcal{L}_R^f$, these rules subsume the approximation and adjunction rules given on Sect. 5.1.1. For example:

- the $\rightarrow$Appr-L rule is a special case of Box-Appr-Neg where $\beta =\rightarrow$ and $j = 2$;
- the $\circ$Appr-R rule is a special case of Diamond-Appr-Pos where $\delta = \circ$ and $i = 2$;
- the rule $\rightarrow$Res rule is a special case of $[\beta]$Pos-Res with $\beta =\rightarrow$ and $i = 2$;
- the rule $\hookrightarrow$Res is a special case of $[\beta]$Pos-Res with $\beta =\rightarrow^{-2}$ and $i = 2$.

We will refer to PEARL augmented with these approximation and residuation rules as PEARL^f.

Example 26 We run PEARL^f on the following flat inductive formula (1.) from Example 19:

$[\rightarrow (\wedge, \rightarrow)]([\rightarrow](p, q), [\rightarrow](q, r), p, r)$.

Applying the first approximation rule to $\mathbf{t} \leq [\rightarrow (\wedge, \rightarrow)]([\rightarrow](p, q), [\rightarrow](q, r), p, r)$ produces

$$\mathbf{i} \leq \mathbf{t},\ [\rightarrow (\wedge, \rightarrow)]([\rightarrow](p, q), [\rightarrow](q, r), p, r) \leq \mathbf{m} \implies \mathbf{i} \leq \mathbf{m}.$$

Now, we apply Box-Appr-Neg three times to get

$$\mathbf{i} \leq \mathbf{t},\ [\to(\wedge,\to)](\mathbf{j}_1, \mathbf{j}_2, \mathbf{j}_3, r) \leq \mathbf{m},$$
$$\mathbf{j}_1 \leq [\to](p, q),\ \mathbf{j}_2 \leq [\to](q, r),\ \mathbf{j}_3 \leq p \implies \mathbf{i} \leq \mathbf{m}.$$

The Ackermann rule RAR is now applicable to eliminate p:

$$\mathbf{i} \leq \mathbf{t},\ [\to(\wedge,\to)](\mathbf{j}_1, \mathbf{j}_2, \mathbf{j}_3, r) \leq \mathbf{m},\ \mathbf{j}_1 \leq [\to](\mathbf{j}_3, q),$$
$$\mathbf{j}_2 \leq [\to](q, r) \implies \mathbf{i} \leq \mathbf{m}.$$

Next, we apply $[\beta]$Pos-Res to solve for q in $\mathbf{j}_1 \leq [\to](\mathbf{j}_3, q)$,

$$\mathbf{i} \leq \mathbf{t},\ [\to(\wedge,\to)](\mathbf{j}_1, \mathbf{j}_2, \mathbf{j}_3, r) \leq \mathbf{m},$$
$$\langle\to^{-2}\rangle(\mathbf{j}_1, \mathbf{j}_3) \leq q,\ \mathbf{j}_2 \leq [\to](q, r) \implies \mathbf{i} \leq \mathbf{m},$$

and then RAR to eliminate q:

$$\mathbf{i} \leq \mathbf{t},\ [\to(\wedge,\to)](\mathbf{j}_1, \mathbf{j}_2, \mathbf{j}_3, r) \leq \mathbf{m},\ \mathbf{j}_2 \leq [\to](\langle\to^{-2}\rangle(\mathbf{j}_1, \mathbf{j}_3), r) \implies \mathbf{i} \leq \mathbf{m}.$$

Now, we apply $[\beta]$Pos-Res again, this time to solve for r in $\mathbf{j}_2 \leq [\to](\langle\to^{-2}\rangle(\mathbf{j}_1, \mathbf{j}_3), r)$, yielding

$$\mathbf{i} \leq \mathbf{t},\ [\to(\wedge,\to)](\mathbf{j}_1, \mathbf{j}_2, \mathbf{j}_3, r) \leq \mathbf{m},\ \langle\to^{-2}\rangle(\mathbf{j}_2, \langle\to^{-2}\rangle(\mathbf{j}_1, \mathbf{j}_3)) \leq r \implies \mathbf{i} \leq \mathbf{m},$$

from which we can eliminate r through one last application of RAR:

$$\mathbf{i} \leq \mathbf{t},\ [\to(\wedge,\to)](\mathbf{j}_1, \mathbf{j}_2, \mathbf{j}_3, \langle\to^{-2}\rangle(\mathbf{j}_2, \langle\to^{-2}\rangle(\mathbf{j}_1, \mathbf{j}_3))) \leq \mathbf{m} \implies \mathbf{i} \leq \mathbf{m}.$$

This simplifies to the inequality:

$$\mathbf{t} \leq [\to(\wedge,\to)](\mathbf{j}_1, \mathbf{j}_2, \mathbf{j}_3, \langle\to^{-2}\rangle(\mathbf{j}_2, \langle\to^{-2}\rangle(\mathbf{j}_1, \mathbf{j}_3))).$$

The FO translation of this inequality may be simplified to $\forall abc(Rabc \Longrightarrow \exists x(Rabx \wedge Raxc))$.

6 Concluding Remarks and Directions for Further Work

The present paper is a continuation of the project started with [4] and [5]. Here we have presented an approach for defining the class of inductive formulae of relevance logic, alternative to the one followed in [4] and in the style of flat polyadic modal languages developed in [11] and [12] where the logical connectives can be composed into new polyadic box- and diamond- connectives, thus eventually flattening the structure of the formulae in a way that enables both a more transparent definition of the inductive formulae and a streamlined version of the algorithm PEARL, applying approximation and residuation rules directly to the composed modalities. All major results, including first-order definability and canonicity transfer readily from [4] and [5] to the flat inductive formulae, though the latter can also be proved directly, following the model-theoretic technique applied in [12].

Further natural extensions of this project include:

- Adding modal operators and extending the methods and results obtained so far to modal relevance logics, in particular covering and extending the class of Sahlqvist formulae in the language of modal relevance logics defined in [14].
- Strengthening PEARL and PEARL^f to cover a few known cases of FO definable relevance formulae, such as the non-inductive axiom A_5 mentioned in Example 19.
- Developing algorithmic correspondence for other semantics for relevance logics, in particular for Urquhart's semilattice semantics.

Acknowledgments We thank Guillermo Badia for encouraging comments and the anonymous reviewer for pointing out a number or typos and other glitches in the submitted version.

References

1. Badia, G. (2018). On Sahlqvist formulas in relevant logic. *Journal of Philosophical Logic*, *47*(4), 673–691.
2. Blackburn, P., de Rijke, M., & Venema, Y. (2001). *Modal logic*. Cambridge University Press.
3. Conradie, W., Ghilardi, S., & Palmigiano, A. (2014). Unified correspondence. In A. Baltag & S. Smets (Eds.), *Johan van Benthem on logic and information dynamics*, volume 5 of *Outstanding contributions to logic* (pp. 933–975). Springer International Publishing.
4. Conradie, W., & Goranko, V. (2021). Algorithmic correspondence for relevance logics. I. The algorithm PEARL. In I. Düntsch & E. Mares (Eds.), *Alasdair Urquhart on nonclassical and algebraic logic and complexity of proofs* (pp. 163–209). Springer. Available from https://www.springer.com/gp/book/9783030714291
5. Conradie, W., Goranko, V., & Jipsen, P. (2021). Algorithmic correspondence for relevance logics, bunched implication logics, and relation algebras via an implementation of the algorithm PEARL. In U. Fahrenberg et al. (Eds.), *Proceedings of relational and algebraic methods in computer science (RAMICS 2021)*, volume 13027 of *Lecture notes in computer science* (pp. 126–143). Springer.
6. Conradie, W., Goranko, V., & Vakarelov, D. (2006). Algorithmic correspondence and completeness in modal logic. I. The core algorithm SQEMA. *Logical Methods in Computer Science*, *2*(1:5).
7. Conradie, W., & Palmigiano, A. (2012). Algorithmic correspondence and canonicity for distributive modal logic. *Annals of Pure and Applied Logic*, *163*(3), 338–376.
8. Conradie, W., & Palmigiano, A. (2019). Algorithmic correspondence and canonicity for non-distributive logics. *Annals of Pure and Applied Logic*, *170*(9), 923–974.

9. Dunn, J. M., & Restall, G. (2002). Relevance logic. In D. M. Gabbay & F. Guenthner (Eds.), *Handbook of philosophical logic* (2nd ed., Vol. 6, pp. 1–128). Springer - Science+Business Media, B.V.
10. Gehrke, M., Nagahashi, H., & Venema, Y. (2005). A Sahlqvist theorem for distributive modal logic. *Annals of Pure and Applied Logic*, *131*, 65–102.
11. Goranko, V., & Vakarelov, D. (2001). Sahlqvist formulae in hybrid polyadic modal languages. *Journal of Logic and Computation*, *11*(5), 737–754.
12. Goranko, V., & Vakarelov, D. (2006). Elementary canonical formulae: Extending Sahlqvist's theorem. *Annals of Pure and Applied Logic*, *141*(1–2), 180–217.
13. Routley, R., Meyer, R. K., Plumwood, V., & Brady, R. (1982). *Relevant logics and their rivals (Volume I)*. Ridgeview, CA.
14. Seki, T. (2003). A Sahlqvist theorem for relevant modal logics. *Studia Logica*, *73*(3), 383–411.

Quantified Modal Relevant Logics II: Welcome to the Neighbourhood

Nicholas Ferenz and Andrew Tedder

Abstract In this paper we define a neighbourhood semantics for quantified modal relevant logics. These models employ the Mares-Goldblatt [16] interpretation of the quantifiers, and are the natural combination of the semantics found in [27] and [9]. Modular soundness and completeness proofs are given in combination with an procedure for producing semantic conditions from axiom and rule schemes. We further show a model-wise equivalence between fully augmented neighbourhood models and relational models. Finally, we employ the constructed semantics to obtain an independence result surrounding the Barcan formulas, and their relation to various Augmentation principles.

1 Introduction

This is the third in a series of papers on neighbourhood semantics for extensions of relevant logics. In [27], a general neighbourhood semantics for first-order relevant logics was given, and completeness proofs were provided for a wide range of such logics; in [9], such a semantics was given for modal relevant logics, alongside completeness proofs and comparisons to *non-general* neighbourhood models of those logics. In this paper, we combine these projects and provide a general neighbourhood semantics for quantified modal relevant logics. The title reflects the

We'd like to thank audiences at the Non-Classical Modal and Predicate Logics conference in Bochum and at the New Directions in Relevant Logic online conference, in addition to an anonymous referee. Ferenz gratefully acknowledges support by RVO 67985807 and by the Czech Science Foundation project GA22-01137S. Tedder gratefully acknowledges the Humboldt foundation for fellowship funding.

N. Ferenz
Centre of Philosophy, University of Lisbon, Lisbon, Portugal

A. Tedder (✉)
Department of Philosophy I, Ruhr University Bochum, Bochum, Germany

I. Sedlár et al. (eds.), *New Directions in Relevant Logic*, Trends in Logic 63,
https://doi.org/10.1007/978-3-031-69940-5_7

fact that this paper is also a sequel to [6, 8], in which the *relational* semantics of modal relevant logics was extended by quantifiers. Throughout these works, the Mares-Goldblatt treatment of the quantifiers, developed in [11, 16], has been employed, as it is here.

First-order modal logic has been one of the great battlegrounds, and sources of progress, in philosophical logic in the twentieth century. Issues arising there, such as transworld identity, necessitism, and others, have influenced work in metaphysics, ethics, and many other areas of mainstream (i.e., non-logic) philosophy. Since relevant logics capture an essential feature of logical entailment, namely *relevance* between premises and conclusions, it's only natural that the question of how first-order modal relevant logics behave should arise, since the notion of entailment is in such widespread currency in these philosophical debates.

The relational semantic development of such logics in Ferenz's work is a natural starting point, as the relational semantics is the most well known, but it introduces some complications and covers only a portion of the ground. Indeed, when we venture deeper into the terrain of weaker relevant logics, those which are not complete with respect to classes of relational frames, we find a range of interesting systems with properties which are better suited to certain views of the central philosophical issues. In this connection, neighbourhood semantics for modal logics have been discussed, for instance in [17, 24], and it seems that similar considerations may well suggest that a more general treatment of the relevant conditional also holds some promise. One such direction, motivating a relevant conditional for which a neighbourhood semantic treatment is necessary, is [25], and other considerations have previously been discussed in work by Lavers [14, 15].

To that end, we'll set out some definitions and investigate some of the interesting properties of neighbourhood semantics for quantified (first-order) modal relevant logics (NQMRL).[1] The plan of the paper is as follows. After settling some preliminaries, we define the neighbourhood semantics and axiomatic proof systems for a wide class of quantified modal relevant logics in, respectively, Sects. 2 and 3. In the latter, modular soundness is proved for these logics, with an procedure for obtaining the correct semantic conditions for a given axiom/rule scheme. Section 4 complements the soundness with a modular completeness proof. In Sect. 5, we show an equivalence between fully-augmented neighbourhood models and relational models. Then, in Sect. 6, we employ the developed framework to study variants on the Barcan formulas, including their relation (or lack thereof) to augmentation properties of frames.

[1] Here we are focusing merely on *static* modalities, though there is also a growing literature on adding dynamic modalities to relevant and substructural logics, for instance in [21, 22, 26]. For simplicity's sake, we'll leave these to the side here, though there are certainly interesting questions in the neighbourhood of such systems.

1.1 Languages

The propositional language can be defined in BNF as follows (t is a propositional constant):

$$\mathbb{P}|\mathtt{t}|\mathcal{A}|\neg\mathcal{A}|\mathcal{A}\wedge\mathcal{A}|\mathcal{A}\vee\mathcal{A}|\mathcal{A}\to\mathcal{A}|\Box\mathcal{A}|\Diamond\mathcal{A}$$

and will occasionally be extended to include two family members of $\to$, namely $\circ$ and $\leftarrow$ (both of which are also arity 2), and family members, namely duals (with respect to $\neg$), of the modalities $\diamonddot$, $\boxdot$ (i.e., $\diamonddot\mathcal{A} := \neg\Box\neg\mathcal{A}$ and $\boxdot\mathcal{A} := \neg\Diamond\neg\mathcal{A}$). This is extended to a first-order language by the addition, on the logical side, of a denumerable set of variables $Var = \{x_i\}_{i\in\omega}$, and two quantifier symbols $\forall, \exists$. On the extra-logical side, a *signature* comprises a set Con of individual constant symbols (written $c_0, \ldots, c_n, \ldots$) and a set $Pred$ of predicate symbols ($Pred^n \subseteq Pred$ is the set of n-ary predicate symbols, and we'll sometimes write a predicate letter with a number superscript indicating its arity). We can then define $\mathbb{P} = Pred^0$, fix $Term = Var \cup Con$, and define the set of atomic formulas in terms of $Term$ and $Pred$, and the set of complex formulas, as usual. We'll not be interested here in *specific* languages constructed from specific signatures, so generally we'll just use $\mathcal{L}_\Phi$, for $\Phi \subseteq \{\circ, \leftarrow\}$ (taking $\mathcal{L} = \mathcal{L}_\varnothing$) as a generic name for languages.

We'll adopt the convention that $\to$ binds most weakly and $\neg$ most strongly of all the connectives. Otherwise, binding will be marked explicitly with parentheses. Furthermore, we define "$\mathcal{A}[\tau/x]$", where $\tau \in Term$, as usual, to mean that instance of $\mathcal{A}$ in which every free occurrence of x is replaced by τ.

2 NQMRL Frames and Models

Since our focus is on models, we'll start with the semantics, and introduce the logics axiomatically in the next section.

2.1 Basic Definitions

Definition 2.1.1 A frame is a tuple $F = \langle W, N, R, *, S_\Box, S_\Diamond, Prop, D, PropFun\rangle$ where: $\varnothing \neq N \subseteq W$, $* : W \longrightarrow W$, $R \subseteq W \times \mathscr{P}(W) \times \mathscr{P}(W)$, $S_\Box, S_\Diamond \subseteq W \times \mathscr{P}(W)$, $Prop \subseteq \mathscr{P}(W)$, $D \neq \varnothing$, and $PropFun \subseteq \{\varphi \mid \varphi : D^\omega \longrightarrow Prop\}$. Given F, we define the following operations on $X, Y \subseteq W$:

$$\neg X = \{\alpha \in W \mid \alpha^* \notin X\} \qquad \Box X = \{\alpha \in W \mid S_\Box\alpha X\}$$

$$X \to Y = \{\alpha \in W \mid R\alpha XY\} \qquad \Diamond X = \{\alpha \in W \mid S_\Diamond\alpha X\}$$

With these, F is required to satisfied the following constraints:

(c0) $N \in Prop$ and $Prop$ is closed under $\cap, \cup, \neg, \to, \Box, \Diamond$ (and $\leftarrow, \circ$, if present)

(c0.0) There is a $\varphi_N \in PropFun$ s.t. for any $f \in D^\omega$, $\varphi_N f = N$

(c0.1) $(\oplus\varphi)f = \oplus(\varphi f)$ for any $f \in D^\omega$, $\oplus \in \{\neg, \Box, \Diamond\}$

(c0.2) $(\varphi \otimes \psi)f = \varphi f \otimes \psi f$ for any $f \in D^\omega$, $\otimes \in \{\cap, \cup, \leftarrow, \circ, \to\}$

(c0.3) For any $\varphi \in PropFun, n \in \omega, f \in D^\omega$:

$$(\forall_n\varphi)f = \bigsqcap_{f' \sim_x f} \varphi f' = \bigcup\{X \in Prop \mid X \subseteq \bigcap_{f' \sim_x f} \varphi f'\}$$

(c0.4) For any $\varphi \in PropFun, n \in \omega, f \in D^\omega$:

$$(\exists_n\varphi)f = \bigsqcup_{f' \sim_x f} \varphi f = \bigcap\{X \in Prop \mid \bigcup_{f' \sim_x f} \varphi f' \subseteq X\}$$

(c1) $X \subseteq Y \iff N \subseteq X \to Y$ for any $X, Y \in Prop$

Definition 2.1.2 A model M on a frame F consists of a multi-type function M s.t. $M : Con \longrightarrow D$ and $M : Pred^n \times D^n \longrightarrow Prop$. Then, given $f : \omega \longrightarrow D$, we fix:

- $M_f(c) = M(c)$
- $M_f(x_n) = fn$

With these, we extend M to a valuation $[\![\cdot]\!]^M : \mathcal{L} \longrightarrow PropFun$ s.t., for any $f \in D^\omega$—we'll write $[\![\mathcal{A}]\!]^M_f$ instead $([\![\mathcal{A}]\!]^M)f$, where convenient:

- $[\![R^n(\tau_1, \ldots, \tau_n)]\!]^M_f = M(R^n)(M_f(\tau_1), \ldots, M_f(\tau_n))$
- $[\![\mathtt{t}]\!]^M_f = N$
- $[\![\oplus\mathcal{A}]\!]^M_f = \oplus[\![\mathcal{A}]\!]^M_f$ for $\oplus \in \{\neg, \Box, \Diamond\}$
- $[\![\mathcal{A} \wedge \mathcal{B}]\!]^M_f = [\![\mathcal{A}]\!]^M_f \cap [\![\mathcal{B}]\!]^M_f$
- $[\![\mathcal{A} \vee \mathcal{B}]\!]^M_f = [\![\mathcal{A}]\!]^M_f \cup [\![\mathcal{B}]\!]^M_f$
- $[\![\mathcal{A} \otimes \mathcal{B}]\!]^M_f = [\![\mathcal{A}]\!]^M_f \otimes [\![\mathcal{B}]\!]^M_f$ for $\otimes \in \{\leftarrow, \circ, \to\}$
- $[\![\forall x_n\mathcal{A}]\!]^M_f = (\forall_n[\![\mathcal{A}]\!]^M)f$
- $[\![\exists x_n\mathcal{A}]\!]^M_f = (\exists_n[\![\mathcal{A}]\!]^M)f$

A formula $\mathcal{A}$ is true in M, written $\vDash_M \mathcal{A}$ iff $N \subseteq [\![\mathcal{A}]\!]^M_f$ holds for every $f \in D^\omega$. $\mathcal{A}$ is true on a frame F, $\vDash_F \mathcal{A}$ iff $\vDash_M \mathcal{A}$ holds for every F on M. Finally, $\mathcal{A}$ is valid for a class $\mathbf{F}$ of frames just in case it is true on every $F \in \mathbf{F}$.

Definition 2.1.3 F is *right R-augmented* iff for any $\{X\} \cup \{Y_i\}_{i \in I} \subseteq Prop$:

$$\bigcap_{i \in I}(X \to Y_i) = X \to \bigcap_{i \in I} Y$$

F is *left R-augmented* iff for any $\{X\} \cup \{Y_i\}_{i \in I} \subseteq Prop$:

$$\bigcap_{i \in I}(X_i \to Y) = \bigcup_{i \in I} X_i \to Y$$

F is *R-augmented* iff it is both right and left R-augmented.
F is *$S_\Box$-augmented* iff for any $\{X_i\}_{i \in I} \subseteq Prop$:

$$\bigcap_{i \in I} \Box X_i = \Box \bigcap_{i \in I} X_i$$

F is *$S_\Diamond$-augmented* iff for any $\{X_i\}_{i \in I} \subseteq Prop$:

$$\bigcup_{i \in I} \Diamond X_i = \Diamond \bigcup_{i \in I} X_i$$

F is *fully* augmented iff it is R, $S_\Box$, and $S_\Diamond$-augmented.[2]

Definition 2.1.4 We characterise frames and models appropriate for the various language expansions as follows:

(1) A frame F is *apt for* $\circ$ just in case it is right R-augmented and for any $X, Y \in Prop$, $X \circ Y = \bigcap\{Z \in Prop \mid X \subseteq Y \to Z\} \in Prop$. In a model M on such an F, $[\![\mathcal{A} \circ \mathcal{B}]\!]^M_f = [\![\mathcal{A}]\!]^M_f \circ [\![\mathcal{B}]\!]^M_f$.
(2) A frame F is *apt for* $\leftarrow$ just in case it is left R-augmented, and for any $X, Y \in Prop$, $Y \leftarrow X = \bigcup\{Z \in Prop \mid X \subseteq Z \to Y\} \in Prop$. In a model M on such an F, $[\![\mathcal{B} \leftarrow \mathcal{A}]\!]^M_f = [\![\mathcal{B}]\!]^M_f \leftarrow [\![\mathcal{A}]\!]^M_f$.

We're particularly interested in $\leftarrow$, $\circ$, but we could also add residuals for $\Box$, $\Diamond$—say, respectively, $\blacklozenge$, $\blacksquare$, where:

$$X \subseteq \Box Y \iff \blacklozenge X \subseteq Y \qquad\qquad \Diamond X \subseteq Y \iff X \subseteq \blacksquare Y$$

[2] The reader may note that augmentation is here defined only relative to elements of $Prop$. A similar definition can be given for all subsets of $\mathscr{P}(W)$ (or just upwardly closed sets in general). Proving the results of [27], the independence results of Sect. 6, or the frame-equivalence results of Sect. 5 using these alternative definitions constitutes an interesting line or research into what results hold when we move from general to non-general frames, as discussed in [9].

We won't have much call to discuss these operations, so we'll ignore them from here on in. The only point to note is that, in general, none of the following triples need to collapse:

Modals	Negation-Duals	Residuals
$\Box A$	$\boxdot A$	$\blacksquare A$
$\Diamond A$	$\diamond\!\!\!\cdot\, A$	$\blacklozenge A$

absent further frame constraints.

3 Quantified Modal Relevant Logics: Axiomatics

3.1 Propositional Relevant Logics

The propositional logic of the non-modal language we'll call **F**, and it can be axiomatised as follows (throughout we take logics to be sets of formulas, extending the set of theorems of **F**, closed under the rule of *substitution*, allowing one to perform uniform substitutions of some piece of non-logical vocabulary by others of the same time)—below $\Rrightarrow$ is a separator of a rule of *proof* in the sense of [12, 23], so $\mathcal{A}_1, \ldots, \mathcal{A}_n \Rrightarrow \mathcal{B}$ should be read to mean that if each $\mathcal{A}_i$ is derivable, then so is $\mathcal{B}$:

(Id) $\mathcal{A} \to \mathcal{A}$
($\wedge$E) $\mathcal{A} \wedge \mathcal{B} \to \mathcal{A}, \mathcal{A} \wedge \mathcal{B} \to \mathcal{B}$
($\vee$I) $\mathcal{A} \to \mathcal{A} \vee \mathcal{B}, \mathcal{B} \to \mathcal{A} \vee \mathcal{B}$
(Dist) $\mathcal{A} \wedge (\mathcal{B} \vee \mathcal{C}) \to (\mathcal{A} \wedge \mathcal{B}) \vee \mathcal{C}$
(rMP) $\mathcal{A} \to \mathcal{B}, \mathcal{A} \Rrightarrow \mathcal{B}$
(rAdj) $\mathcal{A}, \mathcal{B} \Rrightarrow \mathcal{A} \wedge \mathcal{B}$
(rTrans) $\mathcal{A} \to \mathcal{B}, \mathcal{B} \to \mathcal{C} \Rrightarrow \mathcal{A} \to \mathcal{C}$
(rCong) $\mathcal{A} \leftrightarrow \mathcal{A}', \mathcal{B} \leftrightarrow \mathcal{B}' \Rrightarrow (\mathcal{A} \to \mathcal{B}) \leftrightarrow (\mathcal{A}' \to \mathcal{B}')$
(r$\wedge$I) $\mathcal{A} \to \mathcal{B}, \mathcal{A} \to \mathcal{C} \Rrightarrow \mathcal{A} \to \mathcal{B} \wedge \mathcal{C}$
(r$\vee$E) $\mathcal{A} \to \mathcal{C}, \mathcal{B} \to \mathcal{C} \Rrightarrow \mathcal{A} \vee \mathcal{B} \to \mathcal{C}$
(rCont) $\mathcal{A} \to \mathcal{B} \Rrightarrow \neg\mathcal{B} \to \neg\mathcal{A}$
(rt) $\mathcal{A} \Lleftarrow\!\Rrightarrow \mathtt{t} \to \mathcal{A}$

An axiomatic derivation, as usually defined, is a sequence of formulas each of which is either an instance of an axiom or follows from other formulas in the sequence by one of the rules. Logics will be identified with their set of theorems (formulas provable from the axioms alone), and we will write $\vdash_{\mathbf{L}} \mathcal{A}$ for $\mathcal{A}$ is a theorem of **L**, dropping the subscript when doing so introduces no ambiguity.[3]

To this a wide range of further axioms and rules can be added—lists of such can be found in [18, 20, 27] and elsewhere. Let's just consider a couple of milestones, taking us up to the relational semantics. First, let's consider the 'arrow' form of **FDE**, we call **FD(N)E**, which can be obtained by adding a double negation equivalence axiom—fixing $\mathcal{A} \leftrightarrow \mathcal{B} := (\mathcal{A} \to \mathcal{B}) \wedge (\mathcal{B} \to \mathcal{A})$:

(DNE) $\mathcal{A} \leftrightarrow \neg\neg\mathcal{A}$

To obtain the logic **BB**, investigated by Lavers [14], we need add to **FD(N)E** the following two rules:

(rB) $\mathcal{A} \to \mathcal{B} \Rrightarrow (\mathcal{C} \to \mathcal{A}) \to (\mathcal{C} \to \mathcal{B})$
(rB′) $\mathcal{A} \to \mathcal{B} \Rrightarrow (\mathcal{B} \to \mathcal{C}) \to (\mathcal{A} \to \mathcal{C})$

and from here to **B**, the basic logic of the ternary relation semantics introduced by Sylvan (née Routley) and Meyer, we simply upgrade (r∧I) and (r∨E) to axioms, which requires $\to$ to have a *distribution type* in the sense of [3]:[4]

(∧I) $(\mathcal{A} \to \mathcal{B}) \wedge (\mathcal{A} \to \mathcal{C}) \to (\mathcal{A} \to \mathcal{B} \wedge \mathcal{C})$
(∨E) $(\mathcal{A} \to \mathcal{C}) \wedge (\mathcal{B} \to \mathcal{C}) \to (\mathcal{A} \vee \mathcal{B} \to \mathcal{C})$

Note, however, that having a distribution type in this sense does not ensure that the operation is *R-augmented*, in the sense defined earlier.[5]

It is well known how to get from here to stronger systems like **DW**, **T**, **E**, and **R** (for example, see [27]), so we'll not rehearse those details again.

3.2 Adding Modalities

As with the non-modal part, there is a well-known jungle of modal axioms one can add to classical or intuitionistic logic, and the only major difference here is replacing the underlying propositional vocabulary with their relevant versions, though minor additional differences accrue depending on your goals.[6] To that end, we'll just list a handful of particularly important modal axioms/rules, leaving the interested

[3] To use Humberstone's terminology [12], we present logics as FMLA systems. It is possible to present these as SET-FMLA or SEQ-FMLA, or some other setting, but doing so introduces further complexities and is of uncertain philosophical value.

[4] See [9] for details.

[5] This is demonstrated, in effect, in §5 of [27].

[6] The treatment of modalities in relevant logics, and many other non-classical logics, has been heavily influenced by [5], as is our work here.

reader to consult their favourite treatment of modal logic to get inspiration for other principles they might add (one place to check is [9]):

(rRE) $\mathcal{A} \leftrightarrow \mathcal{B} \Rrightarrow \Box\mathcal{A} \leftrightarrow \Box\mathcal{B}$
(rRE$_\Diamond$) $\mathcal{A} \leftrightarrow \mathcal{B} \Rrightarrow \Diamond\mathcal{A} \leftrightarrow \Diamond\mathcal{B}$
(rRM) $\mathcal{A} \rightarrow \mathcal{B} \Rrightarrow \Box\mathcal{A} \rightarrow \Box\mathcal{B}$
(rRM$_\Diamond$) $\mathcal{A} \rightarrow \mathcal{B} \Rrightarrow \Diamond\mathcal{A} \rightarrow \Diamond\mathcal{B}$
(rNec) $\mathcal{A} \Rrightarrow \Box\mathcal{A}$

(C) $\Box\mathcal{A} \wedge \Box\mathcal{B} \rightarrow \Box(\mathcal{A} \wedge \mathcal{B})$
(C$_\Diamond$) $\Diamond(\mathcal{A} \vee \mathcal{B}) \rightarrow \Diamond\mathcal{A} \vee \Diamond\mathcal{B}$
(BD) $\Box(\mathcal{A} \vee \mathcal{B}) \rightarrow \Box\mathcal{A} \vee \Diamond\mathcal{B}$
(DB) $\Diamond\mathcal{A} \wedge \Box\mathcal{B} \rightarrow \Diamond(\mathcal{A} \wedge \mathcal{B})$
(K) $\Box(\mathcal{A} \rightarrow \mathcal{B}) \rightarrow (\Box\mathcal{A} \rightarrow \Box\mathcal{B})$

The basic propositional logic extending **F** by modalities $\Box$, $\Diamond$, what we'll call **F**.**E**$_{\Box\Diamond}$, is obtained by adding (rRE) and (rRE$_\Diamond$), and relevant versions of well-known classical and intuitionistic modal logics can be obtained by adding further axioms and rules. Note that, as above, the addition of (rRM), (rRM$_\Diamond$), (C), and (C$_\Diamond$) require $\Box$, $\Diamond$ to have a distribution type though, as again, not necessarily to be $S_\Box$- or $S_\Diamond$-augmented in the sense defined above.[7]

Let us rehearse how to obtain general neighbourhood frame conditions from arbitrary axioms and rules. Essentially, because we represented propositions directly in the frame we can just use the following simplistic translation scheme to find canonical frame constraints—a translation of this form was used by Lavers [14] and in [9], and in Sect. 3.4.1 we'll adapt it to the quantified language:

3.2.1 Algorithm for Getting Frame Constraints from Axioms/Rules

(1) First replace each metavariable $\mathcal{A}$ with a variable X over $Prop$ (e.g., in $\mathcal{A} \leftrightarrow \neg\neg\mathcal{A}$, obtain $X \leftrightarrow \neg\neg X$). For the next step, do you have an axiom or a rule?

  (a) If an axiom, is the main connective an arrow or not?

    (i) If so, replace the main connective with $\subseteq$ (i.e., in $X \rightarrow Y$, obtain $X \subseteq Y$)
    (ii) If not, simply prefix the whole thing with "$N \subseteq$" (i.e., for X, obtain $N \subseteq X$)

  (b) Then, replace each of the remaining connectives with the associated frame operations (the only place where this isn't a homophone 'translation' is for $\wedge$, $\vee$, which are then replaced by $\cap$, $\cup$, where the 'identity' between the symbols expressing the salient operations requires squinting) and replace every occurrence of $\mathtt{t}$ with N.

(3) If a rule, replace $\Rrightarrow$ by $\Rightarrow$, i.e., the metalanguage implication symbol. Then proceed as if each remaining formula were an axiom, as above, applying steps (a) and (b) as needed.

[7] In a later section, Sect. 5, we'll demonstrate that this is the case, and discuss some upshots.

(4) When none of the above rules are applicable, stop. For an axiom, the resulting expression will either be of the form $N \subseteq X$ where X is some complex element in $Prop$, with basic parts $\{X_i\}_{i \leq n} \subseteq Prop$, or will be of the form $X \subseteq Y$ with basic parts $\{X_i\}_{i \leq n}, \{Y_j\}_{j \leq m} \subseteq Prop$. In the former case, prefix the expression by "$\underset{i \leq n}{\forall} X_i \in Prop($" and write ")" at the end of the expression; in the latter, do so with "$\underset{i \leq n}{\forall} X_i \in Prop \underset{j \leq m}{\forall} Y_j \in Prop($", and write ")" at the end of the expression. For a rule, while each formula in the rule will have the forms indicated above, the prepended quantifier expressions have full scope over $\Rrightarrow$.

When applied to a formula $\mathcal{A}$, call the resulting translation $\mathcal{A}^t$. When applied to a rule (rR) $\mathcal{A}_1, \ldots, \mathcal{A}_n \Rrightarrow \mathcal{B}$, call the resulting constraint (rR^t). We'll call either of these a "t-transform" of the associated axiom/rule. When convenient, we will also denote t-transform of a rule or axiom (cφ), where (φ) denotes the rule or axiom in question.

As an example, this procedure will take the formula $\mathcal{A} \to ((\mathcal{A} \to \mathcal{B}) \to \mathcal{B})$ and generate the frame constraint $\forall X, Y \in Prop(X \subseteq (X \to Y) \to Y)$ and will take the rule $\mathcal{A} \Rrightarrow (\mathcal{A} \to \mathcal{B}) \to \mathcal{B}$ and generate the frame constraint $\forall X, Y \in Prop(N \subseteq X \Rrightarrow X \to Y \subseteq Y)$. Doing this for any additional axiom/rule one likes will produce a canonical frame constraint—we'll verify this as part of the propositional soundness. We will often suppress the prefixed universal quantifiers when displaying a t-transformed axiom or rule. Thus, for the axiom earlier in this paragraph, we would write $X \subseteq (X \to Y) \to Y$.

One easy shortcut: if you want to consider an axiom or rule using $\leftrightarrow$ as the main connective, you can simplify the process by, from (a.i), replacing $\leftrightarrow$ with $=$.

Now we'll prove the soundness and completeness results, which essentially combine the moves employed in [9, 27].

3.3 Propositional Soundness

In the first place, it can be shown that $\mathbf{F.E}_{\Box\Diamond}$ is sound w.r.t. the frames introduced above (ignoring the quantifier machinery). Indeed, the proof can be found in [9], so we just state it here:

Lemma 3.3.1 *If* $\vdash_{\mathbf{F.E}_{\Box\Diamond}} \mathcal{A}$ *then* $\vDash_F \mathcal{A}$ *holds for any frame* F.

Furthermore, in [9, 19, 27] the lever was cranked on certain axioms/rules to obtain further soundness results, but here let's give a more general result, relying on the translation algorithm specified earlier—this is straightforwardly extended to incorporate the modalities.

Lemma 3.3.2 *In any frame satisfying* $\mathcal{A}^t$, *any formula of the form* $\mathcal{A}$ *is satisfied. Furthermore, in any frame satisfying* (*rR*^{t}), *any instance of the rule* (*rR*) *is obeyed.*

Proof The proof proceeds by a straightforward induction on the complexity of the formula(s) involved. The key points to note are just that, first, an implication formula $\mathcal{A} \to \mathcal{B}$ is satisfied in a model M on a frame F just in case $[\![\mathcal{A}]\!]^M \subseteq [\![\mathcal{B}]\!]^M$, justifying step (2.a.i). Otherwise, the result just relies on the rewriting of operations in accordance with (ii) and, in the case for rules, condition (3).

As an example, consider (DB); so suppose that in each frame, for all $X, Y \in Prop$, $\Diamond X \cap \Box Y \subseteq \Diamond(X \cap Y)$. Clearly, then, for any formulas $\mathcal{A}, \mathcal{B}$, $[\![\Diamond\mathcal{A} \wedge \Box\mathcal{B}]\!]^M = \Diamond[\![\mathcal{A}]\!]^M \cap \Box[\![\mathcal{B}]\!]^M \subseteq \Diamond([\![\mathcal{A}]\!]^M \cap [\![\mathcal{B}]\!]^M) = [\![\Diamond(\mathcal{A} \wedge \mathcal{B})]\!]^M$, and the result follows from this by (c1). □

With these, we immediately obtain the following general propositional soundness fact:

Theorem 3.3.3 *Any extension of* $\mathbf{F.E}_{\Box\Diamond}$ *by propositional (modal) axioms and rules is sound w.r.t. the associated class of frames—i.e., that obeying the t-transform of all its axioms and rules.*

This result is, perhaps, not terribly surprising, as the translation essentially just *brute forces* the axiom/rule in question. We mainly note the result here for generality, and as a set up for the proof, later, that every t-transform is canonical w.r.t. its associated axiom/rule.

3.4 Adding Quantifiers

Given any propositional (modal) logic **L**, we can extend it in a few different ways, that we'll consider. First, the most basic extension, which we'll call $\mathbf{QL}^-$, adds the following axioms and rules—below, we'll write a formula with a superscript variable, like $\mathcal{A}^x$, to indicate that the displayed variable is not free in the formula—for (∀E) and (∃I), we require that τ be free for x in $\mathcal{A}$:

(∀E) $\forall x\mathcal{A} \to \mathcal{A}[\tau/x]$ (r∀I) $\mathcal{A}^x \to \mathcal{B} \Rrightarrow \mathcal{A}^x \to \forall x\mathcal{B}$
(∃I) $\mathcal{A}[\tau/x] \to \exists x\mathcal{A}$ (r∃E) $\mathcal{A} \to \mathcal{B}^x \Rrightarrow \exists x\mathcal{A} \to \mathcal{B}^x$

The logic **QL** extends $\mathbf{QL}^-$ by strengthening these two rules to axioms:

(∀I) $\forall x(\mathcal{A}^x \to \mathcal{B}) \to (\mathcal{A}^x \to \forall x\mathcal{B})$
(∃E) $\forall x(\mathcal{A} \to \mathcal{B}^x) \to (\exists x\mathcal{A} \to \mathcal{B}^x)$

Finally, we obtain the system $\mathbf{LQ}/\mathbf{LQ}^-$ by adding to $\mathbf{QL}/\mathbf{QL}^-$ the following axioms:[8]

(EC) $\forall x(\mathcal{A} \vee \mathcal{B}^x) \to \forall x\mathcal{A} \vee \mathcal{B}^x$

[8] Over many logics **L**, (EC) and (dEC) are not equivalent. By considering these axiom schemes separately, we could define two logics between $\mathbf{LQ}/\mathbf{LQ}^-$ and $\mathbf{QL}/\mathbf{QL}^-$. Rather, we leave this task for another time.

(dEC) $\exists x\mathcal{A} \wedge \mathcal{B}^x \to \exists x(\mathcal{A} \wedge \mathcal{B}^x)$

Note that any system extending **QFDE**$^-$ will prove the following:

(dual 1) $\forall x\mathcal{A} \leftrightarrow \neg\exists x\neg\mathcal{A}$
(dual 2) $\exists x\mathcal{A} \leftrightarrow \neg\forall x\neg\mathcal{A}$

In light of this last fact, it is worth noting that without the distribution of negation over quantifiers (given by Dual 1, 2, contraposition, and DNE) the formulas (EC) and (dEC) are not provable from each other in a logic contained in **QFDE**$^-$. Similarly, one may note that, on the semantic side of the fence, the operators $\bigsqcap$ and $\bigsqcup$ are only dual (with respect to negation) in models validating (DNE) (and (rCont), which is assumed here). In [8, Lemma 4.8], it is shown that the duality of $\bigsqcap$ and $\bigsqcup$ with respect to negation requires the usual set theoretic properties of $\bigcap$ and $\bigcup$, that $*$ that is an involution (here $X = X^{**}$, given by (DNE)), and that $*$ is order-inverting (here, by (rCont), as $X^* \subseteq Y$ iff $Y^* \subseteq X$).

A particularly interesting logic, due to its being the a sort of minimal logic for ternary relational semantics, is **BM.C**, which is defined as the logic **F**, together with ($\wedge$I), ($\vee$E), (rB), (rB$'$), (C), (C$_\Diamond$) and the rules (rRM) and (rRM$_\Diamond$). The logic **QMB.C** is of course a quantified extension of **BM.C**, as defined above.

One further fact that comes in handy later is that the addition of $\mathtt{t}$ ensures that the rule form of *universal generalisation* is derivable from (rt) and (r$\forall$I), given that no variable occurs free in $\mathtt{t}$:

(rUG) $\mathcal{A} \Rrightarrow \forall x\mathcal{A}$

Moreover, a universal generalization on constants is also a derivable rule.

Lemma 3.4.1 *The rule (rUG^{con}), that is:*

(rUG)^{con} $\mathcal{A}[c/x] \Rrightarrow \forall x\mathcal{A}$*, where c does not occur in* $\mathcal{A}$*,*

is derivable in any logic that includes ***QF.E***$^-_{\Box\Diamond}$.

This may be proved using the fact that (rUG) is a derivable rule, using the arguments in Mares and Goldblatt [16]. That is, we may rewrite the proof, replacing every instance of the constant c with x.

For the interaction between quantifiers and modalities, we'll be interested in a few particular (well-known) formulas, the *Barcan* and *Buridan* formulas (and their converses):[9]

(BF) $\forall x\Box\mathcal{A} \to \Box\forall x\mathcal{A}$	(BF$_\Diamond$) $\Diamond\exists x\mathcal{A} \to \exists x\Diamond\mathcal{A}$
(CBF) $\Box\forall x\mathcal{A} \to \forall x\Box\mathcal{A}$	(CBF$_\Diamond$) $\exists x\Diamond\mathcal{A} \to \Diamond\exists x\mathcal{A}$
(BuF) $\exists x\Box\mathcal{A} \to \Box\exists x\mathcal{A}$	(BuF$_\Diamond$) $\Diamond\forall x\mathcal{A} \to \forall x\Diamond\mathcal{A}$
(CBuF) $\Box\exists x\mathcal{A} \to \exists x\Box\mathcal{A}$	(CBuF$_\Diamond$) $\forall x\Diamond\mathcal{A} \to \Diamond\forall x\mathcal{A}$

[9] The latter are so-called by Garson [10], though this name is less well-known than that for the former.

A related, interesting, formula is:

$$\forall x \Box(\mathcal{A} \to \mathcal{B}) \to \Box(\forall x \mathcal{A} \to \forall x \mathcal{B})$$

as this motivated Barcan Marcus [2] to introduce the formulas bearing her name: in the context of strict implication logic, this amounts to $\forall x(\mathcal{A} \multimap \mathcal{B}) \multimap (\forall x \mathcal{A} \multimap \forall x \mathcal{B})$. Of course, related reasons suggest that we consider similar formulas, modalised versions of ($\forall$I) and ($\exists$E):

$$\forall x \Box(\mathcal{A}^x \to \mathcal{B}) \to \Box(\mathcal{A}^x \to \forall x B)$$

$$\forall x \Box(\mathcal{A} \to \mathcal{B}^x) \to \Box(\exists x \mathcal{A} \to \mathcal{B}^x)$$

For reasons of space, we'll not go into these and their behaviours relative to the usual Barcan formulas, but leave the question for future work.

3.4.1 Extending the Algorithm to Translate Quantifiers

In the case of quantifiers, we need to be slightly more careful than we do in the propositional case, because now we need to concern ourselves not just with propositions but with restrictions concerning variable occurrences. The translation manual for the quantifiers themselves is obvious: we take $\forall x_n$ to $\forall_n$ and $\exists x_n$ to $\exists_n$.

The side conditions, like those in ($\forall$I), place restrictions on where in the formulas concerned the instantial variable must not occur free. These have been, and will continue to be, stated by writing the variable in question as a superscript on the subformula in which it may no occur free: for instance, we write "$\mathcal{A}^{x_n} \to \mathcal{B}$" rather than "$\mathcal{A} \to \mathcal{B}$, where x_n may not occur free in $\mathcal{A}$". In generating frame constraints, we can state these side conditions explicitly by clauses like: "for all $f, f' \in D^\omega$, if $f \sim_x f'$ then $\varphi f = \varphi f'$", for instance obtaining the condition for ($\forall$I) given in [27]:

$$\text{If, for all } f' \sim_{x_n} f, \varphi f = \varphi f', \text{ then } (\forall_n(\varphi \to \psi))f \subseteq \varphi f \to (\forall_n \psi)f.$$

The trick, then, just to generalise this to accommodate conditions that impose restrictions on the freedom of variables in subformulas. In particular, we can adapt the algorithm in Sect. 3.2.1 with a few simple steps, as below (we also simplify the presentation a bit). One thing to note first is that we might have chosen to write an axiom, or a formula occurring in a rule, with reuse of item variables x_n. So when producing the transform of a formula, first we must comb through the formula and ensure that distinct variables are truly distinct. This is to ensure we don't unnecessarily identify variables that occur as distinct in some instances of an axiom or rule; it's clear that such 'combing' is possible, so we don't go into details.

Given a formula $\mathcal{A}$, we'll provide a translation scheme producing expressions of the form $\mathcal{A}^t$:="if Λ then $\Phi f \subseteq \Psi f$" where Λ is a (meta-language) conjunction of expressions of the form "if $f' \sim_{x_n} f$ then $\varphi f' = \varphi f$" where $f, f' \in D^\omega$ and $\varphi \in PropFun$, and Φ, Ψ are both in $PropFun$. We'll construct this kind of expression as follows:

(1) Is $\to$ the main connective of $\mathcal{A}$? If yes, then replace this occurrence with "$\subseteq$". If no, prefix the formula with "$N \subseteq$."
(2) Enumerate the most basic subformulas of $\mathcal{A}$, as they occur from left to right, $\mathcal{A}_1, \ldots, \mathcal{A}_m$. To each $\mathcal{A}_i$ assign $\varphi^{\mathcal{A}_i} \in PropFun$. If $\mathcal{A}_i$ is marked $\mathcal{A}_i^{x_n}$ for $n \in \omega$, then add to Λ the expression "if $f' \sim_{x_n} f$ then $\varphi^{\mathcal{A}_i} f' = \varphi^{\mathcal{A}_i} f$."
(3) Replace each connective with the associated frame operation: i.e., rewrite $\wedge \mapsto \cap$, $\neg \mapsto \neg$, and so on. Next, where a quantified variable "$\forall x_n$" or "$\exists x_n$" occurs, rewrite this with "$\forall_n$" or "$\exists_n$", respectively.
(4) At this point, we have an expression of the form $\Phi \subseteq \Psi$; replace this with the expression "$\Phi f \subseteq \Psi f$."
(5) Prefix "$\Lambda \Rightarrow \Phi f \subseteq \Psi f$" with "for all $f \in D^\omega$", and "for all $\varphi^{\mathcal{A}'} \in PropFun$" for every basic subformula $\mathcal{A}'$ of $\mathcal{A}$. Stop.

So this has taken a formula $\mathcal{A}$ (with atomic subformulas $\mathcal{A}_1, \ldots, \mathcal{A}_n$ and generated an expression of the form "for all $\varphi^{\mathcal{A}_1}, \ldots, \varphi^{\mathcal{A}_n} \in PropFun$, if Λ then $\Phi f \subseteq \Psi f$" (in the future, we'll leave the "for all $\varphi^{\mathcal{A}_1}, \ldots, \varphi^{\mathcal{A}_n} \in PropFun$" prefix tacit). In the case of a rule (rR) $\mathcal{A}_1, \ldots, \mathcal{A}_n \Rrightarrow \mathcal{B}$, *where no constant occurs in its formulation*, we take the metalanguage implication "if Λ, then $\mathcal{A}_1^t$ & ... & $\mathcal{A}_n^t \Rightarrow \mathcal{B}^t$", where Λ is the collection of Λ_i from each $\mathcal{A}_i^t$ and $\mathcal{B}^t$ and the individual Λ_i are removed from each $\mathcal{A}_i^t$ and $\mathcal{B}^t$, which we'll call (rR^t).

Note that we implicitly prefix each transformation with "For each $n_i \in \omega$". When dealing with notational variants (of every instance) of an axiom or rule scheme, we will deal with each instance individually. In effect, we treat t-transform of an instance to be given by the algorithm, and the t-transform of a scheme to be the set of t-transforms of every instance obtained from the scheme by the renaming of bound variables—that is, the universal closure under each $n_i \in \omega$, as left implicit.

Let's briefly consider an informative example. Take the formula $\forall x_n \exists x_m P$ as an axiom scheme. We produce the t-transform of this instances as "For all $f \in D^\omega$ and for all $\varphi^P \in PropFun$, $N \subseteq (\forall_n \exists_m \varphi^P) f$". (Note the resemblance to formulas such as "for every natural number, there is a greater natural number." This is misleading. As an axiom scheme, the particular predicate letter of the atomic P is not important, and each substitution instances of it is acceptable, no matter which admissible propositional function is used.) The key features to note here are that (1) all instances of the axiom scheme will be made true by a model, even when n and m are not distinct in that model, and that (2) the canonical model construction below will not identify n and m, and therefore will give us the t-transform above.

Note that the algorithm above and the facts proved using it rely on the fact that both rules schemes and axiom schemes do not have occurrences of constants. For rules, this is because many rules with constants do not preserve truth with respect to single models, but rather for an entire frame. Indeed, the admissible rule (rUG^{con}) above only preserves truth with respect to frames (cf. Goldblatt's [11, Theorem

1.7.7]). For axioms, consider the axiom $P(c) \to (P(c) \to P(c))$. For such as axiom we want only models that assign the right type of object to c, but this is a model condition and not a frame condition of the type constructed from the algorithm above. The key point, however is the following: by preventing constants from appearing in the statement of axiom and rule schemes we preserve the reasoning in the proof of the admissibility of (rUG^{con}), which is essential for the completeness proof.[10]

The definition of the algorithm above is quite simple, and there are reasons one might want to add complexity. For example, one might want to add restrictions on the set of predicates for which an axiom scheme is formulated, which we do not consider. Indeed, these kinds of axioms strictly preclude a full, non-sorted substitution of predicate letters in theorems. However, these kinds of axioms have precedent in relevant logics, as in the relevant indiscernibility (RI) approach to identity [7, 13]. There, the exclusion of substitution on predicate letters is a feature, not a bug, for identity is related only to a special subset of predicates. On the (RI) approach, indiscernibility is restricted to a subset of predicates called the relevant predicates: $x = y \to (Gx \to Gy)$, where G is a relevant predicate.[11] We leave the treatment of axioms and rules with restrictions on the predicates as a future exercise, and turn our attention solely to restrictions on free occurrences of variables.

3.5 Soundness

As before, we need the basic part, to be extended in accordance with the translation manual—we'll go into a bit more detail in this proof than in the above ones:

Lemma 3.5.1 *If* $\vdash_{\mathbf{QF.E}^{-}_{\Box\Diamond}} \mathcal{A}$ *then* $\vDash_F \mathcal{A}$ *holds for every frame* F.

Proof As usual, this proceeds by induction on proofs, and this can easily be compiled by comparing the proofs in [9, 27]. We leave the proof to the interested reader. □

The next step concerns the translation manual, as before:

Lemma 3.5.2 *If a frame* F *satisfies* $\mathcal{A}^t$*, then* $\vDash_F \mathcal{A}$*; similarly, if it satisfies* (rR^t) *then it obeys the rule* (rR).

Proof This proof just extends that of Lemma 3.3.2, with modifications appropriate for the atomic and quantifier cases. In addition, we have to consider how the

[10] Alternatively it seems we could allow constants in the schemes, but ensure all scheme instances are closed under universal substitution of all terms that are not bound by a quantifier (including constants).

[11] The theory of *relevant predication* was introduced by Dunn [4]. Roughly, a predicate is relevant to an object (in a context) iff *being that object* relevantly implies *having that predicate*.

translation affects $PropFun$ and elements of D^{ω}. There are two main cases: axioms and rules.

Axioms Suppose we have an axiom-scheme $\mathcal{A}$ with t-transform $\mathcal{A}^t$ = 'if Λ then $\Phi f \subseteq \Psi f$'. We will show that every (restriction-obeying) instance of $\mathcal{A}$ is true in every model $\mathfrak{M}$ satisfying $\mathcal{A}^t$.

Choose an arbitrary (restriction-obeying) instance of $\mathcal{A}$. This is provably and semantically equivalent to a conditional $\mathcal{B} \to \mathcal{C}$, which is a fact made use of in the transformation algorithm. First, for every atomic subformula of $\mathcal{B}$ and $\mathcal{C}$, identify the propositional functions assigned to these formulas in $\mathfrak{M}$. Instantiate the Φ and Ψ in $\mathcal{A}^t$ with these propositional functions, assigning every function to its corresponding atomic sub-formula of $\mathcal{A}$. The result is $\Phi^{\mathcal{A}} \subseteq \Psi^{\mathcal{A}}$, which will automatically satisfy Λ by construction, and thus be an *instance* of the more general $\mathcal{A}^t$ (which the model satisfies in full generality). Then the desired result follows by (c1) and establishing the following:

$$[\![\mathcal{B}]\!] = \Phi^{A} \subseteq \Psi^{A} = [\![\mathcal{C}]\!]$$

The subset relation is established by its being an instance of $\mathcal{A}^t$. Each identity is established by applying the valuation identities, in the definition of a model, for the logical operators down to atomics. By construction, the resulting atomics are assigned the same propositional functions as their corresponding parts in Φ and Ψ. Thus, the result is established and the instance of the axiom scheme $\mathcal{A}$ is true in this model. However, the same arguments apply to any instance of the axiom scheme, and in any model, so every instance of $\mathcal{A}$ is true in every model.

Rules Assume we have an instance of the rule of the form $\mathcal{A}_1, \ldots, \mathcal{A}_n \Rightarrow \mathcal{B}$, which we will call R, where our t-transform (suppressing Λ) is $\mathcal{A}_1^t, \ldots, \mathcal{A}_n^t \Rightarrow \mathcal{B}^t$, or R^t. For some arbitrary model $\mathfrak{M}$ satisfying$\mathcal{A}_1^t, \ldots, \mathcal{A}_n^t \Rightarrow \mathcal{B}^t$, assume that $\mathfrak{M} \models \mathcal{A}_i$ for each for $1 \leq i \leq n$ (of the instance of R). Call $\mathcal{A}_i^R$ the instantiation of $\mathcal{A}_i^t$ obtain by instantiating each propositional function in $\mathcal{A}_i^t$ with the propositional function in $\mathfrak{M}$ corresponding to the atomic formulas in $\mathcal{A}_i$, and similarly for B^t. Then we have $\mathcal{A}_1^R, \ldots, \mathcal{A}_n^R \Rightarrow \mathcal{B}^R$, which is an instance of R^t, and which satisfies Λ by construction.

Now, the model satisfies each $\mathcal{A}_i^R$, for $1 \leq i \leq n$, given that $\mathfrak{M} \models \mathcal{A}_i$, and the definition of $\mathcal{A}_i^R$. Thus, the model satisfies B^R. But by construction this entails that $\mathfrak{M} \models \mathcal{B}$, as desired. □

This gives us the grist to immediately prove a general soundness theorem:

Theorem 3.5.3 (Soundness) *Let* $\mathbf{L}$ *be any extension of* $\mathbf{QF.E}_{\Box\Diamond}$, *and let* $\mathbb{L}$ *be the set of frames obeying the frame constraints of any axioms/rules distinguishing* $\mathbf{L}$. *Then it follows that for any formula* $\mathcal{A}$, $\vdash_{\mathbf{L}} \mathcal{A} \Rightarrow \models_{\mathbb{L}} \mathcal{A}$.

4 Completeness

The completeness argument proceeds by a canonical model construction.

4.1 Canonical Model Definition

Definition 4.1.1 Let **L** be an expansion/extension of $\mathbf{QF.E}^{-}_{\Box\Diamond}$. We define the *canonical model* for **L**, $M^{\mathbf{L}}$. Let $\mathcal{L}^C$ be the set of formulas from the language $\mathcal{L}$ which are *closed*—i.e., those in which no free variables occur.

- $W^{\mathbf{L}}$ is the set of prime theories in **L**: i.e., the set of those $\alpha \subseteq \mathscr{P}(\mathcal{L})$ s.t. (a) if $\vdash_{\mathbf{L}} \mathcal{A} \to \mathcal{B}$ and $\mathcal{A} \in \alpha$ then $\mathcal{B} \in \alpha$, (b) if $\mathcal{A}, \mathcal{B} \in \alpha$ then $\mathcal{A} \wedge \mathcal{B} \in \alpha$, and (c) if $\mathcal{A} \vee \mathcal{B} \in \alpha$ then either $\mathcal{A} \in \alpha$ or $\mathcal{B} \in \alpha$. We define the canonical valuation as usual: $[\![\mathcal{A}]\!]^{\mathbf{L}} := \{\alpha \in W^{\mathbf{L}} \mid \mathcal{A} \in \alpha\}$, where $\mathcal{A}$ is a sentence.
- $N^{\mathbf{L}} = \{\alpha \in W^{\mathbf{L}} \mid \vdash_{\mathbf{L}} \mathcal{A} \Rightarrow \mathcal{A} \in \alpha\}$
- $\alpha^{*^{\mathbf{L}}} = \{\mathcal{A} \mid \neg\mathcal{A} \notin \alpha\}$
- $R^{\mathbf{L}}\alpha XY \iff$ for some $\mathcal{A}, \mathcal{B} \in \mathcal{L}^C$, $X = [\![\mathcal{A}]\!]^{\mathbf{L}}$ & $Y = [\![\mathcal{B}]\!]^{\mathbf{L}}$ & $\mathcal{A} \to \mathcal{B} \in \alpha$
- $S^{\mathbf{L}}_{\Box}\alpha X \iff$ for some $\mathcal{A} \in \mathcal{L}^C$, $X = [\![\mathcal{A}]\!]^{\mathbf{L}}$ & $\Box\mathcal{A} \in \alpha$
- $S^{\mathbf{L}}_{\Diamond}\alpha X \iff$ for some $\mathcal{A} \in \mathcal{L}^C$, $X = [\![\mathcal{A}]\!]^{\mathbf{L}}$ & $\Diamond\mathcal{A} \in \alpha$
- $D^{\mathbf{L}} = Con$
- $Prop^{\mathbf{L}} = \{X \subseteq W^{\mathbf{L}} \mid$ for some $\mathcal{A} \in \mathcal{L}^C, X = [\![\mathcal{A}]\!]^{\mathbf{L}}\}$

Given any formula $\mathcal{A}$, let A^f be $\mathcal{A}$ with every free variable x_n replaced by $fx_n \in Con$. Furthermore, Let $\varphi_{\mathcal{A}} : (D^{\mathbf{L}})^{\omega} \longrightarrow \mathscr{P}(W)$ be defined by $\varphi_{\mathcal{A}} f = [\![\mathcal{A}^f]\!]^{\mathbf{L}}$. With this, we can go on to define:

- $PropFun^{\mathbf{L}} = \{\varphi_{\mathcal{A}} : \mathcal{A} \in \mathcal{L}\}$
- $M_f(c) = c \in Con$
- $M_f(x_n) = fx_n \in Con$
- $M^{\mathbf{L}}(P)(M^{\mathbf{L}}_f(t_1), \ldots, M^{\mathbf{L}}_f(t_n)) = [\![P(t_1, \ldots, t_n)]\!]^{\mathbf{L}}$

This concludes the definition of $M^{\mathbf{L}}$. Now we proceed to the, more or less standard, proof method.

4.2 The Proof of Completeness

The basic completeness result for $\mathbf{QF.E}^{-}_{\Box\Diamond}$ is a straightforward variation on those we've given in previous papers, so we'll just state some of the key lemmas, proofs of which can be found in [9, 27], or reconstructed from those therein.

Lemma 4.2.1 (Pair Extension) *Fix* Γ, Δ, *sets of formulas, and* $\mathbf{L}$ *an extension of* $\mathbf{QF.E}^{-}_{\Box\Diamond}$. *If there are no finite sets* $\{\mathcal{A}_i\}_{i\in I} \subseteq \Gamma$, $\{\mathcal{B}_j\}_{j\in J} \subseteq \Delta$ *s.t.* $\vdash_{\mathbf{L}} \bigwedge_{i\in I} \mathcal{A}_i \to \bigvee_{j\in J} \mathcal{B}_j$, *then there is a prime theory* $\alpha \supseteq \Gamma$ *s.t.* $\alpha \cap \Delta = \varnothing$.

Lemma 4.2.2 *For any closed formulas* $\mathcal{A}$, $\mathcal{B}$, *and logic* $\mathbf{L}$ *extending* $\mathbf{QF.E}^{-}_{\Box\Diamond}$*:*

$$\vdash_{\mathbf{L}} \mathcal{A} \to \mathcal{B} \iff [\![\mathcal{A}]\!]^{\mathbf{L}} \subseteq [\![\mathcal{B}]\!]^{\mathbf{L}}$$

$$\vdash_{\mathbf{L}} \mathcal{A} \leftrightarrow \mathcal{B} \iff [\![\mathcal{A}]\!]^{\mathbf{L}} = [\![\mathcal{B}]\!]^{\mathbf{L}}$$

Lemma 4.2.3 *Let* $\mathbf{L}$ *be any logic extending* $\mathbf{QF.E}^{-}_{\Box\Diamond}$. *If* $\forall x\mathcal{A}$ *is a closed formula, then* $[\![\forall x\mathcal{A}]\!]^{\mathbf{L}} = \bigsqcap_{c\in Con} [\![\mathcal{A}[c/x]]\!]^{\mathbf{L}}$.

Lemma 4.2.4 *Let* $\mathbf{L}$ *be any logic extending* $\mathbf{QF.E}^{-}_{\Box\Diamond}$. *Then* $M^{\mathbf{L}}$ *satisfies the conditions in Def. 2.1.1, and so is a frame.*

Lemma 4.2.5 *Let* $\mathbf{L}$ *be any logic extending* $\mathbf{QF.E}^{-}_{\Box\Diamond}$. *If* $\vdash_{\mathbf{L}} \mathcal{A}$ *then* $\vdash_{\mathbf{L}} \mathcal{A}^f$, *for any* $f \in D^\omega$.

Proof This proof is simple a relies on the kinds of restrictions we are considering. Take any derivation of $\mathcal{A}$. Extend this proof by following these two steps as many times as needed:

(1) Universally quantify each free variable in $\mathcal{A}$, obtaining $\forall x_1 \ldots \forall x_n\mathcal{A}$ for the variables $x_1, \ldots, x_n$ occurring free in $\mathcal{A}$.
(2) Use modus ponens and the axiom ($\forall$E) as many times as possible, replacing each bound variable x_n with fx_n.

Given that our restrictions are on free variable occurrences, the result trivially satisfies these restrictions. Thus, we have a derivation showing that $\vdash_{\mathbf{L}} \mathcal{A}^f$. □

Lemma 4.2.6 *Let* $\mathbf{L}$ *be any logic extending* $\mathbf{QF.E}^{-}_{\Box\Diamond}$. *For any conditional formula* $\mathcal{A} \to \mathcal{B}$*:*

$$\vdash_{\mathbf{L}} \mathcal{A} \to \mathcal{B} \textit{ if and only if for every } f \in D^\omega, \varphi_{\mathcal{A}}f \subseteq \varphi_{\mathcal{B}}f$$

Proof Left-to-right: Assume $\vdash_{\mathbf{L}} \mathcal{A} \to \mathcal{B}$, with a (possible empty) set of restrictions on free occurrences of variables. For any $\alpha \in W$ and $f \in D^\omega$, if $\alpha \in \varphi_{\mathcal{A}}f$, then it follows that $\mathcal{A}^f \in \alpha$. But since $\mathcal{A} \to \mathcal{B}$ is a theorem, so is the sentence $\mathcal{A}^f \to \mathcal{B}^f$ (by Lemma 4.2.5.). So then $\mathcal{B}^f \in \alpha$, which gives $\alpha \in \varphi_{\mathcal{B}}f$. As f and α were arbitrary, we get that for every $f \in D^\omega$, $\varphi_{\mathcal{A}}f \subseteq \varphi_{\mathcal{B}}f$.

Right-to-left: Suppose that $\varphi_{\mathcal{A}}f \subseteq \varphi_{\mathcal{B}}f$ holds for every $f \in D^\omega$. For reductio, suppose that $\nvdash_{\mathbf{L}} \mathcal{A} \to \mathcal{B}$. It follows that $\nvdash_{\mathbf{L}} (\mathcal{A} \to \mathcal{B})^f$, for some $f \in D^\omega$. For suppose that every $g \in D^\omega$ was such that $\vdash_{\mathbf{L}} (\mathcal{A} \to \mathcal{B})^g$. In particular there is a one-to-one mapping g' which maps each variable onto a unique constant not occurring in the formula $\mathcal{A} \to \mathcal{B}$. By universal generalization on constants (Lemma 3.4.1) on the new constants in $(\mathcal{A} \rightarrow \mathcal{B})^g$, no constant symbols from $\mathcal{A} \rightarrow \mathcal{B}$ are removed. Then, by (∀E), each of the new quantifiers can be eliminated, replacing the variables with the original variables in $\mathcal{A} \rightarrow \mathcal{B}$. The resulting sentence is thus derived from the proof of $\vdash_{\mathbf{L}} (\mathcal{A} \rightarrow \mathcal{B})^g$, and satisfies the restriction (of course, since it is syntactically identical to $\mathcal{A} \rightarrow \mathcal{B}$).

Thus, $\nvdash_{\mathbf{L}} (\mathcal{A} \rightarrow \mathcal{B})^f$, for some $f \in D^\omega$. Therefore, by Lemma 4.2.1, there is a prime theory γ extending $\{\mathcal{A}^f\}$ that does not contain $\mathcal{B}^f$. Now, $\mathcal{A}^f \in \gamma$, so $\gamma \in \varphi_{\mathcal{A}} f$, but $\gamma \notin \varphi_{\mathcal{B}} f$. Thus $\varphi_{\mathcal{A}} f \nsubseteq \varphi_{\mathcal{B}} f$, a contradiction. So then $\vdash_{\mathbf{L}} \mathcal{A} \rightarrow \mathcal{B}$.

□

In the previous proof, one may be concerned about the restrictions on free variable occurrences in $\mathcal{A} \rightarrow \mathcal{B}$. However, The canonical model handles these restrictions automatically, since by definition the atomic subformulas are mapped onto propositional functions that ensure the semantic restrictions are met. If x_n is not to occur free in $\mathcal{A}$ but it in fact did, then $\varphi_{\mathcal{A}} f$ is not $\varphi_{\mathcal{A}} f'$, with $f \sim_{x_m} f'$, because $\nvdash_{\mathbf{L}} \mathcal{A}(c) \rightarrow \mathcal{A}(d)$ (and so $[\![\mathcal{A}(c)]\!]^{\mathbf{L}} \neq [\![\mathcal{A}(d)]\!]^{\mathbf{L}}$).

This leaves the new, more general, lemma we'll use here to obtain the general completeness result below:

Lemma 4.2.7 (Canonicity for *t*-Transforms) *Let* $\mathbf{L}$ *be an expansion/extension of* $\mathbf{QF.E}^{-}_{\Box\Diamond}$ *by a selection of axioms* $\mathcal{A}_1, \ldots, \mathcal{A}_m, \ldots$ *and rules* $(rR_1), \ldots, (rR_n), \ldots$. *Then* $M^{\mathbf{L}}$ *satisfies* $\mathcal{A}^t_i$ *and* (rR^t_j) *for* $i, j \in \mathbb{Z}^+$.

Proof **Case: Axioms** Suppose the axiom scheme is $\mathcal{A} \rightarrow \mathcal{B}$, with certain restrictions on free variable occurrences. (With $\mathcal{A} = \mathtt{t}$ possible, this is with no loss of generality.) Our corresponding condition is of the form 'Λ implies $\Phi \subseteq \Psi$' where Λ is obtained from the said restrictions. By the definition of the canonical model, the atomic subformulas of any instance of $\mathcal{A} \rightarrow \mathcal{B}$ are mapped onto propositional functions that automatically satisfy Λ.[12] It thus suffices to show that $\varphi_{\mathcal{A}} f \subseteq \varphi_{\mathcal{B}} f$, given how the translation preserves logical form. By Lemma 4.2.6, it indeed follows that $\varphi_{\mathcal{A}} f \subseteq \varphi_{\mathcal{B}} f$, as desired.

Case: Rules Suppose the rule is $\mathcal{A}_1, \ldots \mathcal{A}_n \Rrightarrow \mathcal{D}$, with certain restrictions. Without loss of generality, we treat the $\mathcal{A}_i$ and $\mathcal{D}$ as conditionals: if they are conditionals, great; if they are not, then we treat them with the antecedent $\mathtt{t}$. That is, we will take $\mathcal{A}_i = \mathcal{B}_i \rightarrow \mathcal{C}_i$ and $\mathcal{D} = \mathcal{E} \rightarrow \mathcal{F}$. We presume Λ holds of the propositional functions in question, and that each of $\mathcal{A}^t_i$ holds of the canonical model, and want to show that $\mathcal{D}^t$ does as well. Lem. 3.5.2 guarantees that the canonical model satisfies each $\mathcal{A}_i$, and then also satisfies $\mathcal{D}$. Thus, Lem. 4.2.6 guarantees that $\mathcal{D}^t$ is satisfied.

□

[12] For example, if $\mathcal{D}$ is a subformula for which x_n should not occur, then $\varphi_{\mathcal{D}} f = [\![\mathcal{D}^f]\!]^{\mathbf{L}} = [\![\mathcal{D}^{f'}]\!]^{\mathbf{L}} = \varphi_{\mathcal{D}} f'$, when $f \sim_{x_n} f'$, since $\mathcal{D}^f$ is syntactically identical to $\mathcal{D}^{f'}$.

Theorem 4.2.8 (Completeness) *Let* $\mathbf{L}$ *be an expansion/extension of* $\mathbf{QF.E}^{-}_{\Box\Diamond}$*, and let* $\mathbb{L}$ *be the class of NQMRL frames satisfying the t-transforms of all axioms/rules distinguishing* $\mathbf{L}$ *from* $\mathbf{QF.E}^{-}_{\Box\Diamond}$*. Then* $\vDash_{\mathbb{L}} \mathcal{A}$ *only if* $\vdash_{\mathbf{L}} \mathcal{A}$*.*

Proof For this, suppose that $\nvdash_{\mathbf{L}} \mathcal{A}$, we can just take $\Gamma = \{\mathcal{B} \mid \vdash_{\mathbf{L}} \mathcal{B}\}$ and $\{\mathcal{A}\}$ and apply Lemma 4.2.1 to obtain a prime theory $\Gamma' \supseteq \Gamma$ s.t. $\mathcal{A} \notin \Gamma'$—Lemma 4.2.7 guarantees that the canonical model for $\mathbf{L}$ has a frame in $\mathbb{L}$, so it provides the desired counterexample, showing that $\nvDash_{\mathbb{L}} \mathcal{A}$, as desired. □

Note that there is a sense of consequence available in this semantics for which a strong completeness result is available, following the discussion in [1, §48.4]. We leave it to the reader so motivated to check that we can obtain a similar kind of completeness result here (going through compactness).

5 Augmentation and Relational Semantics

In the realm of modal extensions of classical logic, relational models are often equivalent to *augmented* neighbourhood models, which is shown by establishing a point-wise equivalence between the models. In the setting of propositional modal relevant logics, there is an equivalence between the relational and the fully augmented neighbourhood models, as demonstrated in [9], which relies on showing a model-wise equivalence. Here we show that this result carries over to quantified extensions, where (1) augmentation is similarly defined, and (2) the models built on the corresponding frames also agree on the extensions of predicates and constant symbols.

But first, let's define the relational models for which the result will hold.

5.1 Relational Models for Quantified Modal Relevant Logics

Definition 5.1.1 (General Relational Frame) A *general relational frame* for a quantified modal relevant logic is a tuple

$$\mathfrak{R} = \langle W, N, R, *, S_{\Box}, S_{\Diamond}, Prop, D, PropFun\rangle$$

where $\emptyset \neq N \subseteq W$, $R \subseteq W^3$, $* : W \longrightarrow W$, $S_{\Box}, S_{\Diamond} \subseteq W^2$, and while fixing $\mathscr{P}(W)^{\uparrow} =_{df} \{X \subseteq W \mid \forall a, b \in W((a \leq b \;\&\; a \in X) \Rightarrow b \in X)\}$, where $a \leq b =_{df} \exists x \in N(Rxab)$, we have that $Prop \subseteq \mathscr{P}(W)^{\uparrow}$, with $N \in Prop$, D is a non-empty set of individuals, and $PropFun \subseteq \{\varphi \mid \varphi : D^{\omega} \longrightarrow Prop\}$. In addition, the following conditions are satisfied:

(r1) $\leq$ is a preorder on W;
(r2) if $a \leq a'$, $b \leq b'$, $c' \leq c$ and $Ra'b'c'$, then $Rabc$;

(r3) if $a \leq b$ then $b^* \leq a^*$;
(r4) if $a \leq b$ and $S_\Box bc$ then $S_\Box ab$;
(r5) if $a \leq b$ and $S_\Diamond ac$ then $S_\Diamond bc$;
(r6) if $X, Y \in Prop$, then so are $X \cup Y$, $X \cap Y$, X, $X \to Y$, $X \circ Y$, $X \leftarrow Y$, $\neg X$, $\Box X$ and $\Diamond X$;[13]
(r7) if $\varphi, \psi \in PropFun$, then so are $\varphi \cup \psi, \varphi \cap \psi, \varphi, \varphi \to \psi, \varphi \circ \psi, \varphi \leftarrow \psi, \neg\varphi, \Box\varphi, \Diamond\varphi, \forall_n\varphi$ and $\exists_n\varphi$ (for every $n \in \omega$);[14] the constant function $\varphi_N f = N$ for every $f \in D^\omega$ is in PropFun.

Definition 5.1.2 (General Relational Model) A *general relational pre-model* $\mathfrak{N}$ *on a frame* $\mathfrak{R}$ is obtained by means of a valuation function $|{-}|^{\mathfrak{N}}$ which assigns:

(i) an individual $|c|^{\mathfrak{N}} \in D$ to each constant symbol c;
(ii) for every term τ and $f \in D^\omega$, if τ is a constant $|\tau|^{\mathfrak{N}} f = |c|^{\mathfrak{N}}$, and if τ is the variable x_n, then $|\tau|^{\mathfrak{N}} f = fn$;
(iii) an function $|P|^{\mathfrak{N}} : D^n \longrightarrow \mathscr{P}(W)$ to each n-ary predicate symbol P;
(iv) a propositional function $|\mathcal{A}|^{\mathfrak{N}} : D^\omega \longrightarrow \mathscr{P}(W)$ to each formula $\mathcal{A}$ such that, when $\mathcal{A}$ is the atomic $P\tau_n, \ldots \tau_n$, the propositional function assigned to it is given by, for every $f \in D^\omega$,

$$|P\tau_1, \ldots, \tau_n|f = |P|(|\tau_1|f, \ldots |\tau_n|f)$$

This is extended to the all formulas by the following:

$$|\mathtt{t}|f = \varphi_N f$$

$$|\oplus \mathcal{A}|f = \oplus|\mathcal{A}|f \qquad \text{for } \oplus \in \{\neg, \Box, \Diamond\}$$

$$|\mathcal{A} \otimes \mathcal{B}|f = |\mathcal{A}|f \otimes |\mathcal{B}|f \qquad \text{for } \otimes \in \{\to, \circ, \leftarrow\}$$

$$|\mathcal{A} \wedge \mathcal{B}|f = |\mathcal{A}|f \cap |\mathcal{B}|f \qquad |\forall x_n \mathcal{A}|f = \forall_n|\mathcal{A}|f$$

$$|\mathcal{A} \vee \mathcal{B}|f = |\mathcal{A}|f \cup |\mathcal{B}|f \qquad |\exists x_n \mathcal{A}|f = \exists_n|\mathcal{A}|f$$

---

[13] In the relational case, some of these operations have new definitions, as in the following:

(i) $\neg X =_{df} \{a \in W \mid \alpha^* \notin X\}$
(ii) $\Box X =_{df} \{a \in W \mid \forall b(S_\Box ab \Rightarrow b \in X)\}$
(iii) $\Diamond X =_{df} \{a \in W \mid \forall b(S_\Diamond ab \mathbin{\&} b \in X)\}$
(iv) $X \to Y =_{df} \{a \in W \mid \forall b, c(Rabc \mathbin{\&} b \in X \Rightarrow c \in Y)\}$
(v) $X \circ Y =_{df} \{a \in W \mid \exists b, c(Rbca \mathbin{\&} b \in X \mathbin{\&} c \in Y)\}$
(vi) $X \leftarrow Y =_{df} \{a \in W \mid \forall b, c(Rbac \mathbin{\&} b \in X \Rightarrow c \in Y)\}$

[14] Where these operators are defined as in the neighbourhood case, by (i) lifting the propositional connectives and (ii) by the same definition of $\forall_n$ and $\exists_n$.

A *general relational model* $\mathfrak{N}$ *on a frame* $\mathfrak{R}$ is a pre-model that assigns each atomic propositional to an element of $PropFun$.

It is provable that every formula is assigned to an element of $PropFun$, provided that the atomic propositions (including $\mathtt{t}$) are assigned, as in a model, to elements of $PropFun$.

For satisfaction by a variable assignment, we write $\mathfrak{R}, \mathfrak{N}, f \models \mathcal{A}$ iff $N \subseteq |\mathcal{A}|^{\mathfrak{N}} f$ for $f \in D^{\omega}$ and $\mathfrak{N}$ based on $\mathfrak{R}$. For truth in a model (i.e., satisfied by every variable assignment) we write $\mathfrak{R}, \mathfrak{N} \models \mathcal{A}$ iff $N \subseteq |\mathcal{A}|^{\mathfrak{N}} f$ for every $f \in D^{\omega}$. Further, we write $\mathfrak{R} \models \mathcal{A}$ iff $\mathfrak{R}, \mathfrak{N} \models \mathcal{A}$ for every model $\mathfrak{N}$ built on $\mathfrak{R}$.

Lemma 5.1.3 *Any relational model for quantified modal relevant logic verifies the non-modal axioms* ($\wedge I$), ($\vee E$), *and the non-modal rules* (rB), (rB′) *and the modal axioms* (C) *and* ($C_{\Diamond}$) *and modal rules* (rRM) *and* ($rRM_{\Diamond}$).

The proof of the above lemmas is as in [9]. The upshot, implicit in the proof of model-wise equivalence, is that the logics determined by fully augmented neighbourhood models each contain the logic **QBM.C**, the basic logic of the ternary relational semantics.

It is worth noting that augmentation only refers to elements of $Prop$. The construction below uses filters of sets of propositions to construct relational models. Augmentation gives us what is required for the squeeze lemma; that is, distribution types. So we only need to define augmentation using elements of $Prop$, at least while we have general frames in the picture.

Theorem 5.1.4 *Every general relational frame* $\mathfrak{R}$ *can be transformed into a fully augmented general neighbourhood frame* $\mathfrak{F}^{\mathfrak{M}}$ *with the same domain D. Moreover, for any model* $\mathfrak{N}$ *with valuation* $|-|$ *on* $\mathfrak{R}$ *and model* $\mathfrak{M}$ *with valuation M on* $\mathfrak{F}^{\mathfrak{M}}$, *if* $|-|$ *and M agree on the assignments to every constants on every predicate symbol—that is,* $|c|^{\mathfrak{N}} = M(c)$ *and* $|P|^{\mathfrak{N}} d_1, \ldots d_n = M(P) d_n \ldots d_n$ *for* $d_1, \ldots d_n \in D$*—then for every formula* $\mathcal{A}$ *and every* $f \in D^{\omega}$,

$$a \in |\mathcal{A}|^{\mathfrak{N}} f \text{ iff } a \in [\![\mathcal{A}]\!]^{M}_{f}$$

Proof Given a relational frame $\mathfrak{R} = \langle W, N, R, *, S_{\Box}, S_{\Diamond}, Prop, D, PropFun \rangle$. We construct a neighbourhood frame

$$\mathfrak{F}^{\mathfrak{M}} = \langle W^{\mathfrak{M}}, N^{\mathfrak{M}}, R^{\mathfrak{M}}, *^{\mathfrak{M}}, S^{\mathfrak{M}}_{\Box}, S^{\mathfrak{M}}_{\Diamond}, Prop^{\mathfrak{M}}, D^{\mathfrak{M}}, PropFun^{\mathfrak{M}} \rangle$$

as follows:

(i) $W^{\mathfrak{M}} = W$; $N^{\mathfrak{M}} = N$; $*^{\mathfrak{M}} = *$; $Prop^{\mathfrak{M}} = Prop$; $D^{\mathfrak{M}} = D$; $PropFun^{\mathfrak{M}} = PropFun$;
(ii) $R^{\mathfrak{M}} = \{\langle a, X, Y \rangle \in W \times Prop \times Prop | \forall b, c((Rabc \,\&\, b \in X) \Rightarrow c \in Y)\}$;

(iii) $S_{\Box}^{\mathfrak{M}} = \{\langle a, X\rangle \in W \times Prop | \forall b(S_{\Box}ab \Rightarrow b \in X)\}$
(iv) $S_{\Diamond}^{\mathfrak{M}} = \{\langle a, X\rangle \in W \times Prop | \exists b(S_{\Diamond}ab \;\&\; b \in X)\}$

To show this model is fully augmented, we rely on the arguments of [9], for a similar model construction in the propositional case. (That is, the proof of this fact relies primarily on the definition of $R^{\mathfrak{M}}$, $S_{\Box}^{\mathfrak{M}}$, and $S_{\Diamond}^{\mathfrak{M}}$.)

For the second part of the lemma, we choose an arbitrary pair of models $\mathfrak{N}$ built on $\mathfrak{R}$ with valuation $|-|^{\mathfrak{N}}$ and $\mathfrak{M}$ built on $\mathfrak{F}^{\mathfrak{M}}$ with valuation M that agree on the extensions of predicate symbols and constants. We show that $a \in |\mathcal{A}|^{\mathfrak{N}} f$ iff $a \in [\![\mathcal{A}]\!]_f^M$ by induction on the complexity of $\mathcal{A}$. For the base case, note the following list of equalities:

$$\begin{aligned}|P\tau_1, \ldots \tau_n|^{\mathfrak{N}} f &= |P|^{\mathfrak{N}}(|\tau_1|^{\mathfrak{N}} f, \ldots, |\tau_n|^{\mathfrak{N}} f)\\ &= |P|^{\mathfrak{N}}(M_f(\tau_1), \ldots, M_f(\tau_n))\\ &= M(P)(M_f(\tau_1), \ldots, M_f(\tau_n))\\ &= [\![P\tau_1, \ldots, \tau_n]\!]_f^M\end{aligned}$$

The first equality is a statement of definition. The second uses the identity of the f function, the definitions of M and $|-|^{\mathfrak{N}}$, and the agreement on constant symbols. The third follows by the agreement on the extensions of predicate symbols, and the fourth by definition. To complete the induction, we show only the case for $\mathcal{A} = \forall x_n \mathcal{B}$, leaving the remaining cases to the reader. Note that, since quantified sentences are modeled by the exact same operation in both cases, unlike for the lifted propositional connectives, the result is straightforward from the equality $(\forall_n |\mathcal{B}|^{\mathfrak{N}}) f = (\forall_n [\![\mathcal{B}]\!]^M) f$, which is given by the induction hypothesis. □

For the more involved and interesting direction, we detour through an algebraic construction, enabling us to employ the gaggle-theoretic framework to construct an equivalent relational model, while making requisite modifications to deal with the internal structure of atomic formulas and quantification.

Definition 5.1.5 Given a fully augmented general neighbourhood frame

$$\mathfrak{F}^{\mathfrak{M}} = \langle W^{\mathfrak{M}}, N^{\mathfrak{M}}, R^{\mathfrak{M}}, *^{\mathfrak{M}}, S_{\Box}^{\mathfrak{M}}, S_{\Diamond}^{\mathfrak{M}}, Prop^{\mathfrak{M}}, D^{\mathfrak{M}}, PropFun^{\mathfrak{M}}\rangle$$

define its *underlying propositional complex algebra* as

$$\mathcal{A}_p^{\mathfrak{M}} = \langle Prop^{\mathfrak{M}}; N^{\mathfrak{M}}, \neg, \cap, \cup, \rightarrow, \Box, \Diamond\rangle$$

where the operations are as defined on $Prop^{\mathfrak{M}}$, and define the *complex first-order structure* as

$$\langle \mathcal{A}_p^{\mathfrak{M}}, PropFun^{\mathfrak{M}}, D^{\mathfrak{M}}, \varphi_N, \neg, \cap, \cup, \rightarrow, \Box, \Diamond, \{\forall_n\}_{n\in\omega}, \{\exists_n\}_{n\in\omega}\rangle$$

where φ_N is the constant function returning N, and the operations are defined on $PropFun^{\mathfrak{M}}$.

On an underlying propositional complex algebra, a *filter* is a subset of $Prop^{\mathfrak{M}}$ closed under $\cap$ and, in the upwards direction, under $\subseteq$. A filter is prime when, if $X \cup Y \in a$, then either $X \in a$ or $Y \in a$. The principle filter of a set $X \in Prop^{\mathfrak{M}}$ is the set $[X) = \{Y \in Prop^{\mathfrak{M}} | X \subseteq Y\}$.

Lemma 5.1.6 (Squeeze) *The underlying propositional complex algebras of any fully augmented general neighbourhood frame is such that each of* $\Box$, $\Diamond$, *and* $\rightarrow$ *has a distribution type, in the sense of [3, Def. 1.3.2] (indeed, each is a multi-gaggle, even if we also add* $\leftarrow, \circ$ *in the obvious way). Therefore, we have the following special cases of the generalized Squeeze Lemma [3, Lem. 2.3.19]:*

(i) If b, c *are prime filters on* $Prop^{\mathfrak{M}}$ *and* a' *is a filter thereon, such that* $\forall X, Y \in Prop^{\mathfrak{M}}((X \rightarrow Y \in a' \;\&\; X \in b) \Rightarrow Y \in c)$, *then there is a prime* $a \supseteq a'$ *such that* $\forall X, Y \in Prop^{\mathfrak{M}}((X \rightarrow Y \in a \;\&\; X \in b) \Rightarrow Y \in c)$.

(ii) If a *is a prime filter on* $Prop^{\mathfrak{M}}$, *and* b' *is a filter thereon, such that* $X \in b'$ *and* $\forall Y \in Prop^{\mathfrak{M}}(Y \in b' \Rightarrow \Diamond Y \in a)$, *then there is a prime* $b \supseteq b'$ *such that* $X \in b$ *and* $\forall Y \in Prop^{\mathfrak{M}}(Y \in b \Rightarrow \Diamond Y \in a)$.

(iii) If a *is a prime filter on* $Prop^{\mathfrak{M}}$, *and* b' *is a filter thereon, such that* $X \notin b'$ *and* $\forall Y \in Prop^{\mathfrak{M}}(\Box Y \in a \Rightarrow Y \in b')$, *then there is a prime* $b \supseteq b'$ *such that* $X \notin b$ *and* $\forall Y \in Prop^{\mathfrak{M}}(\Box Y \in a \Rightarrow Y \in b)$.

(iv) If a, c *are prime filters on* $Prop^{\mathfrak{M}}$, *and* b' *is a principal filter thereon, say* $[X) = \{Y \mid X \subseteq Y\}$, *and* $b^- = \{Z \mid \exists U \notin c(Z \rightarrow Y \in a)\}$, *then there is a prime* $b \supseteq b'$ *such that* $b \cap b^- = \varnothing$.

Proof See [3, Ch. 2] for details. □

Definition 5.1.7 Given a *complex first-order structure*

$$\langle \mathcal{A}_p^{\mathfrak{M}}, PropFun^{\mathfrak{M}}, D^{\mathfrak{M}}, \varphi_N, \neg, \cap, \cup, \rightarrow, \Box, \Diamond, \{\forall_n\}_{n\in\omega}, \{\exists_n\}_{n\in\omega}\rangle$$

based on a frame $\mathfrak{F}^{\mathfrak{M}}$, and with the underlying propositional complex algebra with the carrier set $Prop'$, define a corresponding general relational model $\mathfrak{R}$ by

(i) $W =$ the set of prime filters of $Prop'$ (written $a, b, c \ldots$);
(ii) $N = \{a \in W | N^{\mathfrak{M}} \in a\}$;
(iii) $R = \{\langle a, b, c\rangle \in W^3 | \forall X, Y \in Prop'((X \rightarrow Y \in a \;\&\; X \in b) \Rightarrow Y \in c)\}$;
(iv) $a* = \{X \in Prop' | \neg X \notin a\}$, for all $a \in W$;

(v) $S_{\Box} = \{\langle a, b\rangle \in W^2 | \forall X \in Prop'(\Box X \in a \Rightarrow X \in b)\}$;
(vi) $S_{\Diamond} = \{\langle a, b\rangle \in W^2 | X \in b \Rightarrow \Diamond X \in a\}$;
Given $X \in Prop'$, let $\Sigma_X = \{a \in W | X \in a\}$;
(vii) $Prop = \{\Sigma_X | X \in Prop'\}$;
(viii) $D = D^{\mathfrak{M}}$;
(ix) For each $\varphi^{\mathfrak{M}} \in PropFun^{\mathfrak{M}}$, define $PropFun$ to include φ defined by

$$\varphi f = \Sigma_{(\varphi^{\mathfrak{M}} f)}$$

Lemma 5.1.8 *The frame general relational frame $\mathfrak{R}$ corresponding to a* complex first-order structure *obtained from a fully augmented neighbourhood frame $\mathfrak{F}$ is a general relational frame as in definition 5.1.1.*

Proof It is clear from the definition that $\emptyset \neq N \subseteq W$, and that the relations are appropriately defined in terms of arity on W. The $\leq$ is suitable, and such that $\leq = \subseteq$. $N = \Sigma_N \in Prop$, and the elements of $PropFun$ or of the right type. It remains to show that (r1)–(r7) are satisfied.

Because $\subseteq$ is a partial ordering, (r1) is satisfied. Using the squeeze lemma, it is fairly straightforward to show that (r2)–(r5) are satisfied. The interested reader is directed to [9] for sufficient arguments.

For (r6), we may use the arguments of [9]. The relevant parts of the structure are defined similarly, and the same squeeze lemma is proved for a similarly defined complex algebra based on a set of admissible propositions. In particular, we obtain the following equalities, for each f:[15] $\Sigma_{X\cap Y} = \Sigma_X \cap \Sigma_Y$; $\Sigma_{X\cup Y} = \Sigma_X \cup \Sigma_Y$; $\Sigma_{X\to Y} = \Sigma_X \to \Sigma_Y$; $\Sigma_{X\circ Y} = \Sigma_X \circ \Sigma_Y$; $\Sigma_{X\leftarrow Y} = \Sigma_X \leftarrow \Sigma_Y$; $\Sigma_{\neg Y} = \neg\Sigma_Y$; $\Sigma_{\Box Y} = \Box\Sigma_Y$; $\Sigma_{\Diamond Y} = \Diamond\Sigma_Y$.

The real meat of this proof comes from showing that (r7) is satisfied. We show three illustrative cases: The constant function (1) $\varphi_N \in PropFun$, (2) $\varphi_{X\cap Y} f = \varphi_X f \cap \varphi_Y f$, and (3) $(\varphi_{\forall x_n \mathcal{B}}) f = \forall_n(\varphi_{\mathcal{B}}) f$. For (1), note that $\varphi_N^{\mathfrak{M}} \in PropFun^{\mathfrak{M}}$, which constantly returns N. We have that $\Sigma_{N^{\mathfrak{M}}} = N$, and so $\varphi_N f = N$, for every $f \in U^\omega$, as required.

For (2), assume that $\varphi_X, \varphi_Y \in PropFun$. We know that $\varphi_X^{\mathfrak{M}}, \varphi_Y^{\mathfrak{M}}, (\varphi_X^{\mathfrak{M}} \cap \varphi_Y^{\mathfrak{M}}) \in PropFun^{\mathfrak{M}}$, by the closure of $PropFun^{\mathfrak{M}}$ under $\cap$ and the definition of $PropFun$. It is sufficient to show that, for every $f \in D^\omega$, $(\varphi_X \cap \varphi_Y) f = \Sigma_{(\varphi_X^{\mathfrak{M}} \cap \varphi_Y^{\mathfrak{M}}) f}$.

Assume an arbitrary $f \in D^\omega$. By an equality given in the proof of (r6), $a \in (\varphi_X \cap \varphi_Y) f$ iff $a \in \varphi_X f \cap \varphi_Y f$. That is, $a \in \varphi_X f$ and $a \in \varphi_Y f$, which is the case iff $a \in \Sigma_{(\varphi_X^{\mathfrak{M}} f)}$ and $a \in \Sigma_{(\varphi_Y^{\mathfrak{M}} f)}$. The latter are equivalent to a is an upset of $Prop'$

[15] The arguments in [9] do not explicitly contain the cases for $\circ$ and $\leftarrow$, but these cases are straightforward (or at least similar enough to previous cases) and are also omitted here.

filers contains both $\varphi_X^{\mathfrak{M}} f$ and $\varphi_Y^{\mathfrak{M}} f$. Being a filter, this means that $\varphi_X^{\mathfrak{M}} f \cap \varphi_Y^{\mathfrak{M}} f = (\varphi_X^{\mathfrak{M}} \cap \varphi_Y^{\mathfrak{M}}) f \in a$, which is just that $a \in \Sigma_{(\varphi_X^{\mathfrak{M}} \cap \varphi_Y^{\mathfrak{M}}) f}$.

For (3), assume that $\varphi_X \in PropFun$ and $n \in \omega$. This assume means that $\varphi_X^{\mathfrak{M}} \in PropFun^{\mathfrak{M}}$. It is sufficient to show that, for every $f \in D^{\omega}$, $(\forall_n \varphi_X) f = \Sigma_{(\forall_n \varphi_X^{\mathfrak{M}}) f}$.

For an $a \in W$, $a \in (\forall_n \varphi_X) f$ iff $a \in \Sigma_Y \in Prop$ (which entails that $Y \in Prop'$) such that $\Sigma_Y \subseteq \bigcap_{j \in D} \varphi_X f[j/x_n]$. This is iff $Y \in a$ and thus $\varphi_X^{\mathfrak{M}} f[j/x_n] \in a$, for each $j \in D$. Starting from the other side of the 'iff', $a \in \Sigma_{(\forall_n \varphi_X^{\mathfrak{M}}) f}$ iff $(\forall_n \varphi_X^{\mathfrak{M}} f) \in a$. The latter is equivalent to the existing of a $Z \in Prop'$ such that $Z \in a$ and $Z \subseteq \bigcap_{j \in D} \varphi_X^{\mathfrak{M}} f[j/x_n]$. With the following fact, and by identifying Y and Z, we can obtain the result.

It is now sufficient for us to prove the following, for any $Y \in Prop'$, $n \in \omega$, $\varphi_X^{\mathfrak{M}} \in PropFun^{\mathfrak{M}}$ and $f \in D^{\omega}$:

$$Y \subseteq \bigcap_{j \in D} \varphi_X^{\mathfrak{M}} f[j/x_n] \text{ iff } \Sigma_Y \subseteq \bigcap_{j \in D} \varphi_X f[j/x_n]$$

For the left-to-right direction, suppose that $a \in \Sigma_Y$, which is that $Y \in a$. By the hypothesis, this entails that $\bigcap_{j \in D} \varphi_X^{\mathfrak{M}} f[j/x_n] \in a$, since a is a filter. Moreover, we have that $\varphi_X^{\mathfrak{M}} f[j/x_n] \in a$, for each $j \in D$. This is equivalent to $a \in \Sigma_{\varphi_X^{\mathfrak{M}} f[j/x_n]}$, for each $j \in D$, and so $a \in \bigcap_{j \in D} \varphi_X f[j/x_n]$.

For the converse, consider a $Y \in Prop'$. Then $Y \in \Sigma_Y \subseteq \varphi_X f[j/x_n]$, for each $j \in D$. In particular, $Y \in \Sigma_{\varphi_X^{\mathfrak{M}} f[j/x_n]}$, for each $j \in D$. Since Y is in every prime filter of $Prop'$ containing each $\varphi_X^{\mathfrak{M}} f[j/x_n]$ (for $j \in D$), the proof is complete.

With this last fact in place, Y and Z can be identified in the above argument completing the equivalence: that is: for every $f \in D^{\omega}$, $(\forall_n \varphi_X) f = \Sigma_{(\forall_n \varphi_X^{\mathfrak{M}}) f}$. □

Lemma 5.1.9 *Given a model $\mathfrak{M}$ based on a fully augmented neighbourhood frame $\mathfrak{F}$, construct a relational model $\mathfrak{N}$ by taking the corresponding general relational frame $\mathfrak{R}$ as in the previous Lemma, and by defining a valuation function $|-|$ by setting $|c|^{\mathfrak{N}} = M(c)$ and $|P|^{\mathfrak{N}} d_1, \ldots d_n = M(P) d_n \ldots d_n$ for $d_1, \ldots d_n \in D$—. Then for every formula $\mathcal{A}$ and every $f \in D^{\omega}$,*

$$N \subseteq |\mathcal{A}|^{\mathfrak{N}} f \text{ iff } N^{\mathfrak{M}} \subseteq [\![\mathcal{A}]\!]_f^M$$

Proof The proof is by induction on $\mathcal{A}$. The bases cases are determined by the valuation function defined in the statement of the lemma. The inductive cases are given by identities in the previous lemma. We show a single case. Suppose $\mathcal{A} = \forall x \mathcal{B}$. Then, by the previous lemma, $|\forall x \mathcal{B}|^{\mathfrak{N}} f = \Sigma_{(\forall x \mathcal{B})^{\mathfrak{M}} f}$. Thus $N \subseteq |\forall x \mathcal{B}|^{\mathfrak{N}} f$

iff $N \subseteq \{a \in W : (\forall x\mathcal{B})^{\mathfrak{M}} f \in a\}$. By the definition of N, this is equivalent to $N^{\mathfrak{M}} \in a$ implies $(\forall x\mathcal{B})^{\mathfrak{M}} f \in a$, which is sufficient for this case. □

6 Putting NQMRL Frames to Work

In this section we'll present an independence result one can obtain using our general neighbourhood models.

6.1 Dependence and Independence

Let's start by pulling apart our desired constraints. We'll proceed here to study models *without* negation, as we can produce quite simple models that way (basically, negation is mildly annoying). In addition, the models constructed in this section often display the independence of properties where such independence is highlighted by the lack of negation: the similarity between $S_\Box$-augmentation and the Barcan Formula make it notable that such an independence holds (even) with the lack of negation. We conjecture that the same results hold for models including negation, but leave that question for future work or the interested reader.

Theorem 6.1.1 *Each of the following properties of frames are independent; that is, for each property below, we can construct a* **QF.E**$_{\Box\Diamond}$ *frame which fails to satisfy that property, while satisfying all of the others, and for each we can construct a frame which satisfies the property while failing to satisfy all the others:*[16]

(1) left R-augmentation
(2) right R-augmentation
(3) $S_\Box$-augmentation
(4) $S_\Diamond$-augmentation
(5) Satisfying BF
(6) Satisfying $BF_\Diamond$

Proof We'll go through each of these properties, showing that there are models in which the property holds, while every other fails, and that there are models in which the property fails, while every other holds.

First note that Goldblatt [11, pp. 82–83] proves that **QS4**+UI doesn't prove BF, which is equivalent to $\text{BF}_\Diamond$. **RQ4** is a subsystem of this logic, and so BF and $\text{BF}_\Diamond$ are not provable in it, but **RQ4** is complete w.r.t. relational Mares-Goldblatt frames, and so we can build a general neighbourhood Mares-Goldblatt frame which is fully augmented and which falsifies BF and $\text{BF}_\Diamond$.

Independence of (1), (2) To show that there is a model where (1) fails and (2)–(6) hold, we can just take the model defined in [27, §5.2], fixing $\Box X = X = \Diamond X$. In

[16] A similar independence result should apply for the Buridan formulas introduced earlier.

that model, we have $Prop = \mathscr{P}(W)$, so the fact that (3) and (4) hold is equivalent to that (5) and (6) hold. To see that (3) holds, note that $\Box\bigcap_{i\in I} X_i = \bigcap_{i\in I} X_i$, and since $\Box X_i = X_i$ for all $i \in I$, it follows that $\bigcap_{i\in I}\Box X_i = \bigcap_{i\in I} X_i$. The case for (4) is similar. To find a model where (2) fails and (1), (3)–(6) hold, we can just employ [27, §5.1] as in the case of (1).

It remains to show that there is a model where (1) holds and all the rest fail, and similarly for (2). We'll just spell out (2), using the model defined in [27, §5.1]. Note that in this model, right R augmentation fails, but left R augmentation succeeds, so it just suffices to add interpretations of $\Box$, $\Diamond$, and in fact we can do this simply as follows—we presume we're working in a language with just one unary predicate letter F:[17]

$$\Box X = \begin{cases} \varnothing & \text{if } X = \varnothing \\ \{n \mid 16 \lneq n\} & \text{else} \end{cases}$$

For note then that

$$\bigcap_{n\in D} \Box M(F)n = \bigcap\{n \mid 16 \lneq n\} = \{n \mid 16 \lneq n\}$$

and

$$\Box\bigcap_{n\in D} M(F)n = \Box\varnothing = \varnothing$$

and $\{n \mid 16 \lneq n\} \nsubseteq \varnothing$, so (3) fails. Now for $\Diamond$, we may use a very similar model construction and take

$$\Diamond X = \begin{cases} W & \text{if } X = W \\ \{n \mid 16 \lneq n\} & \text{else} \end{cases}$$

We leave it to the reader to verify that this does not satisfy (4).

Independence of (3) To show that (3) can fail while (1)–(2) and (4)–(6) hold, we can once again adapt the models employed in [27, §5]. We give a frame, and note that it fails on some valuation

- $W = \mathbb{N} \cup \{\infty\}$ with $\infty > n$ for all $n \in \mathbb{N}$
- $N = W$
- $R = \{\langle n, X, Y\rangle \mid n \in W \;\&\; X \subseteq Y\}$

---

[17] We have chosen a constant upset, $\{n \mid 16 \lneq n\}$, more or less at random (though we like the number of 17). The reader is encouraged to choose some other such set if they prefer.

- $\Box X = \begin{cases} \{n \mid 16 \leq n\} & \text{if } X = \varnothing \\ \{n \mid 16 \leq n\} & \text{if } \exists m \in \omega(X = \{n \mid m \leq n\}) \\ \varnothing & \text{else} \end{cases}$
- $\Diamond X = X$
- $D = \mathbb{N}$
- $Prop = \{X \subseteq W \mid \exists n \in \mathbb{N} \forall x \in W(x \in X \iff n \leq x)\} \cup \{\varnothing\}$
- $PropFun = \{\varphi \mid \varphi : D^\omega \longrightarrow Prop\}$

First off, note that this is a frame. To obtain a model we can just define M so that $M(F)n = \{x \mid n \leq x\}$, and then note that:

$$\Box \bigsqcap M(F)n = \Box\varnothing = \{x \mid 16 \leq x\}$$

$$\bigsqcap \Box M(F)n = \{x \mid 16 \leq x\}$$

so we obtain the satisfaction of the BF, but:

$$\Box \bigcap M(F)n = \Box\{\infty\} = \varnothing$$

$$\bigcap \Box M(F)n = \bigcap \{x \mid 16 \leq x\} = \{x \mid 16 \leq x\}$$

and so $\bigcap \Box M(F)n \nsubseteq \Box \bigcap M(F)n$, and so we're not in a $S_\Box$-augmented frame as desired. It's straightforward to verify that all the other augmentation properties hold, and that $BF_\Diamond$ is satisfied.

It remains to show that (3) can hold while the other constraints fail. Here we rely on an argument sketch, which goes as follows. There are models of classical quantified modal logic that do not validate the Barcan Formula, and each of the logics in question are sublogics of classical quantified modal logic (with universal domains and hence unrestricted universal instantiation). One may take a (classical) countermodel of any instance of the Barcan Formula (or its converse) and turn it into a fully augmented model (of the relevant kind) which does not validate the corresponding instance of the Barcan Formula (with relevant instance of classical implication). One may consult Goldblatt [11] for the construction of such models. These models can be straightforwardly given as ternary relational models (as each classical model can so be re-described). Finally, using the augmentation results of the previous section, one may turn this ternary relational model into a fully augmented neighbourhood model which does not validate the Barcan Formula. This model is used in the case for (5) below. However, to show the current case we must further break the R-augmentation in this model. This is accomplished via the methods used in other cases, such as case (1) and (2). (The reader will note that the differences between Box and Diamond augmentation in this construction has been glossed over, the reason is that their independence is quite easy to show using the arguments found in other cases.)

Independence of (4) To show that (4) can fail while the others hold, a model construction similar to that for showing that (3) can fail, above, suffices. The trick essentially just adapts the same model construction, and we leave this to the refer to check to their satisfaction.

Independence of (5) First, we'll show that (5) can fail while all the other conditions hold by means of sketched construction in case (3). Prior to the breaking of R-augmentation in that construction (modulo the differences between the box and diamond) we have just the model we are searching for.

To show that (5) can hold while all the other constraints fail, we can simply adapt the proof that (3) can fail while all other constraints hold. We alter the model given there by the following changes, writing $[n) = \{x \in W \mid n \leq x\}$:

- $RnXY$ iff either $n \geq 0$ and $X \subseteq Y$ or
  $n \geq 1$ and $X = W$ and $\exists m \in \mathbb{N}([m) \subseteq Y)$ or
  $n \geq 2$ and $\exists m \in \mathbb{N}([m) \subseteq X)$ and $\varnothing = Y$

Note then that $\bigcap_{n \in \mathbb{N}} (W \to [m)) = \bigcap [1) = [1) \not\subseteq \varnothing = W \to \{\infty\} = W \to \bigcap_{m \in \mathbb{N}} [m)$. A similar argument goes for left-augmentation, which is left to the interested reader, as is verifying that this is a frame.

Independence of (6) To show both cases for (6), we can employ a tactic like we did for the proof that (5) can hold while the other fail; we leave this to the fastidious reader.

□

References

1. Anderson, A. R., Belnap, N. D., & Dunn, J. M. (1992). *Entailment: The logic of relevance and necessity* (Vol. 2). Princeton University Press.
2. Barcan, R. C. (1946). A functional calculus of first order based on strict implication. *Journal of Symbolic Logic*, *11*, 1–16.
3. Bimbó, K., & Dunn, J. M. (2008). *Generalized Galois logics: Relational semantics for nonclassical logical calculi*. CSLI.
4. Dunn, J. M. (1987). Relevant predication 1: The formal theory. *Journal of Philosophical Logic*, *16*, 347–381.
5. Dunn, J. M. (1995). Positive modal logic. *Studia Logica*, *55*(2), 301–317.
6. Ferenz, N. (2019). *Quantified modal relevant logics*, Ph.D. Dissertation, University of Alberta.
7. Ferenz, N. (2020). Identity in relevant logics: A relevant predication approach. In M. Blicha & I. Sedlar (Eds.), *Logica yearbook 2020* (pp. 49–64). College Publications.
8. Ferenz, N. (2023). Quantified modal relevant logics. *Review of Symbolic Logic*, *16*, 210–240.
9. Ferenz, N., & Tedder, A. (2023). Neighbourhood semantics for modal relevant logics. *Journal of Philosophical Logic*, *52*, 145–181.
10. Garson, J. W. (2001). Quantification in modal logic. In D. Gabbay & F. Guenthner (Eds.), *Handbook of philosophical logic* (Vol. 3, pp. 267–323).
11. Goldblatt, R. (2011). *Quantifiers, propositions and identity: Admissible semantics for quantified modal and substructural logics*. Cambridge University Press (2011).

12. Humberstone, L. (2011). *The connectives*. MIT Press.
13. Kremer, P. (1999). Relevant identity. *Journal of Philosophical Logic*, *28*, 199–222.
14. Lavers, P. (1985). *Generating intensional logics*, M.A. Thesis, University of Adelaide.
15. Lavers, P. (1994). Generalising tautological entailment. *Logique & Analyse*, *146*, 367–377.
16. Mares, E. D., & Goldblatt, R. (2006). An alternative semantics for quantified relevant logic. *Journal of Symbolic Logic*, *71*(1), 163–187 (2006).
17. Pacuit, E. (2017). *Neighborhood semantics for modal logic*. Springer.
18. Robles, G., & Méndez, J. M. (2018). *Routley-Meyer ternary relational semantics for intuitionistic-type negations*. Elsevier.
19. Routley, R., & Meyer, R. K. (1975). Towards a general semantical theory of implication and conditionals. I. Systems with normal conjunctions and disjunctions and aberrant and normal negations. *Reports on Mathematical Logic*, *4*, 67–89.
20. Routley, R., Meyer, R. K., Plumwood, V., & Brady, R. T. (1982). *Relevant logics and their rivals* (Vol. 1). Ridgeview Publishing.
21. Sedlár, I. *Relational semantics for propositional dynamic logics*. Unpublished Manuscript.
22. Sedlár, I., Punčochář, V., & Tedder, A. (2023). Relevant epistemic logic with public announcements and common knowledge. *Journal of Logic and computation*, *33*(2), 436–461.
23. Smiley, T. (1963). Relative necessity. *Journal of Symbolic Logic*, *28*(2), 113–134.
24. Standefer, S. (2019). Tracking reasons with extensions of relevant logics. *Logic Journal of the IGPL*, *27*(4), 543–569.
25. Tedder, A. (2021). Information flow in logics in the vicinity of **BB**. *Australasian Journal of Logic*, *18*(1), 1–24.
26. Tedder, A., & Bilková, M. (2022). Relevant propositional dynamic logics. *Synthese*, *200*, 235.
27. Tedder, A., & Ferenz, N. (2022). Neighbourhood semantics for quantified relevant logics. *Journal of Philosophical Logic*, *51*, 457–484.

Semantics for Second-Order Relevant Logics

Shay Allen Logan

Abstract In this paper we extend Kit Fine's semantics for first-order relevant logic so as to treat dyadic second-order logic. Along the way we discuss how to understand the semantics, problems one encounters in extending it to account for various sorts of comprehension, and partial solutions to those problems.

1 Preamble

Here's the thing: when you look at it from just the right angle, it's entirely obvious how semantics for second-order relevant logics ought to go. Or at least, if you've understood how semantics for *first*-order relevant logics ought to go (those seeking a gentle introduction might find the discussions in [10] or [12] helpful), there are perspectives like this. What's more is that from any such angle, the metatheory that needs doing can be summed up in one line: everything is just as in the first-order case, but with more indices.

Of course, it's no small matter finding the magical angle from which everything becomes obvious. And even having found this perspective, one cannot assume one's audience will find things as obvious as oneself. All that to say this: if the results in the paper below strike you as obvious, pay attention to the perspective that makes that possible. And if they don't, feel free to ignore this preamble in its entirety.

S. A. Logan (✉)
Kansas State University, Manhattan, KS, USA
e-mail: salogan@ksu.edu

I. Sedlár et al. (eds.), *New Directions in Relevant Logic*, Trends in Logic 63,
https://doi.org/10.1007/978-3-031-69940-5_8

2 The Language

We will work in a somewhat idiosyncratic setting: dyadic second-order logic. I've chosen this setting not for technical reasons but for reasons of pedagogical expediency. I've explained this material to a number of people and have found the following:

- Monadic second-order logic, while easiest to understand, doesn't leave everyone clear on what to do when it comes to extending yet higher.
- Third-(or higher-)order logic has too much machinery for any but the devout to make it through. Those that do, though, are left able to see their way anywhere they want.
- Dyadic second-order logic is a middle ground—even the apostate are usually able to tolerate working through it if they decide they really care. And seeing one's way from dyadic second-order logic to third-order logic and higher is usually doable in a matter of days.

That said, we explicitly define the language $\mathcal{L}$ as follows:

Vocabulary The set of symbols of $\mathcal{L}$ consists of

- Countably many individual constants ($c_1, c_2, \ldots$), the set of which we denote by Con_0;
- Countably many unary predicate constants ($P_1, P_2, \ldots$), the set of which we denote by Con_1;
- Countably many binary predicate constants ($Q_1, Q_2, \ldots$), the set of which we denote by Con_2;
- Countably many individual variables ($x_1, x_2, \ldots$), the set of which we denote by Var_0;
- Countably many unary predicate variables ($X_1, X_2, \ldots$), the set of which we denote by Var_1;
- Countably many binary predicate variables ($Y_1, Y_2, \ldots$), the set of which we denote by Var_2;
- The connectives $\neg$, $\wedge$, $\vee$, and $\rightarrow$; and
- The quantifiers $\forall$ and $\exists$.

Grammar $\mathcal{L}$, understood as its set of formulas, is then defined as follows:

- If $T \in \mathsf{Con}_1 \cup \mathsf{Var}_1$ and $t \in \mathsf{Con}_0 \cup \mathsf{Var}_0$, then Tt is a(n atomic) formula;
- If $T \in \mathsf{Con}_2 \cup \mathsf{Var}_2$ and $t_i \in \mathsf{Con}_0 \cup \mathsf{Var}_0$, then Tt_1t_2 is an atomic formula;
- If A and B are formulas, then so are $\neg A$, $(A \wedge B)$, $(A \vee B)$, and $(A \rightarrow B)$;
- If A is a formula and x is an individual variable, then $\forall x A$ and $\exists x A$ are formulas;
- If A is a formula and X is a unary predicate variable, then $\forall X A$ and $\exists X A$ are formulas; and
- If A is a formula and Y is a binary predicate variable, then $\forall Y A$ and $\exists Y A$ are formulas.

I will usually stick to the metavariable conventions implicitly specified in the above definition. We'll also occasionally use $\leftrightarrow$, defined in its usual way. Free and bound occurrences of a variable are also defined in the expected ways; for substitution, we write $A(x/t)$ (resp. $A(X/T)$; $A(Y/S)$) for the formula that results from replacing each free occurrence of x (resp. X; Y) in A with an occurrence of t (resp. T; S). With respect to such substitutions, we define what it means for t (resp. T; S) to be *free for* x (resp. X; Y) in A in the expected ways. Finally, where A is a formula and x_1 is an individual variable (resp. x_1 and x_2 are individual variables), we write $B(X/A(x_1))$ (resp. $B(Y/A(x_1, x_2))$) for the formula that results from replacing, for each individual term t (resp. for each pair of individual terms t_1 and t_2) each occurrence of Xt (resp. Yt_1t_2) in B in which the 'X' (resp. 'Y') is occurring freely with an occurrence of $A(x/t)$ (resp. $A(x_1/t_1, x_2/t_2)$). We extend the notion of 'free for' from variable-substitutions to formula-substitutions in the obvious way.

We write $\mathcal{L}$ for the language so-defined. Where F_0, F_1, and F_2 are pairwise disjoint sets of symbols not found in our language and $\overline{F} = \langle F_0, F_1, F_2\rangle$, we take $\mathcal{L}(\overline{F})$ to be the extension of $\mathcal{L}$ that adds the members of F_0 as additional individual constants, adds the members of F_1 as additional unary predicate constants, and adds the members of F_2 as additional binary predicate constants. An $\mathcal{L}(\overline{F})$-sentence is an $\mathcal{L}(\overline{F})$-formula in which no variables occur free.

3 The Logic

The propositional (or zero-order) fragment of each of the logics we discuss in this paper is the weak relevant logic **B** discussed in (among many other places) §4.3 of [14]. This is also the base logic for which Kit Fine first defined stratified semantics in [8].[1] The reader interested in extending the results of this paper to logics with stronger propositional fragments will find that the tools for doing so introduced by Fine are sufficient. On the other hand, the reader interested in extending my results by including further types of quantification will have to piece together how to do so on their own; I nonetheless think that anyone who takes the time to properly understand what I've done here will find doing so entirely straightforward.

[1] This is not, strictly speaking, correct, as a referee has pointed out. Fine's system included the law of excluded middle, which is not typical for **B**. These days, **B** extended to include excluded middle is usually called **BX**.

Now to business. For each language $\mathcal{L}(\overline{F})$ (including $\mathcal{L} = \mathcal{L}(\langle\emptyset, \emptyset, \emptyset\rangle)$), the dyadic second-order logic **B2Q**$(\overline{F})$ is defined to be the logic axiomatized by the $\mathcal{L}(\overline{F})$-instances of the following axioms and rules:

A1 $A \to A$

A2 $(A \wedge B) \to A$

$(A \wedge B) \to B$

A3 $((A \to B) \wedge (A \to C)) \to (A \to (B \wedge C))$

A4 $A \to (A \vee B)$

$B \to (A \vee B)$

A5 $((A \to C) \wedge (B \to C)) \to ((A \vee B) \to C)$

A6 $(A \wedge (B \vee C)) \to ((A \wedge B) \vee (A \wedge C))$

A7 $\neg\neg A \to A$

A8 $\forall x A \to A(x/t)$

$\forall X A \to A(X/T)$

$\forall Y A \to A(Y/S)$

A9 $\forall x(A \to B) \to (A \to \forall x B)$

$\forall X(A \to B) \to (A \to \forall X B)$

$\forall Y(A \to B) \to (A \to \forall Y B)$

A10 $\forall x(A \vee B) \to (A \vee \forall x B)$

$\forall X(A \vee B) \to (A \vee \forall X B)$

$\forall Y(A \vee B) \to (A \vee \forall Y B)$

A11 $A(t/x) \to \exists x A$

$A(T/X) \to \exists X A$

$A(S/Y) \to \exists Y A$

A12 $(A \wedge \exists x B) \to \exists x(A \wedge B)$

$(A \wedge \exists X B) \to \exists X(A \wedge B)$

$(A \wedge \exists Y B) \to \exists Y(A \wedge B)$

R1 $\dfrac{A \quad A \to B}{B}$

R2 $\dfrac{A \quad B}{A \wedge B}$

R3 $\dfrac{A \to B \quad C \to D}{(B \to C) \to (A \to D)}$

R4 $\dfrac{A \to \neg B}{B \to \neg A}$

R5 $\dfrac{A}{\forall x A}$

R6 $\dfrac{A}{\forall X A}$

R7 $\dfrac{A}{\forall Y A}$

R8 $\dfrac{\forall x(B \to A)}{\exists x B \to A}$

R9 $\dfrac{\forall X(B \to A)}{\exists X B \to A}$

R10 $\dfrac{\forall Y(B \to A)}{\exists Y B \to A}$

Note that we do *not* take $\exists$ to be a defined connective here, which partially explains some of the apparent redundancies in this presentation. Also note that in A8 and A11, we require that t (resp. T; S) be free for x (resp. X; Y), and in A9, A10, A12, and in R8–R10, we require that x (resp. X; Y) not occur free in A. Finally, note that there is an asymmetry in our treatment of intensional confinement for universals in A9 and our treatment of intensional confinement for existentials in R8–R10. This is done to preserve the usual negation duality between the existential and the universal in the presence of rule contraposition (rule R4) rather than axiomatic contraposition.

Recall that a comprehension axiom for A is a formula of one of the following forms:

$$\exists X \forall x(Xx \leftrightarrow A)$$

$$\exists Y \forall x_1 \forall x_2(Yx_1x_2 \leftrightarrow A)$$

We implicitly assume when discussing comprehension axioms that X (resp. Y) does not occur freely in A. Given a set $\mathbb{A}$ of formulas, the logic $\mathbf{B2Q}^{\mathbb{A}}(\overline{F})$ augments $\mathbf{B2Q}(\overline{F})$ with comprehension axioms for each $A \in \mathbb{A}$.

For any of the above logics L, we write $\vdash_L A$ to mean that there is a sequence of formulas $B_1, \ldots, B_n$ with $B_n = A$ so that for $1 \le i \le n$, either B_i is an instance of an L-axiom or B_i follows from previous members of the sequence using one of the L-rules. We write $X \vdash_L A$ to mean that there is a sequence of formulas $B_1, \ldots, B_n$ with $B_n = A$ so that for $1 \le i \le n$, either B_i is a member of X or there are $j < i$ and $k < i$ so that $B_i = B_j \wedge B_k$ or there is $j < i$ such that $\vdash B_j \to B_i$.

Given a set of formulas Γ, we say that Γ is disjunctively closed when $A \in \Gamma$ and $B \in \Gamma$ only if $A \vee B \in \Gamma$. We say that Γ is prime when $A \vee B \in \Gamma$ only if either $A \in \Gamma$ or $B \in \Gamma$. We say that Γ is an L-theory when $\Gamma \vdash_L A$ only if $A \in \Gamma$.

4 Semantics

We define an $\overline{F}$-premodel to be a tuple $\langle T, P, \ell, \sqsubseteq, \cdot, \star, v\rangle$ where $\ell \in T \supseteq P$, $\sqsubseteq$ is a partial ordering of T, $\cdot$ is a binary operation on T, $\star$ is a unary operation on P, and v is a function mapping $t \in T$ to a set of atomic $\mathcal{L}(\overline{F})$ sentences. Note that for a more traditional presentation of things, one can recover from v functions v^1 and v^2 that (respectively) map each unary $\overline{F}$-predicate constant to a function from theories to sets of names and map each binary $\overline{F}$-predicate constant to a function from theories to sets of pairs of names. Explicitly, we have

$$v^1(P, t) = \{c : Pc \in v(t)\}$$

$$v^2(Q, t) = \{\langle c_1, c_2\rangle : Qc_1c_2 \in v(t)\}$$

We define $t \sqsubseteq_P q$ to mean $t \sqsubseteq q$ and $q \in P$. An $\overline{F}$-model is an $\overline{F}$-premodel that satisfies the following conditions:

- Covariance: If $s \sqsubseteq t$, then
  - $u \cdot s \sqsubseteq u \cdot t$,
  - $s \cdot u \sqsubseteq t \cdot u$, and
  - $v(s) \subseteq v(t)$.
- Minimality: If $s \cdot t \sqsubseteq_P p$, then
  - There is $s \sqsubseteq_P q$ so that $q \cdot t \sqsubseteq p$ and
  - There is $t \sqsubseteq_P r$ so that $s \cdot r \sqsubseteq p$.
- If $A \in v(p)$ for all $t \sqsubseteq_P p$, then $A \in v(t)$.
- $\ell \cdot t = t$
- $p^{\star\star} = p$

- If $P \ni p \sqsubseteq_P q$, then $q^\star \sqsubseteq p^\star$.
- For all t there is p so that $t \sqsubseteq_P p$.

As the reader can verify, aside from the fact that we are working in a language with additional structure, an $\overline{F}$-premodel is exactly a model in the sense of Fine's [7], with two modifications: first, we've dropped the requirement that primes above the logic be above their duals (that is, if $\ell \sqsubseteq_P p$, then $p^\star \sqsubseteq p$). Second, we've added (in the final bullet above) the requirement that all theories be extended by some prime. The first change reflects our rejection of $A \vee \neg A$ as an axiom. The second change makes no difference at all in the logic. This isn't hard to see: since it amounts to a restriction on the class of models, the soundness proof still goes through, and since the language $\mathcal{L}(\overline{F})$, thought of as a theory, is obviously prime, the canonical model verifies the condition and the completeness proof will still go through as well.

We intuitively interpret the members of T as theories, the members of P as prime theories, ℓ as the logic, $\sqsubseteq$ as the containment relation among the theories, $\cdot$ as the application operation, $\star$ as the dual of a theory, and v as the function mapping each theory to its set of atomic members. For more detail on these interpretations, see e.g. [7, 11], or [12].

4.1 The Base System

As in [8], so also here a *stratified model* is essentially a poset of $\overline{F}$-models. Fine, in constructing stratified models for first-order quantification, gave models that were 'fibered along' a poset $\mathcal{D}$ of what he called domains. But there was nothing all that 'domain-y' about $\mathcal{D}$—its underlying class was just a class of sets satisfying the following:

- Extendibility: For all $\alpha \in \mathcal{D}$, there is $\beta \in \mathcal{D}$ so that $\beta \supsetneq \alpha$.
- Upper Bound: For all $\alpha \in \mathcal{D}$ and $\beta \in \mathcal{D}$, there is $\gamma \in \mathcal{D}$ so that $\alpha \cup \beta \subseteq \gamma$.
- Reversibility: if $\alpha \in \mathcal{D}$, $\beta \in \mathcal{D}$, $\gamma \in \mathcal{D}$, and $\alpha \subseteq \beta \subseteq \gamma$, then $\alpha \cup (\gamma - \beta) \in \mathcal{D}$ as well.

Thinking of the members of $\mathcal{D}$ as sets of names, we can justify these conditions as follows:

- Extendibility: language is indefinitely extensible; no matter how many names we've added to our vocabulary, we can always add more.
- Upper Bound: No two ways of adding names to our language are incompatible with each other.
- Reversibility: Given any two sets of names we might add to our language, we can add them in either order without consequence.

We can generalize these to the case at hand by 'fibering along' not just a poset of sets of names, but along a poset of *vocabularies*, now understood to include not just *names*, but also predicate symbols and relation symbols. Of course, we'll have to require that such a poset satisfy the obvious analogues of the conditions Fine gave. This does not introduce as much new complexity as one would expect, especially once we take the step of extending set-theoretic notions from sets to triples of sets in the expected ways—that is, by defining e.g. $\overline{F} \subseteq \overline{F'}$ to mean that $F_0 \subseteq F_0'$, $F_1 \subseteq F_1'$, and $F_2 \subseteq F_2'$ and, for $\overline{F} \subseteq \overline{F'}$, by defining $\overline{F} - \overline{F'}$ to be the triple $\langle F_0 - F_0', F_1 - F_1', F_2 - F_2' \rangle$. With this in hand, we define the (oh so) technical term 'appropriate class of vocabularies' by saying that an appropriate class of vocabularies is any class V of triples $\overline{F} = \langle F_0, F_1, F_2 \rangle$ of sets such that

- Extendibility: For each $\overline{F} \in V$, there are (not necessarily distinct) $\overline{F} \subseteq \overline{G} \in V$, $\overline{F} \subseteq \overline{H} \in V$ and $\overline{F} \subseteq \overline{K} \in V$ so that $F_0 \subsetneq G_0$, $F_1 \subsetneq H_1$, and $F_2 \subsetneq K_2$.
- Upper Bound: For all $\overline{F} \in V$ and $\overline{G} \in V$, there is $\overline{H} \in V$ with $\overline{F} \cup \overline{G} \subseteq \overline{H}$.
- Reversibility: if $\overline{F} \in V$, $\overline{G} \in V$, $\overline{H} \in V$, and $\overline{F} \subseteq \overline{G} \subseteq \overline{H}$, then $\overline{F} \cup (\overline{H} - \overline{G}) \in V$ as well.

To define stratified models we begin by settling on a particular appropriate class of vocabularies V such that no member of any component of any member of V occurs as a symbol in $\mathcal{L}$. We then define a V-stratified premodel to be a tuple $\langle M, \Uparrow, \Downarrow, [\![-]\!] \rangle$ where

- M is a function that maps each $\overline{F} \in V$ to a $\overline{F}$-model $M(\overline{F}) = \langle T_{\overline{F}}, P_{\overline{F}}, \ell_{\overline{F}}, \sqsubseteq_{\overline{F}}, \cdot_{\overline{F}}, \star_{\overline{F}}, v_{\overline{F}} \rangle$;
- $\Uparrow$ is a set containing one function $\uparrow_{\overline{F}}^{\overline{F'}} : T_{\overline{F}} \longrightarrow T_{\overline{F'}}$ for each pair $\langle \overline{F}, \overline{F'} \rangle$ with $\overline{F} \subseteq \overline{F'}$.
- $\Downarrow$ is a set containing one function $\downarrow_{\overline{F}}^{\overline{F'}} : T_{\overline{F'}} \longrightarrow T_{\overline{F}}$ for each pair $\langle \overline{F}, \overline{F'} \rangle$ with $\overline{F} \subseteq \overline{F'}$.
- $[\![-]\!]$ is a set containing one function $[-]_{\overline{F}}^{\sigma_1,\sigma_2} : T_{\overline{F}} \longrightarrow T_{\overline{F}}$ for each triple $\langle \overline{F}, \sigma_1, \sigma_2 \rangle$ with $\langle \sigma_1, \sigma_2 \rangle \in (\mathsf{Con}_0 \cup F_0)^2 \cup (\mathsf{Con}_1 \cup F_1)^2 \cup (\mathsf{Con}_2 \cup F_2)^2$.

Intuitively, each model $M(\overline{F})$ is a space of theories in the language $\mathcal{L}(\overline{F})$, each function $\uparrow_{\overline{F}}^{\overline{F'}}$ maps the $\mathcal{L}(\overline{F})$-theory t to the $\mathcal{L}(\overline{F'})$-theory $t \uparrow_{\overline{F}}^{\overline{F'}}$ that t generates under $\ell_{\overline{F'}}$, each function $\downarrow_{\overline{F}}^{\overline{F'}}$ maps the $\mathcal{L}(\overline{F'})$-theory t to the $\mathcal{L}(\overline{F})$-theory $t \cap \mathcal{L}(\overline{F})$, and each function $[-]_{\overline{F}}^{\sigma_1,\sigma_2}$ maps the $\mathcal{L}(\overline{F})$-theory t to the $\mathcal{L}(\overline{F})$-theory $[t]_{\overline{F}}^{\sigma_1,\sigma_2}$ that we get by extending t so as to make σ_1 and σ_2 indistinguishable. We extend all of these functions from functions from theories to theories to functions from sets of theories to sets of theories in the obvious ways.

Given $\langle \sigma_1, \sigma_2 \rangle \in (\mathsf{Con}_0 \cup F_0)^2 \cup (\mathsf{Con}_1 \cup F_1)^2 \cup (\mathsf{Con}_2 \cup F_2)^2$ and a formula $A \in \mathcal{L}(V)$, a $\langle \sigma_1, \sigma_2 \rangle$-variant of A is a formula that results from replacing zero or more occurrences of σ_i in A with σ_j, where $i \neq j \in \{1, 2\}$. Now consider the function $\Sigma_{\langle \sigma_1, \sigma_2 \rangle}$ from subsets of $\mathcal{L}(V)$ to subsets of $\mathcal{L}(V)$ defined by

$$\Sigma_{\langle \sigma_1, \sigma_2 \rangle}(\Gamma) = \bigcup_{A \in \Gamma} \{B : B \text{ is a } \langle \sigma_1, \sigma_2 \rangle\text{-variant of } A\}$$

$\Sigma_{\langle\sigma_1,\sigma_2\rangle}(\Gamma)$ is the *symmetrization* of Γ at $\langle\sigma_1,\sigma_2\rangle$. We say that Γ is symmetric when $\Sigma_{\langle\sigma_1,\sigma_2\rangle}(\Gamma) = \Gamma$.

A V-stratified *model* is a V-stratified premodel that meets the following conditions:

1. Covariance: If $s \sqsubseteq_{\overline{F}} t$, then
   - $s{\uparrow}^{\overline{F'}}_{\overline{F}} \sqsubseteq_{\overline{F'}} t{\uparrow}^{\overline{F'}}_{\overline{F}}$, and
   - $s{\downarrow}^{\overline{F}}_{\overline{F'}} \sqsubseteq_{\overline{F'}} t{\downarrow}^{\overline{F}}_{\overline{F'}}$.
2. Identity: $t{\uparrow}^{\overline{F}}_{\overline{F}} = t{\downarrow}^{\overline{F}}_{\overline{F}} = t$.
3. Transitivity: $t{\uparrow}^{\overline{F'}}_{\overline{F}}{\uparrow}^{\overline{F''}}_{\overline{F'}} = t{\uparrow}^{\overline{F''}}_{\overline{F}}$ and $t{\downarrow}^{\overline{F}}_{\overline{F'}}{\downarrow}^{\overline{F'}}_{\overline{F''}} = t{\downarrow}^{\overline{F}}_{\overline{F''}}$.
4. Extension-Restriction: $t{\uparrow}^{\overline{F'}}_{\overline{F}}{\downarrow}^{\overline{F'}}_{\overline{F}} = t$, but $t{\downarrow}^{\overline{F}}_{\overline{F'}}{\uparrow}^{\overline{F}}_{\overline{F'}} \sqsubseteq t$.
5. Vertical Atomic Heredity: $v_{\overline{F'}}(t{\downarrow}^{\overline{F}}_{\overline{F'}}) = v_{\overline{F}}(t) \cap \mathcal{L}(\overline{F'})$.
6. Primes Down: $P_{\overline{F}}{\downarrow}^{\overline{F}}_{\overline{F'}} \subseteq P_{\overline{F'}}$.
7. Prime Restriction Down: If $q \in P_{\overline{F}}$ and $p \sqsubseteq q{\downarrow}^{\overline{F}}_{\overline{F'}}$, then there is $r \in P_{\overline{F}}$, with $r{\downarrow}^{\overline{F}}_{\overline{F'}} = p$ and $r \sqsubseteq q$.
8. Prime Extension Down: If $t{\downarrow}^{\overline{F}}_{\overline{F'}} \sqsubseteq_{P_{\overline{F'}}} p$, then there is $t \sqsubseteq_{P_{\overline{F}}} q$ with $q{\downarrow}^{\overline{F}}_{\overline{F'}} = p$.
9. Duality Down: $p^{\star_{\overline{F}}}{\downarrow}^{\overline{F}}_{\overline{F'}} = (p{\downarrow}^{\overline{F}}_{\overline{F'}})^{\star_{\overline{F'}}}$.
10. Distribution Up: $(t \cdot_{\overline{F}} u){\uparrow}^{\overline{F'}}_{\overline{F}} = t{\uparrow}^{\overline{F'}}_{\overline{F}} \cdot_{\overline{F'}} u{\uparrow}^{\overline{F'}}_{\overline{F}}$.
11. Distribution Down: $(t \cdot_{\overline{F}} u{\uparrow}^{\overline{F}}_{\overline{F'}}){\downarrow}^{\overline{F}}_{\overline{F'}} = t{\downarrow}^{\overline{F}}_{\overline{F'}} \cdot_{\overline{F'}} u$.
12. Logics Up: $\ell_{\overline{F}}{\uparrow}^{\overline{F'}}_{\overline{F}} = \ell_{\overline{F'}}$.
13. Bracket is a Closure Operator:
    - $t \sqsubseteq_{\overline{F}} [t]^{\sigma_1,\sigma_2}_{\overline{F}}$;
    - If $s \sqsubseteq_{\overline{F}} t$, then $[t]^{\sigma_1,\sigma_2}_{\overline{F}} \sqsubseteq_{\overline{F}} [s]^{\sigma_1,\sigma_2}_{\overline{F}}$; and
    - $[[t]^{\sigma_1,\sigma_2}_{\overline{F}}]^{\sigma_1,\sigma_2}_{\overline{F}} = [t]^{\sigma_1,\sigma_2}_{\overline{F}}$.
14. Bracket Duality: $[([t]^{\sigma_1,\sigma_2}_{\overline{F}})^{\star_{\overline{F}}}]^{\sigma_1,\sigma_2}_{\overline{F}} = ([t]^{\sigma_1,\sigma_2}_{\overline{F}})^{\star_{\overline{F}}}$
15. Bracket Application: $[s \cdot_{\overline{F}} t]^{\sigma_1,\sigma_2}_{\overline{F}} \sqsubseteq_{\overline{F}} [s]^{\sigma_1,\sigma_2}_{\overline{F}} \cdot_{\overline{F}} [t]^{\sigma_1,\sigma_2}_{\overline{F}} \sqsubseteq_{\overline{F}} [s]^{\sigma_1,\sigma_2}_{\overline{F}} \cdot_{\overline{F}} t$
16. Bracket Up: $[t{\uparrow}^{\overline{F}}_{\overline{F'}}]^{\sigma_1,\sigma_2}_{\overline{F}} = [t]^{\sigma_1,\sigma_2}_{\overline{F'}}{\uparrow}^{\overline{F}}_{\overline{F'}}$
17. Bracket Down: If $\sigma_1 \in F_i - F_i'$ and $\sigma_2 \in F_i'$, then $[t{\uparrow}^{\overline{F}}_{\overline{F'}}]^{\sigma_1,\sigma_2}_{\overline{F}}{\downarrow}^{\overline{F}}_{\overline{F'}} \sqsubseteq_{\overline{F'}} t$
18. Bracket Symmetry: $[t]^{\sigma_1,\sigma_2}_{\overline{F}}$ is symmetric in σ_1 and σ_2.
19. Symmetric Prime Extension: If $v(t)$ is symmetric in σ_1 and σ_2 and $t \sqsubseteq_{P_{\overline{F}}} p$, then there is a q so that $v(q)$ is symmetric in σ_1 and σ_2 and $t \sqsubseteq_{P_{\overline{F}}} q \sqsubseteq_{P_{\overline{F}}} p$.

The forcing relation, which holds between triples $\langle S, \overline{F}, t\rangle$—where S is a V-stratified model, $\overline{F} \in V$, and $t \in M_S(\overline{F})$—and $\mathcal{L}(\overline{F})$-*sentences* (not *formulas*!) is then defined as follows:

- $S, \overline{F}, t \models Pt$ iff $Pt \in v_{\overline{F}}(t)$.
- $S, \overline{F}, t \models Qt_1t_2$ iff $Qt_1t_2 \in v_{\overline{F}}(t)$.
- $S, \overline{F}, t \models A \wedge B$ iff $S, \overline{F}, t \models A$ and $S, \overline{F}, t \models B$
- $S, \overline{F}, t \models A \vee B$ iff for all $t \sqsubseteq_{P_{\overline{F}}} p$, $S, \overline{F}, p \models A$ or $S, \overline{F}, p \models B$
- $S, \overline{F}, t \models \neg A$ iff for all $t \sqsubseteq_{P_{\overline{F}}} p$, $S, \overline{F}, p^{\star}_{\overline{F}} \not\models A$.
- $S, \overline{F}, t \models A \rightarrow B$ iff for all $u \in T_{\overline{F}}$, if $S, \overline{F}, u \models A$, then $S, \overline{F}, t \cdot_{\overline{F}} u \models B$.
- $S, \overline{F}, t \models \forall x A$ iff for some $\overline{G} \supsetneq \overline{F}$ and $g \in G_0 - F_0$, $S, \overline{G}, t{\uparrow}^{\overline{G}}_{\overline{F}} \models A(x/g)$.
- $S, \overline{F}, t \models \exists x A$ iff for all $t \sqsubseteq_{P_{\overline{F}}} p$ there are $\overline{G} \supseteq \overline{F}$, $q \in P_{\overline{G}}$, and $g \in G_0$ so that $q{\downarrow}^{\overline{G}}_{\overline{F}} = p$ and $S, \overline{G}, q \models A(x/g)$.
- $S, \overline{F}, t \models \forall X A$ iff for some $\overline{G} \supsetneq \overline{F}$ and $G \in G_1 - F_1$, $S, \overline{G}, t{\uparrow}^{\overline{G}}_{\overline{F}} \models A(X/G)$.
- $S, \overline{F}, t \models \exists X A$ iff for all $t \sqsubseteq_{P_{\overline{F}}} p$ there are $\overline{G} \supseteq \overline{F}$, $q \in P_{\overline{G}}$, and $G \in G_1$ so that $q{\downarrow}^{\overline{G}}_{\overline{F}} = p$ and $S, \overline{G}, q \models A(X/G)$.
- $S, \overline{F}, t \models \forall Y A$ iff for some $\overline{G} \supsetneq \overline{F}$ and $G \in G_2 - F_2$, $S, \overline{G}, t{\uparrow}^{\overline{G}}_{\overline{F}} \models A(Y/G)$.
- $S, \overline{F}, t \models \exists Y A$ iff for all $t \sqsubseteq_{P_{\overline{F}}} p$ there are $\overline{G} \supseteq \overline{F}$, $q \in P_{\overline{G}}$, and $G \in G_2$ so that $q{\downarrow}^{\overline{G}}_{\overline{F}} = p$ and $S, \overline{G}, q \models A(Y/G)$.

Letting $\overline{\Omega} = \langle \Omega_0, \Omega_1, \Omega_2\rangle$, with $\Omega_i = \cup_{\overline{F} \in V} F_i$, we say that an $\mathcal{L}(\overline{\Omega})$-sentence A is valid in S when $S, \overline{F}, \ell_{\overline{F}} \models A$ whenever $A \in \mathcal{L}(\overline{F})$. We say that a formula $A(\Lambda_1, \ldots, \Lambda_n)$ in which the Λ_i occur free is valid in S when all its substitution instances are valid in S. We say that A is V-valid when A is valid in every V-stratified model S and that A is valid when A is V-valid for every appropriate class of vocabularies V. Finally, where $t \in T_{\overline{F}}$ for some $\overline{F} \in V$, we write $\underline{t}$ for $\{A : S, \overline{F}, t \models A\}$. In the remainder, when they are abundantly clear from context, we will drop some of the subscripts to enhance readability.

We now record a few important facts about the semantics:

Lemma 1

- *If $s \sqsubseteq_{\overline{F}} t$ and $S, \overline{F}, s \models A$, then $S, \overline{F}, t \models A$.*
- *If $S, \overline{F}, p \models A$ for all $t \sqsubseteq_{P_{\overline{F}}} p$, then $S, \overline{F}, t \models A$ as well.*
- *If $v(t)$ is symmetric in σ_1 and σ_2, then $\underline{t}$ is symmetric in σ_1 and σ_2.*
- *If $\overline{F} \subseteq \overline{G}$ and $A \in \mathcal{L}(\overline{F})$, then $S, \overline{G}, t \models A$ iff $S, \overline{F}, t{\downarrow}^{\overline{G}}_{\overline{F}}$.*
- *If $\overline{F} \subseteq \overline{G}$ and $A \in \mathcal{L}(\overline{F})$, then $S, \overline{F}, t \models A$ iff $S, \overline{G}, t{\uparrow}^{\overline{G}}_{\overline{F}}$.*
- *If $S, \overline{F}, t \models \forall x A$ (resp. $S, \overline{F}, t \models \forall X A$; $S, \overline{F}, t \models \forall Y A$) and $f \in F_0$ (resp. $F \in F_1$; $F \in F_2$), then $S, \overline{F}, t \models A(x/t)$.*

Proof By induction on A starting at the top of the list and working to the bottom of the list. In each case, the proof is exactly as in [8] except in the case of the existentials and in each the proof for existentials is essentially immediate. □

Theorem 2 **B2Q** *is sound for the semantics.*

Proof By induction on the complexity of the proof. Again, almost everything goes as it did in [8]. We'll look only at one instance each of A11 and A12, since those are newish, and also at R10, just to round things out.

For A11, we examine the individual existential. So let $A(t/x) \to \exists x A \in \mathcal{L}(\overline{F})$ and $S, \overline{F}, s \models A(t/x)$. Then for all $s \sqsubseteq_{P_{\overline{F}}} p$, $S, \overline{F}, p \models A(t/x)$. Thus, letting $\overline{G} = \overline{F}, q = p$ and $g = t$ and recalling that $\downarrow_{\overline{F}}^{\overline{F}}$ is the identity function, we have that for all $t \sqsubseteq_{P_{\overline{F}}} p$ there is $\overline{G} \supseteq \overline{F}$, $q \in P_{\overline{G}}$ and $g \in G_0$ so that $q\downarrow_{\overline{G}}^{\overline{F}} = p$ and $S, \overline{G}, q \models A(g/x)$. So $S, \overline{F}, s \models \exists x A$. If follows that $S, \overline{F}, \ell_{\overline{F}} \models A(t/x) \to \exists x A$.

For A12 we examine the relation existential. So, let $(A \wedge \exists Y B) \to \exists Y(A \wedge B) \in \mathcal{L}(\overline{F})$ and $S, \overline{F}, t \models A \wedge \exists Y B$. Choose $t \sqsubseteq_{P_{\overline{F}}} p$. Clearly $p \models A$ and $p \models \exists Y B$. So there is $\overline{G} \supseteq \overline{F}$, $G \in G_2$, and $q \in P_{\overline{G}}$ with $q\downarrow_{\overline{F}}^{\overline{G}} = p$ and $q \models B(Y/G)$. Since $q\downarrow_{\overline{F}}^{\overline{G}} = p$, $q \supseteq q\downarrow_{\overline{F}}^{\overline{G}}\uparrow_{\overline{F}}^{\overline{G}} = p\uparrow_{\overline{F}}^{\overline{G}}$. Thus since $A \in \mathcal{L}(\overline{F})$, $q \models A$. So $q \models A \wedge B(Y/G) = (A \wedge B)(Y/G)$. So $t \models \exists Y(A \wedge B)$.

For R10, we examine the predicate existential. So, let $\forall X(B \to A) \in \mathcal{L}(\overline{F})$ and suppose $S, \overline{F}, \ell_{\overline{F}} \models \forall X(B \to A)$. To see that $S, \overline{F}, \ell_{\overline{F}} \models \exists X B \to A$, let $S, \overline{F}, t \models \exists X t$. Choose $t \sqsubseteq_{P_{\overline{F}}} p$. Then there is $\overline{G} \supseteq \overline{F}$, $q \in P_{\overline{G}}$, and $G \in G_1$ so that $q\downarrow_{\overline{F}}^{\overline{G}}$ and $S, \overline{G}, q \models B(X/G)$. Since $S, \overline{F}, \ell_{\overline{F}} \models \exists X B \to A$ we also have that $S, \overline{G}, \ell_{\overline{G}} \models \exists X B \to A$. So $S, \overline{G}, \ell_{\overline{G}} \models B(X/G) \to A$ Thus $S, \overline{G}, q \models A$. And since $A \in \mathcal{L}(\overline{F})$, and $q\downarrow_{\overline{F}}^{\overline{G}} = p$, it then follows that $S, \overline{F}, p \models A$. So all prime extensions of t verify A and thus $S, \overline{F}, t \models A$. □

Lemma 3 (Deduction) *If* $A \vdash B$, *and* $A \to B \in \mathcal{L}(\overline{F})$, *then* $A \to B \in$ **B2Q**$(\overline{F})$.

Proof By induction along $\vdash$; see e.g. [8]. □

Lemma 4 (Lindenbaum) *Let* t *be an* $\mathcal{L}(\overline{F})$*-theory,* $\Delta \subseteq \mathcal{L}(\overline{F})$ *be closed under disjunction, and* $t \cap \Delta = \emptyset$. *Then there is a prime* $\mathcal{L}(\overline{F})$*-theory* $p \supseteq t$ *with* $p \cap \Delta = \emptyset$

Proof In the usual way; see e.g. [8]. □

Corollary 5 *If* t *is an* $\mathcal{L}(\overline{F})$*-theory and* $Pr_{\overline{F}}$ *is the set of prime* $\mathcal{L}(\overline{F})$*-theories, then* $t = \bigcap_{t \subseteq p \in Pr_{\overline{F}}} p$.

Theorem 6 **B2Q** *is complete for the semantics.*

Proof By a canonical model construction. In case it's not clear, the key is to use the appropriate class of vocabularies given by the finite triples of sets of variables and then—surprise surprise—to do what Fine did, one more time.

The only interesting thing worth pausing to verify is that in the key lemma (which says that $t \vDash A$ iff $A \in t$) the induction still goes through in the existential cases. We'll prove this for the individual existential since the other cases are exactly parallel.

So, let C be the canonical model, t be an $\mathcal{L}(\overline{F})$-theory, and $t \vDash \exists x A$. Choose $t \subseteq p \in Pr_{\overline{F}}$. Then since $t \vDash \exists x A$, there is $\overline{G} \supseteq \overline{F}$, $g \in G_0$, and $q \in Pr_{\overline{G}}$ so that $q \cap \mathcal{L}(\overline{F}) = p$ and $q \vDash A(x/g)$. But then by the inductive hypothesis, $A(x/g) \in q$. So since $A(x/g) \to \exists x A$ is a theorem, $\exists x A \in q$. But also $\exists x A \in \mathcal{L}(\overline{F})$, so $\exists x A \in q \cap \mathcal{L}(\overline{F}) = p$. Thus, $\exists x A$ is in every prime extension of t and thus in t.

Now suppose that $\exists x A \in t$. Choose $t \subseteq p \in Pr_{\overline{F}}$ and let $p^{c_{\overline{F}}} = \mathcal{L}_{\overline{F}} - p$. Choose $g \in \Omega_0 - F_0$ and let $\overline{G} = \overline{F} \cup \{g\}$. I claim that

$$q = \{B \in \mathcal{L}(\overline{G}) : p \cup \{A(x/g)\} \vdash B\} \cap p^{c_{\overline{F}}} = \emptyset.$$

To see this, suppose to the contrary that for some B, $p \cup \{A(x/g)\} \vdash B$ and $B \in p^{c_{\overline{F}}}$. Then there will be some $C \in p$ so that $C \wedge A(x/g) \vdash B$. Thus $(C \wedge A(x/g)) \to B \in \mathbf{B2Q}(\overline{G})$. It follows that $\forall g(\neg B \to \neg(C \wedge A(x/g))) \in \mathbf{B2Q}(\overline{G})$. But then since A and B are in $\mathcal{L}(\overline{F})$, confining the universal twice (first intensionally, then extensionally) we get that $\neg B \to (\neg C \vee \forall g \neg A(x/g)) \in \mathbf{B2Q}(\overline{G})$. But then we also get that $(C \wedge \exists x A) \to B \in \mathbf{B2Q}(\overline{F})$, and thus since $C \in p$ and $\exists x A \in t \subseteq p$, $B \in p$, which is a contradiction.

Thus, since it is obvious that $p^{c_{\overline{F}}}$ is closed under disjunctions and q is a theory that contains $A(x/g)$ (and thus that, by the inductive hypothesis, $q \vDash A(x/g)$) there is, by the Lindenbaum Lemma, a prime $q' \in Pr_{\overline{G}}$ with $q' \cap \mathcal{L}(\overline{F}) = p$ and $q' \vDash A(x/g)$. So $t \vDash \exists x A$. □

4.2 Adding Comprehension

B2Q is a second-order logic in only the most technical sense. Lacking comprehension axioms, it's really just a many-sorted first-order logic with a complicated vocabulary. In order to add comprehension axioms, we of course have to restrict the class of models we allow.

There's also a complication to consider that arises in logics that don't admit axiomatic transitivity; that is in which the following are not theorems:

$$(A \to B) \to ((B \to C) \to (A \to C))$$

$$(B \to C) \to ((A \to B) \to (A \to C))$$

The problem is this: one of the key uses to which one puts comprehension axioms is in proving that second-order universals are well-behaved. As a particular case of the general phenomenon, it is typically the case in second-order systems that once one has added comprehension axioms for B, one can then derive all instances of

$\forall X A \to A(X/B(y))$ in which $B(y)$ is free for X.[2] This is a fairly natural thing to want of one's second-order universals.

The natural way to go about deriving all such formulas is by first doing something like this:

$$\dfrac{\dfrac{\dfrac{\forall X A \to A \qquad A(X/B) \to A(X/B)}{(A \to A(X/B)) \to (\forall X A \to A(X/B))}}{\forall X[(A \to A(X/B)) \to (\forall X A \to A(X/B))]}}{\exists X(A \to A(X/B)) \to (\forall X A \to A(X/B))}$$

All one has to do then is derive the antecedent; viz. $\exists X(A \to A(X/B))$. Now, if A is atomic, there's no problem here: either X occurs in A (so A has the form Xa for some a) or it doesn't. In the latter case, $A = A(X/B)$, so by a degenerate instance of A11 and the fact that $A \to A$ is an axiom, $\exists X(A \to A(X/B))$ is provable. In the former case, $\exists X(A \to A(X/B))$ is (essentially) one half of an instance of comprehension.

For complex A, the case where X doesn't occur in A goes through as before. But for the other case things are trickier. After poking at the problem a bit, one sees that the right approach is to prove something slightly stronger; viz. that all instances of $\exists X(A \leftrightarrow A(X/B))$ are theorems. And since comprehension gives us this for atoms, the natural way to proceed is by induction on the complexity of A. The inductive hypothesis then gives that all formulas of the form $\exists X(A \leftrightarrow A(X/B))$ are theorems when A has at most k connectives and quantifiers. There are then induction cases for each connective and each quantifier. For most of these, elbow grease is sufficient to get things done. For the conditional, it's not.

To say more, let's make a few assumptions. For simplicity, we'll consider a conditional $A_1 \to A_2$ in which X doesn't occur free in A_1. Our goal is to show that we can derive $\exists X[(A_1 \to A_2) \leftrightarrow (A_1 \to A_2)(X/B)]$. Given our assumption, though, this is just $\exists X[(A_1 \to A_2) \leftrightarrow (A_1 \to A_2(X/B))]$.

Now, as we'll leave the reader to check, if we have axiomatic transitivity on board, we can then prove that the following is a theorem:[3]

$$(A_2 \leftrightarrow A_2(X/B)) \to ((A_1 \to A_2) \leftrightarrow (A_1 \to A_2(X/B)))$$

And from here it's not hard at all to get to the following:

$$\exists X(A_2 \leftrightarrow A_2(X/B)) \to \exists X((A_1 \to A_2) \leftrightarrow (A_1 \to A_2(X/B)))$$

And induction can do the rest of the job from there.

---

[2] See e.g. [13] for a discussion.

[3] Actually, you only need one of the two forms of axiomatic transitivity to do this. But the argument corresponding to the one I'm giving but assuming that X isn't free in A_2 relies on the other form of axiomatic transitivity so the point is moot.

The problem, now returning to the thread, is what to do if (as in the case at hand) we *don't* have the transitivity axioms in the logic we're working with. It's clear (in a highly suggestive comment that we'll be returning to below) that if we were working not with existentials but *instances*, the induction could still be done. That is, if rather than having $\exists X(A_2 \leftrightarrow A_2(X/B))$ as a theorem, we had (e.g.) $A_2(X/P) \leftrightarrow A_2(X/B)(X/P)$ as a theorem, then there would be no problem.[4] Indeed, were this the case, the derivation would be straightforward: just apply R3 to $A_1 \rightarrow A_1$ and the two halves of $A_2(X/P) \leftrightarrow A_2(X/B)(X/P)$ to get the two halves of $(A_1 \rightarrow A_2(X/P)) \leftrightarrow (A_1 \rightarrow A_2(X/B)(X/P))$, which is an instance of what we're looking for. Note that this argument will work given absolutely any *particular* instance, but it doesn't work if all we know is that *there is* an instance—that is if all we know is the existential $\exists X(A_2 \leftrightarrow A_2(X/B))$. The problem, in short is that absent transitivity axioms we just lack any resources that would let us 'pass' this fact about A_2 outside the scope of the existential.

We'll be returning below to the strange and suggestive bits of the above discussion. For now we'll note that what we can conclude is the following: if we indeed want a decently well-behaved universal on hand—one for which we can prove $\forall X(A \rightarrow A(X/B))$, then we need fuller-than-full comprehension; not just comprehension *for* every formula in the language, but also comprehension *within* every formula of the language. More helpfully, what's needed to ensure a well-behaved universal is *something* that ensures all of the following comprehension-esque sentences are theorems:

$$\exists X \forall y_1 \ldots \forall y_n (B \leftrightarrow B(X/A(z)))$$

Where the y_i are the variables that occur free in B. Call such a sentence a comprehension axiom *for A along B*. Note that in the case where $B = Xy$, this just is ordinary comprehension for A, so ordinary comprehension is just comprehension along atoms.

It remains unclear to me exactly how to model, semantically, a system incorporating comprehension not only *for* all formulas but also *along* all formulas. So, for the time being, we'll focus on what needs to be done to semantically model comprehension along atoms. And since there's nothing particularly interesting that's different between how things go in the unary case and how things go in the binary case, we'll present only the former.

To that end, let $\mathbb{A}$ be the set of formulas one wishes to adopt comprehension axioms for. Choose an appropriate class of vocabularies V and for each $A \in \mathbb{A}$, each variable y, and each $\overline{F} \in V$, let $in_{\overline{F}}(A, y)$ be the set of formulas that result from replacing free variables in A with appropriate members of (components of) $\overline{F}$ until at most y remains free in A.

[4] It's worth pointing out that $A_2(X/B)(X/P)$ needn't reduce to $A_2(X/B)$ since X might occur free in B.

The semantic condition we then require is this:

- For all $A \in \mathbb{A}$ and $B \in in_{\overline{F}}(A, y)$, if $\overline{F} \subsetneq \overline{G}$ and $F_1 \subsetneq G_1$, then there is $G \in G_1$ and $\overline{H} \supsetneq \overline{G}$ with $h \in H_0 - G_0$ so that $Gh \in v_{\overline{H}}(t)$ iff $S, \overline{H}, t \vDash B(y/h)$.

The loose idea of the condition is this: at every level $\overline{G}$, and for every instance B of a formula we want comprehension over that is defined at one of the levels $\overline{F}$ preceding $\overline{G}$, we require that there be a predicate $G \in G_1$ and a fresh constant h so that Gh is equivalent to $B(y/h)$.

Call a V-stratified model satisfying the above condition an $\mathbb{A}$, V-stratified model

Theorem 7 *Every theorem of* **B2Q**$^{\mathbb{A}}$ *is valid in every* $\mathbb{A}$, *V-stratified model.*

Proof Clearly the only thing to check is whether the comprehension axioms are valid. To that end, suppose $A \in \mathbb{A}$ and S is an $\mathbb{A}$, V-stratified model. Then $\exists X \forall y(Xy \leftrightarrow A)$ is valid in S just if all of its substitution instances are valid. In turn, this happens just if, for all $\overline{F} \in V$ and $B \in in_{\overline{F}}(A)$, $S, \overline{F}, \ell_{\overline{F}} \vDash \exists X \forall y(Xy \leftrightarrow B)$.

Note by Extendibility there is $\overline{G} \in V$ with $\overline{F} \subsetneq \overline{G}$ and $F_1 \subsetneq G_1$. Thus, by the new semantic condition, there is $G \in G_1$ and $\overline{H} \supsetneq \overline{G}$ with $h \in H_0 - G_0$ so that $Gh \in v_{\overline{H}}(t)$ iff $t \vDash B(y/h)$. Choose $\ell_{\overline{F}} \sqsubseteq_{P_{\overline{F}}} p$. By Prime Extension Down, there is then $\ell_{\overline{G}} \sqsubseteq q$ with $q{\downarrow}^{\overline{G}}_{\overline{F}} = p$. But since $\ell_{\overline{G}} \sqsubseteq q$, $\ell_{\overline{H}} \sqsubseteq q{\uparrow}^{\overline{H}}_{\overline{G}}$.

Now, since $Gh \in v_{\overline{H}}(t)$ iff $S, \overline{H}, t \vDash B(y/h)$, $S, \overline{H}, \ell_{\overline{H}} \vDash Gh \leftrightarrow B(y/h)$. So since $\ell_{\overline{H}} \sqsubseteq q{\uparrow}^{\overline{H}}_{\overline{G}}$, $S, \overline{H}, q{\uparrow}^{\overline{H}}_{\overline{G}} \vDash Gh \leftrightarrow B(y/h)$. Thus $S, \overline{G}, q \vDash \forall y(Gy \leftrightarrow B)$. It follows that $S, \overline{F}, \ell_{\overline{F}} \vDash \exists X(Xy \leftrightarrow B)$. □

Theorem 8 *If C is a nontheorem of* **B2Q**$^{\mathbb{A}}$, *then there is an* $\mathbb{A}$, *V-stratified model in which C isn't valid.*

Proof Extend the usual canonical model by adding to the language designated predicates that do what we need. □

4.3 The 'Full' Semantics

Call a function $T_{\overline{F}} \longrightarrow 2^{F_0}$ a generalized formula. Loosely, a generalized formula represents a way an atomic formula might have been interpreted. In the full semantics, at every level, every generalized formula in fact represents an atom; more to the point, we require that the following be satisfied:

- If $\phi : T_{\overline{F}} \longrightarrow 2^{F_0}$ is a function, then there is $F \in F_1$ so that $Fh \in v_{\overline{F}}(t)$ iff $h \in \phi(t)$.

Clearly full models are $\mathbb{A}$,V-stratified models for all $\mathbb{A}$. Equally clear is that, qua models of language, they're quite odd: each language at each level has uncountably many predicates, and if $F_0 \subsetneq G_0$, then $\overline{G}$ is uncountably enriched over $\overline{F}$. But such are the wages of sin.

Let's write $\mathbf{B2Q}^{full}$ for the set of formulas that are valid in all full models. Whether $\mathbf{B2Q}^{full}$ admits a recursive axiomatization at all is not clear to me. There are well-known reasons for suspecting it isn't that I won't rehearse. But there is also surprising evidence on the other side; e.g. in [9] it's shown that the second-order version of FDE (which many take to be a relevant logic) *is* recursively axiomatizable.

In fact, [9] proves a range of interesting things about second-order FDE (and about second-order LP) and it would be a fun project to see exactly which of them remain true about second-order **B**—or other second-order relevant logics, for that matter. But those are jobs for the future.

5 Adding Metarules

Let's return to the funny business involving comprehension and look at a concrete example of what goes wrong. So, suppose we wanted to prove the following was a theorem of the system we get by adding $\exists X \forall y(Xy \leftrightarrow Ray)$ as our lone comprehension axiom:

$$\exists X \forall y_1 \forall y_2((Xy_1 \rightarrow Xy_2) \leftrightarrow (Ray_1 \rightarrow Ray_2))$$

A natural thought is that we would prove this by showing it follows from the only thing it really could follow from (namely the comprehension axiom we've added) and that we'd accomplish this goal by proving the following:

$$\exists X \forall y(Xy \leftrightarrow Ray) \rightarrow \exists X \forall y_1 \forall y_2((Xy_1 \rightarrow Xy_2) \leftrightarrow (Ray_1 \rightarrow Ray_2))$$

This, in turn would most naturally be proved by using R9 applied to the following:

$$\forall X[\forall y(Xy \leftrightarrow Ray) \rightarrow \exists X \forall y_1 \forall y_2((Xy_1 \rightarrow Xy_2) \leftrightarrow (Ray_1 \rightarrow Ray_2))]$$

Of course, the right way to prove a universal is by universalization, so we expect to first prove the following:

$$\forall y(Xy \leftrightarrow Ray) \rightarrow \exists X \forall y_1 \forall y_2((Xy_1 \rightarrow Xy_2) \leftrightarrow (Ray_1 \rightarrow Ray_2))$$

And, since the existential consequent here follows from an instance, we expect to prove *that* by proving *this*:

$$\forall y(Xy \leftrightarrow Ray) \rightarrow \forall y_1 \forall y_2((Xy_1 \rightarrow Xy_2) \leftrightarrow (Ray_1 \rightarrow Ray_2))$$

But this is really just a universalized version of axiomatic transitivity. Thus, one obvious way to get from ordinary comprehension to all the fancy comprehension we really want is by adding axiomatic transitivity to our logic.

But this isn't the only way to go; an alternative way to do things is to add Bradian metarules.

Before doing so, a confession: I've always been a bit befuddled by Brady's metarules. So what I'd like to do here is describe how I've come to understand what they mean. And since we're at the confessing game, I'll also go ahead and tell you that since what I'm about to say is a bit complicated and what Brady usually says (see e.g. [1] or [2] or [3]) isn't, I worry that maybe I'm getting things wrong. I suppose if that *is* the case, then what we need isn't Bradian metarules, but pseudoBradian metarules.

Whichever way it is, we need some definitions. We define the terms 'derivation of A' and 'metaderivation of R in which $x/X/Y$ has/hasn't been universalized' by simultaneous recursion as follows:

- Each instance of one of A1–A12 is a derivation of itself.
- Each instance of one of R1–R4 and R8–R10 is a metaderivation of itself in which nothing has been universalized.
- Each instance of R5 is a metaderivation of itself in which x has been universalized.
- Each instance of R6 is a metaderivation of itself in which X has been universalized.
- Each instance of R7 is a metaderivation of itself in which Y has been universalized.
- If Δ is a metaderivation of $\dfrac{A_1 \ldots A_n}{B}$ and $\delta_1, \ldots, \delta_n$ are derivations of $A_1, \ldots, A_n$, then $\Delta[\delta_1, \ldots, \delta_n] := \dfrac{\delta_1 \cdots \delta_n}{B}$ is a derivation of B.
- If Δ_1 is a metaderivation of $\dfrac{A_1 \cdots A_n}{B}$ and Δ_2 is a metaderivation of $\dfrac{B \qquad C_1 \cdots C_m}{D}$ then $\Delta_3 := \left(\dfrac{\dfrac{A_1 \cdots A_n}{B} \qquad C_1 \cdots C_m}{D}\right)$ is a metaderivation of $\dfrac{A_1 \cdots A_n \qquad C_1 \cdots C_m}{D}$ x (resp. X; Y) has been universalized in Δ_3 iff it has been universalized in Δ_1 or in Δ_2.
- If δ is a derivation of A and Δ_1 is a metaderivation of $\dfrac{A \qquad B_1 \cdots B_n}{C}$ then $\Delta_2 := \left(\dfrac{\delta \qquad B_1 \cdots B_n}{C}\right)$ is a metaderivation of $\dfrac{B_1 \cdots B_n}{C}$ x (resp. X; Y) has been universalized in Δ_2 iff it has been universalized in Δ_1.[5]

[5] Note that using the generalization rules in the derivation δ does *not* make it the case that one has used generalization in Δ_2.

- If Δ is a metaderivation of $\dfrac{A}{B}$ in which x (resp. X; Y) has not been universalized, then $\dfrac{\exists x A}{\exists x B}$ (resp. $\dfrac{\exists X A}{\exists X B}$; $\dfrac{\exists Y A}{\exists Y B}$) is a metaderivation of itself in which x (resp. X; Y) has not been universalized.

We say that A is a theorem just if there exists a derivation of A.

The 'metarule' bit is in the last bullet in the above list. The idea is that we close the set of rules under a metarule that roughly looks like this:

$$\text{If } A \Rightarrow B, \text{ then } \exists x A \Rightarrow \exists x B$$

Recall now, if you will, the funny business about comprehension from the previous section. The issue was this: to get a well-behaved universal, we need our logic to contain as theorems not only all instances of comprehension *for* all formulas, but also *along* all formulas. We also saw that, provided we had axiomatic transitivity along, giving ourselves access to comprehension for all formulas along atoms—which is to say, giving ourselves what's usually called 'full' comprehension—sufficed for ensuring all instances of comprehension both for and along all formulas.

The argument establishing this broke if we didn't have transitivity. But as we noted, there was something a bit fishy going on—a slightly different argument *did* work given any *particular* instance, but not if all we knew was that *there was* an instance.

Examining the explanation of the metarule above, one suspects that it directly gets around this issue. Indeed, what the metarule seems to say is that anytime you have an argument that works when given an instance, you're allowed to conclude that its existential analogue works.

Happily, this suspicion turns out to be correct: **B2Q**$(\overline{F})$ contains a metaderivation of the following in which no predicate universalization at all occurs:

$$\frac{\forall y(Xy \leftrightarrow A(y))}{\forall y_1 \forall y_2((Xy_1 \to Xy_2) \leftrightarrow (A(y_1) \to A(y_2)))}$$

Thus, the metarule on board, the following is a metaderivation of itself:

$$\frac{\exists X \forall y(Xy \leftrightarrow A(y))}{\exists X \forall y_1 \forall y_2((Xy_1 \to Xy_2) \leftrightarrow (A(y_1) \to A(y_2)))}$$

So if we write **B2QE** for **B2Q** plus the existential metarule, then if $A \in \mathbf{A}$, then $\exists X \forall y_1 \forall y_2((Xy_1 \to Xy_2) \leftrightarrow (A(y_1) \to A(y_2)))$ is a theorem of $\mathbf{B2QE}^{\mathbb{A}}$. And from here, we can very easily finish the job we started in §4.2 of showing that $\forall X(A \to A(X/B))$ is a theorem.

In short, adding the metarule lets us bootstrap our way from comprehension along atoms to comprehension along formulas of arbitrary complexity, and thus to a well-behaved universal.

Here are two more fun facts about Bradian metarules. Fun fact the first: I don't know how to modify my semantic story to accommodate this metarule. Fun fact the second: it's entirely clear how to modify the semantics to accommodate the other Bradian metarule; viz. the metarule 'if $A \Rightarrow B$, then $C \vee A \Rightarrow C \vee B$'. I'll leave the proof to the reader, but all it takes to model this is adding the condition $\ell \in P$ to the semantics. So I suspect that there is a natural way to accommodate the above metarule in the semantics, I just haven't figured out what it is yet.

6 Conclusion

Some natural options for next steps include adding λ terms, exploring second-order analogues of various relevant theories—arithmetic comes to mind, given [6]—and a rethinking of identity using higher-order resources. Of course, the most glaring thing to do as a next step is to figure out exactly what to do about comprehension.

An alternative next step would be to attempt to work *backwards* from an appropriate second (or higher) order relevant logic to a relevant set theory of some sort. It's been shown (see [5]) that such theories aren't going to work in exactly the way one might have hoped. But there's also good reason (see e.g. [4]) to think such theories have some promise.

All of this points to a general thing worth keeping in mind: there are a range of interesting *expressive* extensions of the usual languages that relevance logicians concern themselves with that are massively, embarrassingly underexplored. The problem seems to be essentially the one we encountered with comprehension above: in many cases, the classical logicians have set the syllabus and decided not just how things ought to be explored, but what there is to explore in the first place. But we're not in school anymore, and thus there's no real reason to stick to the syllabus. So go on; explore the strange rich expressive potential we've been gifted. And when you come back, tell us how the semantics goes.

References

1. Brady, R. T. (1984). Natural deduction systems for some quantified relevant logics. *Logique et Analyse*, *27*(108), 355–377.
2. Brady, R. T. (1989). A content semantics for quantified relevant logics. II. *Studia Logica*, *48*, 243–257.
3. Brady, R. T. (2006). *Universal logic*. CSLI Publications.
4. Brady, R. T. (2014). The simple consistency of naive set theory using metavaluations. *Journal of Philosophical Logic*, *43*, 261–281.

5. Dunn, J. M. (1988). The impossibility of certain higher-order non-classical logics with extensionality. In D. F. Austin (Ed.), *Philosophical analysis: A defense by example* (pp. 261–279). Springer.
6. Ferguson, T. M., & Priest, G. (2021). Robert Meyer's publications on relevant arithmetic. *Australasian Journal of Logic*, *18*(5), 146–149.
7. Fine, K. (1974). Models for entailment. *Journal of Philosophical Logic*, *3*(4), 347–372.
8. Fine, K. (1988). Semantics for quantified relevance logic. *Journal of Philosophical Logic*, *17*(1), 27–59.
9. Hazen, A. P., & Pelletier, F. J. (2018). Second-order logic of paradox. *Notre Dame Journal of Formal Logic*, *59*(4), 547–558.
10. Logan, S. A. (2019). Notes on stratified semantics. *Journal of Philosophical Logic*, *48*(4), 749–786.
11. Logan, S. A. (2020). Putting the stars in their places. *Thought: A Journal of Philosophy*, *9*(3), 188–197.
12. Logan, S. A. (2024). *Relevance logic*. Cambridge University Press.
13. Rossberg, M. (2006). *Second-order logic: Ontological and epistemological problems*. PhD thesis, St Andrews.
14. Routley, R., Plumwood, V., Meyer, R. K., & Brady, R. T. (1982). *Relevant logics and their rivals*. Ridgeview.

Implying and Containing in Truthmaker Semantics

Vít Punčochář and Thomas M. Ferguson

Abstract Kit Fine's truthmaker semantics allows us to distinguish between implying and containing: *A implies B* iff every truthmaker of *A* includes (as a part) a truthmaker of *B*, and *A contains B* iff every truthmaker of *B* is included (as a part) in a truthmaker of *A*. These are two different notions that can both be transformed into a proper object language connective. While the former notion amounts to the usual notion of implication, the latter one leads to a novel connective with interesting logical properties. In this paper we put forward a truthmaker semantics for the implication of the relevant logic R and we accompany it with the corresponding "containing" operator. This framework will allow us to relate two seemingly very different kinds of relevant implication that has so far been studied separately, namely the implication of the logic R and Angell's analytic containment. Our approach semantically determines a logic that we call RAC. We explore its connection to R, the logic of analytic containment AC, Urquhart's operational semantics and inquisitive semantics.

1 Introduction

The general aim of relevant logic is to capture a notion of implication that encompasses relevance: Such an implication can hold only if there is a relevant connection between the antecedent and the consequent. One can also consider a notion of containment that imposes an additional requirement that the meaning of the consequent is in a sense contained in the meaning of the antecedent. This is inspired by Kant's characterization of analyticity according to which a judgement of the form *S is P* is analytic if the concept of the predicate *P* is contained in the concept

V. Punčochář (✉)
Institute of Philosophy, Czech Academy of Sciences, Prague, Czech Republic
e-mail: puncochar@flu.cas.cz

T. M. Ferguson
Department of Cognitive Science, Rensselaer Polytechnic Institute, Troy, NY, USA

I. Sedlár et al. (eds.), *New Directions in Relevant Logic*, Trends in Logic 63,
https://doi.org/10.1007/978-3-031-69940-5_9

of the subject S. Following Parry [22, 23], Angell [2, 3] and Fine [16], we take this criterion, which was originally formulated for subject-predicate judgements, and apply it to implications.

In various relational semantic frameworks, e.g. in Kripke semantics for modal and intuitionistic logic or in Routley-Meyer semantics for relevant logic, we obtain the following property of implication: An implication is valid in a model if and only if

(A) every state in which the antecedent holds is a state in which the consequent holds.

This does not apply to Kit Fine's [15] truthmaker semantics for intuitionistic logic and to its generalization [20] for substructural logics, where the characterization of implication is a bit more subtle: An implication is valid in a model if and only if

(B) every "exact" truthmaker of the antecedent includes (as a part) an "exact" truthmaker of the consequent.

The more refined logical structure of the condition (B) invites us to consider its symmetric counterpart that is not available in Kripke and Routley-Meyer semantics (simply because (A) does not have any interesting symmetric counterpart of this kind):

(C) every "exact" truthmaker of the consequent is included (as a part) in an "exact" truthmaker of the antecedent.

The conditions (B) and (C) were used together in [16] for a semantic characterization of Angell's [2, 3] notion of "analytic containment".

In this paper, we consider two different logical connectives, $\rightharpoondown$ and $\rightharpoonup$, respectively related to the conditions (B) and (C). While $\rightharpoondown$ amounts to the usual implication, $\rightharpoonup$ is novel, and, as far as we know, it has not been considered before as a separate logical connective. The condition (C) suggests that $\varphi \rightharpoonup \psi$ can be interpreted as meaning that φ *contains* ψ. This notion of containment does not however directly coincide with Angell's analytic containment. Only when $\rightharpoonup$ is combined with $\rightharpoondown$ we obtain a connective $\Rightarrow$ related to analytic containment, so that *A analytically contains B* in Angell's sense if and only if *A implies* and *contains B* in our sense ($\varphi \Rightarrow \psi =_{def} (\varphi \rightharpoondown \psi) \wedge (\varphi \rightharpoonup \psi)$).

A merit of our approach is that it allows us to treat analytic containment as a proper object language connective that can be arbitrarily embedded into more complex formulas. In this respect, our approach goes beyond Angell's and Fine's treatment of analytic containment that does not allow nesting and embedding of analytic implication.

Our approach reconciles two different treatments of relevant implication and captures them both in a uniform framework. The first approach, represented here by the paradigmatic relevant implication of the logic R [6], is nowadays more standard. The second approach, related to Parry's [22, 23] and Angell's [2, 3] notions of analytic implication and containment, is perhaps a bit less familiar. Generally speaking, as Sylvan et al. point out in [26], the two schools differ with respect to the idioms through which such relevance is enforced; in particular, [26] argue that the Parry/Angell approach to relevance rests on ad hoc devices rather than emerging naturally from prior semantic intuitions. Thus far there has been little investigation into the degree to which these two approaches are connected or whether they can be reconciled. In the one case of which we are aware—Sylvan's "synthesis" in [29]—the synthesis between mainstream relevance and Parry/Angell-style relevance emerges only artificially. In what follows, we will show that the two treatments can interact in an intriguing, uniform, and highly natural way.

2 Truthmaker Semantics for RAC

The semantic framework that we use is a variation of Fine's *truthmaker semantics*. Kit Fine formulated a truthmaker semantics for Angell's system AC in [16]. In [15] he also introduced truthmaker semantics for intuitionistic logic which was generalized to various substructural logics including the implicational fragment of R [20].[1]

The framework has a strong algebraic flavour. The structures that will serve as frames of our semantics are distributive complete partially ordered residuated commutative and contractive monoids (turned upside down). We will call them simply E-frames (exact frames). Let us first properly introduce these structures.

We say that a partial order $\langle S, \sqsubseteq \rangle$ is *complete* if every $T \subseteq S$ has the least upper bound $\bigsqcup T \in S$. In each complete partial order there is also the greatest lower bound $\bigsqcap T$ for each $T \subseteq S$ that can be defined as the least upper bound of all lower bounds of T. Each complete partial order has the least element $\bigsqcup \emptyset$, and the greatest element $\bigsqcap \emptyset$. As usual, we write $s \sqcup t$ and $s \sqcap t$ instead of $\bigsqcup \{s, t\}$ and $\bigsqcap \{s, t\}$. We say that a complete partial order is *distributive* if the following two equations generally hold:[2]

$$s \sqcup (t \sqcap u) = (s \sqcup t) \sqcap (s \sqcup u) \text{ and } s \sqcap (t \sqcup u) = (s \sqcap t) \sqcup (s \sqcap u).$$

A *monoid* is a structure $\langle S, *, l \rangle$, where $S \neq \emptyset$, $*$ is an associative binary operation on S, and l is a neutral element with respect to $*$, i.e. $l * s = s * l = s$, for every $s \in S$. A monoid $\langle S, *, l \rangle$ is called *commutative* if $*$ is commutative.

Definition 1 An *E-frame* is a structure $\langle S, \sqsubseteq, *, l \rangle$, where $\langle S, \sqsubseteq \rangle$ is a distributive complete partial order and $\langle S, *, l \rangle$ is a commutative monoid satisfying the following three conditions:

---

[1] A different approach to the truthmaker semantics for R was developed in [19]. An interesting approach to substructural logics via truthmaker semantics is developed also in [17].

[2] As is well-known, one can require only one of these distributivity conditions and the other follows.

(a) for every $s, t \in S$ there is $u \in S$ such that $s \sqsubseteq t * u$;
(b) for every $s \in S$ and $T \subseteq S$, $s * \bigsqcap T = \bigsqcap\{s * t \mid t \in T\}$;
(c) for every $s \in S$, $s * s \sqsubseteq s$.

We view the elements of S in an E-model as *situations* in the sense of situation semantics [5]. The ordering $\sqsubseteq$ is viewed as the part-whole relation. There are two different notions of fusion in the semantics. First, $s \sqcup t$ is understood as a fusion of the situations s and t in the sense that we just put together the two situations by which we create a bigger one, which includes both. The operation $*$ can also be viewed as a fusion but in a slightly different sense. The situation $s * t$ is more like a result of applying a function to its argument. This will be further illustrated in the following paragraphs and in Lemma 1 below. The situation l is interpreted as logical situation that contains the truthmakers of logical laws.

The conditions (a) and (b) from Definition 1 give us residuation, or, more precisely, the dual of residuation. This aspect of truthmaker semantics might be a bit confusing and we should clarify it here. Truthmaker semantics, like algebraic semantics, employs certain algebraic structures, but these structures are dual to those of algebraic semantics. What is treated as meet in algebraic semantics is regarded as join (fusion) in truthmaker semantics. This makes good sense because in algebraic semantics meet corresponds to conjunction but in truthmaker semantics conjunction is characterized in terms of fusion, i.e. algebraic join. Given this dual perspective, we define, in accordance with [15], or more precisely with the generalization in [20], for every $s, t \in S$ in a given E-frame:

$$s \Rightarrow t = \bigsqcap\{u \in S \mid t \sqsubseteq s * u\}.$$

Fine [15] defines $s \Rightarrow t$ in terms of $\sqcup$ instead of $*$ and calls the state $s \Rightarrow t$ *conditional connection*. It represents a path from s to t. By Fine's own words $s \Rightarrow t$ amounts to "the state of s's leading to t". We think that $s \Rightarrow t$ can also be viewed as an *information channel* in the sense of [4]. The most important feature of $\Rightarrow$ is described in the following lemma which shows that applying a channel $s \Rightarrow t$ to the situation s will result in a situation that includes t.

Lemma 1 *In every E-frame the following law holds for every $s, t \in S$:*

$$t \sqsubseteq s * (s \Rightarrow t).$$

Proof $t \sqsubseteq \bigsqcap\{s * u \in S \mid t \sqsubseteq s * u\} = s * \bigsqcap\{u \in S \mid t \sqsubseteq s * u\} = s * (s \Rightarrow t)$. □

We will also need the following properties of E-frames, which will be used without an explicit reference.

Lemma 2 *Let $\langle S, \sqsubseteq, *, l\rangle$ be an E-frame. Then it holds for all $s, t, u \in S$:*

(a) $s \Rightarrow t \sqsubseteq l$ *if and only if* $t \sqsubseteq s$;
(b) *if* $s \sqsubseteq t$ *then* $s * u \sqsubseteq t * u$;
(c) *if* $s \sqsubseteq t$ *then* $t \Rightarrow u \sqsubseteq s \Rightarrow u$;
(d) $s \Rightarrow t \sqsubseteq (s \Rightarrow u) \sqcup (u \Rightarrow t)$.

Now we introduce our main object language L. It is like a common propositional language which, besides the connectives $\neg$, $\vee$, $\wedge$ includes two implication-like and two further conjunction-like connectives the role of which is explained below:

$$\varphi ::= p \mid \neg\varphi \mid \varphi \vee \varphi \mid \varphi \wedge \varphi \mid \varphi \underset{\cdot}{-} \varphi \mid \varphi \rightharpoondown \varphi \mid \varphi \dot{-} \varphi \mid \varphi \rightharpoonup \varphi \text{ (the language } L\text{).}$$

Later on, $\rightharpoondown$ and $\rightharpoonup$ will be respectively related to $\underset{\cdot}{-}$ and $\dot{-}$. We define "B-equivalence" (where "B" stands for "Belnap" because this equivalence will correspond to the equivalence of the logic R)

$$\varphi \leftrightharpoons \psi =_{def} (\varphi \rightharpoondown \psi) \wedge (\psi \rightharpoondown \varphi)$$

and two more connectives that we call "A-implication" and "A-equivalence" (where "A" can stand for "analytic" or for "Angell" because these connectives will be respectively related to Angell's notions of analytic implication and analytic equivalence)

$$\varphi \Rightarrow \psi =_{def} (\varphi \rightharpoondown \psi) \wedge (\varphi \rightharpoonup \psi) \text{ and } \varphi \Leftrightarrow \psi =_{def} (\varphi \Rightarrow \psi) \wedge (\psi \Rightarrow \varphi).$$

Definition 2 An *E-model* is a tuple $\langle S, \sqsubseteq, *, l, V^+, V^- \rangle$, where $\langle S, \sqsubseteq, *, l \rangle$ is an E-frame and both V^+ and V^- are functions assigning to each atomic formula a non-empty subset of S. An E-model is called *simple* if $V^+(p)$ and $V^-(p)$ are singletons, for every atomic formula p.

Now, relative to a given E-model $\mathcal{M} = \langle S, \sqsubseteq, *, l, V^+, V^- \rangle$, we define (using Fine's terminology) an *exact truthmaking relation* $\Vdash^+_{\mathcal{M}}$ and an *exact falsemaking relation* $\Vdash^-_{\mathcal{M}}$ relating elements of S and L-formulas. So, $s \Vdash^+_{\mathcal{M}} \varphi$ says that s is an exact truthmaker of φ, and $s \Vdash^-_{\mathcal{M}} \varphi$ says that s is an exact falsemaker of φ.

The meaning of "exactness" in this context is the following one: A state exactly verifies (falsifies) a statement only if the state is wholly (or exactly) relevant to establishing the truth (or falsity) of the statement. So, typically, a state exactly verifies a statement even if it is contained (as a part) in a bigger state that does not exactly verify the statement (because the bigger state is no longer *wholly* relevant). In this respect, exact truthmaker semantics is unlike Kripke semantics for intuitionistic logic or Routley-Meyer semantics for relevant logic because it does not require any persistence conditions. This notion of exact verification/falsification itself suggests an intuitive connection to the broad project of relevant logic.

We will denote the set of (exact) truthmakers and falsemakers of φ respectively as $[\varphi]^+_{\mathcal{M}}$ and $[\varphi]^-_{\mathcal{M}}$. So,

$$[\varphi]^+_{\mathcal{M}} = \{s \in S \mid s \Vdash^+_{\mathcal{M}} \varphi\} \text{ and } [\varphi]^-_{\mathcal{M}} = \{s \in S \mid s \Vdash^-_{\mathcal{M}} \varphi\}.$$

The subscript $\mathcal{M}$ will be omitted if no confusion arises. The relations $\Vdash^+$ and $\Vdash^-$ are defined recursively in Table 1 and explained below.

Table 1 Exact truthmaking and falsemaking conditions

1. $s \Vdash^+ p$ iff $s \in V^+(p)$,
2. $s \Vdash^- p$ iff $s \in V^-(p)$,
3. $s \Vdash^+ \neg\varphi$ iff $s \Vdash^- \varphi$,
4. $s \Vdash^- \neg\varphi$ iff $s \Vdash^+ \varphi$,
5. $s \Vdash^+ \varphi \vee \psi$ iff $s \Vdash^+ \varphi$ or $s \Vdash^+ \psi$,
6. $s \Vdash^- \varphi \vee \psi$ iff $\exists t, u \in S$: $t \Vdash^- \varphi$, $u \Vdash^- \psi$ and $s = t \sqcup u$,
7. $s \Vdash^+ \varphi \wedge \psi$ iff $\exists t, u \in S$: $t \Vdash^+ \varphi$, $u \Vdash^+ \psi$ and $s = t \sqcup u$,
8. $s \Vdash^- \varphi \wedge \psi$ iff $s \Vdash^- \varphi$ or $s \Vdash^- \psi$,
9. $s \Vdash^+ \varphi \mathbin{\underset{\cdot}{-}} \psi$ iff $\exists t, u \in S$: $t \Vdash^+ \varphi$, $u \Vdash^+ \psi$ and $s = t * u$,
10. $s \Vdash^- \varphi \mathbin{\underset{\cdot}{-}} \psi$ iff $\exists f : [\varphi]^+ \to [\psi]^-$ such that $s = \bigsqcup\{t \Rightarrow f(t) \mid t \in [\varphi]^+\}$,
11. $s \Vdash^+ \varphi \rightharpoondown \psi$ iff $\exists f : [\varphi]^+ \to [\psi]^+$ such that $s = \bigsqcup\{t \Rightarrow f(t) \mid t \in [\varphi]^+\}$,
12. $s \Vdash^- \varphi \rightharpoondown \psi$ iff $\exists t, u \in S$: $t \Vdash^+ \varphi$, $u \Vdash^- \psi$ and $s = t * u$,
13. $s \Vdash^+ \varphi \dot{-} \psi$ iff $\exists t, u \in S$: $t \Vdash^+ \varphi$, $u \Vdash^+ \psi$ and $s = t * u$,
14. $s \Vdash^- \varphi \dot{-} \psi$ iff $\exists f : [\psi]^- \to [\varphi]^+$ such that $s = \bigsqcup\{f(t) \Rightarrow t \mid t \in [\psi]^-\}$,
15. $s \Vdash^+ \varphi \rightharpoonup \psi$ iff $\exists f : [\psi]^+ \to [\varphi]^+$ such that $s = \bigsqcup\{f(t) \Rightarrow t \mid t \in [\psi]^+\}$,
16. $s \Vdash^- \varphi \rightharpoonup \psi$ iff $\exists t, u \in S$: $t \Vdash^+ \varphi$, $u \Vdash^- \psi$ and $s = t * u$.

Definition 3 An L-formula φ is valid in $\mathcal{M}$ if the logical state of $\mathcal{M}$ contains a truthmaker of φ, i.e. there is $s \sqsubseteq l$ (in $\mathcal{M}$) such that $s \Vdash^+ \varphi$.

The clauses for atomic formulas, negation, conjunction and disjunction are the usual clauses of truthmaker semantics that were originally taken from [31]. The positive clause for the implication $\rightharpoondown$ mirrors the clause for intuitionistic implication from [15] with the difference that in [15] the conditional connections are defined in terms of $\sqcup$ and we define them in terms of $*$. By this modification we achieve to obtain that the implication is relevant rather than intuitionistic. We will see that it corresponds to the implication of the relevant logic R. Note that any truthmaker of an implication encodes a function from the truthmakers of the antecedent to the truthmakers of the consequent. Such a truthmaker is a fusion of the conditional connections leading from the antecedent proposition to the consequent proposition.

The conjunction $\mathbin{\underset{\cdot}{-}}$ amounts to the intensional conjunction of the logic R. This determines its positive semantic clause which should be compared with the clause for $\wedge$. The negative clauses for $\rightharpoondown$ and $\mathbin{\underset{\cdot}{-}}$ are defined so that we obtain, in correspondence to R, the following equations:[3]

$$[\neg(\varphi \rightharpoondown \psi)]^+ = [\varphi \mathbin{\underset{\cdot}{-}} \neg\psi]^+ \text{ and } [\neg(\varphi \mathbin{\underset{\cdot}{-}} \psi)]^+ = [\varphi \rightharpoondown \neg\psi]^+$$

The connective $\rightharpoonup$ is introduced by a semantic symmetry with $\rightharpoondown$. To describe its intended meaning we will employ the following notions.

Definition 4 For each E-frame $\mathcal{F} = \langle S, \sqsubseteq, *, l\rangle$ we define two preorders, $\preccurlyeq$ and $\dot{\preccurlyeq}$, on the power set of S. Let $T, U \subseteq S$. Then

[3] One may note that in this case the style of bilateralism espoused in [16] dovetails neatly with the Nelsonian style of bilateralism described in [21].

(a) $U \preccurlyeq T$ iff $\forall t \in T\ \exists u \in U : u \sqsubseteq t$;
(b) $U \curlyeqprec T$ iff $\forall u \in U\ \exists t \in T : u \sqsubseteq t$.

We say that (in $\mathcal{F}$ or in an E-model based on $\mathcal{F}$) (a) T *implies* U iff $U \preccurlyeq T$; (b) T *contains* U iff $U \curlyeqprec T$; (c) T *analytically implies* U iff T implies and contains U; (d) T and U are *synonymous* iff T analytically implies U and U analytically implies T.

Let us make two brief comments concerning Definition 4. First, we acknowledge some need for caution with respect to the notation of $\preccurlyeq$ and $\curlyeqprec$. Brink's work on *power structures* in e.g. [7] and [8] uses a dual notation in which these relations between sets are symbolized as $\lessdot$ and $\leqslant$, respectively. Second, it is worth mentioning that the use of the term "synonymity" is aligned with Angell's goals for AC, which is defended as a logic capturing synonymy in [3].

So, a proposition T implies a proposition U if every exact truthmaker of T includes an exact truthmaker of U. The notion of analytic implication requires further that T contains U, which means that every exact truthmaker of U is included in an exact truthmaker of T. Now we can observe how these notions are internalized in the object language by the connectives $\rightharpoondown$ and $\rightharpoonup$.

Lemma 3 *In every E-model $\mathcal{M}$, and for all L-formulas φ, ψ:*

(a) $\varphi \rightharpoondown \psi$ *is valid in* $\mathcal{M}$ *iff* $[\varphi]^+$ *implies* $[\psi]^+$ *in* $\mathcal{M}$;
(b) $\varphi \rightharpoonup \psi$ *is valid in* $\mathcal{M}$ *iff* $[\varphi]^+$ *contains* $[\psi]^+$ *in* $\mathcal{M}$;
(c) $\varphi \Rightarrow \psi$ *is valid in* $\mathcal{M}$ *iff* $[\varphi]^+$ *analytically implies* $[\psi]^+$ *in* $\mathcal{M}$;
(d) $\varphi \Leftrightarrow \psi$ *is valid in* $\mathcal{M}$ *iff* $[\varphi]^+$ *and* $[\psi]^+$ *are synonymous in* $\mathcal{M}$.

Proof

(a) We can observe that the following claims are equivalent:

1. $\varphi \rightharpoondown \psi$ is valid in $\mathcal{M}$, i.e. there is $s \sqsubseteq l$ such that $s \Vdash^+ \varphi \rightharpoondown \psi$,
2. there is an $f : [\varphi]^+ \to [\psi]^+$ such that $\bigsqcup\{t \Rightarrow f(t) \mid t \in [\varphi]^+\} \sqsubseteq l$,
3. there is an $f : [\varphi]^+ \to [\psi]^+$ such that $f(t) \sqsubseteq t$, for all $t \in [\varphi]^+$,
4. for every $t \in [\varphi]^+$ there is $u \in [\psi]^+$ such that $u \sqsubseteq t$,
5. $[\psi]^+ \preccurlyeq [\varphi]^+$.

(b) Similarly, one can observe that the following are equivalent:

1. $\varphi \rightharpoonup \psi$ is valid in $\mathcal{M}$, i.e. there is $s \sqsubseteq l$ such that $s \Vdash^+ \varphi \rightharpoonup \psi$,
2. there is an $f : [\psi]^+ \to [\varphi]^+$ such that $\bigsqcup\{f(t) \Rightarrow t \mid t \in [\psi]^+\} \sqsubseteq l$,
3. there is an $f : [\psi]^+ \to [\varphi]^+$ such that $t \sqsubseteq f(t)$, for all $t \in [\psi]^+$,
4. for every $t \in [\psi]^+$ there is $u \in [\varphi]^+$ such that $u \sqsubseteq t$,
5. $[\psi]^+ \curlyeqprec [\varphi]^+$.

(c) and (d) follow from (a) and (b).

□

We have not discussed yet the negative clause for $\rightharpoonup$. There are more possible alternative ways of how it can be reasonably defined. Intuition gives us no strong clue in this case but we have decided to define the negative clause so that some symmetries between the positive and negative clauses are preserved. The disadvantage of this approach is that we had to introduce yet another conjunction $\dot{-}$ that interacts with $\rightharpoonup$ in analogy to how $\underset{\cdot}{-}$ interacts with $\rightharpoondown$. The advantage is that when we fix the clauses as in Table 1 we obtain the following symmetries:

$$[\varphi \underset{\cdot}{-} \psi]^- = [\varphi \rightharpoondown \neg\psi]^+ \text{ and } [\varphi \dot{-} \psi]^- = [\varphi \rightharpoonup \neg\psi]^+$$

$$[\varphi \rightharpoondown \psi]^- = [\varphi \rightharpoonup \psi]^- = [\varphi \underset{\cdot}{-} \neg\psi]^+ = [\varphi \dot{-} \neg\psi]^+$$

In general, we have $[\varphi \underset{\cdot}{-} \psi]^+ = [\varphi \dot{-} \psi]^+$ but not $[\varphi \underset{\cdot}{-} \psi]^- = [\varphi \dot{-} \psi]^-$, and we have $[\varphi \rightharpoondown \psi]^- = [\varphi \rightharpoonup \psi]^-$ but not $[\varphi \rightharpoondown \psi]^+ = [\varphi \rightharpoonup \psi]^+$.

Now we are ready to introduce our logic RAC that, as we will see, incorporates the implication of the logic R as well as Angell's analytic implication of the logic AC. For the sake of comparison, we define, besides the logic RAC, also its generalized version that we call GenRAC and that is also of some interest. Note that it follows immediately from the definition that RAC is at least as strong as GenRAC. It will turn out later that it is properly stronger.

Definition 5 We say that an L-formula φ is logically valid in GenRAC, and we write $\vDash_{\mathsf{GenRAC}} \varphi$, if φ is valid in every E-model. We say that an L-formula φ is logically valid in RAC, and we write $\vDash_{\mathsf{RAC}} \varphi$, if φ is valid in every simple E-model.

Let us discuss some logical features of the novel containing connective $\rightharpoonup$. It holds that for every positive (i.e. negation-free) and disjunction-free formulas α, β, we obtain the following A-equivalence:

$$\vDash_{\mathsf{RAC}} (\alpha \rightharpoonup \beta) \Leftrightarrow (\alpha \rightharpoondown \beta).$$

Let us verify this fact. One can check by induction that $[\alpha]^+$ is a singleton, in every simple E-model and for every positive and disjunction-free formula α. In particular, if $[\alpha]^+$ and $[\beta]^+$ are singletons then $[\alpha \rightharpoondown \beta]^+$ and $[\alpha \rightharpoonup \beta]^+$ are singletons too and $[\alpha \rightharpoondown \beta]^+ = [\alpha \rightharpoonup \beta]^+$. Assume that $[\alpha]^+ = \{s\}$ and $[\beta]^+ = \{t\}$. Then there is just one function $f : [\alpha]^+ \to [\beta]^+$ assigning t to s and just one function $g : [\beta]^+ \to [\alpha]^+$ assigning s to t. Then we obtain:

$$[\alpha \rightharpoondown \beta]^+ = \{s \Rightarrow f(s)\} = \{s \Rightarrow t\} = \{g(t) \Rightarrow t\} = [\alpha \rightharpoonup \psi]^+.$$

This observation (which does not hold for GenRAC) gives us many logical principles governing $\rightharpoonup$, given that $\rightharpoondown$ is the implication of the logic R (as we will show in the next section). However, the way how $\rightharpoonup$ and $\rightharpoondown$ interact with disjunction is symmetric but inverted. For example, we can observe in both GenRAC and RAC that p implies but does not contain $p \vee q$ (i.e. $p \rightharpoondown (p \vee q)$ is valid but $p \rightharpoonup (p \vee q)$ is not). On the other hand, $p \vee q$ contains but does not imply p (i.e. $(p \vee q) \rightharpoonup p$ is valid but $(p \vee q) \rightharpoondown p$ is not). Moreover, we obtain the following B-equivalences illustrating nicely the inverted interaction with disjunction:

$$\vDash_{\mathsf{GenRAC}} ((\varphi \vee \psi) \rightharpoondown \chi) \leftrightharpoons ((\varphi \rightharpoondown \chi) \wedge (\psi \rightharpoondown \chi)),$$

$$\vDash_{\mathsf{GenRAC}} (\chi \rightharpoonup (\varphi \vee \psi)) \leftrightharpoons ((\chi \rightharpoonup \varphi) \wedge (\chi \rightharpoonup \psi)).$$

The former B-equivalence is the standard principle of R (and many other logics). Let us check that the latter B-equivalence holds as well. In the light of Lemma 3, we need to show that (a) for any truthmaker s of the formula $\chi \rightharpoonup (\varphi \vee \psi)$ there is a truthmaker $t \sqsubseteq s$ of $(\chi \rightharpoonup \varphi) \wedge (\chi \rightharpoonup \psi)$, and (b) for any truthmaker s of the formula $(\chi \rightharpoonup \varphi) \wedge (\chi \rightharpoonup \psi)$ there is a truthmaker $t \sqsubseteq s$ of $\chi \rightharpoonup (\varphi \vee \psi)$.

(a) Let $s \Vdash^+ \chi \rightharpoonup (\varphi \vee \psi)$. So, there is a function $f : [\varphi \vee \psi]^+ \to [\chi]^+$ such that $s = \bigsqcup\{f(t) \Rightarrow t \mid t \in [\varphi \vee \psi]^+\}$. We can split f into two functions $g_1 : [\varphi]^+ \to [\chi]^+$ and $g_2 : [\psi]^+ \to [\chi]^+$ such that $g_1(t) = f(t)$ for each $t \in [\varphi]^+$, and $g_2(t) = f(t)$ for each $t \in [\psi]^+$. Let $u_1 = \bigsqcup\{g_1(t) \Rightarrow t \mid t \in [\varphi]^+\}$ and $u_2 = \bigsqcup\{g_2(t) \Rightarrow t \mid t \in [\psi]^+\}$. Clearly, $u_1 \sqcup u_2 \Vdash^+ (\chi \rightharpoonup \varphi) \wedge (\chi \rightharpoonup \psi)$ and $u_1 \sqcup u_2 = s$.

(b) Now assume $s \Vdash^+ (\chi \rightharpoonup \varphi) \wedge (\chi \rightharpoonup \psi)$. So, there are u_1, u_2 such that $u_1 \Vdash^+ \chi \rightharpoonup \varphi$, $u_2 \Vdash^+ \chi \rightharpoonup \psi$, and $u_1 \sqcup u_2 = s$. Then there are $g_1 : [\varphi]^+ \to [\chi]^+$ and $g_2 : [\psi]^+ \to [\chi]^+$ such that $u_1 = \bigsqcup\{g_1(t) \Rightarrow t \mid t \in [\varphi]^+\}$ and $u_2 = \bigsqcup\{g_2(t) \Rightarrow t \mid t \in [\psi]^+\}$. Now we define a function $f : [\varphi \vee \psi]^+ \to [\chi]^+$ in the following way:

$$f(s) = \begin{cases} g_1(s) & \text{for } s \in [\varphi]^+ \\ g_2(s) & \text{for } s \in [\psi]^+ \setminus [\varphi]^+ \end{cases}$$

and we fix $v = \bigsqcup\{f(t) \Rightarrow t \mid t \in [\varphi \vee \psi]^+\}$. It is clear that $v \sqsubseteq s$ and v is a truthmaker of $\chi \rightharpoonup (\varphi \vee \psi)$.

An axiomatic characterization of RAC and GenRAC is an open problem left for future research. However, in the rest of this paper we will characterize some fragments of RAC. It will be shown that RAC integrates naturally into one framework the implications of R and AC, which is the main goal of this paper. We will also show a connection of RAC to inquisitive logic and make some further comments on GenRAC connecting it to the logic of Urquhart's operational semantics.

3 A Connection to R

In this section, we will show that one can obtain a semantics for R by some simple modifications of the semantics for RAC introduced in the previous section. This should also clarify some peculiarities of our approach.

Let us introduce alternative truthmaking and falsemaking relations $\Vvdash^+$ and $\Vvdash^-$, relative to any simple E-model $\mathcal{M}$. These relations are defined just like $\Vdash^+$ and $\Vdash^-$, except that we modify the positive semantic clause for disjunction and the negative clause for conjunction. Instead of those presented in Table 1 we define:

- $s \Vvdash^{+} \varphi \vee \psi$ iff $\exists t, u \in S\colon t \Vvdash^{+} \varphi, u \Vvdash^{+} \psi$ and $s = t \sqcap u$,
- $s \Vvdash^{-} \varphi \wedge \psi$ iff $\exists t, u \in S\colon t \Vvdash^{-} \varphi, u \Vvdash^{-} \psi$ and $s = t \sqcap u$.

As before, let us define:

$$[\![\varphi]\!]^{+} = \{s \in S \mid s \Vvdash^{+} \varphi\} \text{ and } [\![\varphi]\!]^{-} = \{s \in S \mid s \Vvdash^{-} \varphi\}.$$

We say that an L-formula φ is *simply valid in* $\mathcal{M}$ if there is $s \sqsubseteq l$ (in $\mathcal{M}$) such that $s \Vvdash^{+} \varphi$. Although this modification flattens some features of our semantics from the previous section, other interesting features emerge. One can check by induction that all the semantic clauses in the modified semantics preserve the basic property of atomic formulas, namely that they have a unique truthmaker and a unique falsemaker. In other words, for any simple E-model and any L-formula φ, both $[\![\varphi]\!]^{+}$ and $[\![\varphi]\!]^{-}$ are singletons. As a consequence, the difference between $\rightharpoondown$ and $\rightharpoonup$, and also the difference between $\underset{\cdot}{-}$ and $\dot{-}$, are lost. By inspecting the semantic clauses for $\underset{\cdot}{-}, \rightharpoondown, \dot{-}, \rightharpoonup$ one can observe that if φ and ψ have unique truthmakers and falsemakers then $s \Vvdash^{+(-)} \varphi \rightharpoondown \psi$ iff $s \Vvdash^{+(-)} \varphi \rightharpoonup \psi$, and $s \Vvdash^{+(-)} \varphi \underset{\cdot}{-} \psi$ iff $s \Vvdash^{+(-)} \varphi \dot{-} \psi$.

Let us denote the single element of $V^{+}(p)$ as $\mathcal{V}^{+}(p)$ and the single element of $V^{-}(p)$ as $\mathcal{V}^{-}(p)$. That is $V^{+}(p) = \{\mathcal{V}^{+}(p)\}$ and $V^{-}(p) = \{\mathcal{V}^{-}(p)\}$. Now we can extend $\mathcal{V}^{+}$ and $\mathcal{V}^{-}$, in a given simple E-model $\mathcal{M}$, and assign recursively to each L-formula φ a unique situation $\mathcal{V}^{+}(\varphi)$ as its positive algebraic value and a unique situation $\mathcal{V}^{-}(\varphi)$ as its negative algebraic value:

$$\begin{array}{llll}
\mathcal{V}^{+}(\neg\varphi) & = \mathcal{V}^{-}(\varphi) & \mathcal{V}^{-}(\neg\varphi) & = \mathcal{V}^{+}(\varphi) \\
\mathcal{V}^{+}(\varphi \vee \psi) & = \mathcal{V}^{+}(\varphi) \sqcap \mathcal{V}^{+}(\psi) & \mathcal{V}^{-}(\varphi \vee \psi) & = \mathcal{V}^{-}(\varphi) \sqcup V^{-}(\psi) \\
\mathcal{V}^{+}(\varphi \wedge \psi) & = \mathcal{V}^{+}(\varphi) \sqcup \mathcal{V}^{+}(\psi) & \mathcal{V}^{-}(\varphi \wedge \psi) & = \mathcal{V}^{-}(\varphi) \sqcap V^{-}(\psi) \\
\mathcal{V}^{+}(\varphi \underset{\cdot}{-} \psi) & = \mathcal{V}^{+}(\varphi) * \mathcal{V}^{+}(\psi) & \mathcal{V}^{-}(\varphi \underset{\cdot}{-} \psi) & = \mathcal{V}^{+}(\varphi) \Rightarrow \mathcal{V}^{-}(\psi) \\
\mathcal{V}^{+}(\varphi \rightharpoondown \psi) & = \mathcal{V}^{+}(\varphi) \Rightarrow \mathcal{V}^{+}(\psi) & \mathcal{V}^{-}(\varphi \rightharpoondown \psi) & = \mathcal{V}^{+}(\varphi) * \mathcal{V}^{-}(\psi) \\
\mathcal{V}^{+}(\varphi \dot{-} \psi) & = \mathcal{V}^{+}(\varphi) * \mathcal{V}^{+}(\psi) & \mathcal{V}^{-}(\varphi \dot{-} \psi) & = \mathcal{V}^{+}(\varphi) \Rightarrow V^{-}(\psi) \\
\mathcal{V}^{+}(\varphi \rightharpoonup \psi) & = \mathcal{V}^{+}(\varphi) \Rightarrow \mathcal{V}^{+}(\psi) & \mathcal{V}^{-}(\varphi \rightharpoonup \psi) & = \mathcal{V}^{+}(\varphi) * \mathcal{V}^{-}(\psi)
\end{array}$$

Lemma 4 *For each simple E-model* $\mathcal{M} = \langle S, \sqsubseteq, *, l, V^{+}, V^{-}\rangle$, *and for every* L-*formula* φ, *it holds that*

$$[\![\varphi]\!]^{+} = \{\mathcal{V}^{+}(\varphi)\} \textit{ and } [\![\varphi]\!]^{-} = \{\mathcal{V}^{-}(\varphi)\}.$$

This lemma shows that the (modified) relational semantic clauses for $\Vvdash$ just reduce to the standard algebraic clauses if applied to singleton propositions. Since in this modified semantics the positive as well as the negative clauses for $\underset{\cdot}{-}, \rightharpoondown$ coincide respectively with the positive and negative clauses for $\dot{-}, \rightharpoonup$, we simplify the language and consider just one from these two pairs. So, let us introduce the language of the logic R, denoted as L_{R}, as generated by the following grammar:

$$\varphi ::= p \mid \neg\varphi \mid \varphi \vee \varphi \mid \varphi \wedge \varphi \mid \varphi \rightharpoondown \varphi \mid \varphi \div \varphi \text{ (the language } L_{\mathsf{R}}).$$

The positive part of L_{R}, i.e. when negation is dropped, will be denoted as L_{R}^{p}, and the disjunction-free fragment of L_{R}^{p} will be denoted as L_{R}^{pdf}.

In restriction to L_{R}^{p} the logic of all E-models, relative to the truthmaking relation $\Vdash^{+}$, is the positive fragment of the logic R. We obtain the full R, for the whole language L_{R} involving negation, if we impose the following constraint on valuations.[4]

Definition 6 By an R-*model* we will understand a simple E-model in which the following equation holds for all L_{R}-formulas φ, ψ:

$$\mathcal{V}^{+}(\varphi) \Rightarrow \mathcal{V}^{-}(\psi) = \mathcal{V}^{+}(\psi) \Rightarrow \mathcal{V}^{-}(\varphi).$$

We can show that the logic R is indeed sound and complete in the modified semantics with respect to all R-models. An axiomatic system for the logic R in the language L_{R} is presented in Table 2. If an L_{R}-formula φ is derivable in this system we write $\vdash_{\mathsf{R}} \varphi$. One can check that this system is indeed sound with respect to the class of R-models in the modified truthmaker semantics. For an illustration, let us show soundness of B14. Given any R-model we can reason as follows:

1. $\mathcal{V}^{+}(\varphi) * \mathcal{V}^{+}(\varphi) \sqsubseteq \mathcal{V}^{+}(\varphi)$,
2. $\mathcal{V}^{+}(\varphi) \Rightarrow (\mathcal{V}^{+}(\varphi) * \mathcal{V}^{+}(\varphi)) \sqsubseteq l$,
3. $\mathcal{V}^{+}(\varphi) \Rightarrow \mathcal{V}^{-}(\varphi \rightharpoondown \neg\varphi) \sqsubseteq l$,
4. $\mathcal{V}^{+}(\varphi \rightharpoondown \neg\varphi) \Rightarrow \mathcal{V}^{-}(\varphi) \sqsubseteq l$,
5. $\mathcal{V}^{+}((\varphi \rightharpoondown \neg\varphi) \rightharpoondown \neg\varphi) \sqsubseteq l$.

To demonstrate completeness, we can show how this semantics is related to the common algebraic semantics for R developed in [12]. The algebraic semantics is based on structures called De Morgan monoids. A *De Morgan monoid* is an algebraic structure of the shape $\langle A, \curlywedge, \curlyvee, \cdot, -, 1\rangle$, where $\langle A, \curlywedge, \curlyvee\rangle$ is a distributive lattice with the lattice ordering $\leq$; $\langle A, \cdot, 1\rangle$ is a commutative monoid which is square-increasing, i.e. $x \leq x \cdot x$, involutive, i.e. $- - x = x$, and the following law is satisfied:

$$x \cdot y \leq z \text{ if and only if } x \cdot -z \leq -y.$$

[4] A semantics with this kind of constraint on valuations is quite nonstandard and rather inelegant. One could proceed in an alternative way by extending E-frames with a unary algebraic operation $-$, require that the whole structure is a dual of a De Morgan monoid (which is defined below) and define negation directly in terms of the added operation: $V^{+}(\neg\varphi) = -V^{+}(\varphi)$. In this way the semantics would differ from the usual algebraic semantics only by the requirement that the algebras are complete. However, our aim here is not to present an elegant semantics for R. We are just explaining what distortion of the framework for RAC is needed to obtain R.

Table 2 Axioms and rules of R

B1	$\varphi \rightharpoondown \varphi$
B2	$(\varphi \rightharpoondown \psi) \rightharpoondown ((\chi \rightharpoondown \varphi) \rightharpoondown (\chi \rightharpoondown \psi))$
B3	$(\varphi \rightharpoondown (\varphi \rightharpoondown \psi)) \rightharpoondown (\varphi \rightharpoondown \psi)$
B4	$(\varphi \rightharpoondown (\psi \rightharpoondown \chi)) \rightharpoondown (\psi \rightharpoondown (\varphi \rightharpoondown \chi))$
B5	$((\varphi \rightharpoondown \psi) \wedge (\varphi \rightharpoondown \chi)) \rightharpoondown (\varphi \rightharpoondown (\psi \wedge \chi))$
B6	$(\varphi \wedge \psi) \rightharpoondown \varphi$
B7	$(\varphi \wedge \psi) \rightharpoondown \psi$
B8	$\varphi \rightharpoondown (\varphi \vee \psi)$
B9	$\psi \rightharpoondown (\varphi \vee \psi)$
B10	$((\varphi \rightharpoondown \chi) \wedge (\psi \rightharpoondown \chi)) \rightharpoondown ((\varphi \vee \psi) \rightharpoondown \chi))$
B11	$(\varphi \wedge (\psi \vee \chi)) \rightharpoondown ((\varphi \wedge \psi) \vee (\varphi \wedge \chi))$
B12	$(\varphi \rightharpoondown (\psi \rightharpoondown (\varphi \div \psi))$
B13	$(\varphi \rightharpoondown (\psi \rightharpoondown \chi)) \rightharpoondown ((\varphi \div \psi) \rightharpoondown \chi)$
B14	$(\varphi \rightharpoondown \neg\varphi) \rightharpoondown \neg\varphi$
B15	$(\varphi \rightharpoondown \neg\psi) \rightharpoondown (\psi \rightharpoondown \neg\varphi)$
B16	$\neg\neg\varphi \rightharpoondown \varphi$
C1	$\varphi, \varphi \rightharpoondown \psi / \psi$
C2	$\varphi, \psi / \varphi \wedge \psi$

A *valuation* in a De Morgan monoid $\mathcal{A} = \langle A, \curlywedge, \curlyvee, \cdot, -, 1\rangle$ is a function v assigning to each atomic formula an element of A. This function is extended to the whole language L_{R} in the following way:

$$\begin{array}{lll} v(\neg\varphi) & = -v(\varphi) & \\ v(\varphi \vee \psi) & = v(\varphi) \curlyvee v(\psi) & v(\varphi \wedge \psi) = v(\varphi) \curlywedge v(\psi) \\ v(\varphi \rightharpoondown \psi) & = -(v(\varphi) \cdot -v(\psi)) & v(\varphi \div \psi) = v(\varphi) \cdot v(\psi) \end{array}$$

An L_{R}-formula φ is *v-valid* in $\mathcal{A}$ if $1 \leq v(\varphi)$. We say that φ is *valid* in $\mathcal{A}$ if φ is v-valid in $\mathcal{A}$, for every valuation v. It is well-known that φ is derivable in the system for R if and only if it is valid in every De Morgan monoid. So, if φ is not derivable in the system for R there is a De Morgan monoid $\mathcal{A} = \langle A, \curlywedge, \curlyvee, \cdot, -, 1\rangle$ (e.g. the Lindenbaum-Tarski algebra of R) and a valuation v in $\mathcal{A}$ such that φ is not v-valid in $\mathcal{A}$. To show that the system for R is complete with respect to the modified truthmaker semantics it suffices to show that $\mathcal{A}$ and v can be transformed into an R-model that validates the same formulas. Such a transformation is obtained by the usual duality result for De Morgan monoids exploited typically in the completeness proof for R with respect to its relational (Routley-Meyer) semantics.

A *filter* F in a De Morgan monoid $\mathcal{A} = \langle A, \curlywedge, \curlyvee, \cdot, -, 1\rangle$ is a non-empty upward closed subset of A that is closed under $\curlywedge$. Equivalently, a non-empty $F \subseteq A$ is a filter in $\mathcal{A}$ if it satisfies the following condition:

$$x \curlywedge y \in F \text{ if and only if } x \in F \text{ and } y \in F.$$

A filter is called *prime* if also

$$x \curlyvee y \in F \text{ if and only if } x \in F \text{ or } y \in F.$$

We will denote the set of all prime filters in $\mathcal{A}$ as $Prime(\mathcal{A})$. Now, given any De Morgan monoid $\mathcal{A} = \langle A, \curlywedge, \curlyvee, \cdot, -, 1\rangle$ and a valuation v in $\mathcal{A}$, we construct the structure $E(\mathcal{A}, v) = \langle S, \sqsubseteq, *, l, V^+, V^-\rangle$, where

- S is the set of all upward closed subsets of $Prime(\mathcal{A})$, that is, $X \in S$ iff $X \subseteq Prime(\mathcal{A})$ and for all $F, G \in Prime(\mathcal{A})$, if $F \in X$ and $F \subseteq G$ then $G \in X$;
- $\sqsubseteq$ is the superset relation $\supseteq$;
- $X * Y = \{F \in Prime(\mathcal{A}) \mid \exists G \in X, H \in Y\ \forall x \in G, y \in H, x \cdot y \in F\}$;
- $l = \{F \in Prime(\mathcal{A}) \mid 1 \in F\}$;
- $V^+(p) = \{\{F \in Prime(\mathcal{A}) \mid v(p) \in F\}\}$;
- $V^-(p) = \{\{F \in Prime(\mathcal{A}) \mid -v(p) \in F\}\}$.

Lemma 5 *Let $\mathcal{A} = \langle A, \curlywedge, \curlyvee, \cdot, -, 1\rangle$ be a De Morgan monoid and v a valuation in $\mathcal{A}$. Then $E(\mathcal{A}, v)$ is an* R*-model. Moreover, for every L_{R}-formula φ:*

(a) $\mathcal{V}^+(\varphi) = \{F \in Prime(\mathcal{A}) \mid v(\varphi) \in F\}$;
(b) $\mathcal{V}^-(\varphi) = \{F \in Prime(\mathcal{A}) \mid -v(\varphi) \in F\}$.

As a consequence, for every L_{R}-formula φ, φ is v-valid in $\mathcal{A}$ if and only if φ is simply valid in $E(\mathcal{A}, v)$.

Proof This is the usual duality result for De Morgan monoids. □

Theorem 1 *For every L_{R}-formula φ, φ is simply valid in all* R*-models if and only if $\vdash_{\mathsf{R}} \varphi$.*

Proof This result follows from soundness and Lemma 5. □

As a consequence of this result, we obtain the following connection between (Gen)RAC and R.

Theorem 2 *Both* GenRAC *and* RAC *conservatively extend the disjunction-free fragment of positive* R *(axiomatizable by axioms B1-B7, B12, B13, and the rules C1, C2), i.e. for any L^{pdf}_{R}-formula α, $\vDash_{\mathsf{GenRAC}} \alpha$ if and only if $\vDash_{\mathsf{RAC}} \alpha$ if and only if $\vdash_{\mathsf{R}} \alpha$.*

Proof We have observed that in restriction to simple E-models and positive disjunction-free formulas the semantics of this section coincides with the semantics from the previous section. So, Theorem 1 gives us immediately conservativity of RAC over disjunction-free positive R. It gives us also completeness of this fragment for GenRAC. Moreover, one can check soundness of the disjunction-free fragment of positive R (i.e. axioms B1-B7, B12, B13, and the rules C1, C2 restricted to the language L^{pdf}_{R}) w.r.t. all (not only all simple) E-models, which finishes the proof.

□

In this section we have shown how the semantics that we introduced in the previous section can be modified to obtain a semantics for R. In particular, the semantics for RAC is very close to the algebraic semantics for R but it differs in two important respects: (a) the bilateral clauses do not validate contraposition; (b) the positive semantic clause for disjunction and the negative clause for conjunction generate propositions that have not just one but multiple exact truthmakers. It is this kind of multiplicity of semantic values that is important for a semantic characterization of analytic implication and analytic equivalence that will be discussed in Sect. 5.

4 A Connection to UR and InqR

Theorem 2 cannot be extended to the full language L_{R}. As already remarked, the bilateral treatment of negation does not guarantee contraposition, e.g. in the form of the axiom B15, which is valid in R. So, there are laws of negation valid in R but invalid in RAC (and thus also in GenRAC).

On the other hand, there are laws of disjunction valid in GenRAC (and thus also in RAC) which are not valid in R. Interestingly, it turns out that GenRAC is related to the logic determined by Urquhart's operational semilattice semantics that is known to be stronger than R (see [30], and [28], for a useful overview of Urquhart's semantics and the corresponding logic).

In accordance with [28], we will denote Urquhart's logic as UR. It was shown by [18] that in the operational semantics one can consider two alternative treatments of disjunction, one leading to UR and the other to R (in the context of the language L^p_{R}). These two treatments are closely connected to our two characterizations of disjunction, one w.r.t. $\Vdash^+$ and the other w.r.t. $\Vvdash^+$. Nevertheless, the truthmaker semantics of implication is quite different from how it is treated in the operational semantics, and so the connection is not immediately obvious.

Let us consider the following characteristic principle that holds in UR but not in R:

$$((\varphi \rightharpoondown (\psi \vee \chi)) \wedge (\psi \rightharpoondown \vartheta)) \rightharpoondown (\varphi \rightharpoondown (\vartheta \vee \chi)).$$

It will be illustrative if we show that this principle is indeed valid in GenRAC. Take an arbitrary E-model and assume $s \Vdash^+ (\varphi \rightharpoondown (\psi \vee \chi)) \wedge (\psi \rightharpoondown \vartheta)$, i.e. $s = s_1 \sqcup s_2$, where $s_1 \Vdash^+ \varphi \rightharpoondown (\psi \vee \chi)$ and $s_2 \Vdash^+ \psi \rightharpoondown \vartheta$. We have to show that there is $t \sqsubseteq s$ such that $t \Vdash^+ \varphi \rightharpoondown (\vartheta \vee \chi)$. Our assumption means that there are functions $f_1 : [\varphi]^+ \rightarrow [\psi \vee \chi]^+$ and $f_2 : [\psi]^+ \rightarrow [\vartheta]^+$ and

$$s_1 = \bigsqcup \{u \Rightarrow f_1(u) \mid u \in [\varphi]^+\} \text{ and } s_2 = \bigsqcup \{u \Rightarrow f_2(u) \mid u \in [\psi]^+\}.$$

Now we define a function $g : [\varphi]^+ \rightarrow [\vartheta \vee \chi]^+$ in this way:

$$g(u) = \begin{cases} f_2(f_1(u)) & \text{if } f_1(u) \in [\psi]^+ \\ f_1(u) & \text{otherwise} \end{cases}$$

Let $t = \bigsqcup\{u \Rightarrow g(u) \mid u \in [\varphi]^+\}$. We will show that $t \sqsubseteq s$, which means that $u \Rightarrow g(u) \sqsubseteq s$, for each $u \in [\varphi]^+$. Let $u \in [\varphi]^+$. Assume first that $f_1(u) \in [\psi]^+$. Then

$$u \Rightarrow g(u) = u \Rightarrow f_2(f_1(u)) \sqsubseteq (u \Rightarrow f_1(u)) \sqcup (f_1(u) \Rightarrow f_2(f_1(u))) \sqsubseteq s_1 \sqcup s_2 = s.$$

Now assume $f_1(u) \notin [\psi]^+$. In that case $f_1(u) \in [\chi]^+$, and we obtain

$$u \Rightarrow g(u) = u \Rightarrow f_1(u) \sqsubseteq s_1 \sqsubseteq s.$$

This finishes the proof. The exact relation of GenRAC to UR will be explored in future research.

The logic RAC is stronger than GenRAC even in this fragment. It can be shown that in the L^p_{R}-fragment the logic RAC coincides with positive inquisitive R (i.e. positive InqR) introduced in [25].[5] This logic can be obtained by axioms B1-B13, rules C1, C2 and one extra principle called Split:

$$(\alpha \rightharpoondown (\varphi \vee \psi)) \rightharpoondown ((\alpha \rightharpoondown \varphi) \vee (\alpha \rightharpoondown \psi)),$$

where α is any L^{pdf}_{R}-formula and φ, ψ are arbitrary L^p_{R}-formulas. Split is valid in RAC but not in GenRAC.

Theorem 3 RAC *conservatively extends positive* InqR *(axiomatizable by the axioms B1-B13, Split, and the rules C1, C2), that is,* $\vDash_{\mathsf{RAC}} \varphi$ *if and only if* $\vdash_{\mathsf{InqR}} \varphi$, *for every* L^p_{R}*-formula* φ.

Proof We give only a sketch of the proof. One can mechanically check, using the axiomatic system, that every formula of positive InqR is valid in RAC. This gives us one half of the theorem. For the other half, we first show a disjunctive normal form: in the system for positive InqR one can prove that every L^p_{R}-formula is B-equivalent to a disjunction of L^{pdf}_{R}-formulas. One also needs to show that the disjunction in RAC has the disjunction property: $\varphi \vee \psi$ is valid in RAC only if φ or ψ is valid in RAC. Finally, assume that $\vDash_{\mathsf{RAC}} \varphi$, for some L^p_{R}-formula φ, which is B-equivalent (in InqR and thus also in RAC) to $\alpha_1 \vee \ldots \vee \alpha_n$. Then, by the disjunction property,

[5] Inquisitive logic is a logic of questions but here we do not rely on this particular interpretation of the system. For more details on the standard framework of inquisitive logic including its interpretation in terms of questions, see [10, 11]. As far as we know a close connection between truthmaker semantics and inquisitive semantics was originally discovered and described by Ivano Ciardelli in an unpublished manuscript [9].

$\vDash_{\mathsf{RAC}} \alpha_i$, for some $1 \leq i \leq n$. But then, by Theorem 2 and the fact that α_i is an L^{pdf}_{R}-formula, $\vdash_{\mathsf{R}} \alpha_i$, and thus also $\vdash_{\mathsf{InqR}} \alpha_i$. It follows that $\vdash_{\mathsf{InqR}} \varphi$. □

5 A Connection to Analytic Containment

Now we finally show how is RAC related to the logic of analytic containment.

By L_0 we understand the $\{\neg, \wedge, \vee\}$-fragment of L. So, L_0-formulas are generated by the following grammar:

$$\alpha ::= p \mid \neg\alpha \mid \alpha \wedge \alpha \mid \alpha \vee \alpha \text{ (the language } L_0\text{).}$$

An equivalential formula (or L^e_0-formula, for short) is an expression of the form $\alpha \Leftrightarrow \beta$ where α and β are L_0-formulas.

The logic of analytic containment AC was introduced by Angell in [2, 3]. In Table 3 we present an axiomatic system for AC that is taken from [15] and that is based on analytic equivalence rather than implication. Note that the system allows us to derive only equivalential formulas. The implication $\varphi \Rightarrow \psi$ is definable in terms of the equivalence as $\varphi \Leftrightarrow (\varphi \wedge \psi)$. Our semantics can be viewed as a modified and expanded version of Fine's [15] semantics for AC and we will show that AC can be embedded into RAC. However, this connection is not completely straightforward and there are some issues that need to be clarified.

We say that a set of states $U \subseteq S$, in a given E-model, is *complete* if it is closed under arbitrary fusions, that is, if $\bigsqcup T \in U$, for every non-empty $T \subseteq U$. Fine obtained a correspondence between AC and his semantics by imposing a

Table 3 Axioms and rules of AC

E1	$\alpha \Leftrightarrow \alpha \wedge \alpha$
E2	$\alpha \wedge \beta \Leftrightarrow \beta \wedge \alpha$
E3	$(\alpha \wedge \beta) \wedge \gamma \Leftrightarrow \alpha \wedge (\beta \wedge \gamma)$
E4	$\alpha \Leftrightarrow \alpha \vee \alpha$
E5	$\alpha \vee \beta \Leftrightarrow \beta \vee \alpha$
E6	$(\alpha \vee \beta) \vee \gamma \Leftrightarrow \alpha \vee (\beta \vee \gamma)$
E7	$\alpha \wedge (\beta \vee \gamma) \Leftrightarrow (\alpha \wedge \beta) \vee (\alpha \wedge \gamma)$
E8	$\alpha \vee (\beta \wedge \gamma) \Leftrightarrow (\alpha \vee \beta) \wedge (\alpha \vee \gamma)$
E9	$\alpha \Leftrightarrow \neg\neg\alpha$
E10	$\neg(\alpha \wedge \beta) \Leftrightarrow \neg\alpha \vee \neg\beta$
E11	$\neg(\alpha \vee \beta) \Leftrightarrow \neg\alpha \wedge \neg\beta$
D1	$\alpha \Leftrightarrow \beta / \beta \Leftrightarrow \alpha$
D2	$\alpha \Leftrightarrow \beta, \beta \Leftrightarrow \gamma / \alpha \Leftrightarrow \gamma$
D3	$\alpha \Leftrightarrow \beta / (\alpha \wedge \gamma) \Leftrightarrow (\beta \wedge \gamma)$
D4	$\alpha \Leftrightarrow \beta / (\alpha \vee \gamma) \Leftrightarrow (\beta \vee \gamma)$

constraint requiring that, for each L_0-formula, the sets of its exact truthmakers and falsemakers are always complete in this sense. In our semantics for RAC, this is immediately satisfied for atomic formulas because the sets of their exact truthmakers and falsemakers are always singletons. In Fine's semantics this constraint must be imposed explicitly on valuations because they can assign also non-singletons.

The property of being complete is then preserved by negation, positive clause for conjunction, and negative clause for disjunction. However, it is not immediately preserved by positive clause for disjunction and negative clause for conjunction. For this reason, Fine introduces the following "inclusive" interpretation of these two cases:[6]

- $s \Vdash^+ \varphi \vee \psi$ iff $s \Vdash^+ \varphi$ or $s \Vdash^+ \psi$ or $\exists t, u \in S$: $t \Vdash^+ \varphi$, $u \Vdash^+ \psi$ and $s = t \sqcup u$,
- $s \Vdash^- \varphi \wedge \psi$ iff $s \Vdash^- \varphi$ or $s \Vdash^- \psi$ or $\exists t, u \in S$: $t \Vdash^- \varphi$, $u \Vdash^- \psi$ and $s = t \sqcup u$.

These inclusive clauses preserve the property of being complete. In our semantics, we can preserve the "exclusive" clauses from Sect. 2 and just introduce the inclusive disjunction and its dual as defined symbols:

$$\alpha \varovee \beta =_{def} \alpha \vee \beta \vee (\alpha \wedge \beta) \text{ and } \alpha \varowedge \beta =_{def} \neg(\neg\alpha \varovee \neg\beta).$$

The following lemma shows that, when applied to formulas whose sets of truthmakers and falsemakers are closed under fusion, $\varovee$ and $\varowedge$ have exactly the inclusive truthmaking and falsemaking conditions used in [15] in connection with the semantics for AC.

Lemma 6 *Let α, β be L_0-formulas such that $[\alpha]^+, [\alpha]^-, [\beta]^+, [\beta]^-$ are all complete in a given E-model. Then for each state s of this E-model we obtain the following characterization of $\varovee$ and $\varowedge$:*

(a) $s \Vdash^+ \alpha \varovee \beta$ iff $s \Vdash^+ \alpha$ or $s \Vdash^+ \beta$ or $\exists t, u \in S$: $t \Vdash^+ \alpha$, $u \Vdash^+ \beta$ and $s = t \sqcup u$;
(b) $s \Vdash^- \alpha \varovee \beta$ iff $\exists t, u \in S$: $t \Vdash^- \alpha$, $u \Vdash^- \beta$ and $s = t \sqcup u$;
(c) $s \Vdash^+ \alpha \varowedge \beta$ iff $\exists t, u \in S$: $t \Vdash^+ \alpha$, $u \Vdash^+ \beta$ and $s = t \sqcup u$;
(d) $s \Vdash^- \alpha \varowedge \beta$ iff $s \Vdash^- \varphi$ or $s \Vdash^- \beta$ or $\exists t, u \in S$: $t \Vdash^- \alpha$, $u \Vdash^- \beta$ and $s = t \sqcup u$.

Proof (a) is given directly by the definition of $\varovee$ and the semantic clauses for $\vee$ and $\wedge$. (b) $s \Vdash^- \alpha \varovee \beta$ iff $s \Vdash^- \alpha \vee \beta \vee (\alpha \wedge \beta)$ iff $\exists s_1, s_2, s_3 \in S$ such that $s_1 \Vdash^- \alpha$, $s_2 \Vdash^- \beta$, $s_3 \Vdash^- \alpha \wedge \beta$, and $s = s_1 \sqcup s_2 \sqcup s_3$ iff $\exists s_1, s_2, s_3 \in S$ such that $s_3 \Vdash^- \alpha$ or $s_3 \Vdash^- \beta$, and $s_1 \Vdash^- \alpha$, $s_2 \Vdash^- \beta$, and $s = s_1 \sqcup s_2 \sqcup s_3$ iff $\exists t, u \in S$: $t \Vdash^- \alpha$, $u \Vdash^- \beta$ and $s = t \sqcup u$. The last equivalence is given by the assumption that $[\alpha]^-$, $[\beta]^-$ are complete, so, in the left-to-right implication one can take $t = s_1 \sqcup s_3$, and $u = s_2$, if $s_3 \Vdash^- \alpha$, and $t = s_1$, and $u = s_2 \sqcup s_3$, if $s_3 \Vdash^- \beta$. In the right-to-left implication one can take $s_1 = t$ and $s_2 = s_3 = u$. (c) $s \Vdash^+ \alpha \varowedge \beta$ iff $s \Vdash^- \neg\alpha \varovee \neg\beta$ iff, by (b), $\exists t, u \in S$: $t \Vdash^- \neg\alpha$, $u \Vdash^- \neg\beta$ and $s = t \sqcup u$ iff $\exists t, u \in S$: $t \Vdash^+ \alpha$, $u \Vdash^+ \beta$

[6] They are inclusive in the sense that a truthmaker makes a disjunction true iff either it makes true one of the disjuncts, or it makes true the corresponding conjunction. The latter option is not included in the exclusive interpretation.

and $s = t \sqcup u$. (d) $s \Vdash^- \alpha \owedge \beta$ iff $s \Vdash^+ \neg\alpha \ovee \neg\beta$ iff, by (a), $s \Vdash^- \alpha$ or $s \Vdash^- \beta$ or $\exists t, u \in S$: $t \Vdash^- \alpha$, $u \Vdash^- \beta$ and $s = t \sqcup u$. □

Let α be an L_0-formula. We define the *inclusive variant* of α, denoted as α^{in}, as the result of replacing in α all $\vee$ by $\ovee$ and all $\wedge$ by $\owedge$. Then we obtain the following embedding of AC into RAC.

Theorem 4 *For all L_0-formulas α, β it holds that $\vdash_{\mathsf{AC}} \alpha \Leftrightarrow \beta$ if and only if $\vDash_{\mathsf{RAC}} \alpha^{in} \Leftrightarrow \beta^{in}$.*

Proof We can proceed as in [15]. Using Lemma 3, one can check the soundness of the system with respect to our semantics. For the completeness part we can use the same structure that was used as a canonical model also by Fine in his completeness proof. We can present it as a simple E-model in the sense of this paper. So, let us define the structure $\mathcal{M}^c = \langle S, \sqsubseteq, *, l, V^+, V^- \rangle$, where:

- S is the set of all sets of literals (atomic formulas and their negations),
- $\sqsubseteq$ is the subset relation $\subseteq$,
- $*$ is intersection,
- l is the empty set,
- $V^+(p) = \{\{p\}\}$ and $V^-(p) = \{\{\neg p\}\}$.

$\mathcal{M}^c$ is indeed an E-model in our sense and Fine [16] showed that, for every L_0-formulas α, β, $\vdash_{\mathsf{AC}} \alpha \Leftrightarrow \beta$ if and only if $[\alpha]^+$ and $[\beta]^+$ are synonymous in $\mathcal{M}^c$ *in the inclusive interpretation*, that is, according to Lemma 3-d and Lemma 6, if and only if $\alpha^{in} \Leftrightarrow \beta^{in}$ is valid in $\mathcal{M}^c$ according to our definition. So, if $\nvdash_{\mathsf{AC}} \alpha \Leftrightarrow \beta$ then there is a simple E-model in which $\alpha^{in} \Leftrightarrow \beta^{in}$ is not valid and thus $\nvDash_{\mathsf{RAC}} \alpha^{in} \Leftrightarrow \beta^{in}$, which establishes completeness. □

Without taking into account the inclusive interpretation, we would not obtain AC but a weaker logic. In particular, the axioms E1 and E8 do not hold in the exclusive interpretation. The failure of idempotence in the case of conjunction can be demonstrated as follows. Consider an E-model in which t_p and t_q are respectively the unique truthmakers of p and q, and assume that they are incomparable (that is, t_p is not a part of t_q and vice versa). Then $[p \vee q]^+ = \{t_p, t_q\}$ but $[(p \vee q) \wedge (p \vee q)]^+ = \{t_p, t_q, t_p \sqcup t_q\}$. Then $(p \vee q) \rightharpoonup ((p \vee q) \wedge (p \vee q))$, and thus also $(p \vee q) \Leftrightarrow ((p \vee q) \wedge (p \vee q))$, are not valid in the model. The failure of E8 can be demonstrated in a similar way.

Since there is no restriction on $V^+(p)$ in the general semantics for GenRAC, the argument above can be used to show that even $p \Leftrightarrow (p \wedge p)$ is not valid in GenRAC and thus Theorem 4 would not hold if we replaced RAC with GenRAC.

6 Conclusion

In this paper, we defined an analytic implication and equivalence in terms of the implication of the logic R. Analytic implication and equivalence could be analogously defined also in terms of different implicational connectives, for example in terms of intuitionistic implication. Nevertheless, we find interesting that one can achieve such a synthesis of two approaches to relevant logic that are seemingly very different.

The availability of such syntheses opens space for a number of directions for future research.

Firstly, the foregoing work provides some preliminary progress towards a solution to a pair of challenges facing the Parry/Angell tradition in relevant logics, namely, the expansion of Angell's intuitions regarding synonymy to a higher-degree language and the refining Parry's intuitions regarding containment between higher-degree sentences beyond the mere inclusion of atomic formulas criticized so heavily by Sylvan in [26].

Some of the consequences of **RAC** may intuitively agree or disagree with these sets of intuitions. E.g., that **R** proves *Assertion* $p \rightharpoondown ((p \rightharpoondown q) \rightharpoondown q)$ implies that the formula $p \rightharpoonup ((p \rightharpoondown q) \rightharpoondown q)$ is valid in **RAC**. This violates the respective variable inclusion requirements of Angell and Parry's systems, as well as the refinements thereto presented in [13] and [14], respectively.[7] Despite this, the uniformity of the semantics for **RAC** undoubtedly provides an advantage over the more *ad hoc* treatments to these challenges and suggests the investigations into whether refinements to the truthmaker semantics will provide more satisfying resolutions to the difficulties between the Parry/Angell idioms and higher-degree languages.

Secondly, the foregoing work on **RAC** opens up some interesting and worthwhile interpretative questions. That s acts as a truthmaker for $\varphi \rightharpoondown \psi$ admits a distinctly Brouwer-Heyting-Kolmogorov-style interpretation in which s encodes a function that computes for any truthmaker t for φ some truthmaker $f(t)$ for ψ. Such a computational interpretation is available in the case of truthmakers for $\varphi \rightharpoonup \psi$ as well, by encoding a function that finds for each truthmaker t for ψ a truthmaker $f(t)$ for φ from which ψ may be computed.

This suggests several intriguing avenues to pursue, especially in the context of domain theory. For one, a truthmaker for $\varphi \rightharpoonup \psi$ might be read as standing in for a *reversible computation* (see [1] for a survey), i.e., a computation in which solutions for ψ can be constructively traced back to the particular input φ from which it was calculated. Moreover, following Plotkin's [24] and Smyth's [27] interpretations of the Egli-Milner ordering in powerdomain constructions, $\varphi \rightharpoonup \psi$ may be read as representing a nondeterministic computation (exemplified e.g. by the principle $(\chi \rightharpoonup (\varphi \vee \psi)) \rightharpoondown ((\chi \rightharpoonup \varphi) \wedge (\chi \rightharpoonup \psi))$). We view the foregoing remarks provide an opportunity to examine the relationship between truthmaker semantics, relevant logic, and Scott domains in a novel fashion and plan to develop this thread further in future work.

[7] It is not clear at this moment whether these variable inclusion principles are violated also in **GenRAC**. We leave this as an open question for further research.

Acknowledgments This paper is an outcome of the project Logical Structure of Information Channels, no. 21-23610M, supported by the Czech Science Foundation and carried out at the Institute of Philosophy of the Czech Academy of Sciences.

References

1. Aman, B., et al. (2020). Foundations of reversible computation. In I. Ulidowski, I. Lanese, U. Schultz, & C. Ferreira, (Eds.), *Reversible computation: Extending horizons of computing* (pp. 1–40). Cham: Springer.
2. Angell, R. B. (1977). Three systems of first degree entailment. *The Journal of Symbolic Logic, 42*, 147.
3. Angell, R. B. (1989). Deducibility, entailment and analytic containment. In J. Norman & R. Sylvan (Eds.), *Directions in relevant logic*. Kluwer.
4. Barwise, J. (1993). Constraints, channels, and the flow of information. In: P. Aczel et al. (Eds.), *Situation theory and its applications, Volume 3* (pp. 3–28). Stanford: CSLI Publications.
5. Barwise, J., & Perry, J. (1983). *Situations and attitudes*. Cambridge, MA: The MIT Press.
6. Belnap, N.D. Jr. (1967). Intensional models for first degree formulas. *The Journal of Symbolic Logic, 32*, 1–22.
7. Brink, C. (1986). Power structures and logic. *Quaestiones Mathematicae, 9*, 69–94.
8. Brink, C. (1993). Power structures. *Algebra Universalis, 30*, 177–216.
9. Ciardelli, I. *Inquisitive semantics as a truth-maker Semantics*. Unpublished manuscript.
10. Ciardelli, I. (2022). *Inquisitive logic. Consequence and inference in the realm of questions*. Springer.
11. Ciardelli, I., Groenendijk, J., & Roelofsen, F. (2019). *Inquisitive semantics*. Oxford: Oxford University Press.
12. Dunn, J. M. (1966). *The algebra of intensional logics*. PhD thesis, University of Pittsburgh.
13. Ferguson, T. M. (2014). Logics of nonsense and Parry systems. *Journal of Philosophical Logic, 44*, 65–80.
14. Ferguson, T. M. (2023). Subject-matter and intensional operators I: Conditional-agnostic analytic implication. *Philosophical Studies, 180*, 1849–1879.
15. Fine, K. (2014). Truth-maker semantics for intuitionistic logic. *Journal of Philosophical Logic, 43*, 221–246.
16. Fine, K. (2016). Angellic content. *Journal of Philosophical Logic, 45*, 199–226.
17. Hlobil, U. (2022). The laws of thought and the laws of truth as two sides of one coin. *Journal of Philosophical Logic, 52*, 313–343.
18. Humberstone, L. (1988). Operational semantics for positive R. *Notre Dame Journal of Formal Logic, 29*, 61–80.
19. Jago, M. (2020). Truthmaker semantics for relevant logic. *Journal of Philosophical Logic, 49*, 681–702.
20. Majer, O., Punčochář, V., & Sedlár, I. (2023). Truth-maker semantics for some substructural logics. In F. Faroldi & F. Van De Putte (Eds.), *Outstanding contributions to logic: Kit Fine on truthmakers, relevance, and non-classical logic*. Springer.
21. Nelson, D. (1949). Constructible falsity. *Journal of Symbolic Logic, 14*, 16–26.
22. Parry, W. T. (1968). The logic of C. I. Lewis. In P. A. Schilpp (Ed.), *The philosophy of C. I. Lewis*. Open Court.
23. Parry, W. T. (1972). Comparison of entailment theories. *Journal of Symbolic Logic, 37*, 441–442.
24. Plotkin, G. (1976). A powerdomain constructions. *SIAM Journal of Computing, 5*, 452–486.
25. Punčochář, V. (2020). A relevant logic of questions. *Journal of Philosophical Logic, 49*, 905–939.

26. Routley, R., Meyer, R. K., Plumwood, V., & Brady, R. (1982). *Relevant logics and their rivals* (Vol. I). Atascadero: Ridgeview Publishing.
27. Smyth, M. B. (1978). Power domains. *Journal of Computer and System Sciences, 16*, 23–36.
28. Standefer, S. (2021). Revisiting semilattice semantics. In I. Duntsch & E. Mares (Eds.), *Outstanding contributions to logic: Alasdair Urquhart on nonclassical and algebraic logic and complexity of proofs* (pp. 241–258). Springer.
29. Sylvan, R. (1988). Relevant containment logics and certain frame problems of AI. *Logique et Analyse, 31*, 11–24.
30. Urquhart, A. (1972). Semantics for relevant logics. *The Journal of Symbolic Logic, 37*, 159–169.
31. Van Fraassen, B. (1969). Facts and tautological entailments. *Journal of Philosophy, 66*, 477–487.

The Only 3-Valued Logic Which Is a Natural Implication Expansion with the Variable-Sharing Property of Kleene's Strong Logic

Gemma Robles and José M. Méndez

Abstract Let us refer by MK3 to Kleene's strong 3-valued matrix. An implicative expansion of MK3 is natural if the conditional function defining it verifies *modus ponens*, assigns a designated value to a conditional whenever it assigns the same value to its antecedent and its consequent, and, finally, it coincides with the classical conditional function when restricted to the "classical" values $\mathfrak{t}$ and $\mathfrak{f}$. Two are the main results of this paper. (1) It is proven that, from the viewpoint of functional strength, there is only one 3-valued natural implication expansion of MK3 with the variable-sharing property, the logic we dub $L3^{VSP}$. (2) It is shown that $L3^{VSP}$ is a significant and strong logic that can be seen from different perspectives, one of them being to consider it an expansion of classical positive propositional logic.

MSC Classification 03B47

1 Introduction

A propositional logic L has the variable-sharing property (VSP) if in all L-theorems of implication form antecedent and consequent share at least a propositional variable. Given that in propositional logic the non-logical content is conveyed by propositional variables, if L is a propositional logic with the VSP, then it is free from "paradoxes of relevance" in the sense that L does not contain theorems of implication form where the semantical content of antecedent and consequent is disjoint. Anderson and Belnap consider the VSP a necessary property a relevant logic has to fulfill (cf. [1, 4]), but some authors go so far as to consider that "the

G. Robles (✉)
Dpto. de Psicología, Sociología y Filosofía, Universidad de León, León, Spain
e-mail: gemma.robles@unileon.es

J. M. Méndez
Universidad de Salamanca, Edificio FES, Campus Unamuno, Salamanca, Spain
e-mail: sefus@usal.es

I. Sedlár et al. (eds.), *New Directions in Relevant Logic*, Trends in Logic 63,
https://doi.org/10.1007/978-3-031-69940-5_10

concept of relevant logic is coextensional with that of having the variable-sharing property" (cf. [19, p. 28]).

On the other hand, Kleene's strong 3-valued matrix MK3 (our label) was defined in [16] in the context of the treatment of partial recursive functions. The matrix MK3 can be rendered as shown in Definition 2.3 below, The connectives are conjunction, disjunction and negation. We take 1 and 2 as designated values (cf. Remark 2.4). Then "1" represents "both true and false", while "2" represents "true only" and "0" stands for "false only".

Finally, the notion of a "natural conditional" is here understood as an extension of that introduced in [27] and can be described as shown in Definition 2.5. That is, a conditional function, $f_{\rightarrow}$, expanding MK3 is natural if the three following conditions are fulfilled: (1) $f_{\rightarrow}$ coincides with the classical conditional function when restricted to the values 0 and 2; (2) $f_{\rightarrow}$ satisfies *modus ponens*; (3) $f_{\rightarrow}$ assigns a designated value to a conditional whenever the same value is assigned to its antecedent and consequent (cf. [22]. Of course, there are stricter notions than those defined in [22] and [27]; cf., e.g., [2]).

Now, in [22] it is proved that there are exactly 11 3-valued natural $f_{\rightarrow}$-functions expanding MK3 defining relevant conditionals in the sense that $A \rightarrow B$ (cf. Definition 2.1) is falsified if A and B do no have at least a variable in common.

Well then, two are the main results of this paper. (1) It is proven that, from the standpoint of functional strength, the 11 implicative expansions of MK3 referred to above actually determine only one logic, since they are functionally equivalent to each other. This unique logic (but cf. the concluding remarks to the paper), unique in the sense just remarked (that is, it is the only 3-valued implication expansion with the VSP of MK3, from the functional viewpoint), is dubbed $L3^{VSP}$. (2) It is shown that $L3^{VSP}$ is not an artificial construct, but a logic with remarkable properties among which we note, without trying to be exhaustive, the following (cf. also §5 and the appendix to the paper): natural conditionals in the sense of [22]; self-extensionality (i.e., "replacement"); considerable syntactical strength; considerable expressive power (for example, the important 3-valued logics Pac and RM3—cf., e.g., [15]—are definable in $L3^{VSP}$), and finally, its being interpretable in the clear and important two-valued Belnap-Dunn semantics.

The introduction is ended by explaining the structure of the paper. But before doing this, let us point out some notes on many-valued logics with the VSP.

It is known that there are infinitely many logics with the VSP (cf. [12]). Furthermore, some many-valued logics with the VSP have been studied in the literature. For example, the logic determined by Belnap's eight element matrix M_0 (cf. [4], axiomatized in [8]); or the logic characterized by Meyer's six element Crystal lattice, CL, also axiomatized in [8]. But it does not seem easy to interpret the meaning of the logical values in these matrices in an intuitively clear way. However, the meaning of the three truth-values in MK3 and its 3-valued expansions is crystal-clear.

The paper is organized as follows. In Sect. 2, some preliminary notions together with the 11 implicative expansions of MK3 with the VSP are defined. In Sect. 3 it is proved that the aforementioned 11 implicative expansions are functionally equivalent to each other and so determine only one logic, L3^{VSP}. It is also proved that the logics Pac ("Paraconsistency") and RM3 ("the strongest logic in the family of relevance logics"—[2, p. 276]) are functionally included in L3^{VSP} (in addition to [2], cf. [1] and [7] about RM3; [15] and references therein, about Pac). In Sect. 4, L3^{VSP} is presented as an expansion of classical positive propositional logic and, finally, in Sect. 5, the paper is ended with some concluding remarks on the results obtained and some observations on possible future work to be done in the topic. We have added an appendix including some complementary material as well as a proof of some of the properties L3^{VSP} sports that have been referred to throughout this introduction.

2 The Class MI3^{VSP} of Implicative Expansions of MK3

In this section, we define the class MI3^{VSP} of matrices. The label MI3^{VSP} intends to abbreviate "natural implicative expansions of Kleene's strong matrix MK3 with the variable-sharing property (VSP)". We begin by stating some prior concepts.

Definition 2.1 (Some Preliminary Notions) The propositional language consists of a denumerable set of propositional variables $p_0, p_1, \ldots, p_n, \ldots$, and some or all of the following connectives: $\rightarrow$ (conditional or implication[1]), $\wedge$ (conjunction), $\vee$ (disjunction) and $\sim$ (negation). The biconditional ($\leftrightarrow$) and the set of formulas (wffs) are defined in the customary way. A, B, C, etc. are metalinguistic variables. Then the ensuing concepts are understood in a fairly standard sense: logical matrix M, M-interpretation, M-consequence and M-validity. Also, the following notions: functions definable in a matrix, functional inclusion and functional equivalence (cf., e.g., [22, §2] or [23]).

Remark 2.2 (Logics) As suggested in the introduction, in this paper, logics are primarily viewed as M-determined structures, i.e., as structures of the type $(\mathcal{L}, \vDash_M)$ where $\mathcal{L}$ is a propositional language and $\vDash_M$ is a (consequence) relation defined in $\mathcal{L}$ according to the logical matrix M as follows: for any set of wffs Γ and wff A, $\Gamma \vDash_M A$ iff $I(A) \in D$ whenever $I(\Gamma) \in D$ for all M-interpretations I ($I(\Gamma) \in D$ iff $I(A) \in D$ for all $A \in \Gamma$; D is the set of designated values in M). Thus, from this viewpoint, we can safely travel back and forth from matrices to logics, given the aims of this paper.

Nevertheless, logics are sometimes defined as Hilbert-type axiomatic systems, the notions of "theorem" and "proof from premises" being the usual ones. Furthermore, in a derived or secondary sense, we can regard an M-determined logic as a,

[1] We follow Anderson and Belnap's "Grammatical Propaedeutic", Appendix to [1]: "The principal aim of this piece is to convince the reader that it is philosophically respectable to "confuse" implication and entailment with the conditional, and indeed philosophically suspect to harp on the dangers of such "confusion"" [1, p. 473].

say, Hilbert-type system (or a natural deduction system or a Gentzen-type system) L such that $\Gamma \vdash_L A$ iff $\Gamma \vDash_M A$, where $\vDash_M$ is the consequence relation defined above and $\Gamma \vdash_L A$ means "A is provable from Γ in L".

Definition 2.3 (Kleene's Strong 3-Valued Matrix) The propositional language consists of the connectives $\wedge, \vee, \sim$. Kleene's strong 3-valued matrix, MK3 (our label), is the structure $(\mathcal{V}, D, \mathrm{F})$ where (1) $\mathcal{V} = \{0, 1, 2\}$ with $0 < 1 < 2$; (2) $D = \{1, 2\}$; (3) F$= \{f_\wedge, f_\vee, f_\sim\}$ where $f_\wedge$ and $f_\vee$ are defined as the glb (or lattice meet) and the lub (or lattice join), respectively, and $f_\sim$ is an involution with $f_\sim(2) = 0$, $f_\sim(0) = 2$ and $f_\sim(1) = 1$. We display the tables for $\wedge$, $\vee$ and $\sim$:

$\wedge$	0	1	2
0	0	0	0
1	0	1	1
2	0	1	2

$\vee$	0	1	2
0	0	1	2
1	1	1	2
2	2	2	2

$\sim$	0
0	2
1	1
2	0

Remark 2.4 (On Designated Values in MK3) Kleene does not seem to have considered designated values in [16], §64, although he remarks: "The third "truth-value" $\mathfrak{u}$ is thus not on a par with the other two $\mathfrak{t}$ and $\mathfrak{f}$ in our theory. Consideration of its status will show that we are limited to a special kind of truth-value" [16, p. 333]. We use 2, 0 and 1 instead of $\mathfrak{t}$, $\mathfrak{f}$ and $\mathfrak{u}$, respectively, used by Kleene. The former have been chosen in order to use the tester in [13], in case one is needed. Also, to put in connection the results in the present paper with previous work by us. Finally, we note that the set D can be restricted to $\{2\}$.

On the other hand, we set:

Definition 2.5 (Natural Conditionals) Let $\mathcal{V}$ and D be defined as in Definition 2.3. Then, an $f_\rightarrow$-function on $\mathcal{V}$ defines a natural conditional if the following conditions are satisfied:

1. $f_\rightarrow$ coincides with the $f_\rightarrow$-function for the classical conditional when restricted to the subset $\{0, 2\}$ of $\mathcal{V}$.
2. $f_\rightarrow$ satisfies modus ponens, that is, for any $a, b \in \mathcal{V}$, if $a \rightarrow b \in D$ and $a \in D$, then $b \in D$.
3. For any $a, b \in \mathcal{V}$, $a \rightarrow b \in D$ if $a = b$.

Remark 2.6 (Natural Conditionals in Tomova's Original Paper) We note that natural conditionals are defined in [27] exactly as in Definition 2.5 except for condition (3), which reads there as follows: for any $a, b \in \mathcal{V}$, $a \rightarrow b \in D$ if $a \leq b$.

Definition 2.7 (The Class MI3$^{\mathbf{VSP}}$) In [22, Appendix III, Proposition C.9], it is proved that the only natural implicative expansions of MK3 determining logics (in the sense of Remark 2.2) with the VSP are the ones built up with the conditional described by the following truth tables:

(t1)

→	0	1	2
0	2	0	2
1	0	1	0
2	0	0	2

(t2)

→	0	1	2
0	2	0	2
1	0	1	1
2	0	0	2

(t3)

→	0	1	2
0	2	0	2
1	0	1	2
2	0	0	2

(t4)

→	0	1	2
0	2	0	2
1	0	1	0
2	0	1	2

(t5)

→	0	1	2
0	2	0	2
1	0	1	0
2	0	2	2

(t6)

→	0	1	2
0	2	1	2
1	0	1	0
2	0	0	2

(t7)

t7	0	1	2
0	2	1	2
1	0	1	0
2	0	1	2

(t8)

→	0	1	2
0	2	1	2
1	0	1	0
2	0	2	2

(t9)

→	0	1	2
0	2	2	2
1	0	1	0
2	0	0	2

(t10)

→	0	1	2
0	2	2	2
1	0	1	0
2	0	1	2

(t11)

→	0	1	2
0	2	2	2
1	0	1	0
2	0	2	2

Then the class MI3^{VSP} is the class of all 3-valued natural implicative expansions of MK3 with the VSP. It consists of the matrices M1, M2,..., M11. Each Mi ($1 \leq i \leq 11$) is the structure ($\mathcal{V}$, D, F) where $\mathcal{V}$, D and $f_\wedge$, $f_\vee$, $f_\sim$ $\in$ F are as in Definition 2.3, and $f_\rightarrow$ is defined according to table ti.

Finally, we set:

Definition 2.8 (The LM*i*-Logics) The logic LMi ($1 \leq i \leq 11$) is the logic determined by the matrix Mi in the sense explained in Remark 2.2: for any set of wffs Γ and wff A, $\Gamma \vDash_{\text{LM}i} A$ iff $I(A) \in \{1, 2\}$ whenever $I(\Gamma) \in \{1, 2\}$ for all Mi-interpretations I ($I(\Gamma) \in \{1, 2\}$ iff $I(A) \in \{1, 2\}$ for all $A \in \Gamma$).

Then the term LMi-logic (s) can be used to refer to the logics determined by the elements in MI3^{VSP} generally.

In the next section, it is proved that, from the viewpoint of functional strength, the LMi-logics are equivalent different versions of the same logic we name L3^{VSP} (natural 3-valued logic with the VSP).

3 LM1-LM11 Are Functionally Equivalent

In this section, it is proved that the logics LM1 through LM11 introduced in Definition 2.8 are functionally equivalent to each other. Consequently, these 11 expansions of Kleene's strong logic are just different formulations of the same logic we have named L3^{VSP} with a different choice of the primitive implication connective (nevertheless, cf. the concluding remarks to the paper). It will also be proved that the 3-valued paraconsistent logic Pac and "the strongest logic in the family of relevance logics" (cf. [2, p. 276]), RM3, are functionally included in $L3^{VSP}$ (cf. [1, 7] and references therein about RM3; [15] and references therein about Pac).

We begin by investigating the functional relations the logics LM1, LM2, LM3, LM6 and LM9 maintain to each other (Lemmas 3.2–3.5), but firstly, we note a remark on the proofs to follow.

Remark 3.1 (Functions and Truth-Tables. On Displaying Proofs of Definability) Let f_* be a function defined in $\mathcal{V} = \{0, 1, 2\}$. In this paper, f_* is usually represented by means of a truth-table t_* (or simply $*$), as for instance, it is the case with $\wedge$, $\vee$ and $\sim$ in MK3 (Definition 2.3). In addition, by k_* (or simply $*$) we refer to the connective defined by t_*. Now, let M be MK3 or an expansion of it. The proof that a given unary or binary function f_* is definable in M is easily visualized by using the connectives corresponding to the functions in M needed in the proof in question. In general, proofs provided below are simplified as just indicated (A, B refer to any wffs—cf. Definition 2.1) On the other hand, in order to prove that an LMi-logic is functionally included in another one, it is clear that it suffices to show that the implication table of the former is definable in the latter, given that we treat only implicative expansions of MK3. (We will simply say "equivalent logics" instead of "functionally equivalent logics", "inclusion" instead of "functional inclusion", etc.) Finally, by $\stackrel{i}{\to}$ $(1 \leq i \leq 11)$, we refer to the conditional or implication (cf. Note 1) given by table ti. (In case a tester is needed, the one in [13] can be used.)

Lemma 3.2 (LM2 and LM6; LM3 and LM9)

1. *LM2 and LM6 are equivalent to each other.*
2. *LM3 and LM9 are equivalent to each other.*

Proof We set:

(1a) $A \stackrel{6}{\to} B =_{\text{df}} {\sim}B \stackrel{2}{\to} {\sim}A$
(1b) $A \stackrel{2}{\to} B =_{\text{df}} {\sim}B \stackrel{6}{\to} {\sim}A$
(2a) $A \stackrel{9}{\to} B =_{\text{df}} {\sim}B \stackrel{3}{\to} {\sim}A$
(2b) $A \stackrel{3}{\to} B =_{\text{df}} {\sim}B \stackrel{9}{\to} {\sim}A$

□

Lemma 3.3 (LM1, LM2 and LM3)

1. *LM1 is included in LM2.*
2. *LM1 is included in LM3.*

Proof Given Lemma 3.2, we set:

(1) $A \stackrel{1}{\to} B =_{\text{df}} (A \stackrel{2}{\to} B) \wedge (A \stackrel{6}{\to} B)$
(2) $A \stackrel{1}{\to} B =_{\text{df}} (A \stackrel{3}{\to} B) \wedge (A \stackrel{9}{\to} B)$

□

Lemma 3.4 (LM1, LM6 and LM9) *Both LM9 and LM6 are included in LM1.*

Proof We set:

(1) $A \overset{9}{\rightarrow} B =_{\text{df}} A \overset{1}{\rightarrow} (A \overset{1}{\rightarrow} B)$
(2) $A \overset{6}{\rightarrow} B =_{\text{df}} (A \overset{9}{\rightarrow} B) \wedge [B \vee (A \overset{1}{\rightarrow} B)]$

□

The ensuing corollary follows from the lemmas proved above.

Lemma 3.5 (LM1, LM2, LM3, LM6 and LM9) *The logics LM1, LM2, LM3, LM6 and LM9 are equivalent.*

Proof The proof is immediate by Lemmas 3.2, 3.3 and 3.4. □

Next, it is shown that LM1 is included in LM4 and in LM5 (Lemma 3.6). Then we investigate the relations between LM4, LM7 and LM10 (Lemma 3.7), on the one hand, and those between LM5, LM8 and LM11, on the other hand (Lemma 3.8).

Lemma 3.6 (LM1, LM4 and LM5) *LM1 is included in both LM4 and LM5.*

Proof We set:

(1) $A \overset{1}{\rightarrow} B =_{\text{df}} (A \overset{4}{\rightarrow} B) \wedge ({\sim}B \overset{4}{\rightarrow} {\sim}A)$
(2) $A \overset{1}{\rightarrow} B =_{\text{df}} (A \overset{5}{\rightarrow} B) \wedge ({\sim}B \overset{5}{\rightarrow} {\sim}A)$

□

Lemma 3.7 (LM4, LM7 and LM10) *The logics LM4, LM7 and LM10 are equivalent to each other.*

Proof We set:

(1) $A \overset{7}{\rightarrow} B =_{\text{df}} (A \overset{4}{\rightarrow} B) \vee (A \overset{6}{\rightarrow} B)$
(2) $A \overset{4}{\rightarrow} B =_{\text{df}} (A \overset{7}{\rightarrow} B) \wedge [A \vee ({\sim}B \overset{7}{\rightarrow} {\sim}A)]$
(3) $A \overset{10}{\rightarrow} B =_{\text{df}} A \overset{4}{\rightarrow} (A \overset{4}{\rightarrow} B)$
(4) $A \overset{4}{\rightarrow} B =_{\text{df}} (A \overset{10}{\rightarrow} B) \wedge [A \vee ({\sim}B \overset{10}{\rightarrow} {\sim}A)]$

(Notice that table t6 can be used in (1) by virtue of Lemmas 3.5 and 3.6.) □

Lemma 3.8 (LM5, LM8 and LM11) *The logics LM5, LM8 and LM11 are equivalent to each other.*

Proof We set:

(1) $A \overset{11}{\rightarrow} B =_{\text{df}} A \overset{5}{\rightarrow} (A \overset{5}{\rightarrow} B)$
(2) $A \overset{5}{\rightarrow} B =_{\text{df}} (A \overset{11}{\rightarrow} B) \wedge [A \vee ({\sim}B \overset{11}{\rightarrow} {\sim}A)]$
(3) $A \overset{8}{\rightarrow} B =_{\text{df}} (A \overset{11}{\rightarrow} B) \wedge [B \vee (A \overset{5}{\rightarrow} B)]$
(4) $A \overset{5}{\rightarrow} B =_{\text{df}} (A \overset{8}{\rightarrow} B) \wedge [A \vee ({\sim}B \overset{8}{\rightarrow} {\sim}A)]$

□

In what follows, it is shown that LM4 (resp., LM11) is included in LM1 (resp., LM10). Then LM4 and LM5 are proved equivalent (Lemmas 3.9, 3.10, and 3.11).

Lemma 3.9 (LM1, LM4) *LM4 is included in LM1.*

Proof Consider the connective $k12$ defined by the ensuing table:

t12	0	1	2
0	0	0	0
1	0	1	0
2	0	1	0

Given LM1, table 4 is defined as follows: $A \overset{4}{\to} B =_{\mathrm{df}} (A \overset{1}{\to} B) \vee (A\ k12\ B)$. So, let us defined t12. We use the connectives $k13$, $k14$ given by the tables t13, t14.

t13	0	1	2
0	2	0	2
1	0	1	1
2	0	1	2

t14	0	1	2
0	0	1	0
1	2	1	0
2	2	1	0

which are defined as follows: $A\ k13\ B =_{\mathrm{df}} (A \overset{1}{\to} B) \vee (A \wedge B)$; $A\ k14\ B =_{\mathrm{df}} {\sim}[B \vee (A \overset{1}{\to} B)]$. Then t12 is defined as follows: $A\ k12\ B =_{\mathrm{df}} (A\ k13\ B) \wedge (A\ k14\ B)$ □

Lemma 3.10 (LM11 and LM10) *LM11 is included in LM10.*

Proof We need the connective $k12$ defined in Lemma 3.9 by using LM1 (notice that LM1 is included in LM4—Lemma 3.6—and that LM4 and LM10 are equivalent by Lemma 3.7). Also, we use the connectives $k15$, $k16$ and $k17$ given by the ensuing tables:

t15	0	1	2
0	2	0	2
1	0	1	2
2	2	0	2

t16	0	1	2
0	0	0	0
1	2	1	2
2	0	2	0

t17	0	1	2
0	0	0	0
1	0	1	0
2	0	2	0

which are defined as follows: $A\ k15\ B =_{\mathrm{df}} (A\ k12\ B) \overset{4}{\to} (A \vee B)$; $A\ k16\ B =_{\mathrm{df}} {\sim}({\sim}B\ k15\ {\sim}A)$; $A\ k17\ B =_{\mathrm{df}} (A\ k16\ B)\ k16\ (A\ k12\ B)$. Finally, t11 is defined as follows: $A \overset{11}{\to} B =_{\mathrm{df}} (A \overset{10}{\to} B) \vee (A\ k17\ B)$. □

Lemma 3.11 (LM4 and LM5) *LM4 and LM5 are equivalent.*

Proof Given Lemmas 3.7, 3.8, and 3.10, it suffices to prove that LM4 is included in LM5. We set: $A \overset{4}{\to} B =_{\mathrm{df}} (A \overset{5}{\to} B \wedge ({\sim}A \vee B)$. □

Finally, we have:

Theorem 3.12 (LM1 Through LM11) *LM1 through LM11 are equivalent to each other.*

Proof Consider the groups of logics (a) = {LM1, LM2, LM3, LM6, LM9}; (b) = {LM4, LM7, LM10}; (c) = {LM5, LM8, LM11}. By Lemmas 3.5, 3.7, and 3.8, it is proved that each one of these groups consists of logics equivalent to each other. Then, given the lemmas just quoted, the logics in (b) and (c) are equivalent by Lemma 3.11; and logics in (a) and in (b) (so in (c)) are equivalent by Lemmas 3.6 and 3.9. Consequently, the 11 LMi-logics introduced in Definition 2.8 are equivalent to each other. □

It follows from Theorem 3.12 that L3^{VSP} can be defined as the logic determined by any of the matrices in Definition 2.7 (but cf. the concluding remarks to the paper).

The section is ended by showing that the logics Pac and RM3 are included in L3^{VSP}. These logics are determined by the implicative expansions of MK3 given by the following tables

Pac	0	1	2
0	2	2	2
1	0	1	2
2	0	1	2

RM3	0	1	2
0	2	2	2
1	0	1	2
2	0	0	2

We have:

Proposition 3.13 (Pac, RM3 and L3^{VSP}) *The logics Pac and RM3 are included in L3^{VSP}.*

Proof Given Theorem 3.12, the ensuing definitions are sufficient:

(1) $A \overset{\text{Pac}}{\rightarrow} B =_{\text{df}} (A \overset{10}{\rightarrow} B) \vee (A \overset{3}{\rightarrow} B)$

(2) $A \overset{\text{RM3}}{\rightarrow} B =_{\text{df}} (A \overset{9}{\rightarrow} B) \vee (A \overset{3}{\rightarrow} B)$

□

4 L3^{VSP} as an Expansion of Classical Positive Propositional Logic

In this section, it is shown how to use the fact that Pac (so classical positive propositional logic) is definable in L3^{VSP} in order to give easy Hilbert-style formulations (H-formulations) of L3^{VSP}. The H-formulations we define present L3^{VSP} as an expansion of classical positive propositional logic. We use, say, the implication of LM1 for axiomatizing L3^{VSP}. In the appendix to the paper, an alternative axiomatization is sketched by using that of LM10.

In order to give an H-formulation for L1, we rely upon a strategy based upon Belnap-Dunn two-valued semantics introduced by Brady in [7] (cf. also [8, 9, 21]) as illustrated for 3-valued logics in some papers such as [21, 25].

As it is well-known, Belnap-Dunn two-valued semantics (BD-semantics) is characterized by the possibility of assigning T, F, both T and F or neither T not F to the formulas of a given language (cf. [5, 6, 10, 11]); T represents truth, F represents falsity). Concerning 3-valued logics, two variants of BD-semantics, overdetermined BD-semantics (o-semantics) and underdetermined BD-semantics (u-semantics) can be considered. Formulas can be assigned T, F or both T and F in the former; T, F or neither T nor F in the latter (cf. [21, 25]). U-semantics is especially adequate to 3-valued logics determined by matrices with only one designated value; o-semantics, for those determined by matrices where only one value is not designated.

Given an implicative expansion of MK3, M, with 1 and 2 as designated values, the idea for defining an o-semantics, $\mathrm{M_o}$, equivalent to the matrix semantics based upon M is simple: a wff A is assigned both T and F in $\mathrm{M_o}$ iff A is assigned 1 in M. Next, A is assigned T (resp., F) in $\mathrm{M_o}$ iff it is not assigned 0 (resp., 2) in M. (Notice that, unlike in u-semantics, interpretation of formulas cannot be empty in o-semantics.)

Then below an o-semantics for LM1 is introduced by defining the notion of an LM1-model and the accompanying notions of LM1-consequence and LM1-validity.

Definition 4.1 (LM1-Model) An LM1-model is a structure (K, I) where (i) $K = \{\{T\}, \{F\}, \{T, F\}\}$, and (ii) I is an LM1-interpretation from the set of all wffs to K, this notion being defined according to the following conditions for each propositional variable p and wffs A, B: (1) $I(p) \in K$; (2a) $T \in I(\sim A)$ iff $F \in I(A)$; (2b) $F \in I(\sim A)$ iff $T \in I(A)$; (3a) $T \in I(A \wedge B)$ iff $T \in I(A)$ & $T \in I(B)$; (3b) $F \in I(A \wedge B)$ iff $F \in I(A)$ or $F \in I(B)$; (4a) $T \in I(A \vee B)$ iff $T \in I(A)$ or $T \in I(B)$; (4b) $F \in I(A \vee B)$ iff $F \in I(A)$ & $F \in I(B)$; (5a) $T \in I(A \rightarrow B)$ iff $[T \notin I(A)$ & $T \notin I(B)]$ or $[T \notin I(A)$ & $F \notin I(B)]$ or $[T \in I(A)$ & $F \in (A)$ & $T \in I(B)$ & $F \in I(B)]$ or $[F \notin I(A)$ & $F \notin I(B)]$; (5b) $F \in I(A \rightarrow B)$ iff $[T \in I(A)$ & $F \in I(A)]$ or $[T \in I(B)$ & $F \in I(B)]$ or $[T \in I(A)$ & $F \in I(B)$.

Definition 4.2 (LM1-Consequence, LM1-Validity) Let M be an LM1-model. For any set of wffs Γ and wff A:

1. $\Gamma \vDash_{\mathrm{M}} A$ (A is a consequence of Γ in M) iff $T \in I(A)$ whenever $T \in I(\Gamma)$. ($T \in I(\Gamma)$ iff $\forall A \in \Gamma(T \in I(A))$; $F \in I(\Gamma)$ iff $\exists A \in \Gamma(F \in I(A))$.)
2. $\Gamma \vDash_{\mathrm{LM1}} A$ (A is a consequence of Γ in LM1-semantics) iff $\Gamma \vDash_{\mathrm{M}} A$ for each LM1-model M.
3. In particular, $\vDash_{\mathrm{LM1}} A$ (A is valid in LM1-semantics) iff $\vDash_{\mathrm{M}} A$ for each LM1-model M (i.e., iff $T \in I(A)$ for each LM1-model M).

By $\vDash_{\mathrm{LM1}}$ we shall refer to the relation just defined.

Now, given Definition 2.7 together with the adjoined notions of M1-interpretation and M1-validity (cf. Definition 2.1) and Definitions 4.1 and 4.2. We easily prove:

Proposition 4.3 (Coextensiveness of $\vDash_{\mathbf{M1}}$ and $\vDash_{\mathbf{LM1}}$) *For any set of wffs* Γ *and a wff* A, $\Gamma \vDash_{M1} A$ *iff* $\Gamma \vDash_{LM1} A$. *In particular,* $\vDash_{M1} A$ *iff* $\vDash_{LM1} A$.

Proof See the proof of Proposition 7.4 in [21] where the simple proof procedure is exemplified in the case of the 24 3-valued natural implicative logics introduced by Tomova in [27]. □

Proposition 4.3 simply formalizes the intuitive translation (explained above) of the matrix semantics based upon M1 into Belnap and Dunn's two-valued type LM1-semantics. Nevertheless, it is a useful proposition, since it gives us the possibility of proving soundness of LM1 w.r.t. $\vDash_{\mathrm{M1}}$ while proving completeness w.r.t. $\vDash_{\mathrm{LM1}}$ by using a canonical model construction.

But let us now define the H-system HLM1. We use $\stackrel{1}{\rightarrow}$, $\wedge$, $\vee$ and $\sim$ as primitive connectives ($\supset$ is interpreted by table Pac—cf. Proposition 3.13)[2]

Definition 4.4 (The System HLM1) The system HLM1 can be formulated as follows ($A_1, \ldots, A_n \Rightarrow B$ means "if $A_1, \ldots, A_n$, then B"; the superindex "1" above $\rightarrow$ is dropped):

Axioms:

A1. $A \supset (B \supset A)$

A2. $[A \supset (B \supset C)] \supset [(A \supset B) \supset (A \supset C)]$

A3. $[(A \supset B) \supset A] \supset A$

A4. $(A \wedge B) \supset A; \ \ (A \wedge B) \supset B$

A5. $A \supset [B \supset (A \wedge B)]$

A6. $A \supset (A \vee B); \ \ B \supset (A \vee B)$

A7. $(A \supset C) \supset [(B \supset C) \supset [(A \vee B) \supset C]$

A8. $(A \rightarrow B) \supset (A \supset B)$

A9. $\sim\sim A \leftrightarrow A$

A10. $(A \rightarrow B) \leftrightarrow (\sim B \rightarrow \sim A)$

A11. $\sim(A \vee B) \leftrightarrow (\sim A \wedge \sim B)$

A12. $\sim(A \wedge B) \leftrightarrow (\sim A \vee \sim B)$

A13. $A \vee \sim A$

---

[2] We remark that $\supset$ (as interpreted by Pac—cf. Proposition 3.13) could have been used as a primitive connective instead of any of the 11 conditionals given in Definition 2.7—cf. Proposition 4.15.

A14. $(A \vee B) \vee (A \rightarrow B)$

A15. $(A \vee {\sim}B) \vee (A \rightarrow B)$

A16. $[A \wedge {\sim}A) \wedge (B \wedge {\sim}B)] \rightarrow (A \rightarrow B)$

A17. $[(A \rightarrow B) \wedge (A \wedge {\sim}A)] \supset {\sim}B$

A18. $[{\sim}(A \rightarrow B) \wedge (A \wedge B)] \rightarrow ({\sim}A \vee {\sim}B)$

A19. $[{\sim}(A \rightarrow B) \wedge ({\sim}A \wedge B)] \rightarrow (A \vee {\sim}B)$

A20. $(A \wedge {\sim}A) \supset {\sim}(A \rightarrow B)$

A21. $(A \wedge {\sim}B) \supset {\sim}(A \rightarrow B)$

Rules of inference:

Modus ponens (MP $\supset$): $A \supset B, A \Rightarrow B$

Definitions:

$$A \supset B =_{\text{df}} A \overset{\text{Pac}}{\rightarrow} B$$

$$A \leftrightarrow B =_{\text{df}} (A \rightarrow B) \wedge (B \rightarrow A)$$

$$A \equiv B =_{\text{df}} (A \supset B) \wedge (B \supset A)$$

Remark 4.5 (On the H-formulation of LM1) We note that if $\wedge$ and $\vee$ are defined as in Łukasiewicz's 3 valued logic Ł3 (cf. [17]), the resulting tables are different from the corresponding ones in MK3. On the other hand, we note that the H-formulation of LM1 is not more complex (in fact, it is simpler) than, say, those for strong 3-valued logics such as G3 (cf. [3] and references therein).

Below, we remark some proof-theoretical properties of LM1.

Proposition 4.6 (Some Basic Theorems and Rules of HLM1) *The following theorems and rules are provable in HLM1:*

1. *Modus ponens for $\rightarrow$ (MP$\rightarrow$): $A \rightarrow B, A \Rightarrow B$*
2. *Adjunction (Adj): $A, B \Rightarrow A \wedge B$*
3. *Elimination of $\wedge$ (E$\wedge$): $A \wedge B \Rightarrow A, B$*
4. *Deduction theorem for $\supset$ (DT): If $\Gamma, A \vdash_{HLM1} B$, then $\Gamma \vdash_{HLM1} A \supset B$*
5. *If A is a classical positive propositional tautology, then $\vdash_{HLM1} A$.*
6. *(t1) $(A \rightarrow B) \equiv ({\sim}B \rightarrow {\sim}A)$; (t2) ${\sim}{\sim}A \equiv A$; (t3) ${\sim}({\sim}A \rightarrow {\sim}B) \leftrightarrow {\sim}(B \rightarrow A)$; (t4) ${\sim}({\sim}A \rightarrow {\sim}B) \equiv {\sim}(B \rightarrow A)$.*

Proof It is immediate. (1) By A8 and MP$\supset$. (2) By A5 and MP$\supset$. (3) By A4 and MP$\supset$. (4) By A1 and A2 since MP$\supset$ is the only rule of inference. (5) By A1 through A7 and MP$\supset$, as these theses and rule axiomatize classical positive propositional logic. (6) t1-t4 are immediate by A8, A9, A10, E$\wedge$, Adj and definitions of $\leftrightarrow$ and $\equiv$. □

By using this proposition we can prove some easy theorems of HLM1 which are instrumental in the completeness proof.

Proposition 4.7 (More Theorems of HLM1) *The following theorems are provable in HLM1. (t5)* $(\sim A \vee \sim B) \vee (A \rightarrow B)$*; (t6)* $[\sim(A \rightarrow B) \wedge (\sim A \wedge \sim B)] \supset (A \vee B)$*; (t7)* $(B \wedge \sim B) \supset \sim(A \rightarrow B)$*; (t8)* $[(A \rightarrow B) \wedge \sim B] \supset \sim A$*; (t9)* $[(A \rightarrow B) \wedge (B \wedge \sim B)] \supset A$.

Proof It is easy by using Proposition 4.6. In particular, we have: t5: by A14 and t1; t6: by A18, A8 and t2; t7: by A20, t2 and t4; t8: by t1; t9: by A17, t1 and t2. □

In what follows, we proceed to the proofs of soundness and completeness.

Theorem 4.8 (Soundness of HLM1) *For any set of wffs* Γ *and a wff* A*, if* $\Gamma \vdash_{HLM1} A$*, then (1)* $\Gamma \vDash_{M1} A$ *and (2)* $\Gamma \vDash_{LM1} A$.

Proof (1) It is immediate: the axioms of HLM1 are M1 valid and MP$\supset$ preserves M1-validity (recall that the material conditional is understood according to the table $\overset{\text{Pac}}{\rightarrow}$: cf. Proposition 3.13; in case a tester is needed, the one in [13] can be used). (2) By (1) and Proposition 4.3. □

Concerning completeness, it is proved by a canonical model construction, as suggested above. Let us see how this proof proceeds. We begin by stating a couple of definitions and a remark.

Definition 4.9 (HLM1-Theories) An HLM1-theory is a set of wffs containing all HLM1-theorems and closed under MP$_{\supset}$. An HLM1-theory t is prime if whenever $A \vee B \in t$, then $A \in t$ or $B \in t$; and t is non-trivial if it does not contain all wffs.

Remark 4.10 (Complete HLM1-Theories) An HLM1-theory t is complete if for any wff A, $A \in t$ or $\sim A \in t$. Now, prime HLM1-theories are complete by virtue of A13.

Definition 4.11 (Canonical HLM1-Models) Let $\mathcal{T}$ be a non-trivial prime HLM1-theory. A canonical HLM1-model is the structure $(K, I_{\mathcal{T}})$ where (i) K is defined as in Definition 4.1 and (ii) $I_{\mathcal{T}}$ is a function from the set of all wffs to K defined as follows: For each wff A, $T \in I_{\mathcal{T}}(A)$ iff $A \in \mathcal{T}$ and $F \in I_{\mathcal{T}}(A)$ iff $\sim A \in \mathcal{T}$.

Then, in order to prove completeness, we have to prove the ensuing two facts:

1. An HLM1-theory without a given wff can be extended to a prime HLM1-theory without the same wff.
2. Let $\mathcal{T}$ be a non-trivial prime HLM1-theory. Then $I_{\mathcal{T}}$ (as defined in Definition 4.11) fulfills clauses (2a), (2b), (3a), (3b) (4a), (4b), (5a) and (5b) (it is immediate that $I_{\mathcal{T}}$ fulfills clause (1)). That is, we have to prove that the canonical translations of clauses (1) through (5b) are provable in $\mathcal{T}$.

We proceed to the proofs of facts 1 and 2.

Lemma 4.12 (Primeness) *Let A be a wff and t an HLM1-theory such that $A \notin t$. Then there is a prime HLM1-theory $\mathcal{T}$ such that $t \subseteq \mathcal{T}$ and $A \notin \mathcal{T}$.*

Proof It is easy by using classical positive propositional logic (cf., e.g., Lemma 5.9 in [24]). In case that $\wedge$ and $\vee$ are defined by $\supset$ and $\sim$ —which is not the case here—, it suffices to use classical implicative propositional logic (cf., e.g., Lemma 3.9 in [18]). □

Lemma 4.13 (Canonical HLM1-Models are HLM1-Models) *Let M_c be a canonical HLM1-model. Then M_c is indeed an HLM1-model.*

Proof Let $\mathcal{T}$ be a non-trivial prime HLM1-theory and M_c be the canonical HLM1-model built upon it as indicated in Definition 4.11. In order to prove that M_c is indeed an HLM1-model it suffices to prove that $I_{\mathcal{T}}$ fulfills clauses (2a) through (5b). We have:

- Clause (2a). It is trivial.
- Clause (2b). By using A9.
- Clause (3a). By A4 and A5.
- Clause (3b). By A12.
- Clause (4a). By primeness of $\mathcal{T}$ and A6.
- Clause (4b). By A11.
- Clause (5a). ($\Rightarrow$) Suppose $A \rightarrow B \in \mathcal{T}$. We have to show that at least one of the following alternative obtains: $[A \notin \mathcal{T}$ & $B \notin \mathcal{T}]$ or $[A \notin \mathcal{T}$ & $\sim B \notin \mathcal{T}]$ or $[A \in \mathcal{T}$ & $\sim A \in \mathcal{T}$ & $B \in \mathcal{T}$ & $\sim B \in \mathcal{T}]$ or $[\sim A \notin \mathcal{T}$ & $\sim B \notin \mathcal{T}]$. For *reductio*, suppose that there are wffs A, B such that (1) $A \in \mathcal{T}$ or $B \in \mathcal{T}$ and (2) $A \in \mathcal{T}$ or $\sim B \in \mathcal{T}$ and (3) $A \notin \mathcal{T}$ or $\sim A \notin \mathcal{T}$ or $B \notin \mathcal{T}$ or $\sim B \notin \mathcal{T}$ and (4) $\sim A \in \mathcal{T}$ or $\sim B \in \mathcal{T}$. We have 32 possibilities to consider, but each one of them contains at least one of the following alternatives: (a) a contradiction, e.g., $A \in \mathcal{T}$ & $A \notin \mathcal{T}$; (b) $A \in \mathcal{T}$ and $B \notin \mathcal{T}$; (c) $\sim B \in \mathcal{T}$ & $\sim A \notin \mathcal{T}$; (d) $A \in \mathcal{T}$, $\sim A \in \mathcal{T}$ and $\sim B \notin \mathcal{T}$ or (e) $B \in \mathcal{T}$, $\sim B \in \mathcal{T}$ and $A \notin \mathcal{T}$. But (b)–(e) are also impossible: (b): by MP$\rightarrow$; (c): by t8; (d): by A17; (e): by t9.

  ($\Leftarrow$) Suppose $[A \notin \mathcal{T}$ & $B \notin \mathcal{T}]$ or $[A \notin \mathcal{T}$ & $\sim B \notin \mathcal{T}]$ or $[A \in \mathcal{T}$ & $\sim A \in \mathcal{T}$ & $B \in \mathcal{T}$ & $\sim B \in \mathcal{T}]$ or $[\sim A \notin \mathcal{T}$ & $\sim B \notin \mathcal{T}]$. We have to prove that $A \rightarrow B \in \mathcal{T}$ follows from each one of these four alternatives. Now, this is immediate by using A14, A15, A16 and t5, respectively.
- Clause (5b). ($\Rightarrow$) Suppose $\sim(A \rightarrow B) \in \mathcal{T}$. We have to prove that at least one of the following alternatives follows: $[B \in \mathcal{T}$ & $\sim B \in \mathcal{T}]$ or $[A \in \mathcal{T}$ & $\sim A \in \mathcal{T}]$ or $[A \in \mathcal{T}$ & $\sim B \in \mathcal{T}]$. Suppose for reductio that there are wffs A, B such that (1) $B \notin \mathcal{T}$ or $\sim B \notin \mathcal{T}$ and (2) $A \notin \mathcal{T}$ or $\sim A \notin \mathcal{T}$ and (3) $A \notin \mathcal{T}$ or $\sim B \notin \mathcal{T}$. We have 8 possibilities to consider, but each one of them contains at least one of the following alternatives: (a) $A \notin \mathcal{T}$ and $\sim A \notin \mathcal{T}$; (b) $B \notin \mathcal{T}$ and $\sim B \notin \mathcal{T}$; (c) $A \notin \mathcal{T}$ and $B \notin \mathcal{T}$; (d) $A \notin \mathcal{T}$ and $\sim B \notin \mathcal{T}$ or (e) $\sim A \notin \mathcal{T}$ and $\sim B \notin \mathcal{T}$. But these five situations are impossible: (a), (b): $\mathcal{T}$ is complete. Then, (c)–(d) are shown untenable by using the completeness of $\mathcal{T}$ and t6, A19 and A18, respectively.

($\Leftarrow$) Suppose $[A \in \mathcal{T} \;\&\; {\sim}A \in \mathcal{T}]$ or $[B \in \mathcal{T} \;\&\; {\sim}B \in \mathcal{T}]$ or $[A \in \mathcal{T} \;\&\; {\sim}B \in \mathcal{T}]$. We have to prove that ${\sim}(A \to B) \in \mathcal{T}$, follows from each one of these three alternatives, which is immediate by using A20, t7 and A21, respectively.

□

Once Lemmas 4.12 and 4.13 proved, the proof of completeness is straightforward.

Theorem 4.14 (Completeness of HLM1) *For any set of wffs* Γ *and wff* A, *(1) if* $\Gamma \vDash_{M1} A$, *then* $\Gamma \vdash_{HLM1} A$; *(2) if* $\Gamma \vDash_{LM1} A$, *then* $\Gamma \vdash_{HLM1} A$.

Proof Firstly, case (2) is proved. (2) Suppose $\Gamma \nvdash_{\mathrm{HLM1}} A$, i.e., that A is not included in the set of consequences derivable in HLM1 from Γ (in symbols, $A \notin \mathrm{Cn}\Gamma[\mathrm{HLM1}]$). Then, $\mathrm{Cn}\Gamma[\mathrm{HL1}]$ is extended to a prime HLM1-theory $\mathcal{T}$ such that $A \notin \mathcal{T}$. Next, the canonical HLM1-model $\mathrm{M_c} = (K, I_{\mathcal{T}})$ based upon $\mathcal{T}$ is defined, and we have $\Gamma \nvDash_{\mathrm{M_c}} A$, since $T \in I_{\mathcal{T}}(\Gamma)$ (as $T \in I_{\mathcal{T}}(\mathrm{Cn}\Gamma[\mathrm{HLM1}])$ but $T \notin I_{\mathcal{T}}(A)$), whence $\Gamma \nvDash_{\mathrm{LM1}} A$ (by Definitions 4.1 and 4.2), as was to be proved.

(1) It is immediate by (2) and Proposition 4.3. □

The section is ended by proving that $\overset{1}{\to}$, in fact, $\overset{i}{\to}$ $(1 \le i \le 11)$, can be replaced by $\overset{\mathrm{Pac}}{\to}$ as a primitive connective.

Proposition 4.15 ($\overset{\mathrm{Pac}}{\to}$ as a Primitive Connective) *Let Pac be the result of expanding MK3 with the conditional table Pac defined in Proposition 3.13. Then table 6 (so,* t_i $(i \le 1 \le 11)$ *is definable from Pac.*

Proof Consider the connectives $k18, k19, k20, k21, k22$ and $k23$ defined in the following tables:

	t18
0	2
1	1
2	2

t19	0	1	2
0	2	2	2
1	0	1	2
2	0	0	2

t20	0	1	2
0	0	0	0
1	0	1	0
2	0	0	2

t21	0	1	2
0	2	1	0
1	0	1	0
2	0	0	0

t22	0	1	2
0	2	1	0
1	0	1	0
2	0	0	2

t23	0	1	2
0	0	0	2
1	0	1	0
2	0	0	0

These connectives are defined as follows:

- $k18\, A =_{\mathrm{df}} {\sim}(A \wedge {\sim}A)$
- $A\, k19\, B =_{\mathrm{df}} (A \overset{\mathrm{Pac}}{\to} B) \wedge ({\sim}B \overset{\mathrm{Pac}}{\to} {\sim}A)$
- $A\, k20\, B =_{\mathrm{df}} k18\, (A\, k19\, B)\, k19\, (A \wedge B)$
- $A\, k21\, B =_{\mathrm{df}} {\sim}(A \vee B) \wedge (A\, k19\, B)$

- $A\ k22\ B =_{df} (A\ k20\ B) \vee (A\ k21\ B)$
- $A\ k23\ B =_{df} k18\ (A\ k19\ B)\ k19\ ({\sim}A \wedge B)$

Then, the table $\overset{6}{\rightarrow}$ is defined as follows: $A \overset{6}{\rightarrow} B =_{df} (A\ k22\ B) \vee (A\ k23\ B)$. So the 11 tables in Definition 2.7 are definable from Pac (i.e., the implicative expansion of MK3 with table $\overset{\text{Pac}}{\rightarrow}$). □

5 Concluding Remarks

The paper is ended with some brief concluding remarks on the results obtained and on possible future work to be done on the topic.

The most known 3-valued logics such as Łukasiewicz's Ł3, Gödel G3, the quasi-relevant logic RM3 or the paraconsistent logic Pac (cf., e.g., [15, 21] and references in these papers) lack the VSP. In fact, this property is not even predicable of the natural implicative logics defined in [27]. To the best of our knowledge, [22] is the first item in the literature presenting some instances of 3-valued logics with the VSP. These logics are obtained by extending the notion of a "natural conditional" defined in [27].

It is trivial to build up binary expansions of MK3 with the VSP. Consider the ensuing general table t$*$ (a_i $(1 \leq i \leq 4) \in \{0, 1, 2\}$)

$*$	0	1	2
0	0	a_1	0
1	a_2	2	a_3
2	0	a_4	0

Suppose now that A and B do not share some propositional variables in $A * B$. Let M be any expansion of MK3 built up by adding any of the $*$-functions described in t$*$ and let I be an M-interpretation such that $I(p) = 2$ (resp. $I(p) = 0$) for each propositional variable p in A (resp., in B). Then $I(A * B) = 0$. But it is obvious that none of the $*$-functions of t$*$ can represent a reasonable notion of a conjunction, disjunction or implication connective. However, the 11 implicative expansions of MK3 introduced in [22] cannot be deemed artificial constructs as, we think, it has been shown throughout the paper.

In this paper it has been proved that, from the viewpoint of functional strength, there is only one 3-valued natural (in the sense of [22]) implication expansion with the VSP of Kleene's strong logic, the logic we have dubbed L3^{VSP}. It is our opinion that it has also been shown that L3^{VSP} is a significant and strong logic that, we hope, can be of some use in contexts where relevance, paraconsistency and 3-valued decidability are needed.

There is a number of ways in which the research reported in the present paper could be pursued. We shall limit ourselves to remark two of them, but before doing this, let us stress the point that the fact that LM1 through LM11 are functionally equivalent logics does not necessarily make 10 of them dismissable in favor of a given chosen one, for the same reasons that Łukasiewicz's Ł3 is not dismissable in favor of one of the wealth of functional logics equivalent to it defined in [22].

The two possible paths for developing the results here obtained we propose are these:

1. Investigate the functional relations $L3^{VSP}$ maintain with some significant 3-valued logics such as those treated in [21].
2. Investigate the class of 3-valued non C-extending implication expansions of MK3 with the VSP. The members in this class will be, of course, *contraclassical* logics in the sense of [14] and it is well-known that there are very interesting contraclassical logics such as, e.g., connexive logics (cf., e.g., [28]). (We note that, as a way of an example, a 3-valued non C-extending implication expansion of MK3 with the VSP is briefly treated in the appendix.)

Acknowledgments This work is funded by the Spanish Ministry of Science and Innovation (MCIN/AEI/10.13039/501100011033) under Grant [PID2020-116502GB-I00].

Declarations

- Funding: The authors declare that this work is funded by the Spanish Ministry of Science and Innovation (MCIN/AEI/10.13039/501100011033) (Grant number [PID2020-116502GB-I00]).
- Conflict of interest/Competing interests: The authors declare that there are no conflict of interests/competing interests.
- Ethics approval: Not applicable.
- Consent to participate: Not applicable.
- Consent for publication: Not applicable.
- Availability of data and materials: Not applicable.
- Code availability: Not applicable.
- Authors' contributions (applicable for works with multiple authors): Not applicable.

Appendix A: The System HLM10

In this appendix, we include some complementary material to the topics investigated in the preceding sections. In particular, the ensuing points are treated: as an additional illustration of how to provide H-formulations for the LMi-logics, LM10 is given such a type of axiomatization; some proof-theoretical properties of LM1, LM10 and LMi-logics in general are noted; a comparison between LM1 and the important relevant logics B and DW is made; it is proved that all LMi-logics lack the Ackermann Property, and finally, we briefly present an instance of a 3-valued non C-extending implication expansion of MK3 with the VSP.

Definition A.1 (LM10-Models) An LM10-model is a structure (K, I) where K and I are defined like in LM1-models (cf. Definition 4.1), except for clauses (5a) and (5b), which now read as follows.

(5a) $T \in I(A \rightarrow B)$ iff $T \notin I(A)$ or $[T \in I(B) \ \& \ F \in I(B)]$ or $[F \notin I(A) \ \& \ T \in I(B)]$.
(5b) $F \in (A \rightarrow B)$ iff $[T \in I(A) \ \& \ F \in I(A)]$ or $[T \in I(A) \ \& \ F \in I(B)]$.

The notions of LM10-consequence and LM10-validity are defined similarly as in LM1 (cf. Definition 4.2).

Then HLM10 is introduced in the next definition ($\wedge, \vee, \sim$ and $\stackrel{10}{\rightarrow}$ are the primitive connectives—cf. note 2).

Definition A.2 (The System HLM10) The system HLM10 can be formulated as follows ($A_1, \ldots, A_n \Rightarrow B$ means "if $A_1, \ldots, A_n$, then B"; the superindex "10" above $\rightarrow$ is dropped):

Axioms:

A1. $A \supset (B \supset A)$

A2. $[A \supset (B \supset C)] \supset [(A \supset B) \supset (A \supset C)]$

A3. $[(A \supset B) \supset A] \supset A$

A4. $(A \wedge B) \supset A; \ \ (A \wedge B) \supset B$

A5. $A \supset [B \supset (A \wedge B)]$

A6. $A \supset (A \vee B); \ \ B \supset (A \vee B)$

A7. $(A \supset C) \supset [(B \supset C) \supset [(A \vee B) \supset C)]$

A8. $(A \rightarrow B) \supset (A \supset B)$

A9. $\sim\sim A \leftrightarrow A$

A10. $\sim(A \vee B) \leftrightarrow (\sim A \wedge \sim B)$

A11. $\sim(A \wedge B) \leftrightarrow (\sim A \vee \sim B)$

A12. $A \vee \sim A$

A13. $(A \rightarrow B) \vee A$

A14. $B \supset [\sim A \vee (A \rightarrow B)]$

A15. $[(A \rightarrow B) \wedge (A \wedge \sim A)] \rightarrow \sim B$

A16. $(B \wedge \sim B) \supset (A \rightarrow B)$

A17. $(A \wedge {\sim}A) \supset {\sim}(A \rightarrow B)$

A18. $(A \wedge {\sim}B) \supset {\sim}(A \rightarrow B)$

A19. $[{\sim}(A \rightarrow B) \wedge {\sim}A] \rightarrow A$

A20. $[{\sim}(A \rightarrow B) \wedge (A \wedge B)] \rightarrow ({\sim}A \vee {\sim}B)$

A21. $[{\sim}(A \rightarrow B) \wedge ({\sim}A \wedge B)] \rightarrow (A \vee {\sim}B)$

Rules of inference:

$$\text{Modus ponens (MP} \supset) : A \supset B, A \Rightarrow B$$

Definitions: $\supset$, $\equiv$ and $\leftrightarrow$ are defined as in HLM1 (cf. Definition 4.4.).

Below, some proof-theoretical properties of HLM10 are remarked.

Remark A.3 (Some Rules of HLM10) Items (1)–(5) in Proposition 4.6 for HLM1 are also provable (in a similar way) in the case of HLM10.

Now, following the pattern developed for HLM1 in §4, it is not difficult to prove that HLM10 is sound and complete w.r.t. both M10-semantics and LM10-semantics. In what follows, we note some proof-theoretical properties of the LMi-logics.

Remark A.4 (Some Proof-Theoretical Properties of the LMi-Logics) Consider now the ensuing theses and rules:

(a) $A \rightarrow A$; (b) $(A \wedge B) \leftrightarrow (B \wedge A)$; (c) $(A \vee B) \leftrightarrow (B \vee A)$; (d) $[A \wedge (B \wedge C)] \leftrightarrow [(A \wedge B) \wedge C]$; (e) $[A \vee (B \vee C)] \leftrightarrow [(A \vee B) \vee C]$; (f) $[A \wedge (B \vee C)] \leftrightarrow [(A \wedge B) \vee (A \wedge C)]$; (g) $[A \vee (B \wedge C)] \leftrightarrow [(A \vee B) \wedge (A \vee C)]$; (h) $[A \vee (A \wedge B)] \leftrightarrow A$; (i) $[A \wedge (A \vee B)] \leftrightarrow A$; (j) $A \leftrightarrow (A \wedge A)$; (k) $A \leftrightarrow (A \vee A)$; (l) $A \leftrightarrow {\sim\sim}A$; (m) ${\sim}(A \vee B) \leftrightarrow ({\sim}A \wedge {\sim}B)$; (n) ${\sim}(A \wedge B) \leftrightarrow ({\sim}A \vee {\sim}B)$; (o) $A \vee {\sim}A$; (E$\wedge$) $A \wedge B \Rightarrow A, B$; (Adj) $A, B \Rightarrow A \wedge B$; (CI$\wedge$) $A \rightarrow B, A \rightarrow C \Rightarrow A \rightarrow (A \wedge B)$; (I$\vee$) $A \Rightarrow A \vee B, B \vee A$; (E$\vee$) $A \rightarrow C, B \rightarrow C \Rightarrow (A \vee B) \rightarrow C$; (Trans$\rightarrow$) $A \rightarrow B, B \rightarrow C \Rightarrow A \rightarrow C$; (Con$\leftrightarrow$) $A \leftrightarrow B \Rightarrow {\sim}B \leftrightarrow {\sim}A$; (Fac$\leftrightarrow$) $A \leftrightarrow B \Rightarrow (A \wedge C) \leftrightarrow (B \wedge C)$; (Fac$'$ $\leftrightarrow$) $A \leftrightarrow B \Rightarrow (C \wedge A) \leftrightarrow (C \wedge B)$; (Sum$\leftrightarrow$) $A \leftrightarrow B \Rightarrow (A \vee C) \leftrightarrow (B \vee C)$; (Sum$'$ $\leftrightarrow$) $A \leftrightarrow B \Rightarrow (C \vee A) \leftrightarrow (C \vee B)$; (Trans$\leftrightarrow$) $A \leftrightarrow B, B \leftrightarrow C \Rightarrow A \leftrightarrow C$; (Suf$\leftrightarrow$) $A \leftrightarrow B \Rightarrow (B \rightarrow C) \leftrightarrow (A \rightarrow C)$; (Pref$\leftrightarrow$) $A \leftrightarrow B \Rightarrow (C \rightarrow A) \leftrightarrow (C \rightarrow B)$.

Notice that (a) is the self-identity axiom; (b) through (k) formalize the commutative, associative, absorption, idempotence of $\wedge$ and $\vee$ and distribution between the two connectives. Furthermore, (l) contains the two double negation axioms, (m) and (n) are the De Morgan laws and, finally, (o) is the principle of excluded middle. On the other hand, the abbreviations preceding the rules summarize the following labels. E$\wedge$: elimination of $\wedge$; Adj: adjunction; CI$\wedge$: conditioned introduction of $\wedge$; I$\vee$: introduction of $\vee$; E$\vee$: elimination of $\vee$; Trans$\rightarrow$: transitivity of $\rightarrow$; Fac$\leftrightarrow$, Fac$'$ $\leftrightarrow$ (resp. Sum$\leftrightarrow$, Sum$'$ $\leftrightarrow$): (two versions of) factor w.r.t. $\leftrightarrow$ (resp., summation w.r.t. $\leftrightarrow$); Con$\leftrightarrow$: contraposition w.r.t. $\leftrightarrow$; Trans$\leftrightarrow$: transitivity w.r.t. $\leftrightarrow$; Suf$\leftrightarrow$: suffixing w.r.t. $\leftrightarrow$, and, finally Pref$\leftrightarrow$: prefixing w.r.t. $\leftrightarrow$.

We have:

Proposition A.5 (Theses and Rules Provable in LM1, LM10) *The theses and rules displayed in Remark A.4 are provable in LM1 and LM10.*

Proof It is easy. It suffices to show that the theses are M1-valid and M10-valid, while the rules preserve M1-validity and M10-validity (if needed, the tester in [13] can be used). Then we apply the completeness theorem. □

Remark A.6 (Theses and Rules Provable in the LM*i*-Logics) It is easy to check that all the theses and rules listed in Remark A.4 are provable in each one of the LM*i*-logics, except for LM4 and LM5, where everything is provable save Trans (LM4 and LM5 are non-transitive logics w.r.t. its characteristic implication).

Now, given some of the properties noted in Remark A.4, the ensuing proposition is provable in the LM*i*-logics:

Proposition A.7 (Replacement) *For any wffs A, B, $A \leftrightarrow B \Rightarrow C[A] \leftrightarrow C[A/B]$, where $C[A]$ is a wff in which A appears and $C[A/B]$ is the result of substituting A by B in $C[A]$ in one or more place where A occurs.*

Proof Induction on the length of $C[A]$ by using Trans$\leftrightarrow$, Fac$\leftrightarrow$, Fac$'$ $\leftrightarrow$, Sum$\leftrightarrow$, Sum$'$ $\leftrightarrow$, Pref$\leftrightarrow$, Suf$\leftrightarrow$ and Con$\leftrightarrow$. □

The facts noted in Remark A.6 and Proposition A.7 allow us to compare the LM*i*-logics with "implicative logics" as understood in the classical Polish logical tradition.

Definition A.8 (Implicative Logics) A logic is implicative if it fulfills the ensuing conditions for any wffs A, B, C (cf. [20, pp. 179–180]; [29, p. 228])

C1. $A \to A$	Reflexivity
C2. $A \to B, A \Rightarrow B$	Modus ponens
C3. $A \Rightarrow B \to A$	Veq
C4. $A \to B, B \to C \Rightarrow A \to C$	Transitivity
C5. $A \leftrightarrow B \Rightarrow C[A] \leftrightarrow C[A/B]$	Replacement

(Veq abbreviates "Verum e quodlibet"—"A true proposition follows from any proposition".)

Now, it follows from Remark A.6 and Proposition A.7 that, leaving aside LM4 and LM5, all LM*i*-logics comply with all conditions listed in Definition A.8, except, of course, C3, since Veq encapsulates an infinity of paradoxes of relevance (LM4 and LM5 fail C3 and C4).

Remark A.9 (LM1, B and DW) Routley and Meyer's basic logic B can be axiomatized with the ensuing axioms and rules of inference:

(a1) $A \rightarrow A$; (a2) $(A \wedge B) \rightarrow A$; $(A \wedge B) \rightarrow B$; (a3) $A \rightarrow (A \vee B)$; $B \rightarrow (A \vee B)$; (a4) $[(A \rightarrow B) \wedge (A \rightarrow C)] \rightarrow [A \rightarrow (B \wedge C)]$; (a5) $[(A \rightarrow C) \wedge (B \rightarrow C)] \rightarrow [(A \vee B) \rightarrow C]$; (a6) $[A \wedge (B \vee C)] \rightarrow [(A \wedge B) \vee (A \wedge C)]$; (a7) $A \rightarrow {\sim}{\sim}A$; (a8) ${\sim}{\sim}A \rightarrow A$; (r1 –Adj) $A, B \Rightarrow A \wedge B$; (r2 –MP) $A \rightarrow B, A \Rightarrow B$; (r3 –Suf) $A \rightarrow B \Rightarrow (B \rightarrow C) \rightarrow (A \rightarrow C)$; (r4 –Pref) $B \rightarrow C \Rightarrow (A \rightarrow B) \rightarrow (A \rightarrow C)$; (r5 –Con) $A \rightarrow B \Rightarrow {\sim}B \rightarrow {\sim}A$.

Then, the logic DW is the result of replacing in B the rule Con by the contraposition axiom $(A \rightarrow B) \rightarrow ({\sim}B \rightarrow {\sim}A)$ (cf. [26, Chapter 4] about B and DW). Now, let B′ (resp., DW′) be the result of replacing in B (resp., in DW) the axioms a2 and a3 by the respective corresponding rule E$\wedge$ and I$\vee$ (cf. Remark A.4). Well then, we note that LM1 is an expansion of DW′.

Remark A.10 (Ackermann Property) A logic L has the Ackerman Property (AP) if A contains at least an implicative connective in all L-theorems of the form $A \rightarrow (B \rightarrow C)$. As pointed out above, the VSP is, according to Anderson and Belnap, a necessary property of any *relevant* logic, but, in addition to the VSP, a logic L has to comply with the AP in order to be an *entailment logic* (cf. [1]). LM1-LM11 lack the AP: the wff $p \rightarrow (p \rightarrow p)$ (an instance of the "mingle axiom" $A \rightarrow (A \rightarrow A)$) is provable in each one of them.

Remark A.11 (A 3-Valued Non C-Extending Implication Expansion of MK3 with the VSP) Consider the non C-extending implication expansion of MK3 with the $f_{\rightarrow}$-function given by the ensuing table

$\rightarrow$	0	1	2
0	2	2	0
1	0	1	0
2	0	1	2

Let us name LM12 the logic determined by this expansion. We note that LM12 has the VSP: if A and B do not share propositional variables in $A \rightarrow B$, then $I(A \rightarrow B) = 0$ for any M12-interpretation I such that $I(p) = 1$ (resp., $I(p) = 0$) for each variable p in A (resp., in B). Finally, Pac (so classical positive propositional logic—cf. Proposition 3.13) is definable as follows: $A \overset{\text{Pac}}{\rightarrow} B =_{df} B \vee (A \overset{12}{\rightarrow} B)$. Then, keeping to the pattern set up for axiomatizing LM1 and LM10, LM12 can be given the ensuing H-formulation as an expansion of classical positive propositional logic (cf. note 2):

Axioms: A1–A12 of HLM10 plus (A13) $(B \wedge {\sim}B) \supset (A \rightarrow B)$; (A14) $(A \wedge {\sim}A) \supset {\sim}(A \rightarrow B)$; (A15) $(A \wedge B) \supset [{\sim}A \vee (A \rightarrow B)]$; (A16) $({\sim}A \wedge {\sim}B) \supset [A \vee (A \rightarrow B)]$; (A17) $(A \wedge {\sim}B) \supset [{\sim}A \vee {\sim}(A \rightarrow B)]$; (A18) $({\sim}A \wedge B) \supset [{\sim}B \vee {\sim}(A \rightarrow B)]$; (A19) $(A \rightarrow B) \supset ({\sim}A \vee B)$; (A20) $(A \rightarrow B) \supset (A \vee {\sim}B)$; (A21) $[(A \rightarrow B) \wedge {\sim}A] \supset {\sim}B$; (A22) ${\sim}(A \rightarrow B) \supset (A \vee B)$; (A23) $[{\sim}(A \rightarrow B) \wedge {\sim}B] \supset A$; (A24) ${\sim}(A \rightarrow B) \supset ({\sim}A \vee {\sim}B)$.

The only rule of inference is *modus ponens* for ⊃. Notice that A18, A20 and A21 are *contraclassical* theses (cf. [14]).

References

1. Anderson, A. R., & Belnap, N. D. Jr. (1975). *Entailment. The logic of relevance and necessity* (Vol. I). Princeton, NJ: Princeton University Press.
2. Avron, A. (1991). Natural 3-valued logics–characterization and proof theory. *Journal of Symbolic Logic*, *56*(1), 276–294. https://doi.org/10.2307/2274919
3. Avron, A. (2022). Proof systems for 3-valued logics based on Gödel's implication. *Logic Journal of the IGPL*, *30*(3), 437–453. https://doi.org/10.1093/jigpal/jzab013
4. Belnap, N. D. Jr. (1960). Entailment and relevance. *The Journal of Symbolic Logic*, *25*(2), 144–146. https://doi.org/10.2307/2964210
5. Belnap, N. D. Jr. (1977a). A useful four-valued logic. In G. Epstein & J. M. Dunn (Eds.), *Modern uses of multiple-valued logic* (pp. 5–37). Dordrecht: D. Reidel Publishing Co. https://doi.org/10.1007/978-94-010-1161-7_2
6. Belnap, N. D. Jr. (1977b). How a computer should think. In G. Ryle (Ed.), *Contemporary aspects of philosophy* (pp. 30–55). Stocksfield: Oriel Press Ltd.
7. Brady, R. T. (1982). Completeness proofs for the systems RM3 and BN4. *Logique et Analyse*, *25*, 9-32. https://www.jstor.org/stable/44084001
8. Brady, R. T. (Ed.). (2003). *Relevant logics and their rivals* (Vol. II). Aldershot: Ashgate.
9. Brady, R. T. (2006). *Universal logic*. Stanford, CA: CSLI.
10. Dunn, J. M. (1976). Intuitive semantics for first-degree entailments and "coupled trees." *Philosophical Studies*, *29*, 149–168. https://doi.org/10.1007/BF00373152
11. Dunn, J. M. (2000). Partiality and its dual. *Studia Logica*, *66*(1), 5-40. https://doi.org/10.1023/A:1026740726955
12. Dziobiak, W. (1983). There are $2^{\aleph_0}$ logics with the relevance principle between R and RM. *Studia Logica*, *42*(1), 49–61. https://doi.org/10.1007/BF01418759
13. González, C. (2011). *MaTest*. https://sites.google.com/site/sefusmendez/matest
14. Humberstone, L. (2000). Contra-classical logics. *Australasian Journal of Philosophy*, *78*(4), 438–474. https://doi.org/10.1080/00048400012349741
15. Karpenko, A. S. (1999). Jaśkowski's criterion and three-valued paraconsistent logics. *Logic and Logical Philosophy*, *7*, 81–86. https://doi.org/10.12775/LLP.1999.006
16. Kleene, S. C. (1952). *Introduction to metamathematics* (1st ed.). Amsterdam/Groningnen: North-Holland Publishing Co./P. Noordhoff.
17. Łukasiewicz, J. (1920). On three-valued logic. In L. Borkowski (Ed.), J. Łukasiewicz, *Selected works* (pp. 87–88). Amsterdam: North-Holland Publishing Co., 1970.
18. Méndez, J. M., & Robles, G. (2015). A strong and rich 4-valued modal logic without Łukasiewicz-type paradoxes. *Logica Universalis*, *9*(4), 501–522. https://doi.org/10.1007/s11787-015-0130-z
19. Øgaard, T. F. (2021). Non-Boolean classical relevant logics I. *Synthese*, *198*(8), 6993–7024. https://doi.org/10.1007/s11229-019-02507-z
20. Rasiowa, H. (1974). *An Algebraic approach to non-classical logics* (Vol. 78). Amsterdam: North-Holland Publishing Company.
21. Robles, G., & Méndez, J. M. (2019). Belnap-Dunn semantics for natural implicative expansions of Kleene's strong three-valued matrix with two designated values. *Journal of Applied Non-Classical Logics*, *29*(1), 37–63. https://doi.org/10.1080/11663081.2018.1534487
22. Robles, G., & Méndez, J. M. (2020). The class of all natural implicative expansions of Kleene's strong logic functionally equivalent to Łukasiewicz's 3-valued logic Ł3. *Journal of Logic, Language and Information*, *29*(3), 349–374. https://doi.org/10.1007/s10849-019-09306-2

23. Robles, G., & Méndez, J. M. (2022). A remark on functional completeness of binary expansions of Kleene's strong 3-valued logic. *Logic Journal of the IGPL, 30*(1), 21–33. https://doi.org/10.1093/jigpal/jzaa028
24. Robles, G., & Méndez, J. M. (2023). A Class of implicative expansions of Belnap-Dunn logic in which Boolean negation is definable. *Journal of Philosophical Logic, 52*, 915–938. https://doi.org/10.1007/s10992-022-09692-2
25. Robles, G., Salto, F., & Méndez, J. M. (2019). Belnap-Dunn semantics for natural implicative expansions of Kleene's strong three-valued matrix II. Only one designated value. *Journal of Applied Non-Classical Logics, 29*(3), 307-325. https://doi.org/10.1080/11663081.2019.1644079
26. Routley, R., Meyer, R. K., Plumwood, V., & Brady, R. T. (1982). *Relevant logics and their rivals* (Vol. 1). Atascadero, CA: Ridgeview Publishing Co.
27. Tomova, N. (2012). A Lattice of implicative extensions of regular Kleene's logics. *Reports on Mathematical Logic, 47*, 173–182. http://doi.org/10.4467/20842589RM.12.008.0689
28. Wansing, H. (2023). Connexive logic. In E. N. Zalta & U. Nodelman (Eds.), *The Stanford encyclopedia of philosophy* (Summer 2023 Edition). Metaphysics Research Lab, Stanford University. https://plato.stanford.edu/archives/sum2023/entries/logic-connexive/
29. Wójcicki, R. (1988). *Theory of logical calculi: Basic theory of consequence operations*. Springer Netherlands. https://doi.org/10.1007/978-94-015-6942-2

Part III
Proof Theory

A Conceptual Approach to Restricted Quantification in Relevant Logics

Ross T. Brady

Abstract There are three differing accounts of restricted quantification in relevant logics under the author's name, singly or jointly. The first was Brady (Relevant logics and their rivals, vol 2. Ashgate, Aldershot, 2003), in which an axiomatization was given for the restricted quantifiers (∀xP) and (∃xP), where the respective quantifiers have sub-domains restricted by the predicate P which must satisfy a non-emptiness requirement. The second was in the group work, Beall, Brady, Hazen, Priest and Restall (J Philos Log 35:587–598, 2006), where criteria were set up for restricted quantifiers to satisfy in relevant logics with the appropriate restricted quantifiers being introduced accordingly. The third was in (Brady RT, Logic – the big picture. In: Beziau J-Y, Chakraborty M, Dutta S (eds) New directions in paraconsistent logic, Springer, pp 353–373, 2015), where the classical account, using ⊃ and & to respectively restrict the universal and existential quantifiers, was briefly given within a big-picture view of logic. We will rectify this situation by presenting a novel and simpler account of restricted quantification in accordance with its conceptual analysis. We introduce a new connective > which will replace both the classical ⊃ and &, but behave as a pure positive restrictor without the negative requirements on the restricting predicate which indeed separate the two classical restrictors. We axiomatize it with a basic axiom and rule and a number of general rules with existential premises. We then determine how this new connective fits into the context of the standard connectives and quantifiers of a range of relevant logics, including the logic MCQ of meaning containment. Using this, we then provide a simple account of the square of opposition. We also relate > to the rule $\Longrightarrow$, which is also interpreted positively with no negative requirements on its antecedent/premise. We then relate it as best as we can to the classical account. We finish by examining the extent of the application of >, focussing on the Curry Paradox of naive set theory.

R. T. Brady (✉)
La Trobe University, Melbourne, Australia
e-mail: ross.brady@latrobe.edu.au

I. Sedlár et al. (eds.), *New Directions in Relevant Logic*, Trends in Logic 63,
https://doi.org/10.1007/978-3-031-69940-5_11

1 Introduction

We start by applauding my students over the years for giving an intuitive response to various logical questions that arose in my teaching. Many have been insistent in objecting to Cantor's diagonal argument, some have asked for the real semantics, as opposed to the usual truth-theoretic semantics, and most have restricted universal quantification using & instead of $\supset$. I must admit to fobbing them off on each of these issues at the time, subsequently making amends through my published work. See [10, 12] for my response on Cantor's diagonal argument and [8] for my response on truth-theoretic semantics. I retired from teaching at the end of 2011 and so most of my responses have occurred too late for the students. Here, we will finally deal with the remaining issue, again belatedly.

Briefly, deductive logic is at base conceptual, as conceptualization is essentially what is left when ruling out experimentation and social surveys to provide basic data. Such concepts focus on those of the logical words: and, or, not, implies/entails, for some, for all, and identity, with the possible addition of necessity and possibility and other modal concepts. Deductive logic proceeds through the application of rules, initially to a set of axioms and then to derived theorems in the Hilbert-style approach, which is used in key applications such as set theory and arithmetic and in mathematics generally. Such axioms and rules capture the basic concepts of logic, with derived theorems also capturing these concepts. Thus, logic captures meanings in a proof-theoretic setting.

Here, we focus on the two quantifiers, universal and existential, but restricted to a non-empty subdomain by a predicate. This is what largely occurs in practice as the universal domain is too wide-ranging to apply to specific considerations. Indeed, in practice, almost all uses of the quantifiers are restricted, as evidenced by what occurs in our student translation exercises. The important conceptual observation regarding restricted quantification is its lack of negative requirements. It is just a pure restriction of a usually large domain to a subdomain using a predicate that enables the quantifier to focus on a specific context at hand. There is no need to specify what is not in the subdomain, where such a predicate would fail to hold. Such a specification could be achieved by applying the negative properties, the Law of Excluded Middle and the Disjunctive Syllogism, to the restricting predicate.

The paper will be sectioned as follows. In Sect. 2, we examine the author's three differing published accounts of restricted quantification, to provide a logical background for our account.[1] These consist of [5], where an axiomatization was

[1] We should also mention the earlier works of Belnap and Cohen. Belnap pioneered the work on conditional assertion and its usage to restrict quantification in his [4]. Like our argument given below, when the antecedent of a conditional assertion is not satisfied, the conditional becomes what is called "nonassertive", with it just reducing to its consequent alone. Cohen in Anderson, Belnap and Dunn [2], pp.472–487, then goes on to set up a proof theory for conditional assertion, based on the logic R, which is sound and complete with respect to a 3-valued Routley-Meyer style semantics.

given for the restricted quantifiers, the group work, Beall, Brady, Hazen, Priest and Restall [3], where criteria were set up for restricted quantifiers to satisfy in relevant logics, and [7] with the classical account. In Sect. 3, we introduce the restricting connective > and give its axiomatization, from which we establish all of Brady's axiomatization of restricted quantification in [5] and almost all of the criteria listed in Beall, Brady, Hazen, Priest and Restall [3]. We will apply this axiomatization of > to determine the interaction of its restricted quantification with the standard connectives and quantifiers of relevant logics. We are then able to proceed with a straight-forward account of the Square of Opposition. In Sect. 4, we examine the relationship between the restrictor > and the rule $\Longrightarrow$and also determine the appropriate relationships of our restricted quantification to the corresponding classical account, which uses $\supset$ and & to respectively restrict the universal and existential quantifiers. In Sect. 5, we discuss the extent of the application of the connective > with reference to Curry's Paradox.

2 The Three Published Accounts

We start with the [5] account, which sets out the axiomatization below for the restricted quantifiers ($\forall$xP) and ($\exists$xP), under the condition that the restricting predicate P is non-empty. This condition is built into each of the uses of the two restricted quantifiers though often this needs to be established through a specific instance such as Pa, which would ensure that any instantiation of the variable a lies in the subdomain determined by the predicate P. Such a non-emptiness condition is assumed to hold in the case of the universal domain, this being projected down into its subdomain restricted by the predicate P. Such a condition enables $\forall$xA $\rightarrow$ $\exists$xA to hold universally and ($\forall$xP)A $\rightarrow$ ($\exists$xP)A to hold restrictedly, both of which easily follow from the axiomatizations below. Note that we are not proposing to deal with the so-called free semantics, where the universal domain can be empty allowing $\forall$xA $\rightarrow$ $\exists$xA to fail, and correspondingly allowing ($\forall$xP)A $\rightarrow$ ($\exists$xP)A to fail for such a restricted domain.

Though the author's particular focus has been on the weak relevant logic MC of meaning containment, we will also cover some key relevant logics because what we say will apply to relevant logics in general, except for the fact that MC drops distribution sententially, which will then flow onto its quantified forms as well. So, the quantificational extensions fall into two groups: the standard relevant logics with distribution and the logic MC without distribution. We will start with MC and then present a few key relevant systems.

In accordance with Brady's earlier works, we set out our sentential logic MC of meaning containment and its quantificational extension MCQ, with standard bracketing conventions from Anderson and Belnap [1]. This was initially presented and argued for in Brady [6] in the form of the logic DJ^d, with a one-premise meta-rule (see MR1 below), and then modified to MC in [11], by dropping the distribution axiom, A&(B$\vee$C) $\rightarrow$ (A&B) $\vee$ (A&C). Then, in [7], distribution in rule-form was essentially added through a strengthening of the one-premise meta-rule to a two-premise form. However, subsequently in [9], the rule-form of distribution was also dropped, noting that it still is an admissible rule which holds for a wide range of logical applications. This reinstates just the one-premise meta-rule MR1 below. This is used for a variety of purposes in the sequel and is justified by its deductive equivalence with the rule-form of disjunction elimination: if A ⇒ C and B ⇒ C then A∨B ⇒ C, which is standardly used in mathematical applications. This is the form of MC we present.

MC.
Primitives: ~, &, ∨, →.
Axioms.
A1. A→A.
A2. A&B → A.
A3. A&B → B.
A4. (A→B) & (A→C) → .A→B&C.
A5. A → A∨B.
A6. B → A∨B.
A7. (A→C) & (B→C) → .A∨B→C.
A8. ~~A→A.
A9. A→~B → .B→~A.
A10. (A→B) & (B→C) → .A→C.
Rules.
R1. A, A→B ⇒ B.
R2. A, B ⇒ A&B.
R3. A→B, C→D ⇒ B→C → .A→D.
Meta-Rule.
MR1. If A ⇒ B then C∨A ⇒ C∨B.

We now add the quantifiers to yield MCQ. As in the author's previous presentations, we separate free and bound variables to simplify the conditions on the axioms.

MCQ.
Primitives: ∀, ∃,
a,b,c, ... (free variables)
x,y,z, ... (bound variables)
Axioms.
QA1. ∀xA → A^a/x.
QA2. ∀x(A→B) → .A→∀xB. (Note that x cannot be free in A.)
QA3. A^a/x → ∃xA.
QA4. ∀x(A→B) → . ∃xA→B. (Note that x cannot be free in B.)
Rule.
QR1. A^a/x ⇒ ∀xA, where a does not occur in A. (This ensures that the generalization is just upon the substituted occurrences of a in A^a/x.)
Meta-Rule.

QMR1. If $A^a/x \Rightarrow B^a/x$ then $\exists xA \Rightarrow \exists xB$, where a does not occur in A nor in B, and

where, in the derivation $A^a/x \Rightarrow B^a/x$, QR1 does not generalize on any free variable in the premise A^a/x. The same condition then applies to the premise A of $A \Rightarrow B$ in the meta-rule MR1 of MC.

The quantificational meta-rule QMR1 is very useful in this context and is justified by its deductive equivalence with the rule-form of existential elimination: If $A^a/x \Rightarrow B$ then $\exists xA \Rightarrow B$, where a does not occur in A nor in B. Note that x cannot be free in B, as it can in QMR1 prior to the substitution $^a/x$ made on it.

The author's first account of restricted quantification, given in [5], is an axiomatization of restricted quantification, capturing the restricted versions of the standard quantificational axiomatization. This was set out as follows, in pp. 326–7 of [5], with adjustments made for our distinction between free and bound variables and our omission of distribution, as set out in the axiomatization of MCQ above. Note that we replace the existential premise $\exists xPx$ by Pa when a occurs elsewhere in the rule.

Axioms for Restricted Quantifiers.

AQR1. $(\forall xPx)A \longleftrightarrow A$.

AQR2. $A \longleftrightarrow (\exists xPx)A$.

Note that, for ARQ1 and ARQ2, x cannot occur in A in a free position, as it does in RRQ1 and RRQ3 below.

Rules for Restricted Quantifiers.

RRQ1. $Pa \Rightarrow (\forall xPx)A \rightarrow A^a/x$.

RRQ2. $\exists xPx \Rightarrow (\forall xPx)(A \rightarrow B) \rightarrow .(\forall xPx)A \rightarrow (\forall xPx)B$.

RRQ3. $Pa \Rightarrow A^a/x \rightarrow (\exists xPx)A$.

RRQ4. $\exists xPx \Rightarrow (\forall xPx)(A \rightarrow B) \rightarrow .(\exists xPx)A \rightarrow (\exists xPx)B$.

RRQ5. Pa, $A^a/x \Rightarrow (\forall xPx)A$, where a does not occur in A. (This corrects a mistake in [5], where Pa was omitted as a premise of the rule.)

Meta-rule for Restricted Quantification.

MRRQ1. If Pa and $A^a/x \Rightarrow B^a/x$ then $(\exists xPx)A \Rightarrow (\exists xPx)B$, where a does not occur in either A or B.

MRRQ1 and QMR1 (of MCQ) have the proviso that neither QR1 nor RRQ5 generalize on any free variable in the premise A^a/x in the derivation $A^a/x \Rightarrow B^a/x$. The same proviso also applies to the premise A of the meta-rule MR1 of MC. (We add Pa to ensure that a satisfies the restricting predicate.)

We apply these considerations to the following key distributive relevant logics, though with MR1 included, which we indicate with the superscript d:

B^d: MC - A10 - A9 + A11. $A\&(B \vee C) \rightarrow (A\&B) \vee (A\&C)$ + R4. $A \rightarrow \sim B \Rightarrow B \rightarrow \sim A$.

DW^d: B^d + A9 - R4.

TW^d: DW^d + A12. $A \rightarrow B \rightarrow .B \rightarrow C \rightarrow .A \rightarrow C$ + A13. $A \rightarrow B \rightarrow .C \rightarrow A \rightarrow .C \rightarrow B$ - R3.

RW^d: TW^d + A14. $A \rightarrow .A \rightarrow B \rightarrow B$.

R: RW^d + A15. (A → .A→B) → .A→B. (MR1 is not needed in R, as it is derivable.)

The quantificational extensions of each of these systems are obtained by adding the following distributive axioms to the axioms, rule and meta-rule, given for MCQ above:

QA5. ∀x(A∨B) → A∨∀xB.
QA6. A&∃xB → ∃x(A&B).

Note that x cannot occur free in A in QA5 and QA6.

The restricted quantificational extensions of each of these systems are obtained with the addition of the following distributive rules:

RRQ6. ∃xPx ⇒ (∀xPx)(A∨B) → A∨(∀xPx)B.
RRQ7. ∃xPx ⇒ A&(∃xPx)B → (∃xPx)(A&B).

Note that x cannot occur free in A in RRQ6 and RRQ7.

These restricted quantificational axioms and rules will provide the central justification for our introduction of the new connective > in Sect. 3 below.

The author's second account is in a joint work with Beall, Hazen, Restall and Priest in [3], which is largely due to the co-authors. This account was reliant on five key constraints of restricted quantification, two negative requirements and eight further constraints required to determine the square of opposition, where we assume that domains are non-empty in accordance with the central position of traditional logic. In setting up these restricted quantifications, a conditional connective was designed for use with the universal quantifier and the standard conjunction used with the existential quantifier.

The first five constraints were as follows, adjusted to fit our presentation above:

A1. Pa, (∀xPx)A ⇒ A^a/x.
A2. ∃xPx, ∀xA ⇒ (∀xPx)A.
A3. ∃xPx, ∀x(Px→A) ⇒ (∀xPx)A.
A4. Pa, A^a/x ⇒ (∃xPx)A.
A5. ∃xPx, ~∃xA ⇒ ~(∃xPx)A.

The two negative requirements were:

B1. ∃xPx, ∃x~Ax ⇒ not - {(∀xPx)Ax ⇒ (∀x~Ax)~Px}.
B2. ∃xPx ⇒ not- {(∀xPx)(∀xPx)A ⇒ (∀xPx)A}.

The eight constraints for the Square of Opposition were:

C1. ∃xPx, (∀xPx)A ⇒ (∃xPx)A.
C2. ∃xPx, ~(∃xPx)A ⇒ (∃xPx)~A.
C3. ∃xPx, (∀xPx)A, ~(∃xPx)A ⇒ #,

where # can be thought of as the disjunction of all contradictions of the form B&~B.

C4. ∃xPx, * ⇒ (∃xPx)A ∨ (∃xPx)~A,

where * can be thought of as the conjunction of all formulae of the form B∨~B, this being unnecessary in a relevant logic such as R which contains the Law of Excluded Middle.

C5. ∃xPx, (∀xPx)A & (∃xPx)~A ⇒ #.
C6. ∃xPx, ~(∃xPx)A & (∃xPx)A ⇒ #.

Note that we replaced ∨ by &, as there was clearly a mistake in the above [3].

C7. ∃xPx, * ⇒ (∀xPx)A ∨ (∃xPx)~A.
C8. ∃xPx, * ⇒ ~(∃xPx)A ∨ (∃xPx)A.

Unlike in the joint paper, we include the premise ∃xPx in C5–8 to ensure the non-emptiness of Px, and we also replace their C9 by C8, fixing up a typo.

The paper goes on to establish a special restricting connective ↦ for the universal quantifier, whilst retaining the use of conjunction & for restricting the existential quantifier. It is then shown that all of the 13 constraints above are satisfied by these two restricting connectives. The connective ↦ can be captured in the Routley-Meyer semantics by replacing the 3-place relation R, used to capture →, by R′ where R'abc iff Rabc and a⊑c, where if a⊑c then any formula true at a is true at c, for all worlds a, b and c. The truth condition for ↦ is then given as follows: A ↦ B is true at a iff, for all b,c such that R'abc, if A is true at b then B is true at c. However, for the two negative requirements, there is also a need for a falsity condition for ↦, which is as follows: A ↦ B is false at a iff A is true at a and B is false at a. This is within a four-valued semantics which allows truth and falsity to be independently determined. However, this does make the semantics look somewhat ad hoc in trying to satisfy all the constraints and in the sequel there will clearly be no need for such an artificial separation between truth and falsity.

The third account occurs, very briefly, on p.369 in [7], which we elucidate as follows. This is the familiar classical account, based on the assumption that the restricting predicate P satisfies the negative requirements of the Law of Excluded Middle (LEM), Pa ∨ ~Pa, and the Disjunctive Syllogism (DS), ~Pa, Pa ∨ A ⇒ A. These two laws have the effect of ensuring that the restricting predicate is meta-theoretically classical, viz. that exactly one of Pa and ~Pa is provable, if it is also assumed that priming holds for the two disjunctions and that the system is non-trivial. Moreover, the LEM and the DS do enable all classical tautologies to be proved in relevant logics with the help of distribution in rule-form, which can be shown to be admissible for the logic MC. (See [10] for discussion on this.)

The idea here was to carry the classicality of the universal domain, where all elements are in the domain and no elements are not, down to the restricted domain, which then needs to be specified both positively and negatively, that is, each element is either in the domain or not in it and no element is both in it and not in it. Obviously, this is an ideal situation, but restricted quantification does not generally say anything about what is not in the restricted domain. That is, one only needs to know what is in the restricted domain in order for the restriction to take place. Conceptually, this is all we need to consider, and we take this forward into the next section in introducing the appropriate restricting connective, called the restrictor.

3 The Introduction of the Restrictor >

What students did in using & as a restricting connective for both the universal and existential quantifiers was somewhat intuitive in that one connective is used for both, this being appropriate as the same restricting process is used no matter which quantifier is involved, in accordance with the concept of restricted domain. However, if we take the negative side of the restricted domain into account, then & does not work for the universal quantifier as any element outside the restricted domain {x:Px} will make the whole restricted universal statement, ∀x(Px & A), false, whilst ⊃ will also not work for the existential quantifier as any element outside the restricted domain {x:Px} will make the whole restricted existential statement, ∃x(Px ⊃A), true.

So, what we need to do is to just consider the positive side of the restricted domain, as argued conceptually above, and try to find a single restricting connective, as neither & nor ⊃ will do because of their valuations when their left component Px is false. Indeed, if we just consider the positive side of the restricted domain, then both & and ⊃ agree, that is, when their left component Px is true the value of Px ⊃A and Px &A are the same as their right-hand component A. So, in determining our new restricting connective we just need to capture the value of the right-hand component A when the value of the left-hand component Px is true. This will make our new connective, which we will denote by >, similar in this regard to the rule ⇒, as it only applies when its premise has been assumed. The proof of a premise allows the rule to activate, yielding its conclusion, and similarly the truth of an antecedent Px allows Px >A to be applied, yielding A. Here, in restricted quantification, A would normally contain the bound variable x. If it does not, then the above axioms ARQ1 and ARQ2 of [5] apply to the respective restricted quantifiers, making them redundant.

Here, we call '>' a connective, but it is called as such because it occurs within a formula even though it is unlike any other connective in that its antecedent does not have any negative requirements, just like a premise of a rule. So, restricted quantification restricts the domain of quantification requiring it to satisfy a predicate P and in so doing creates a rule-like assumption which is placed within a formula because of its context rather than be placed as a premise of a rule.

We proceed to axiomatize > with an axiom and a series of rules, the first rule being based on the meaning equivalence of Pa > A with A, on the assumption of Pa, replacing the bound x of the restricted quantifiers by the free a, as the quantifiers are yet to be introduced. In accordance with our understanding of restricted quantification, when Pa is assumed, the restriction of A^a/x to satisfy Pa then applies, leaving just A^a/x by itself to freely engage in this restricted context just as it normally would in a predicate calculus with a universal domain. Thus, the restriction 'Pa >' is redundant once Pa is assumed to be true. As above, this is also supported at the level of truth by the line of the truth-table for Pa ⊃ A^a/x and Pa & A^a/x determined by Pa being true, which shows that Pa ⊃ A^a/x, Pa &A^a/x and A^a/x all have the same truth-values within this classical account. What we do is to replace each restricted quantifier by the corresponding unrestricted universal or existential quantifier, but with the use of the restrictor >, in the manner discussed above.

As will be shown below, this rule with premise Pa, together with subsequent rules with existential premises ∃xPx, can then be used, in conjunction with the properties of the two unrestricted quantifiers, to determine the corresponding properties of restricted quantification, in accordance with their axiomatization in [5] and with respect to the 13 constraints given in Beall, Brady, Hazen, Priest and Restall [3]. However, in the process, one of the 2 negative requirements of [2006] will need to be deleted and one of the 13 constraints recast.

The use of the free variable a in the premise Pa in the rules below ensures that the occurrence of a in the respective rule represents an object in the restricted domain. The predicate P is non-empty here with a as a free variable applying to at least one element of the domain restricted by P.

Primitive: > (restrictor)
Definitions.
1. $(\forall xPx)A =_{df} \forall x(Px > A)$.
2. $(\exists xPx)A =_{df} \exists x(Px > A)$.

Restrictor Axiom.
RA1. $(Pa > A) \longleftrightarrow A$, where a is not free in A.

The restricting predicate Pa is redundant here, as a is not free.

Restrictor Rules.
RR1. $Pa \Rightarrow (Pa > A) \longleftrightarrow A$, where a is free in A.

This rule shows the collapse of the Pa restriction once Pa is satisfied.

We now have a framework for adding the two unrestricted quantifiers in accordance with their properties as applied to a universal domain. In particular, one can straightforwardly show $(\forall xPx)A \longleftrightarrow \forall xA$ and $(\exists xPx)A \longleftrightarrow \exists xA$, when Pa is a theorem, thus showing that restricted quantification reduces to unrestricted quantification, for both quantifiers, upon proof of the restricting predicate.

RR2. $\exists xPx \Rightarrow (\forall xPx)(A\to B) \to .(\forall xPx)A \to (\forall xPx)B$.
RR3. $\exists xPx \Rightarrow (\forall xPx)(A\to B) \to .(\exists xPx)A \to (\exists xPx)B$.

RR2 and RR3 are the more general versions in comparison with the corresponding axioms of MCQ and align with those of [5]. Unfortunately, neither RR2 nor RR3, are derivable from $Pa \Rightarrow (Pa >.A\to B) \to .(Pa > A) \to (Pa > B)$, as the Pa needs to be understood universally.

RR4. $\exists xPx \Rightarrow \forall xA \to (\forall xPx)A$.
RR5. $\exists xPx \Rightarrow (\exists xPx)A \to \exists xA$.

RR4 and RR5 show that the restricted domain is a subset of the universal domain. Their obvious proofs using RR1 fail because QMR1 cannot be applied to introduce the premise ∃xPx because QR1 would be used to generalize on the free variable a, which replaces the bound variable x. The same point applies to RR2 and RR3.

Restriction Distribution

By repeatedly applying RR1, we can derive the following forms of what we will call Restriction Distribution, each relating to a particular connective or quantifier:

RD1. $Pa \Rightarrow Pa > {\sim}Aa \longleftrightarrow {\sim}(Pa > Aa)$.
RD2. $Pa \Rightarrow Pa > A\&B \longleftrightarrow (Pa > A) \& (Pa > B)$.
RD3. $Pa \Rightarrow Pa > A \vee B \longleftrightarrow (Pa > A) \vee (Pa > B)$.
RD4. $Pa \Rightarrow (Pa >.A \rightarrow B) \longleftrightarrow .(Pa > A) \rightarrow (Pa > B)$.
RD5. $Pa \Rightarrow Pa > \forall yA \longleftrightarrow \forall y(Pa > A)$.
RD6. $Pa \Rightarrow Pa > \exists yA \longleftrightarrow \exists y(Pa > A)$.

We now boost up these distribution principles to apply to the respective restricted universal and existential quantifiers, as appropriate, but those listed immediately below will need to be added as further rules, as for RR2-5 above, because their proofs would require generalization on the free variable a, making QMR1 inapplicable.

RD7a. $\exists xPx \Rightarrow (\forall xPx){\sim}Ax \longleftrightarrow {\sim}(\exists xPx)Ax$.
RD7b. $\exists xPx \Rightarrow (\exists xPx){\sim}Ax \longleftrightarrow {\sim}(\forall xPx)Ax$.
RD8a. $\exists xPx \Rightarrow (\forall xPx)A \& (\forall xPx)B \rightarrow (\forall xPx)(A\&B)$.
RD9a. $\exists xPx \Rightarrow (\exists xPx)(A \vee B) \rightarrow (\exists xPx)A \vee (\exists xPx)B$.
RD10a. $\exists xPx \Rightarrow \forall y(\forall xPx)A \rightarrow (\forall xPx)\forall yA$.
RD11a. $\exists xPx \Rightarrow (\exists xPx)\exists yA \rightarrow \exists y(\exists xPx)A$.

The following usual relations between the two restricted quantifiers can then be shown, by first generalizing on an unrestricted variable and then using RR4 to transfer the generalization to the restricted variable, subject just to the assumption $\exists xPx$.

RD8b. $\exists xPx \Rightarrow (\forall xPx)(A\&B) \rightarrow (\forall xPx)A \& (\forall xPx)B$.
RD8c. $\exists xPx \Rightarrow (\exists xPx)(A\&B) \rightarrow (\exists xPx)A \& (\exists xPx)B$.
RD9b. $\exists xPx \Rightarrow (\forall xPx)A \vee (\forall xPx)B \rightarrow (\forall xPx)(A \vee B)$.
RD9c. $\exists xPx \Rightarrow (\exists xPx)A \vee (\exists xPx)B \rightarrow (\exists xPx)(A \vee B)$.
RD10b. $\exists xPx \Rightarrow (\forall xPx)\forall yA \rightarrow \forall y(\forall xPx)A$.
RD10c. $\exists xPx \Rightarrow (\exists xPx)\forall yA \rightarrow \forall y(\exists xPx)A$.
RD11b. $\exists xPx \Rightarrow \exists y(\forall xPx)A \rightarrow (\forall xPx)\exists yA$.
RD11c. $\exists xPx \Rightarrow \exists y(\exists xPx)A \rightarrow (\exists xPx)\exists yA$.

We can now prove the axioms, rules and meta-rule for restricted quantifiers from [5]. We indicate what is used in the proofs in square brackets to the right, given the use of MCQ in each case. However, use of MR1 and QMR1 (and possibly QR1) are indicated when they are employed. (We maintain the numbering from [5])

Axioms for Restricted Quantifiers

1. $(\forall xPx)A \longleftrightarrow A$, where x does not occur free in A. [RA1]
2. $A \longleftrightarrow (\exists xPx)A$, where x does not occur free in A. [RA1]

Rules for Restricted Quantifiers

1. $Pa \Rightarrow (\forall xPx)A \rightarrow A^a/x$. [RR1]
2. $\exists xPx \Rightarrow (\forall xPx)(A \rightarrow B) \rightarrow .(\forall xPx)A \rightarrow (\forall xPx)B$. [RR2]
3. $Pa \Rightarrow A^a/x \rightarrow (\exists xPx)A$. [RR1]
4. $\exists xPx \Rightarrow (\forall xPx)(A \rightarrow B) \rightarrow .(\exists xPx)A \rightarrow (\exists xPx)B$. [RR3]
5. $Pa, A^a/x \Rightarrow (\forall xPx)A$, where a does not occur in A. [RR1, QR1]

Meta-rule for Restricted Quantification

1. If Pa and $A^a/x \Rightarrow B^a/x$ then $(\exists xPx)A \Rightarrow (\exists xPx)B$, where a does not occur in A or B.

QMR1 (of MCQ) and MRRQ1 (immediately above) have the proviso that neither QR1 nor RRQ5 (above) generalize on any free variable in the premise A^a/x in the derivation $A^a/x \Rightarrow B^a/x$. [By RR1, if $A^a/x \Rightarrow B^a/x$ then $Pa > A^a/x \Rightarrow Pa > B^a/x$, with the above proviso, upon which QMR1 can be applied.] The same proviso also applies to the premise A of the meta-rule MR1 of MC.

For those relevant logics with distribution, as set out in Sect. 2 above, the following two rules are also included, as in [5]. These would also need to be included in the above axiomatization for the restrictor, thus proving them here.

6. $\exists xPx \Rightarrow (\forall xPx)(A \vee B) \rightarrow A \vee (\forall xPx)B$.
7. $\exists xPx \Rightarrow A \& (\exists xPx)B \rightarrow (\exists xPx)(A \& B)$.

In 6 and 7, x does not occur free in A.

The first five constraints of Beall, Brady, Hazen, Priest and Restall [3] were as follows, adjusted to fit our presentation above. We will indicate proofs in square brackets, where applicable. (We maintain the rule names of [3].)

A1. $Pa, (\forall xPx)A \Rightarrow A^a/x$. [RR1]
A2. $\exists xPx, \forall xA \Rightarrow (\forall xPx)A$. [RR4]
A3. $\exists xPx, \forall x(Px \rightarrow A) \Rightarrow (\forall xPx)A$.

Since Px is restricted to its positive interpretation, let us consider $Px \rightarrow A$ in rule-form, replacing $\forall x(Px \rightarrow A)$ by $Pa \Rightarrow A^a/x$ and hence $\exists xPx$ by Pa, making it a meta-rule. So, we replace A3 by A3′, capturing the essence of A3.

A3′. $Pa, (Pa \Rightarrow A^a/x) \Rightarrow (\forall xPx)A$. [RR1, QR1]
A4. $Pa, A^a/x \Rightarrow (\exists xPx)A$. [RR1]
A5. $\exists xPx, {\sim}\exists xA \Rightarrow {\sim}(\exists xPx)A$. [RR5]

The two negative requirements were:

B1. $\exists xPx, \exists x{\sim}Ax \Rightarrow$ not - $\{(\forall xPx)Ax \Rightarrow (\forall x{\sim}Ax){\sim}Px\}$.

$\{(\forall xPx)Ax \Rightarrow (\forall x{\sim}Ax){\sim}Px\}$ holds if the negation is classically interpreted, as can be seen by Venn Diagrams. This is not ruled out by the logic. However, non-classical interpretations of negation can make it fail, such as one which fails Disjunctive Syllogism, allowing some of the elements satisfying ~A not to satisfy ~P, whilst satisfying P. So, B1 holds with $\{(\forall xPx)Ax \Rightarrow (\forall x{\sim}Ax){\sim}Px\}$ failing as a general rule.

B2. $\exists xPx \Rightarrow$ not- $\{(\forall xPx)(\forall xPx)A \Rightarrow (\forall xPx)A\}$. [The outer $(\forall xPx)$ is redundant, yielding $(\forall xPx)A \Rightarrow (\forall xPx)A$, which rejects B2.]

The eight constraints for the square of opposition were:

C1. $\exists xPx, (\forall xPx)A \Rightarrow (\exists xPx)A$. [RR1, MR1]
C2. $\exists xPx, {\sim}(\exists xPx)A \Rightarrow (\exists xPx){\sim}A$. [RD7a, C1]
C3. $\exists xPx, (\forall xPx)A, {\sim}(\exists xPx)A \Rightarrow \#$. [C1, yielding a contradiction.]
C4. $\exists xPx, * \Rightarrow (\exists xPx)A \vee (\exists xPx){\sim}A$.

Start with $A^a/x \vee {\sim}A^a/x$, due to the general LEM, then apply RRQ3 of [5] to each disjunct, followed by MR1.

C5. $\exists xPx, (\forall xPx)A \,\&\, (\exists xPx){\sim}A \Rightarrow \#$. [RD7b yields a contradiction]
C6. $\exists xPx, {\sim}(\exists xPx)A \,\&\, (\exists xPx)A \Rightarrow \#$. [RD7a yields a contradiction.]
C7. $\exists xPx, * \Rightarrow (\forall xPx)A \vee (\exists xPx){\sim}A$.

Start with $(\forall xPx)A \vee {\sim}(\forall xPx)A$ and apply RD7b.

C8. $\exists xPx, * \Rightarrow {\sim}(\exists xPx)A \vee (\exists xPx)A$. [Clearly, an instance of the LEM.]

However, we can simplify our treatment of the Square of Opposition as follows. We use the restricting predicate Px to express the key universal component in 'All Ps are As', that is $(\forall xPx)Ax$, this being appropriate because no attention is paid to the elements a that do not satisfy Pa. Traditionally, this is symbolized as PaA and represented by A on the Square below. This same consideration then passes to the other points of the Square: 'Some Ps are As', that is $(\exists xPx)Ax$, traditionally PiA and represented by I; 'No Ps are As', that is ${\sim}(\exists xPx)Ax$, traditionally PeA and represented by E; and 'Some Ps are not As', that is $(\exists xPx){\sim}Ax$, traditionally PoA and represented by O.

The traditional relationships of the Square are, using these capital letters, as set out in the diagram below:

```
A ------ E
|        |
|        |
I ------ O
```

A and E are contraries, which cannot both be true, this being represented using a horizontal line from A to E. Further, A entails I and E entails O, which is represented by the two vertical lines. I and O are subcontraries, which cannot both be false, which is represented by the lower horizontal line. A and O are contradictories, as are E and I, which cannot both be true and cannot both be false, that is, A and O take opposite truth values, as do E and I.

These relationships need to be expressed and then proved within the above relevant logics. The two entailments can be clearly expressed using →, that is A→I and E→O, upon assumption ∃xPx. Here, MC is conceptually more sensitive regarding entailments, in comparison with the other relevant logics. The two contradictories can also be expressed as A⟷~O and E⟷~I, as co-entailments, in the relevant logics, also upon assumption ∃xPx. The entailment-based relations between A and E and between I and O can thus be obtained by transitivity: A→~E and ~I→O, given ∃xPx. These relationships are listed as follows, with indications of proof:

A→I: ∃xPx ⇒ (∀xPx)A → (∃xPx)A. [RR1, MR1]
E→O: ∃xPx ⇒ ~(∃xPx)A → (∃xPx)~A. [RR1, MR1, RD7a]
A⟷~O: ∃xPx ⇒ (∀xPx)A ⟷ ~(∃xPx)~A. [RD7b]
E⟷~I: ∃xPx ⇒ ~(∃xPx)A ⟷ ~(∃xPx)A. [Clear]
A→~E: ∃xPx ⇒ (∀xPx)A → ~~(∃xPx)A. [Clear]
~I→O: ∃xPx ⇒ ~(∃xPx)A → (∃xPx)~A. [Clear]

To make conclusions about truth and falsity for contraries, subcontraries and contradictories, it suffices to apply the Law of Excluded Middle (LEM) to the antecedents of the respective entailments. Of the above relevant logics, only the logic R includes the LEM, but for the others individual instances of the LEM can be established through one of its disjuncts. For then, ~A∨A → ~A∨~E and hence ~A∨~E and ~(A&E) and so either A or E are false and they cannot both be true, as required for contraries. Similarly, I∨~I →I∨O and hence I∨O and ~(~I&~O) and so either I or O are true and they cannot both be false, as required for subcontraries. Furthermore, by applying the LEM to the antecedents of the two entailment components of each of the contradictories, A and O take opposite truth values and so do E and I.

These disjunctions will need to be prime to separate out the respective truth and falsity statements. (See [8] for the discussion of primeness in truth-theoretic semantical systems.) Such primeness can be guaranteed through metacompleteness, which holds for all the relevant logics in our list, except for the logic R. (See [9] for an account of metavaluations and metacompleteness.) For such logics, the LEM is not a theorem, but this means that the LEM can only be shown for its instances through one of its disjuncts as its disjunction is prime. However, for R, the LEM is a theorem but primeness is not guaranteed for its instantiations as it is not metacomplete. Here, we talk of the truth of the respective disjunctive formulae, except where priming holds in special cases.

4 The Relationships with Rules and with the Classical Account

Since both the restrictor > and rule ⇒ have no negative requirements, both being based on their antecedent holding, there should be a clear relationship between the two. Let us consider the two following equivalences, for such a relationship.

(i) (Pa ⇒ Pa > A) ⟺ .Pa ⇒A, incorporating the Pa assumption just for the restrictor. This clearly holds in both directions by applying the rule RR1.
(ii) Pa ⇒ .(Pa > A) ⟺ .Pa ⇒A, where the whole equivalence of which is subject to the Pa assumption.

For L⇒R, apply RR1 yielding A and then just deductively repeat Pa. For R⇒L, deductively contract to Pa ⇒A and apply RR1.

So, both these relationships with rules hold for the restrictor >.

We now determine relationships between our restricted quantifiers expressed using the restrictor > and the respective classical accounts for the two restricted quantifiers, that is, ∀x(Px ⊃A) and ∃x(Px&A), each given ∃xPx. This will be performed as a matter of interest rather than as a classical recapture, as we still maintain the use of the restrictor > as being appropriate for general application. Being classical accounts, we can assume the Rule of Disjunctive Syllogism (~A, A∨B ⇒ B) (DS), for application to Pa.

We will establish rules or entailments between Pa ⊃A and Pa >A, between Pa & A and Pa >A, and between ∃x(Px&A) and (∃xPx)A, as appropriate. Unlike for the universals, we can use the meta-rule MR1 to introduce the existential quantifiers. Unfortunately, MR1 cannot be used for the universals because the derivations between the two universals will use universal generalization in the process. Note that Pa cannot be universally generalized at any stage as this would produce an unrestricted quantifier. As usual, key elements of the following proofs are indicated in square brackets.

1: Pa ⇒ .Pa ⊃A ⇒ Pa > A. [DS, RR1]
2: Pa ⇒ .Pa > A → .Pa ⊃A. [RR1]
3: Pa ⇒ . Pa & A → .Pa >A. [RR1]
4: Pa ⇒ . Pa > A ⇒ .Pa & A. [RR1]
5: ∃x(Px & A) ⇒ (∃xPx)A. [3, MR1]
6: ∃x(Px & (Px > A)) ⇒ ∃x(Px & A). [4, MR1]

For 6, we must existentially quantify over both Px and Px > A to ensure that the same element is used for both of these components. Of course, MR1 only applies to a single premise rule.

5 The Extent of Application and Concluding Remarks

Whilst the motivation behind the introduction of the restrictor > is its use in restricting the two quantifiers, we have initially introduced it freely to provide frameworks to which the quantifiers can be applied. Such free introduction may lead to problems associated with its lack of negative requirements. Nevertheless, we must still assume that the restricted domain is non-empty and that any antecedent Pa of the restrictor must hold for any instance of a that the variable a may represent, in order for the restrictor to be applied.

Let us consider the example of Curry's Paradox of naive set theory. We set out the introduction of the Curry set C as follows.

$\forall x(x \varepsilon x)$	$\Rightarrow \forall x(x \varepsilon C \longleftrightarrow .x \varepsilon x > p)$ [$\forall x(x \varepsilon x)$ is needed here to cover all possible instances.]
	$\Rightarrow C \varepsilon C \longleftrightarrow .C \varepsilon C > p$
	$\Rightarrow C \varepsilon C \longleftrightarrow p$ [RR1] (Note that $C \varepsilon C$ follows from $\forall x(x \varepsilon x)$.)
	$\Rightarrow p$.

However, $\phi \varepsilon \phi$ is false for the null set ϕ and the assumption of the rule cannot be activated and Curry's Paradox using > fails to show that triviality is derivable in general. Nevertheless, this is still a case of a simple inconsistency implying triviality, by justifiably adding the analytic truth $\sim\forall x(x \varepsilon x)$ as an additional premise. So, this a case of Ex Falso Quodlibet, easily derivable from the DS, both being admissible rules of MCQ due to its metacompleteness. Though not proven, such metacompleteness would likely extend to the above restricted extension.

This derivation would seem to bring up another problem for the free use of > in that an equivalence in meaning, expressed by $\longleftrightarrow$, may not hold between a formula involving > and a formula without its lack of negative requirements. However, the restricting predicate must be deletable by RR1, since any use of 'Pa >' must be covered by an assumption of Pa, allowing it to be deleted by applying RR1. Such an example is expressed by '$x \varepsilon C \longleftrightarrow .x \varepsilon x > p$' in the first line. The use of RR1 would collapse it down to '$x \varepsilon C \longleftrightarrow p$'. This requirement on free use does not apply to generalized properties of restricted quantification that stand up as analytic truths, subject only to the non-emptiness of the restricting predicate P. These are evidenced by the rules RR2-5 and RD7a-11c.

Note that, as in [5], there is no semantics proposed to provide a completeness argument for the above restricted quantificational logic. There are two points to make about this. First, there are problems with the usual truth-theoretic semantics regarding primeness and distribution, since both of these can fail, and thus the semantics loses its reality as providing a meaning for the logical components. However, it can still be used as a technical tool to reject non-theorems. Second, the yardstick should be determined by the meanings of the logical words, leaving one to do the mental gymnastics to check that each axiom, rule and meta-rule is in accordance with the meanings of its components.

Acknowledgements I would like to thank the three referees for their comments, which have not only made this paper more readable but also helped to correct some of my mistakes in the original text.

References

1. Anderson, A. R., & Belnap, N. D., Jr. (1975). *Entailment. The logic of relevance and necessity* (Vol. 1). Princeton University Press.
2. Anderson, A. R., Belnap, N. D., Jr., & Dunn, J. M. (1992). *Entailment. The logic of relevance and necessity* (Vol. 2). Princeton University Press.
3. Beall, J. C., Brady, R. T., Hazen, A., Priest, G., & Restall, G. (2006). Relevant restricted quantification. *Journal of Philosophical Logic, 35*, 587–598.
4. Belnap, N. D. Jr. (1973). Restricted quantification and conditional assertion. In H. Leblanc (Ed.), *Truth, syntax and modality*, Proceedings of the Temple University conference on alternative semantics, North-Holland, Amsterdam, pp. 48–75.
5. Brady, R. T. (Ed.). (2003). *Relevant logics and their rivals* (Vol. 2). Ashgate.
6. Brady, R.T. (2006). *Universal logic*, CSLI Publs, Stanford.
7. Brady, R. T. (2015). Logic - The big picture. In J.-Y. Beziau, M. Chakraborty, & S. Dutta (Eds.), *New directions in paraconsistent logic* (pp. 353–373). Springer.
8. Brady, R. T. (2017). Some concerns regarding ternary-relation semantics and truth-theoretic semantics in general. *The IfCoLog Journal of Logics and their Applications, 4*, 755–781.
9. Brady, R. T. (2017a). Metavaluations. *The Bulletin of Symbolic Logic, 23*(2017), 296–323.
10. Brady, R. T. (2022). Intension, Extension, Distribution and Decidability. In K. Bimbó (Ed.), *Relevance Logics and other Tools for Reasoning. Essays in Honor of J. Michael Dunn* (pp. 128–149). College Publications.
11. Brady, R. T., & Meinander, A. (2013). Distribution in the logic of meaning containment and in quantum mechanics. In K. Tanaka, F. Berto, E. Mares, & F. Paoli (Eds.), *Paraconsistency: Logic and applications* (pp. 223–255). Springer.
12. Brady, R. T., & Rush, P. A. (2008). What is wrong with Cantor's diagonal argument? *Logique et Analyse, 51*, 185–219.

Fusion, Fission, and Ackermann's Truth Constant in Relevant Logics: A Proof-Theoretic Investigation

Fabio De Martin Polo

Abstract The aim of this paper is to provide a proof-theoretic characterization of relevant logics including fusion and fission connectives, as well as Ackermann's truth constant. We achieve this by employing the well-established methodology of labelled sequent calculi. After having introduced several systems, we will conduct a detailed proof-theoretic analysis, show a CUT-admissibility theorem, and establish soundness and completeness. The paper ends with a discussion that contextualizes our current work within the broader landscape of the proof theory of relevant logics.

1 Introduction and Aim

Relevant logics are a recognized family of non-classical logics designed to tackle paradoxes of material and strict implication. According to relevantists, the symbol $\rightarrow$ represents a more refined and philosophically motivated concept of conditional. Early proponents, like A. R. Anderson and N. D. Belnap, argued that a valid conditional requires a strong connection between the antecedent and consequent, where both are relevant to each other. These logics have garnered significant attention among logicians, leading to the application of various formal structures to provide detailed and systematic characterizations. In this paper, I extend the proof-theoretic investigation of relevant logics initiated in [13] (see also [29]). I explore the inclusion of additional operators in relevant systems, considering the following:

I would like to thank an anonymous referee for having provided valuable suggestions and comments that have significantly improved the overall quality of the paper. For the same reason, I would also like to express my gratitude to Shawn Standefer. The preparation of this article was supported by a Sofja Kovalevskaja Award granted by the Alexander von Humboldt Foundation and financed by the German Ministry for Education and Research.

F. De Martin Polo (✉)
Ruhr University Bochum, Bochum, Germany

I. Sedlár et al. (eds.), *New Directions in Relevant Logic*, Trends in Logic 63,
https://doi.org/10.1007/978-3-031-69940-5_12

1. The incorporation of *fusion*, denoted as $\circ$, as a primitive binary operator. This study is primarily influenced by the work in [61, pp. 365–366] (Sect. 2).
2. The inclusion of *fission*, i.e., intensional disjunction, following some remarks expressed in [56, 61].
3. The inclusion of *connexive fusion*, denoted as $\bullet$, in relevant systems, as proposed by R. Sylvan (né Routley). The motivation behind this addition is rooted in the idea that "connexive and relevant logics are one and the same" [59, 393] (Sect. 3).
4. The introduction of *Ackermann's truth constant*, denoted as t, within the framework of relevant logics. Specifically, I study a conservative extension of Anderson and Belnap's logic of entailment **E**, named $\mathbf{E}^{\mathsf{t}}$, as discussed in [61, pp. 407–424, Appendix 1] (Sect. 4).

All logics resulting from the above additions are defined using reduced Routley-Meyer frames and models, which are structures featuring a ternary relation among states and a distinguished state interpreted as the *real* or *actual* world (see [22, 60, 61, 63]). By employing these semantic structures, I establish corresponding labelled sequent calculi based on the methodology applied to many other non-classical logics (e.g., [16, 24, 41, 42]). To achieve this, I convert the semantic clauses and frame conditions of each logic under consideration into well-constructed schematic rules, demonstrating that the resulting calculi exhibit various desirable properties (Sects. 5 and 6). The paper ends by offering a comparison with other proof-theoretic techniques used to characterize relevant logics, as well as by identifying potential avenues for future research (Sect. 7).

2 Fusion and Fission in Relevant Logics

The analysis of the connective usually referred to as *fusion*, denoted $\circ$, has played a central role in the formal and philosophical development of relevant logics. In [61, p. 366], amongst other, the authors explain that:

> [...] fusion can be added conservatively to *L* sentential systems, and is a useful connective to introduce for algebraic and Gentzenisation exercises.

Let's start by considering a base system of relevant logic **B**, including $\circ$ among its connectives, denoted $\mathbf{B}^{\circ}$. As usual, $p, q, \ldots$ and $A, B, \ldots$ denote atomic and compound formulas, respectively. The set of formulas is defined recursively in the standard way. At the level of Hilbert systems, $\mathbf{B}^{\circ}$ contains all instances of the following axioms and it is closed under the following rules ("$\Rrightarrow$" is employed as a rule-forming operator distinct from both the sequent arrow "$\Rightarrow$" and the meta-level symbol " $\Longrightarrow$ "):

(A1) $A \to A$
(A2) $A_1 \wedge A_2 \to A_i$
(A3) $(A \to B) \wedge (A \to C) \to (A \to (B \wedge C))$
(A4) $A_i \to (A_1 \vee A_2)$
(A5) $(A \to C) \wedge (B \to C) \to ((A \vee B) \to C)$
(A6) $A \wedge (B \vee C) \to (A \wedge B) \vee (A \wedge C)$
(A7) $\sim\sim A \to A$

(R1) $A, A \to B \Rrightarrow B$
(R2) $A, B \Rrightarrow A \wedge B$
(R3) $A \to B \Rrightarrow (C \to A) \to (C \to B)$
(R4) $A \to B \Rrightarrow (B \to C) \to (A \to C)$
(R5) $A \to B \Rrightarrow \sim B \to \sim A$
(Port°) $(A \circ B) \to C \Lleftarrow\Rrightarrow A \to (B \to C)$

Observation 1 To be precise, within the full system **R**, fusion can be precisely defined as $A \circ B =_{\mathrm{df}} \sim(A \to \sim B)$. However, it is important to note that this definition does not hold in weaker relevant logics (as discussed, e.g., in [15, 56]). Hence, the discussions we will engage with may be particularly valuable for researchers and practitioners working with weaker relevant logics.

The two-ways rule "portation", (Port°), encodes the fundamental feature of $\circ$. Indeed, as highlighted in [61, p. 385], "fusion accomplishes 'conjunction' of nested implications". For the semantic characterization of relevant logics including $\circ$, we will use so-called reduced Routley-Meyer frames and models. Within the context of such a frame semantics, fusion and implication are interconnected through the shared ternary relation R. Implication is associated with the universal condition, while fusion corresponds to the existential condition. Let's start by introducing the semantics for the base logic **B**° [61, Ch. 4 and pp. 365–366].

Definition 2.1 A *reduced* **B**° *frame* $\mathscr{F}$ is a quadruple $\langle 0, K, R, * \rangle$, where K is a set of states (possible worlds, points), $0 \in K$ (the *real* or *actual world*), $R \subseteq K^3$, and $* : K \mapsto K$. We impose the following constraints on R:[1]

(p1) $R0aa$.
(p2) $a = a^{**}$.
(p3) If $R0ad$ and $Rdbc$, then $Rabc$.
(p3′) If $R0ad$ and $Rbca$, then $Rbcd$.
(p4) If $R0ab$, then $R0b^*a^*$.
(p5) If $R0ab$ and $R0bc$, then $R0ac$.

[1] Relations of the form $R0ab$ and $R0ab \wedge R0ba$ can be shortened by using the notation $a \leq b$ and $a = b$, respectively. However, given that the latter symbols are precisely defined in terms of the ternary accessibility relation, we can employ only R to characterize relevant logics.

We recall that (p3′) is the frame condition added specifically to deal with the rule (Port°).

Definition 2.2 A *reduced* **B**° *model* $\mathscr{M}$ is a pair $\langle \mathscr{F}, v \rangle$, where $\mathscr{F}$ is a reduced frame and v is a function, called *valuation*, such that $v : \mathsf{At} \mapsto \wp(K)$, and subject to the so-called *heredity condition*. For $p \in \mathsf{At}$ and $a, b \in K$:

(AtHer) If $R0ab$ and $a \in v(p)$, then $b \in v(p)$.

Finally, the valuation v is extended to the whole language as follows. For all $A, B \in$ Frm and $a \in K$:

$\mathscr{M}, a \Vdash p$	iff	$a \in v(p)$, for $p \in \mathsf{At}$
$\mathscr{M}, a \Vdash {\sim}A$	iff	$\mathscr{M}, a^* \nVdash A$
$\mathscr{M}, a \Vdash A \wedge B$	iff	$\mathscr{M}, a \Vdash A$ and $\mathscr{M}, a \Vdash B$
$\mathscr{M}, a \Vdash A \vee B$	iff	$\mathscr{M}, a \Vdash A$ or $\mathscr{M}, a \Vdash B$
$\mathscr{M}, a \Vdash A \circ B$	iff	$\exists b, c \in K$, s.t. $Rbca$ and $\mathscr{M}, b \Vdash A$ and $\mathscr{M}, c \Vdash B$
$\mathscr{M}, a \Vdash A \rightarrow B$	iff	$\forall b, c \in K$, if $Rabc$ and $\mathscr{M}, b \Vdash A$, then $\mathscr{M}, c \Vdash B$

Finally:
A formula B is *satisfied* in a model $\mathscr{M} = \langle \mathscr{F}, v \rangle$ iff $\mathscr{M}, 0 \Vdash B$.
"*A entails B in* $\mathscr{M}$" iff, for all $a \in K$, if $a \Vdash A$, then $a \Vdash B$.
A formula B is *valid* in a frame $\mathscr{F} = \langle K, 0, *, R \rangle$ iff, for all valuations v, the formula B is satisfied in $\mathscr{M}$.

Finally, two important, standard lemmas are the following ones (see, among others, [15, 56, 58]):

Heredity Lemma *If $R0ab$ and $\mathscr{M}, a \Vdash A$, then $\mathscr{M}, b \Vdash A$.*

Verification Lemma *A entails B in a given model $\mathscr{M}$ iff $A \rightarrow B$ is satisfied in that model, i.e., for all $a \in K$, $(\mathscr{M}, a \Vdash A \implies \mathscr{M}, a \Vdash B)$ iff $\mathscr{M}, 0 \Vdash A \rightarrow B$.*

Observation 2 In the preceding definitions, we introduced "reduced" models for relevant logics (see, for example, [22, 63]). These models differ from those referred to as "non-reduced" (or "unreduced") models (see, for example, [60, 61]).[2] There are several key differences to consider. Let $\mathscr{F}'$ and $\mathscr{M}'$ denote non-reduced frames and models, respectively. $\mathscr{F}'$ is defined as the structure $\langle K, 0, T, *, R \rangle$, where 0 is taken to be a subset of K, rather than a singleton, and T is a distinct element $T \in 0$ called the "designated situation". The members of 0 are referred to as "regular situations". A model $\mathscr{M}'$ is defined as the structure $\langle \mathscr{F}', v \rangle$. Finally, satisfaction in a model is defined with respect to regular situations, i.e., A is satisfied in a model $\mathscr{M}'$ if $\mathscr{M}', x \Vdash A$ for all $x \in 0$. Validity on $\mathscr{F}'$ is defined as before.

[2] The original Routley-Meyer models for relevant logics were the reduced ones, whereas the introduction of unreduced models happened later, driven by the interest in exploring weaker relevant systems. (Thanks to an anonymous reviewer for contributing to the clarification of this point.)

In relevant logics, an additional intensional connective, known as fission or intensional disjunction, denoted by the symbol "+", can also be examined. Its presence within the Anderson and Belnap tradition in relevant logic stems from the criticism directed at the inference $\sim A, A \vee B \Rightarrow B$, which is regarded as a fallacy of relevance (see [5, 19]). As a solution to preserve the relevance criterion while upholding disjunctive syllogism, Anderson and Belnap proposed a version of it that incorporates the connective $+$, for which addition principles do not hold, rather than its extensional counterpart $\vee$. Taking a slightly different perspective, Stephen Read [56] argues that the problem lies not with the rule of disjunctive syllogism itself, but rather with our understanding of disjunction in natural language, and suggests that we should interpret natural language disjunctions as fissions. In a previous work [55], Read explores the question of whether the distinction between the two notions of disjunction, $\vee$ and $+$, can be observed in natural reasoning. He puts forth a positive stance on this matter, contending that his analysis corroborates the relevantist standpoint that *or* is ambiguous. Note that Mares raises doubts about the possibility of considering natural language disjunction as fission. According to Mares, the formal treatment of disjunction, particularly within the framework of ternary frame semantics, may have limited similarity to our intuitive understanding of natural language disjunction (see, [33, 617–619]). Before delving into technical aspects, let's consider the following example to gain an informal understanding of intensional disjunction:

> Intuitively, the intensional variety of *or* would be one requiring "relevance" between the disjuncts. On the intensional reading, "*A* or *B*" would entail that *A* and *B* are so related that we are entitled to say "If *A* had not been true, *B* would have been true" or "If *B* had not been true, *A* would have been true" or the like. A disjunction like *Either Napoleon was born in Corsica or else the number of the beast is perfect* clearly fails to have this property and therefore is of the truth-functional kind. Whereas *That is either Drosophilia Melanogaster or D. virilis, I'm not sure which* appears to entail that if it is not the one then it is the other, and thus is of the intensional kind.[3] [2, 17]

Formally, the choice of treating fission as either a primitive or a defined connective can be made. If one opts for the former and formalizes it as $A + B$, the following frame condition and semantic clause may be considered (see, e.g., [33, 618]):

(p3$''$) If $R0da$ and $Rbca$, then $Rbcd$.
$(+^1)$ $\mathscr{M}, a \Vdash A + B$ iff $\forall b, c \in K$, if $Rbca$, then either $\mathscr{M}, b \Vdash A$ or $\mathscr{M}, c \Vdash B$.

More standardly, by following the presentations in [61, 286, 361] and [56, 72], we can consider fission defined in terms of negation and implication, that is $A + B =_{\mathsf{df}} \sim A \rightarrow B$. The semantic condition for $+$ is precisely derived from its definition, i.e.:

$(+^2)$ $\mathscr{M}, a \Vdash A + B$ iff $\forall b, c \in K$, if $Rabc$, then either $\mathscr{M}, b^* \Vdash A$ or $\mathscr{M}, c \Vdash B$.

[3] The mentioned example was also discussed in Anderson and Belnap's [5, 176–177].

Another notion of fission, which is equivalent to the previous one in **R** but differs from it in weaker systems, is defined as $A \oplus B =_{\text{df}} \sim(\sim A \circ \sim B)$ (see [61, 361], [56, 53]). The corresponding semantic condition can be given as follows:

($\oplus$) $\mathcal{M}, a \Vdash A \oplus B$ iff $\forall b, c \in K$, if $Rbca^*$, then either $\mathcal{M}, b^* \Vdash A$ or $\mathcal{M}, c^* \Vdash B$.

Where, again, the clause is derived from the semantic definitions of $\circ$ and $\sim$.

2.1 Labelled Sequent Systems

In this part of the paper, we shall define Gentzen-style calculi for $\mathbf{B}^\circ$, and extensions thereof, by using the methodology of labelled sequent calculi. We extend the vocabulary of sequents with a bunch of labels $(a, b, c, \ldots, x, y, z \ldots)$ denoting states in Routley-Meyer semantics and, intuitively, use the notation $a : A$ to express the forcing relation $a \Vdash A$ via sequents.[4]

Definition 2.3 Let K be a set of labels, including a distinguished label denoted 0. For any $A \in \mathsf{Frm}$ and labels $a, b, c \in K$, the set of well-formed formulas consists of (i) labelled formulas of the form $a : A$ and (ii) relational atoms of the form $Rabc$.[5] Given two multisets Γ, Δ of labelled formulas and relational atoms, a labelled sequent is an object of the following form: $\Gamma \Rightarrow \Delta$.

The labelled rules of our sequent systems are subject to the following *closure condition*. Consider a rule $\mathbf{r}$ of the following form:

$$\frac{A, B_1, \ldots, B_n, B_{n+1}, B_{n+1}, \Gamma \Rightarrow \Delta}{B_1, \ldots, B_n, \Gamma \Rightarrow \Delta}\ \mathbf{r}$$

Applying the closure condition on $\mathbf{r}$ means to substitute the multiple occurrence B_{n+1}, B_{n+1} with a single one to obtain a rule $\mathbf{r}^*$ of the following shape:

$$\frac{A, B_1, \ldots, B_n, B_{n+1}, \Gamma \Rightarrow \Delta}{B_1, \ldots, B_n, \Gamma \Rightarrow \Delta}\ \mathbf{r}^*$$

The base labelled calculus for $\mathbf{B}^\circ$, termed **G3rB°**, is built as follows.

Initial Sequents $R0ab, a : p, \Gamma \Rightarrow \Delta, b : p$ (possibly a^*, b^*).

---

[4] The relationship between the two forcing expressions will be explicitly presented on pp. 303ff, where the proof of soundness is provided.

[5] In this section, the development of our desired labelled calculi solely relies on relational atoms of the form $Rabc$. However, as we proceed to the subsequent sections, where Routley-Meyer semantics are extended to accommodate additional relations between states, we will also introduce new relational symbols in our calculi.

Logical Rules

$$\frac{\Gamma \Rightarrow \Delta, a^* : A}{a : {\sim}A, \Gamma \Rightarrow \Delta}\ L{\sim} \qquad \frac{a^* : A, \Gamma \Rightarrow \Delta}{\Gamma \Rightarrow \Delta, a : {\sim}A}\ R{\sim}$$

$$\frac{a : A, a : B, \Gamma \Rightarrow \Delta}{a : A \wedge B, \Gamma \Rightarrow \Delta}\ L\wedge \qquad \frac{\Gamma \Rightarrow \Delta, a : A \quad \Gamma \Rightarrow \Delta, a : B}{\Gamma \Rightarrow \Delta, a : A \wedge B}\ R\wedge$$

$$\frac{a : A, \Gamma \Rightarrow \Delta \quad a : B, \Gamma \Rightarrow \Delta}{a : A \vee B, \Gamma \Rightarrow \Delta}\ L\vee \qquad \frac{\Gamma \Rightarrow \Delta, a : A, a : B}{\Gamma \Rightarrow \Delta, a : A \vee B}\ R\vee$$

$$(b, c \text{ fresh})\ \frac{Rbca, b : A, c : B, \Gamma \Rightarrow \Delta}{a : A \circ B, \Gamma \Rightarrow \Delta}\ L\circ$$

$$\frac{Rbca, \Gamma \Rightarrow \Delta, a : A \circ B, b : A \quad Rbca, \Gamma \Rightarrow \Delta, a : A \circ B, c : B}{Rbca, \Gamma \Rightarrow \Delta, a : A \circ B}\ R\circ$$

$$\frac{Rabc, a : A \to B, \Gamma \Rightarrow \Delta, b : A \quad c : B, Rabc, a : A \to B, \Gamma \Rightarrow \Delta}{Rabc, a : A \to B, \Gamma \Rightarrow \Delta}\ L\to$$

$$(b, c \text{ fresh})\ \frac{Rabc, b : A, \Gamma \Rightarrow \Delta, c : B}{\Gamma \Rightarrow \Delta, a : A \to B}\ R\to$$

Relational Rules for R

$$\frac{R0aa, \Gamma \Rightarrow \Delta}{\Gamma \Rightarrow \Delta}\ \text{R1} \qquad \frac{R0aa^{**}, R0a^{**}a, \Gamma \Rightarrow \Delta}{\Gamma \Rightarrow \Delta}\ \text{R2}$$

$$\frac{Rabc, R0ad, Rdbc, \Gamma \Rightarrow \Delta}{R0ad, Rdbc, \Gamma \Rightarrow \Delta}\ \text{R3} \qquad \frac{Rbcd, R0ad, Rbca, \Gamma \Rightarrow \Delta}{R0ad, Rbca, \Gamma \Rightarrow \Delta}\ \text{R3}'$$

$$\frac{R0b^*a^*, R0ab, \Gamma \Rightarrow \Delta}{R0ab, \Gamma \Rightarrow \Delta}\ \text{R4} \qquad \frac{R0ac, R0ab, R0bc, \Gamma \Rightarrow \Delta}{R0ab, R0bc, \Gamma \Rightarrow \Delta}\ \text{R5}$$

Observation 3 Initial sequents are stated in their weakening-absorbing form and express, at the calculus level, the heredity condition for atomic formulas. As the heredity rules represent forms of contraction, namely:

$$\frac{b : p, R0ab, a : p, \Gamma \Rightarrow \Delta}{R0ab, a : p, \Gamma \Rightarrow \Delta}\ \text{AtHer-L} \qquad \frac{R0ab, \Gamma \Rightarrow \Delta, b : p, a : p}{R0ab, \Gamma \Rightarrow \Delta, b : p}\ \text{AtHer-R}$$

it is more advantageous to adopt a system where these rules can be shown to be height-preserving admissible rather than primitive (Proposition 5.2). This is why we incorporate heredity into the axioms. Furthermore, given Proposition 5.1, the generalized version of ATHER can be derived using (admissible) CUT and contraction (consider always Proposition 5.2 below). Rules $R\circ$ and $L\rightarrow$ include copies of the principal formulas in the premises to prove that contraction is height-preserving admissible (Lemma 5.4). Rules $L\circ$ and $R\rightarrow$ are subject to the eigenvariable condition, i.e., each application of those rules requires the introduction of fresh (i.e., not previously used) labels. Finally, all relational rules are obtained by converting the conditions of **B**° frames into schematic rules according to the methodology developed and used, amongst others, in [16, 41] for modal and intermediate logics.

Furthermore, if one intends to incorporate fission as a primitive connective in a proof system, the semantic clause ($+^1$) and the frame condition (p3″) stated on page 271 can be converted into the following labelled rules and relational rule, respectively:

$$\frac{b : A, Rbca, a : A + B, \Gamma \Rightarrow \Delta \qquad c : B, Rbca, a : A + B, \Gamma \Rightarrow \Delta}{Rbca, a : A + B, \Gamma \Rightarrow \Delta}\, L{+}^1$$

$$(b, c \text{ fresh})\ \frac{Rbca, \Gamma \Rightarrow \Delta, b : A, c : B}{\Gamma \Rightarrow \Delta, a : A + B}\, R{+}^1 \qquad \frac{Rbcd, R0da, Rbca, \Gamma \Rightarrow \Delta}{R0da, Rbca, \Gamma \Rightarrow \Delta}\, \mathrm{R3''}$$

Alternatively, if fission is defined in terms of $\rightarrow$ and $\sim$, or in terms of $\circ$ and $\sim$, the semantic clause ($+^2$), or ($\oplus$), can be transformed into the following two sets of left and right rules, respectively:

$$\frac{b^* : A, Rabc, a : A + B, \Gamma \Rightarrow \Delta \qquad c : B, Rabc, a : A + B, \Gamma \Rightarrow \Delta}{Rabc, a : A + B, \Gamma \Rightarrow \Delta}\, L{+}^2$$

$$(b, c \text{ fresh})\ \frac{Rabc, \Gamma \Rightarrow \Delta, b^* : A, c : B}{\Gamma \Rightarrow \Delta, a : A + B}\, R{+}^2$$

$$\frac{b^* : A, Rbca^*, a : A \oplus B, \Gamma \Rightarrow \Delta \qquad c^* : B, Rbca, a : A \oplus B, \Gamma \Rightarrow \Delta}{Rbca^*, a : A \oplus B, \Gamma \Rightarrow \Delta}\, L\oplus$$

$$(b, c \text{ fresh})\ \frac{Rbca^*, \Gamma \Rightarrow \Delta, b^* : A, c^* : B}{\Gamma \Rightarrow \Delta, a : A \oplus B}\, R\oplus$$

It is important to note that we shall not include these two latter sets of rules as primitive in our intended labelled calculi. As demonstrated in Proposition 5.2, rules for the two defined notions of fission are admissible.

Modular Extensions Hilbert systems for some common stronger relevant logics can be obtained by the addition of axioms to the system for **B**°. Likewise, frames for $\mathbf{B}^{\circ}$ can be enriched to capture stronger relevant logics by adding some further constraints on R. In what follows, we list some axioms and the conditions needed to validate them.

(Some of these frame conditions appeal to the standard definitions, $Rabcd ::= \exists x(Rabx \wedge Rxcd)$ and $Ra(bc)d ::= \exists x(Raxd \wedge Rbcx)$.)

(A8)	$(A \to B) \to (\sim B \to \sim A)$	(p6)	$Rabc \implies Rac^*b^*$
(A9)	$(A \to B) \wedge (B \to C) \to (A \to C)$	(p7)	$Rabc \implies Ra(ab)c$
(A10)	$(A \to B) \to ((B \to C) \to (A \to C))$	(p8)	$Rabcd \implies Rb(ac)d$
(A11)	$(A \to B) \to ((C \to A) \to (C \to B))$	(p9)	$Rabcd \implies Ra(bc)d$
(A12)	$(A \to (A \to B)) \to (A \to B)$	(p10)	$Rabc \implies Rabbc$
(A13)	$(A \wedge (A \to B)) \to B$	(p11)	$Raaa$
(A14)	$(A \to \sim A) \to \sim A$	(p12)	Raa^*a
(A15)	$(A \to (B \to C)) \to (B \to (A \to C))$	(p13)	$Rabcd \implies Racbd$
(A16)	$A \to ((A \to B) \to B)$	(p14)	$Rabc \implies Rbac$
(A17)	$A \vee \sim A$	(p15)	$R00^*0$
(A18)	$((A \to A) \to B) \to B$	(p16)	$Ra0a$
(A19)	$A \to (A \to A)$	(p17)	$Rabc \implies (R0ac \vee R0bc)$

From the point of view of labelled systems, it is possible to obtain calculi for stronger relevant logics by converting further frame conditions into relational rules. For example, to obtain a calculus for $\mathbf{DW}^{\circ}$, it suffices to convert condition (p6) into the schematic rule:

$$\frac{Rac^*b^*, Rabc, \Gamma \Rightarrow \Delta}{Rabc, \Gamma \Rightarrow \Delta}\ \mathrm{R6}$$

and add it to the rules for $\mathbf{G3rB}^{\circ}$ displayed above.

3 Fusion in Relevant Connexive Logics à la Sylvan

A less standard approach to understanding the notion of fusion was proposed by Sylvan in [59]. To differentiate it from the connectives $\circ$ and $\wedge$, Sylvan used the symbol $\bullet$ to represent fusion, and we will adopt this notation hereafter. In his 1978 article, Sylvan introduced frames and models that allowed him to capture connexive logics while preserving the relevant implication.[6] He stated:

[6] To place Sylvan's work in relation to other important works on connexive logics, refer to Observation 4 provided below.

> [...] if antecedent and consequent enjoy a meaning connexion then they are relevant in meaning to one another, and if they are relevant in meaning to one another then they have through the relevance relation a connexion in meaning. Thus the general classes of connexive and relevant logics are *one and the same*. [59, 393, Emphasis mine]

Connexivity is typically associated with the presence of specific principles within a system. Sylvan discussed three such principles, namely:

Aristotle's thesis	$\sim(A \rightarrow \sim A)$	(Ar)
Boethius' thesis	$(A \rightarrow B) \rightarrow \sim(A \rightarrow \sim B)$[7]	(Bo)
Strawson's thesis	$\sim((A \rightarrow B) \bullet (A \rightarrow \sim B))$	(St)

The choice of using $\bullet$ instead of $\circ$ or $\wedge$ is related to a common feature in many formalizations of connexive logics: the rejection of "simplification" principles. Specifically, laws such as $A_1 \bullet A_2 \rightarrow A_i$ are required to fail. Sylvan's argument, based on the "cancellation account of negation", explains this as follows:

> [...] $(A \bullet \sim A) \rightarrow A$ and $(A \bullet \sim A) \rightarrow \sim A$ fail. For $\sim A$ cancels out A, so that the conjoined content of $A \bullet \sim A$ is less than that of A and $\sim A$. But implication requires content inclusion, so these (degenerate) examples of Simplification fail. This explains, in a sketchy way, the character of the connexivist argument against Simplification. The same argument explains why $(A \bullet \sim A)$ does not imply $\sim(A \bullet \sim A)$, and provides a basis for an argument for Aristotle's thesis. [59, 395]

After introducing a Hilbert calculus for a basic system of relevant connexive logic (referred to as $\mathbf{B}^{\bullet}$), Sylvan proceeds to introduce specific frames and models by stating:

> Connexive modellings do not differ from relevant modellings as to the implication connective: thus the pure entailment theories are the same: It is only when negation and conjunction are introduced that marked differences begin to emerge. [59, 398]

In constructing these models, Sylvan introduced two new elements: a ternary relational symbol S, distinct from the relevance relation R, to define the truth condition for $\bullet$, and the co-called "generation relation" denoted by G, which

[7] On the one hand, it is true that connexive logics are characterized by formulas that reference, among others, two ancient philosophers, namely Aristotle and Boethius. On the other hand, it remains unclear whether the origins of connexivity can legitimately be traced back to ancient times. Aristotle wrote: "[...] when B is not large, the same B must necessarily be large (which is impossible)" (*Prior Analytics*, 57b14-16). By formalizing the quote with contemporary methods, one obtains the formula usually referred to as Aristotle's Thesis, namely $\sim(\sim B \rightarrow B)$, and/or $\sim(B \rightarrow \sim B)$ (see [73]). A similar strategy occurred with Boethius' work *De Syllogismo Hypothetico*, where he argues that "*Si est A, non est B*" is the negation of "*Si est A, est B*", that is $(A \rightarrow B) \rightarrow \sim(A \rightarrow \sim B)$ in contemporary notation. The utilization of this formal approach in the examination of ancient texts has led to the current convention among logicians to view Aristotle and Boethius as historical points of reference within the domain of connexive logics. However, it's important to note that historians of logic often cast various doubts and hold differing perspectives regarding whether Aristotle and Boethius were conscious advocates of connexivity, both philosophically and logically. For divergent historical reconstructions, see, e.g., Łukasiewicz [32] and McCall [35], as well as Lenzen [30, 31], Weiss [75], and Ruge [62] for more contemporary sources.

indicates that a formula A generates a situation b, meaning that everything holding in b is implied by A. The inclusion of G in the proposed structures validates connexive principles [59, 398–399].

By turning to the formal details of [59], let's introduce frames and models for relevant connexive $\mathbf{B}^\bullet$ as follows:[8]

Definition 3.1 A *reduced* $\mathbf{B}^\bullet$ *frame* $\mathscr{F}$ is a sextuple $\langle 0, K, R, S, *, \mathsf{G}\rangle$, where $K, 0, R$ and $*$ are as before. $S \subseteq K^3$ is an additional relational symbol and G is a relation on formulas and worlds (i.e. the *generation* relation). Finally, we impose the following constraints on R and S:

(p1), (p2), (p3), (p4) of Definition 2.1, plus:
(p6) If $R0ab$ and $Scda$, then $Scdb$.

Definition 3.2 A *reduced* $\mathbf{B}^\bullet$ *model* $\mathscr{M}$ is a pair $\langle \mathscr{F}, v\rangle$, where $\mathscr{F}$ is a reduced frame and v is a function, called *valuation*, such that $v : \mathsf{At} \mapsto \wp(K)$, and subject to the so-called *heredity condition* (see Definition 2.2). The valuation v is extended to the whole language as follows. For all $A, B \in \mathsf{Frm}$ and $a \in K$:

$\mathscr{M}, a \Vdash p$	iff	$a \in v(p)$, for $p \in \mathsf{At}$;
$\mathscr{M}, a \Vdash {\sim}A$	iff	$\mathscr{M}, a^* \nVdash A$;
$\mathscr{M}, a \Vdash A \bullet B$	iff	$\exists b, c \in K$, s.t. $Sbca$ and $\mathscr{M}, b \Vdash A$ and $\mathscr{M}, c \Vdash B$;
$\mathscr{M}, a \Vdash A \to B$	iff	$\forall b, c \in K$, if $Rabc$ and $\mathscr{M}, b \Vdash A$, then $\mathscr{M}, c \Vdash B$.

We include the following clause for G:

(Gen) If $A\mathsf{G}a$, then $\mathscr{M}, a \Vdash A$.

Finally, satisfaction in a model $\mathscr{M}$ and validity on a frame $\mathscr{F}$, as well as the notion of entailment, are as in Definition 2.2.

In order to validate connexive principles, Sylvan suggested to add further constraints on $\mathbf{B}^\bullet$ frames. Specifically, we consider the following additions:

for Aristotle's thesis:	$\exists y(R0^*yy^* \wedge A\mathsf{G}y)$	(F.Ar)
for Boethius' thesis:	$\exists s, t, u(Rast \wedge Ra^*su \wedge A\mathsf{G}s \wedge R0tu^*)$	(F.Bo)
for Strawson's thesis:	$Sbc0^* \implies \exists y, z(Rbyz \wedge Rcyz^* \wedge A\mathsf{G}y)$	(F.St)

Observation 4 Sylvan's approach to connexivity is relatively uncommon in the subject. However, C. Mortensen and R. Brady have given serious consideration to Sylvan's approach and proposed additional modifications to his original ideas. Specifically, Mortensen [40] defines Ar as the condition that for every model $\mathscr{M}$, the set $C_A =_{\mathsf{df}} \{w : \mathscr{M}, w \vDash A \text{ and } \mathscr{M}, w \nvDash {\sim}A\}$ is non-empty. He claims that

[8] Similarly to the previous section, our focus will be on reduced models. However, it should be noted that Sylvan [59] also explores non-reduced models and other types of semantic structures in his work.

this characterization aligns with the idea expressed by Ar and emphasizes that for a self-inconsistent proposition A, if A is denied, the set C_A must be empty. Brady [11], instead, presented another relational semantics for relevant logic **B** extended with either Ar or Bo. The semantics introduced a non-empty subset of worlds $0 \subset K$, including a distinguished element T. The extended model structures include a function mapping sets of worlds, including propositions, i.e., interpretations of formulas, to sets of worlds. However, besides the approaches using ternary relational frames, the debate on connexivity is wide and it's challenging to summarize it under a single definition, as there exist various divergent approaches.

The origins of contemporary connexive logic can be traced back to the works of Richard Angell and Storrs McCall. Specifically, Angell's [6] main goal was to create a formal system that embodies what he termed the *principle of subjunctive contrariety*, which asserts the incompatibility of statements like "*If p were true, then q would be true*" and "*If p were true, then q would be false*". Building upon Angell's work, McCall [35] introduced the terminology *connexive* logic and extensively studied Angell's formal system. Additionally, McCall [36] explored the potential of using connexive implication to capture all valid moods of Aristotle's syllogistic within a first-order language.

In recent years, however, we have witnessed the emergence of several notable programs. For instance, C. Pizzi introduced the concept of consequential implication, which is defined using familiar modal concepts and strongly motivated by historical and philosophical inquiries into ancient notions of conditionals (see, for example, [50–53]). Additionally, H. Wansing developed a constructive connexive logic called **C** by working in a bilateral setting and modifying the falsity condition of implication as given for Nelson's constructive logic **N4** (see, for instance, [26, 72] as well as [17]). Similarly, N. Francez proposed *poly-connexivity*, which alters the familiar falsity conditions of conjunctions and disjunctions, in addition to the falsity clause for intuitionistic implication (see, for example, [20, 21]). Interestingly, this approach also discusses how to modify the base logic to transition from constructive to relevant systems. Drawing inspiration from Wansing's approach, Francez combined relevant logic **R** with connexive principles, devising a system of natural deduction that combines the relevant conditional with connexive theses (see [19]). Other approaches to connexivity are discussed in the following surveys [21, 47, 73].

3.1 Labelled Sequent Systems

A labelled calculus based on the models just introduced can be found in Table 1.[9] We remark that both $R \rightarrow$ and $L\bullet$ are subject to the eigenvariable restriction, and that

[9] The notion of labelled sequent is as in Definition 2.3. Also the closure condition applies in this case.

Table 1 G3rB$^\bullet$

Initial sequents:

$$R0ab, a : p, \Gamma \Rightarrow \Delta, b : p$$

(possibly a^*, b^*)

Logical rules:

$$\frac{\Gamma \Rightarrow \Delta, a^* : A}{a : {\sim}A, \Gamma \Rightarrow \Delta} L{\sim} \qquad \frac{a^* : A, \Gamma \Rightarrow \Delta}{\Gamma \Rightarrow \Delta, a : {\sim}A} R{\sim}$$

$$(b, c \text{ fresh}) \ \frac{Sbca, b : A, c : B, \Gamma \Rightarrow \Delta}{a : A \bullet B, \Gamma \Rightarrow \Delta} L\bullet$$

$$\frac{Sbca, \Gamma \Rightarrow \Delta, a : A \bullet B, b : A \quad Sbca, \Gamma \Rightarrow \Delta, a : A \bullet B, c : B}{Sbca, \Gamma \Rightarrow \Delta, a : A \bullet B} R\bullet$$

$$\frac{Rabc, a : A \to B, \Gamma \Rightarrow \Delta, b : A \quad c : B, Rabc, a : A \to B, \Gamma \Rightarrow \Delta}{Rabc, a : A \to B, \Gamma \Rightarrow \Delta} L\to$$

$$(b, c \text{ fresh}) \ \frac{Rabc, b : A, \Gamma \Rightarrow \Delta, c : B}{\Gamma \Rightarrow \Delta, a : A \to B} R\to$$

Relational rules for R and S:

$$\frac{R0aa, \Gamma \Rightarrow \Delta}{\Gamma \Rightarrow \Delta} \text{R1} \qquad \frac{R0aa^{**}, R0a^{**}a, \Gamma \Rightarrow \Delta}{\Gamma \Rightarrow \Delta} \text{R2}$$

$$\frac{Rabc, R0ad, Rdbc, \Gamma \Rightarrow \Delta}{R0ad, Rdbc, \Gamma \Rightarrow \Delta} \text{R3} \qquad \frac{R0b^*a^*, R0ab, \Gamma \Rightarrow \Delta}{R0ab, \Gamma \Rightarrow \Delta} \text{R4}$$

$$\frac{Scdb, R0ab, Scda, \Gamma \Rightarrow \Delta}{R0ab, Scda, \Gamma \Rightarrow \Delta} \text{R6}$$

R6 is the result of converting frame condition (p6) into a rule. Finally, notice that the premises of both rules, $R\bullet$ and $L \to$, are stated in their contraction-absorbing form. Last but not least, we introduce a rule specific rule for the generation relation G and show that it is height-preserving admissible:

Proposition 3.1 *The following rule:*

$$\frac{A\mathsf{G}a, a : A, \Gamma \Rightarrow \Delta}{A\mathsf{G}a, \Gamma \Rightarrow \Delta} L\mathsf{G}$$

is height-preserving admissible.

Proof As the proof requires other definitions and results to be introduced in Sect. 5, we spell out the details in Proposition 5.2. □

To show that our calculus is adequate to deal with the connexive principles mentioned above, we prove the next result:

Proposition 3.2 **G3rB$^\bullet$**, *extended by the addition of the following rules:*

$$\text{(y fresh)}\ \frac{R0^*yy^*, AGy, \Gamma \Rightarrow \Delta}{\Gamma \Rightarrow \Delta}\,\text{AR} \qquad \text{(s, t, u fresh)}\ \frac{Rast, Ra^*su, AGs, R0tu^*, \Gamma \Rightarrow \Delta}{\Gamma \Rightarrow \Delta}\,\text{BO}$$

$$\text{(y, z fresh)}\ \frac{Rbyz, Rcyz^*, AGy, Sbc0^*, \Gamma \Rightarrow \Delta}{Sbc0^*, \Gamma \Rightarrow \Delta}\,\text{ST}$$

is a connexive system. More precisely:

1. **G3rB$^\bullet$** $+ \text{AR} \vdash\Rightarrow 0 : {\sim}(A \to {\sim}A)$;
2. **G3rB$^\bullet$** $+ \text{BO} \vdash\Rightarrow 0 : (A \to B) \to {\sim}(A \to {\sim}B)$;
3. **G3rB$^\bullet$** $+ \text{ST} \vdash\Rightarrow 0 : {\sim}((A \to B) \bullet (A \to {\sim}B))$.

We remark that AR, BO and ST correspond to the conversion into schematic rules of (F.Ar), (F.Bo) and (F.St), respectively.

Proof We proceed by root-first derivation. For ${\sim}(A \to {\sim}A)$ and $(A \to B) \to {\sim}(A \to {\sim}B)$ see Table 2, and for ${\sim}((A \to B) \bullet (A \to {\sim}B))$ consider the derivation in Table 3. □

Finally, Sylvan observes that:

> The semantics given suggests two reductions, of *S* to *R* or vice versa, both of which should also be resisted. [Such reductions] would make connexive logic a, perhaps *bizarre*, branch of relevant logic. [59, 410, Emphasis mine]

The reduction that Sylvan has in mind ca be spelled out by identifying the relevant relation *R* and the newly introduced one *S*. This would make connexive fusion, $\bullet$, identical to relevant fusion, $\circ$, and would allow one to validate the following rule:

(Port$^\bullet$) $(A \bullet B) \to C \Leftarrow\!\Rightarrow A \to (B \to C)$

As a result, by reducing *R* to *S* and viceversa, one misses Sylvan's leading idea to interpret and characterize $\bullet$ as a proper type of connexive fusion connective, different from both, $\circ$ and $\wedge$. However, if the temptation to identify *R* and *S*, cannot be resisted, we introduce the following notion to cope with the resulting logics:[10]

Definition 3.3 We say that:

1. A Hilbert-system for **B$^\bullet$** is *Sylvan-bizarre* iff it is closed under the two-ways rule (Port$^\bullet$).
2. A reduced **B$^\bullet$** frame is *Sylvan-bizarre* iff it satisfies the following frame constraint: $Sabc \Longleftrightarrow Rabc$ (biz).

[10] The terminology is inspired by the words used by Sylvan himself in the previously mentioned quote.

Table 2 Derivations of Ar and Bo in **G3rB•** (and extensions thereof)

(Ar) $\Rightarrow 0 : {\sim}(A \to {\sim}A)$

$$
\dfrac{
\dfrac{
(\text{Prop. } 5.7)\ \dfrac{\dfrac{R0yy, y : A, R0^*yy^*, AGy, 0^* : A \to {\sim}A \Rightarrow y : A}{y : A, R0^*yy^*, AGy, 0^* : A \to {\sim}A, \Rightarrow y : A}\,\text{R1}}{R0^*yy^*, AGy, 0^* : A \to {\sim}A \Rightarrow y : A}\,LG
\qquad
\dfrac{(\text{Prop. } 5.7)\ \dfrac{\dfrac{R0yy^{**}, R0y^{**}y, y : A, R0^*yy^*, AGy, 0^* : A \to {\sim}A \Rightarrow y^{**} : A}{y : A, R0^*yy^*, AGy, 0^* : A \to {\sim}A, \Rightarrow y^{**} : A}\,\text{R2}}{R0^*yy^*, AGy, 0^* : A \to {\sim}A \Rightarrow y^{**} : A}\,LG}{y^* : {\sim}A, R0^*yy^*, AGy, 0^* : A \to {\sim}A \Rightarrow}\,L{\sim}
}{(y \text{ fresh})\ R0^*yy^*, AGy, 0^* : A \to {\sim}A \Rightarrow}\,L{\to}
}{\dfrac{0^* : A \to {\sim}A \Rightarrow}{\Rightarrow 0 : {\sim}(A \to {\sim}A)}\,R{\sim}}\,\text{AR}
$$

(Bo) $\Rightarrow 0 : (A \to B) \to {\sim}(A \to {\sim}B)$

$$
\dfrac{
\dfrac{
(\text{Prop. } 5.7)\ \dfrac{\dfrac{R0ss, s : A, Ra^*su, AGs, \mathcal{A} \Rightarrow s : A}{s : A, Ra^*su, AGs, \mathcal{A} \Rightarrow s : A}\,\text{R1}}{Ra^*su, AGs, \mathcal{A} \Rightarrow s : A}\,LG
\qquad
\dfrac{\dfrac{(\text{Prop. } 5.7)\ \dfrac{\dfrac{R0ss, s : A, Rast, AGs, \mathcal{A}' \Rightarrow u^* : B, s : A}{s : A, Rast, AGs, \mathcal{A}' \Rightarrow u^* : B, s : A}\,\text{R1}}{Rast, AGs, \mathcal{A}' \Rightarrow u^* : B, s : A}\,LG \qquad t : B, Rast, R0tu^*, \mathcal{A}'' \Rightarrow u^* : B}{Rast, Ra^*su, AGs, R0tu^*, a^* : A \to {\sim}B, a : A \to B \Rightarrow u^* : B}\,L{\to}}{u : {\sim}B, Rast, Ra^*su, AGs, R0tu^*, a^* : A \to {\sim}B, a : A \to B \Rightarrow}\,L{\sim}
}{(s, t, u \text{ fresh})\ Rast, Ra^*su, AGs, R0tu^*, a^* : A \to {\sim}B, a : A \to B \Rightarrow}\,L{\to}
}{(\text{Prop. } 5.7)\ \dfrac{\dfrac{a^* : A \to {\sim}B, a : A \to B \Rightarrow}{a : A \to B \Rightarrow a : {\sim}(A \to {\sim}B)}\,R{\sim}}{\Rightarrow 0 : (A \to B) \to {\sim}(A \to {\sim}B)}\,\text{VER}_2}\,\text{BO}
$$

where $\mathcal{A} = Rast, R0tu^*, a^* : A \to {\sim}B, a : A \to B$, $\mathcal{A}' = Ra^*su, R0tu^*, a^* : A \to {\sim}B, a : A \to B$, and $\mathcal{A}'' = Ra^*su, AGs, a^* : A \to {\sim}B, a : A \to B$.

Table 3 Derivation of St in **G3rB•** (and extensions thereof)

$(\mathsf{St}) \Rightarrow 0 : {\sim}((A \to B) \bullet (A \to {\sim}B))$

$$
\dfrac{
\text{(Prop. 5.7)}\;\dfrac{\dfrac{R0yy, y : A, Rcyz^*, \mathsf{AG}y, Sbc0^*, \mathcal{A} \Rightarrow y : A}{y : A, Rcyz^*, \mathsf{AG}y, Sbc0^*, \mathcal{A} \Rightarrow y : A}\,\text{R1}}{Rcyz^*, \mathsf{AG}y, Sbc0^*, \mathcal{A} \Rightarrow y : A}\,LG
\qquad
\dfrac{\dfrac{
\text{(Prop. 5.7)}\;\dfrac{\dfrac{R0yy, y : A, Rbyz, \mathsf{AG}y, \mathcal{A}' \Rightarrow z^{**} : B, y : A}{y : A, Rbyz, \mathsf{AG}y, \mathcal{A}' \Rightarrow z^{**} : B, y : A}\,\text{R1}}{Rbyz, \mathsf{AG}y, \mathcal{A}' \Rightarrow z^{**} : B, y : A}\,LG
\quad
\dfrac{R0zz^{**}, R0z^{**}z, z : B, Rbyz, \mathsf{AG}y, \mathcal{A}' \Rightarrow z^{**} : B}{z : B, Rbyz, \mathsf{AG}y, \mathcal{A}' \Rightarrow z^{**} : B}\,\text{R2}
}{Rbyz, Rcyz^*, \mathsf{AG}y, Sbc0^*, b : A \to B, c : A \to {\sim}B \Rightarrow y^{**} : B}\,L{\to}}{y^* : {\sim}B, Rbyz, Rcyz^*, \mathsf{AG}y, Sbc0^*, b : A \to B, c : A \to {\sim}B \Rightarrow}\,L{\sim}
}{
\text{(b, c fresh)}\;\dfrac{\text{(y, z fresh)}\;\dfrac{Rbyz, Rcyz^*, \mathsf{AG}y, Sbc0^*, b : A \to B, c : A \to {\sim}B \Rightarrow}{Sbc0^*, b : A \to B, c : A \to {\sim}B \Rightarrow}\,\text{ST}}{\dfrac{0^* : (A \to B) \bullet (A \to {\sim}B) \Rightarrow}{\Rightarrow 0 : {\sim}((A \to B) \bullet (A \to {\sim}B))}\,R{\sim}}\,L\bullet
}\,L{\to}
$$

(In the printed tree, the lowest inference marked $L{\to}$ concludes $Rbyz, Rcyz^*, \mathsf{AG}y, Sbc0^*, b : A \to B, c : A \to {\sim}B \Rightarrow$, which is then followed by ST (y, z fresh), $L\bullet$ (b, c fresh), and $R{\sim}$.)

where $\mathcal{A} = Rbyz, c : A \to {\sim}B, b : A \to B$ and $\mathcal{A}' = Sbc0^*, Rcyz^*, b : A \to B, c : A \to {\sim}B$.

At the calculus level we can cope with the notion of *Sylvan-bizarre* by converting each direction of (biz) into a relational rule. Indeed, by adding such newly constructed relational rules to **G3rB$^\bullet$**, we're allowed to extend the notion of *Sylvan-bizarre* to labelled calculi for connexive relevant logics as follows:

Proposition 3.3 **G3rB$^\bullet$**, *extended by the following two rules:*

$$\dfrac{Rabc, Sabc, \Gamma \Rightarrow \Delta}{Sabc, \Gamma \Rightarrow \Delta}\ \text{BIZ}_1 \qquad \dfrac{Sabc, Rabc, \Gamma \Rightarrow \Delta}{Rabc, \Gamma \Rightarrow \Delta}\ \text{BIZ}_2$$

is Sylvan-bizarre, i.e., the two-ways rule (Port$^\bullet$) *is admissible.*

Before displaying the proof, we recall that a sequent-style rule:

$$\dfrac{\mathcal{P}_1 \ldots \mathcal{P}_n}{\mathcal{C}}\ \mathbf{r}$$

is admissible in a sequent calculus **L**, if $\vdash_{\mathbf{L}} \mathcal{P}_1, \ldots, \vdash_{\mathbf{L}} \mathcal{P}_n$, together imply $\vdash_{\mathbf{L}} \mathcal{C}$. Finally, let's move to the proof:

Proof ($\Rrightarrow$) We first show that, in **G3rB$^\bullet$** $+ \text{BIZ}_1 + \text{BIZ}_2$, if $\Rightarrow 0 : A \bullet B \to C$, then $\Rightarrow 0 : A \to (B \to C)$.

$$\dfrac{\dfrac{\text{(Thm. 5.6)}\ \Rightarrow 0 : A \bullet B \to C \qquad \dfrac{\dfrac{\dfrac{\dfrac{\vdots\ \delta_1}{Sacd, R0dd, 0 : A \bullet B \to C, Racd, c : B, a : A \Rightarrow d : C, d : A \bullet B}}{R0dd, 0 : A \bullet B \to C, Racd, c : B, a : A \Rightarrow d : C, d : A \bullet B}\ \text{BIZ}_2 \qquad R0dd, d : C, \mathcal{A} \Rightarrow d : C}{R0dd, 0 : A \bullet B \to C, Racd, c : B, a : A \Rightarrow d : C}\ L\to}{0 : A \bullet B \to C, Racd, c : B, a : A \Rightarrow d : C}\ \text{R1}}{\dfrac{Racd, c : B, a : A \Rightarrow d : C}{a : A \Rightarrow a : B \to C}\ R\to\ (c, d \text{ fresh})}\ \text{CUT}}{\Rightarrow 0 : A \to (B \to C)}\ \text{VER}_2\ \text{(Prop. 5.7)}$$

where δ_1 is concluded by:

$$\dfrac{\dfrac{R0aa, Sacd, R0dd, a : A, \mathcal{A}'' \Rightarrow d : C, d : A \bullet B, a : A}{Sacd, R0dd, a : A, \mathcal{A}'' \Rightarrow d : C, d : A \bullet B, a : A} \qquad \dfrac{R0cc, Sacd, c : B, \mathcal{A}' \Rightarrow d : C, d : A \bullet B, c : B}{Sacd, c : B, \mathcal{A}' \Rightarrow d : C, d : A \bullet B, c : B}\ \text{R1}}{Sacd, R0dd, 0 : A \bullet B \to C, Racd, c : B, a : A \Rightarrow d : C, d : A \bullet B}\ R\bullet$$

where $\mathcal{A}' = R0dd, 0 : A \bullet B \to C, Racd, a : A$, and $\mathcal{A}'' = 0 : A \bullet B \to C, Racd, c : B$.

(⇐) We finish the proof by showing that, if $\Rightarrow 0 : A \to (B \to C)$, then $\Rightarrow 0 : A \bullet B \to C$.

$$
\dfrac{
  \dfrac{}{\Rightarrow 0 : A \to (B \to C)}\,(\text{Thm. 5.6})
  \qquad
  \dfrac{
    \dfrac{
      R0bb, b : A, \mathcal{A} \Rightarrow a : C, b : A
      \qquad
      \overset{\vdots\, \delta_1}{b : B \to C, R0bb, Rbca, c : B, \mathcal{A}' \Rightarrow a : C}
    }{R0bb, 0 : A \to (B \to C), Rbca, Sbca, b : A, c : B \Rightarrow a : C}\,L\!\to
  }{0 : A \to (B \to C), Rbca, Sbca, b : A, c : B \Rightarrow a : C}\,\text{R1}
}{
  \dfrac{
    \dfrac{
      \dfrac{Rbca, Sbca, b : A, c : B \Rightarrow a : C}{Sbca, b : A, c : B \Rightarrow a : C}\,\text{BIZ}_1
    }{a : A \bullet B \Rightarrow a : C}\,L\bullet\ (b, c \text{ fresh})
  }{\Rightarrow 0 : (A \bullet B) \to C}\,\text{VER}_2\ (\text{Prop. 5.7})
}\,\text{CUT}
$$

where, $\mathcal{A} = 0 : A \to (B \to C), Rbca, Sbca, c : B$ and $\mathcal{A}' = 0 : A \to (B \to C), Sbca, b : A$, and δ_1 is concluded by:

$$
\dfrac{
  \dfrac{R0cc, Rbca, c : B, \mathcal{A}'' \Rightarrow a : C, c : B}{Rbca, c : B, \mathcal{A}'' \Rightarrow a : C, c : B}\,\text{R1}
  \qquad
  \dfrac{R0aa, a : C, Rbca, c : B, \mathcal{A}'' \Rightarrow a : C}{a : C, Rbca, c : B, \mathcal{A}'' \Rightarrow a : C}\,\text{R1}
}{b : B \to C, R0bb, Rbca, c : B, \mathcal{A}' \Rightarrow a : C}\,L\!\to
$$

where $\mathcal{A}'' = 0 : A \to (B \to C), Sbca, b : A, b : B \to C, R0bb$. We remark that all applications of CUT and VER$_2$ are admissible, as it will be shown in Sect. 5. □

Finally, modular extensions of **G3rB**$^\bullet$ can be constructed by converting other frame conditions (see list stated in [59, 397–398]) into relational rules. Notice that the methodology we briefly introduced at the end of last section works also in this case.

4 Ackermann's Truth Constant in Relevant Logics: The Case of E^{t}

In this section, I'll consider the proof theory for relevant logics including so-called *Ackermann's truth constant*. To be precise, W. Ackermann, in his 1956 paper [1], introduced a so-called *falsity* (or *negative truth*) *constant* (⅄), sometimes read as "the absurd", in order to define the modal operators of impossibility, necessity and possibility, as ⅄ $\to A$, $\sim A \to$ ⅄ and $\sim (A \to$ ⅄$)$, respectively. In their 1959 paper [4], Anderson and Belnap proved that the addition of ⅄ is conservative, and that the aforementioned modal notions can be actually defined without relying on ⅄. These considerations motivated Anderson and Belnap in preferring the ⅄-free fragment of their modified, albeit theorem-wise identical, version of Ackermann's logic, i.e., the well-known logic of entailment **E**. More precisely, the Anderson and Belnap tradition in relevant logics (see [5]) takes ⅄ (or its positive counterpart, namely t, that is, basically, shorthand for $\sim$ ⅄) as a constant of convenience. In other words, if it can be added conservatively to a logic, it simplifies the presentation and/or proofs.

Accordingly, the authors of [61, Ch. 5] considered several semantic structures to formalize the behaviour of Ackermann's constant t and, mostly, focused their attention on the logic **E** extended by t (see especially, [61, Appendix 1, pp. 407–424]).

The strategy that we will explore, following the work done in [61], consists in the addition of a unary predicate on states interpreted, roughly, as a *possibility* predicate. In other words, if we denote such a predicate with the letter P and let a be a world in K, then Pa tells us that a is a *logically possible* state:

> P can be construed, without too much distortion as a possibility predicate [...] it shows that t marks out the regular worlds, i.e. those where the theorems hold. [61, 351]
> A situation a in K has property P iff each proposition which is necessarily true holds in a, i.e., a is a possible 'world'. [61, 407]

So, P selects regular (or, normal) worlds, i.e., those worlds at which all theorems are true under all interpretations.

Before entering the specific details of such models, let us draw a distinction between the cases which make t either *distinctive* or *non-distinctive*. The characteristic rules, which make t *distinctive*, are:

(Rt1) $\mathsf{t} \to A \Rrightarrow A$
(Rt2) $A \Rrightarrow \mathsf{t} \to A$

At the model-theoretic level, the authors of [61], discuss two possible semantic clauses that may be formulated for t, depending on whether it is distinctive or not. After some inspection, the authors notice that the two different clauses for t, can be always condensed into a single one. Nevertheless, we remark that, in all these cases, reduced structures face some limitations. Roughly, only a limited amount of, rather weak, relevant logics can be successfully characterized by reduced modellings. A notable exclusion is the well-known logic $\mathbf{E}^{\mathsf{t}}$, for which another solution was introduced.

From a closer perspective, $\mathbf{E}^{\mathsf{t}}$ contains all instances of the following axioms and is closed under the following rules [61, 408]:

(T1)	$(\mathsf{t} \to A) \to A$	(T2)	$\mathsf{t} \to (A \to A)$
(A2)	$A_1 \wedge A_2 \to A_i$	(A3)	$(A \to B) \wedge (A \to C) \to (A \to (B \wedge C))$
(A4)	$A_i \to (A_1 \vee A_2)$	(A5)	$(A \to C) \wedge (B \to C) \to ((A \vee B) \to C)$
(A7)	$\sim\sim A \to A$	(A6′)	$(A \wedge (B \vee C)) \to ((A \wedge B) \vee C)$
(A14)	$(A \to \sim A) \to \sim A$	(A8)	$(A \to B) \to (\sim B \to \sim A)$
(R1)	$A, A \to B \Rrightarrow B$	(A10)	$(A \to B) \to ((B \to C) \to (A \to C))$
(R2)	$A, B \Rrightarrow A \wedge B$	(A12)	$(A \to (A \to B)) \to (A \to B)$
(Rt2)	$A \Rrightarrow \mathsf{t} \to A$		

$\mathbf{E}^{\mathsf{t}}$ is a conservative extension of Anderson and Belanp's logic of entailment **E**, and to get the axiomatization of **E**, it suffices to replace (T1) and (T2) by the following two axioms:

(AxE1) $((A \to A) \to B) \to B$
(AxE2) $\Box A \wedge \Box B \to \Box(A \wedge B)$

where $\Box A =_{\mathsf{df}} (A \to A) \to A$. As it was pointed out by the authors of [61], "part of the point of the t-reformulation of **E** is to dispose of the intially troublesome [modal] axiom", that is (AxE2). As the authors of [61, p. 407] remark:

> The chief innovation in the semantical analysis of **E** we offer, [...] is the introduction of a new property P of situations. [...]
> P is required to cope with **E** in view of the way **E** introduces necessity as part of the logic of entailment, most obviously through the postulates $\Box A \to A$ and $\Box A \wedge \Box B \to \Box(A \wedge B)$ and through the (derived) rule of necessitation . (Notation adapted)

Reduced frames and models for $\mathbf{E}^{\mathsf{t}}$ were introduced as follows [61, 411]:

Definition 4.1 A $\mathbf{E}^{\mathsf{t}}$ *reduced frame* is a quintuple $\langle 0, K, \mathsf{P}, R, *\rangle$, where 0, K, and R are as in Definition 2.1. Finally, P denotes a property on K, called the *possibility predicate*. R and P are subject to the following conditions:

(p1)	$R0aa$.
(p2)	$a = a^{**}$.
(p6)	If $Rabc$, then Rac^*b^*.
(p8)	If $Rabcd$, then $Rb(ac)d$.
(p10)	If $Rabc$, then $Rabbc$.
(p12)	Raa^*a.
(t1)	$\exists x(\mathsf{P}x$ and $Raxa)$.
(t2)	If $\mathsf{P}x$ and $Rxab$, then $R0ab$.
(t3)	If $R0bx$ and $Raxc$, then $Rabc$.

Definition 4.2 An $\mathbf{E}^{\mathsf{t}}$ *reduced model* is a structure $\langle \mathscr{F}, v\rangle$, where $\mathscr{F}$ is an $\mathbf{E}^{\mathsf{t}}$ frame and v is as in Definition 2.2. We add the new condition for t:

$\mathscr{M}, a \Vdash \mathsf{t}$ iff $\exists b \in K$, such that $R0ba$ and $\mathsf{P}b$.

Finally, satisfaction in a model $\mathscr{M}$ and validity on a frame $\mathscr{F}$, as well as the notion of entailment, are as in Definition 2.2.

Based on this model, we construct the corresponding labelled proof system, termed $\mathbf{G3rE}^{\mathsf{t}}$ (Table 4, p. 287).[11] T1, T2 and T3 are (t1), (t2) and (t3) converted into rule form, respectively. We remark that other useful relational rules can be considered as well. For example:

$$\frac{Rabd, R0bc, Racd, \Gamma \Rightarrow \Delta}{R0bc, Racd, \Gamma \Rightarrow \Delta}\ \mathrm{R5}'$$

can be obtained by an application of T3. The same holds for:

$$\frac{R0ac, R0ab, R0bc, \Gamma \Rightarrow \Delta}{R0ab, R0bc, \Gamma \Rightarrow \Delta}\ \mathrm{R5}$$

[11] Definition 2.3 and the closure condition always apply.

Table 4 $\mathbf{G3rE^t}$

Initial sequents: $R0ab, a : p, \Gamma \Rightarrow \Delta, b : p$ (possibly a^*, b^*)

Rules for t:

$$\frac{}{\mathsf{P}b, R0ba, \Gamma \Rightarrow \Delta, a : \mathsf{t}}\, Rt \qquad (b \text{ fresh})\ \frac{\mathsf{P}b, R0ba, a : \mathsf{t}, \Gamma \Rightarrow \Delta}{a : \mathsf{t}, \Gamma \Rightarrow \Delta}\, Lt$$

Logical rules:

$$\frac{\Gamma \Rightarrow \Delta, a^* : A}{a : {\sim}A, \Gamma \Rightarrow \Delta}\, L{\sim} \qquad \frac{a^* : A, \Gamma \Rightarrow \Delta}{\Gamma \Rightarrow \Delta, a : {\sim}A}\, R{\sim}$$

$$\frac{a : A, a : B, \Gamma \Rightarrow \Delta}{a : A \wedge B, \Gamma \Rightarrow \Delta}\, L\wedge \qquad \frac{\Gamma \Rightarrow \Delta, a : A \quad \Gamma \Rightarrow \Delta, a : B}{\Gamma \Rightarrow \Delta, a : A \wedge B}\, R\wedge$$

$$\frac{a : A, \Gamma \Rightarrow \Delta \quad a : B, \Gamma \Rightarrow \Delta}{a : A \vee B, \Gamma \Rightarrow \Delta}\, L\vee \qquad \frac{\Gamma \Rightarrow \Delta, a : A, a : B}{\Gamma \Rightarrow \Delta, a : A \vee B}\, R\vee$$

$$\frac{Rabc, a : A \to B, \Gamma \Rightarrow \Delta, b : A \quad c : B, Rabc, a : A \to B, \Gamma \Rightarrow \Delta}{Rabc, a : A \to B, \Gamma \Rightarrow \Delta}\, L\to$$

$$(b, c \text{ fresh})\ \frac{Rabc, b : A, \Gamma \Rightarrow \Delta, c : B}{\Gamma \Rightarrow \Delta, a : A \to B}\, R\to$$

Relational rules for R and P:

$$\frac{R0aa, \Gamma \Rightarrow \Delta}{\Gamma \Rightarrow \Delta}\, \mathrm{R1} \qquad \frac{R0aa^{**}, R0a^{**}a, \Gamma \Rightarrow \Delta}{\Gamma \Rightarrow \Delta}\, \mathrm{R2}$$

$$\frac{Rac^*b^*, Rabc, \Gamma \Rightarrow \Delta}{Rabc, \Gamma \Rightarrow \Delta}\, \mathrm{R6} \qquad (y \text{ fresh})\ \frac{Rbyd, Racy, Rabx, Rxcd, \Gamma \Rightarrow \Delta}{Rabx, Rxcd, \Gamma \Rightarrow \Delta}\, \mathrm{R8}$$

$$(x \text{ fresh})\ \frac{Rabx, Rxbc, Rabc, \Gamma \Rightarrow \Delta}{Rabc, \Gamma \Rightarrow \Delta}\, \mathrm{R10} \qquad \frac{Raa^*a, \Gamma \Rightarrow \Delta}{\Gamma \Rightarrow \Delta}\, \mathrm{R12}$$

$$(x \text{ fresh})\ \frac{\mathsf{P}x, Raxa, \Gamma \Rightarrow \Delta}{\Gamma \Rightarrow \Delta}\, \mathrm{T1} \qquad \frac{R0ab, \mathsf{P}x, Rxab, \Gamma \Rightarrow \Delta}{\mathsf{P}x, Rxab, \Gamma \Rightarrow \Delta}\, \mathrm{T2}$$

$$\frac{Rabc, R0bx, Raxc, \Gamma \Rightarrow \Delta}{R0bx, Raxc, \Gamma \Rightarrow \Delta}\, \mathrm{T3}$$

which can be obtained by an application of R5'. More generally speaking, all relational rules, obtained by converting the frame conditions listed on [61, 411], are derivable by means of the relational rules taken as primitive in Table 4. We observe that, instead of $R\mathsf{t}$, it is possible to incorporate another type of initial sequents, namely structures of the following form: $\mathsf{P}b, R0ba, \Gamma \Rightarrow \Delta, a : \mathsf{t}$. However, to maintain a certain symmetry in the construction of the calculus, we made the choice of incorporating the 0-premise rule $R\mathsf{t}$ alongside $L\mathsf{t}$, which represents the rule-version of the semantic clause for t (Definition 4.2), instead of introducing additional initial sequents.[12]

Observation 5 The section started with an examination of Ackermann's falsity constant and its impact on the debate surrounding relevant logics. It was noted that the standard approach considers truth as a primitive concept, while defining falsity as the negation of truth, i.e., $\mathsf{f} =_{\mathsf{df}} {\sim}\mathsf{t}$.[13] I followed this convention and, as a result, the two following admissible rules for f can be considered:

$$(b \text{ fresh})\ \frac{\mathsf{P}b, R0ba^*, a^* : \mathsf{t}, \Gamma \Rightarrow \Delta}{\Gamma \Rightarrow \Delta, a : \mathsf{f}}\ Rf \qquad \frac{}{a : \mathsf{f}, \mathsf{P}b, R0ba^*, \Gamma \Rightarrow \Delta}\ Lf$$

By proceeding root-first then we get the following derivations:

$$(b \text{ fresh})\ \cfrac{\cfrac{\cfrac{\mathsf{P}b, R0ba^*, a^* : \mathsf{t}, \Gamma \Rightarrow \Delta}{a^* : \mathsf{t}, \Gamma \Rightarrow \Delta}\ Lt}{\Gamma \Rightarrow \Delta, a : {\sim}\mathsf{t}}\ R{\sim}}{\Gamma \Rightarrow \Delta, a : \mathsf{f}}\ \mathsf{df.f} \qquad \cfrac{\cfrac{\cfrac{}{\mathsf{P}b, R0ba^*, \Gamma \Rightarrow \Delta, a^* : \mathsf{t}}\ Rt}{a : {\sim}\mathsf{t}, \mathsf{P}b, R0ba^*, \Gamma \Rightarrow \Delta}\ L{\sim}}{a : \mathsf{f}, \mathsf{P}b, R0ba^*, \Gamma \Rightarrow \Delta}\ \mathsf{df.f}$$

Additionally, one might consider a Routley-Meyer semantic presentation of Ackermann's system from [1], known as Π'', which includes the original ⅄, and attempt to convert its semantic definition, along with the frame conditions, into well-constructed sequent-style rules. A semantics of this nature has been developed in [37]. However, delving into the examination of this approach within our current discussion would require introducing certain modifications and explanations that would make this section excessively lengthy. Therefore, a comprehensive investigation of the proof theory stemming from the semantic frameworks discussed in [37] will be temporarily set aside.

As remarked above, $\mathbf{E}^{\mathsf{t}}$ conservatively extends Anderson and Belnap's logic **E**. Accordingly, all **E** theorems are also $\mathbf{E}^{\mathsf{t}}$ theorems. However, as the modality $\Box$ is difficult to use in proofs,[14] another, more standard modal notion can be defined by relying on the truth constant t:

(df■) $\blacksquare A = \mathsf{t} \rightarrow A$.

[12] Let me also emphasize that in the context of labelled calculi based on Negri's methodology, it is quite common to use rules for constants, such as t or $\bot$, rather than introducing additional initial sequents apart from those that include atomic formulas (see, e.g., [45, 46]).

[13] While historically the falsity constant has been denoted as ⅄, one can use the more contemporary notation and represent it with the small case letter f.

[14] Mares and Standefer remark that: "The operator $\Box$ is extremely difficult to use in proofs. [...]. Written in primitive notation [(AxE2)] is $(((A \rightarrow A) \rightarrow A) \wedge ((B \rightarrow B) \rightarrow B)) \rightarrow (((A \wedge B) \rightarrow (A \wedge B)) \rightarrow (A \wedge B))$. Using this formula to prove other modal theses can be quite difficult" [34, pp. 701–702] (Notation adapted).

Accordingly, (T1), (T2) and (Rt2) can be reformulated as $\blacksquare A \to A$, $\blacksquare(A \to A)$ and $A \Rrightarrow \blacksquare A$, respectively. As remarked by Meyer:

> [...] '$\to$' [is] primitive and '$\blacksquare$' [is] contextually defined by [df$\blacksquare$]. ([df$\blacksquare$] is essentially the definition of '$\blacksquare$' of Ackermann's [1], modified by the insights of [Anderson and Belnap's] [4].) [They] prefer the essentially equivalent $\Box A = (A \to A) \to A$ [...]. My view is that the sentential constant [t] is well-motivated and ought to be introduced in the name of elegance; theirs seems to be that it is superfluous and ought to be thrown out in the name of Ockham. [38, 196] (Notation adapted)

In virtue of our observations, we can prove the following result:

Proposition 4.1 *If A is a theorem of Anderson and Belnap's* **E**, *then* $\mathbf{G3rE^t} \vdash \Rightarrow 0 : A$.

Proof We proceed by showing that (AxE1) and (AxE2) can be derived in $\mathbf{G3rE^t}$.

$\mathbf{G3rE^t} \vdash \Rightarrow 0 : ((A \to A) \to B) \to B$

$$
\dfrac{
  (a,b \text{ fresh})\;
  \dfrac{
    (x \text{ fresh})\;
    \dfrac{
      \dfrac{
        (c,d \text{ fresh})\;
        \dfrac{
          \dfrac{R0cd, Rxcd, c : A, \mathsf{P}x, Raxa, \mathcal{A} \Rightarrow b : B, d : A}{Rxcd, c : A, \mathsf{P}x, Raxa, \mathcal{A} \Rightarrow b : B, d : A}\,\text{T2}
        }{\mathsf{P}x, Raxa, \mathcal{A} \Rightarrow b : B, x : A \to A}\,R\!\to
        \qquad
        a : B, R0ab, \mathcal{A}' \Rightarrow b : B
      }{\mathsf{P}x, Raxa, R0ab, a : (A \to A) \to B \Rightarrow b : B}\,L\!\to
    }{R0ab, a : (A \to A) \to B \Rightarrow b : B}\,\text{T1}
  }{\Rightarrow 0 : ((A \to A) \to B) \to B}\,R\!\to
}{}
$$

where $\mathcal{A} = R0ab, a : (A \to A) \to B$ and $\mathcal{A}' = \mathsf{P}x, Raxa, a : (A \to A) \to B$.

$\mathbf{G3rE^t} \vdash \Rightarrow 0 : \blacksquare A \wedge \blacksquare B \to \blacksquare(A \wedge B)$

$$
\dfrac{
  (a,b \text{ fresh})\;
  \dfrac{
    \dfrac{
      (c,d \text{ fresh})\;
      \dfrac{
        \dfrac{
          \dfrac{
            \dfrac{}{\mathsf{P}x, R0xc, Racd, c : \mathsf{t}, \mathcal{A} \Rightarrow d : A \wedge B, c : \mathsf{t}}\,Rt
          }{Racd, c : \mathsf{t}, \mathcal{A} \Rightarrow d : A \wedge B, c : \mathsf{t}}\,Lt
          \qquad
          \begin{matrix}\vdots\;\delta_1 \\ d : A, Racd, Rbcd, c : \mathsf{t}, R0ab, a : \mathsf{t} \to A, a : \mathsf{t} \to B \Rightarrow d : A \wedge B\end{matrix}
        }{Racd, Rbcd, c : \mathsf{t}, R0ab, a : \mathsf{t} \to A, a : \mathsf{t} \to B \Rightarrow d : A \wedge B}\,L\!\to
      }{Rbcd, c : \mathsf{t}, R0ab, a : \mathsf{t} \to A, a : \mathsf{t} \to B \Rightarrow d : A \wedge B}\,\text{R3}
    }{R0ab, a : \mathsf{t} \to A, a : \mathsf{t} \to B \Rightarrow b : \mathsf{t} \to (A \wedge B)}\,R\!\to
  }{R0ab, a : (\mathsf{t} \to A) \wedge (\mathsf{t} \to B) \Rightarrow b : \mathsf{t} \to (A \wedge B)}\,L\wedge
}{\dfrac{\Rightarrow 0 : ((\mathsf{t} \to A) \wedge (\mathsf{t} \to B)) \to (\mathsf{t} \to (A \wedge B))}{\Rightarrow 0 : \blacksquare A \wedge \blacksquare B \to \blacksquare(A \wedge B)}\,\text{df}\blacksquare \text{ (dashed)}}\,R\!\to
$$

and δ_1 is concluded by:

$$
\dfrac{
  \dfrac{
    \dfrac{}{\mathsf{P}x, R0xc, Racd, c : \mathsf{t}, \mathcal{A} \Rightarrow d : A \wedge B, c : \mathsf{t}}\,Rt
  }{Racd, c : \mathsf{t}, \mathcal{A}' \Rightarrow d : A \wedge B, c : \mathsf{t}}\,Lt
  \qquad
  \begin{matrix}\vdots\;\delta_2 \\ d : B, d : A, Racd, \mathcal{A}'' \Rightarrow d : A \wedge B\end{matrix}
}{d : A, Racd, Rbcd, c : \mathsf{t}, R0ab, a : \mathsf{t} \to A, a : \mathsf{t} \to B \Rightarrow d : A \wedge B}\,L\!\to
$$

and δ_2 is obtained as follows:

$$\dfrac{\dfrac{R0dd, d : B, d : A, Racd, \mathcal{A}'' \Rightarrow d : A}{d : B, d : A, Racd, \mathcal{A}'' \Rightarrow d : A}\,\text{R1} \qquad \dfrac{R0dd, d : B, d : A, Racd, \mathcal{A}'' \Rightarrow d : B}{d : B, d : A, Racd, \mathcal{A}'' \Rightarrow d : B}\,\text{R1}}{d : B, d : A, Racd, \mathcal{A}'' \Rightarrow d : A \wedge B}\,R\wedge$$

where $\mathcal{A} = Rbcd, R0ab, a : \mathsf{t} \rightarrow A, a : \mathsf{t} \rightarrow B,, \mathcal{A}' = d : A, Rbcd, R0ab, a : \mathsf{t} \rightarrow A, a : \mathsf{t} \rightarrow B$ and $\mathcal{A}'' = Rbcd, c : \mathsf{t}, R0ab, a : \mathsf{t} \rightarrow A, a : \mathsf{t} \rightarrow B$. □

Extensions The authors of [61, 422] mention some interesting extensions that can be obtained by adding axioms and their related conditions to Hilbert systems and frames for $\mathbf{E^t}$, respectively. They are given as follows:

(B3)	$B \rightarrow (A \rightarrow A)$	(q3)	$Rabc \Longrightarrow R0bc$
(B4)	$A \rightarrow (B \rightarrow A)$	(q4)	$Rabc \Longrightarrow R0ac$
(B5)	$B \rightarrow (A \vee {\sim}A)$[15]	(q5)	$R0a^*a$

The following extensions can be built:

S4:	$\mathbf{E^t}$ + (B3)	$\mathscr{F}_{\mathbf{S4}} = \mathscr{F}_{\mathbf{E^t}}$ + (q3)	(Lewis' modal logic **S4**)
CL:	$\mathbf{E^t}$ + (B4)	$\mathscr{F}_{\mathbf{CL}} = \mathscr{F}_{\mathbf{E^t}}$ + (q4)	(Classical logic)
EDS:	$\mathbf{E^t}$ + (B5)	$\mathscr{F}_{\mathbf{EDS}} = \mathscr{F}_{\mathbf{E^t}}$ + (q5)	

where **EDS** is a proper subsystem of Lewis' logic **S3** (and hence of **S4**), and it is "of importance in the debate on entailment since it includes Disjunctive Syllogism B5′ without conceding the higher degree of strict implication" [61, 422]. Finally, let us remark that by adding the rules obtained by converting the frame conditions displayed above to $\mathbf{G3rE^t}$, namely:

$$\dfrac{R0bc, Rabc, \Gamma \Rightarrow \Delta}{Rabc, \Gamma \Rightarrow \Delta}\,\text{RQ3} \qquad \dfrac{R0ac, Rabc, \Gamma \Rightarrow \Delta}{Rabc, \Gamma \Rightarrow \Delta}\,\text{RQ4} \qquad \dfrac{R0a^*a, \Gamma \Rightarrow \Delta}{\Gamma \Rightarrow \Delta}\,\text{RQ5}$$

one obtains also labelled calculi for the logics just mentioned, i.e., **S4**, **CL** and **EDS**.

5 Proof Analysis

The main aim of this section is to prove the CUT-admissibility theorem for our labelled sequent systems. The general proof presented here is similar to the proof

[15] Or, equivalently, (B5′), namely, $(A \wedge ({\sim}A \vee B)) \rightarrow B$.

for labelled systems for other non-classical logics (see, e.g., [13, 16, 24, 29, 41]). A list of the results proved in this section is as follows:

- Height-preserving admissibility of substitution of labels (Lemma 5.1).
- Derivability of generalized initial sequents (Proposition 5.1).
- Height-preserving admissibility of the weakening rules (Lemma 5.2).
- Height-preserving invertibility of logical and relational rules (Lemma 5.3).
- Height-preserving admissibility of the contraction rules (Lemma 5.4).
- CUT-admissibility (Theorem 5.1).
- Other useful admissibility results (Proposition 5.2).

As there are many cases to be analysed in these proofs, we only outline the important parts here.

Let's start by fixing some standard notions.

Definition 5.1 Let $\mathcal{A}$ be any labelled formula of the form $a : A$. We denote by $l(\mathcal{A})$ the label of a formula $\mathcal{A}$, and by $p(\mathcal{A})$ the pure part of the formula, that is, the unlabelled formula. The *weight* (or *complexity*) of a labelled formula is defined as a lexicographically ordered pair: $\langle \mathsf{w}(p(\mathcal{A})), \mathsf{w}(l(\mathcal{A}))\rangle$, where:

1. for all state labels $a \in W$, $\mathsf{w}(a) = 1$;
2. for all $p \in \mathsf{At}$, $\mathsf{w}(p) = 1$;
3. $\mathsf{w}(\mathsf{t}) = 1$;
4. $\mathsf{w}({\sim}A) = \mathsf{w}(A) + 1$;
5. $\mathsf{w}(A \heartsuit B) = \mathsf{w}(A) + \mathsf{w}(B) + 1$, for $\heartsuit \in \{\circ, \bullet, +, \rightarrow\}$.[16]

Definition 5.2 A rule $\mathbf{r}$ is *height-preserving admissible* just in case: if there is a derivation of the premise(s) of $\mathbf{r}$, then there is a derivation of the conclusion of $\mathbf{r}$ that contains no application of $\mathbf{r}$ (with the height at most n, where n is the maximal height of the derivation of the premise(s)).

Definition 5.3 Let $\mathbf{q}$ be standing for either R or S. We define substitution as follows:

- $\mathbf{q}abc(d/e) \equiv \mathbf{q}abc$, if $e \neq a$, $e \neq b$ and $e \neq c$.
- $\mathbf{q}abc(d/a) \equiv \mathbf{q}dbc$, if $a \neq b$ and $a \neq c$.
- $\mathbf{q}abc(d/b) \equiv \mathbf{q}adc$, if $b \neq a$ and $b \neq c$.
- $\mathbf{q}abc(d/c) \equiv \mathbf{q}abd$, if $c \neq a$ and $c \neq b$.
- $\mathbf{q}aac(d/a) \equiv \mathbf{q}ddc$, if $a = b$ and $a \neq c$.
- $\mathbf{q}abb(d/b) \equiv \mathbf{q}add$, if $b = c$ and $b \neq a$.
- $\mathbf{q}cbc(d/c) \equiv \mathbf{q}dbd$, if $c = a$ and $c \neq b$.
- $\mathbf{q}aaa(d/a) \equiv \mathbf{q}ddd$, if $a = b$ and $a = c$.
- $A\mathsf{G}a(d/e) \equiv A\mathsf{G}a$, if $e \neq a$.

[16] As consequence of our definition, we obtain (i) $\mathsf{w}(A + B) = \mathsf{w}({\sim}A \rightarrow B) = \mathsf{w}({\sim}A) + \mathsf{w}(B) + 1 = \mathsf{w}(A) + \mathsf{w}(B) + 2$, (ii) $\mathsf{w}(A \oplus B) = \mathsf{w}({\sim}({\sim}A \circ {\sim}B)) = \mathsf{w}({\sim}A \circ {\sim}B) + 1 = \mathsf{w}({\sim}A) + \mathsf{w}({\sim}B) + 2 = \mathsf{w}(A) + \mathsf{w}({\sim}B) + 3 = \mathsf{w}(A) + \mathsf{w}(B) + 4$, (iii) $\mathsf{w}(\mathsf{f}) = \mathsf{w}({\sim}\mathsf{t}) = \mathsf{w}(\mathsf{t}) + 1 = 2$ and (iv) $\mathsf{w}(\blacksquare A) = \mathsf{w}(\mathsf{t} \rightarrow A) = \mathsf{w}(\mathsf{t}) + \mathsf{w}(A) + 1 = \mathsf{w}(A) + 2$.

- $A\mathsf{G}a(d/a) \equiv A\mathsf{G}d$.
- $\mathsf{P}a(d/e) \equiv \mathsf{P}a$, if $e \neq a$.
- $\mathsf{P}a(d/a) \equiv \mathsf{P}d$.
- $a : A(d/e) \equiv a : A$, if $e \neq a$.
- $a : A(d/a) \equiv d : A$.

We are now in a position that allows us to extend this definition to multisets.

Lemma 5.1 *Let the variable e stand for either a, b or c. If* $\vdash^n \Gamma \Rightarrow \Delta$ *and, provided d is free for e in* Γ, Δ, *then* $\vdash^n \Gamma(d/e) \Rightarrow \Delta(d/e)$ *(allowing* $*$*-variables to be substituted to variables as well).*

Proof By induction on the height n of the derivation of $\Gamma \Rightarrow \Delta$.

Let $n = 0$. If $\Gamma \Rightarrow \Delta$ is an initial sequent and (d/e) is not a vacuous substitution, then the substitution $\Gamma(d/e) \Rightarrow \Delta(d/e)$ is also an initial sequent. Similarly for conclusions of the $\mathbf{G3rE^t}$ rule $R\mathsf{t}$.

Let $n > 0$. We consider the last rule applied in the derivation.

1. If it is a rule for $\wedge$, $\vee$, or $\sim$, we apply the inductive hypothesis to the premise(s) of the rule, and then the rule.
2. We proceed similarly if the last rule is either $R\circ$, $L{+}^1$, $R\bullet$ or $L\rightarrow$.
3. If the last rule is either $R\rightarrow$, $L\circ$, $R{+}^1$ or $L\bullet$ and the substitution is vacuous ($e \neq a, b, c$), then there's nothing to do.
   Else, if d is not an eigenvariable, and the substitution $d/e)$ is not vacuous, we apply the inductive hypothesis to the derivation of the premise and, finally, the rule.
   If d is the eigenvariable, we first apply the inductive hypothesis in order to replace the eigenvariable d with a fresh variable (by the variable condition the substitution does not affect neither Γ nor Δ). Finally, we apply the inductive hypothesis to the derivation of the premise(s) and then the rule.

A more detailed case analysis can be found and readapted to the systems of this paper from [13]. □

Importantly, the heredity condition can be reflected at the calculus level by means of formal derivations:

Proposition 5.1 *Sequents of the following form:* $R0ab, a : A, \Gamma \Rightarrow \Delta, b : A$ *are derivable.*

Proof By induction on the structure of A.

If $A = B \circ C$, then:

$$
\dfrac{\dfrac{\dfrac{\dfrac{R0cc, Rcdb, c : B, \mathcal{S}, \Gamma \Rightarrow \Delta, b : B \circ C, c : B}{Rcdb, c : B, \mathcal{S}, \Gamma \Rightarrow \Delta, b : B \circ C, c : B}\,\mathrm{R1} \qquad \dfrac{R0dd, Rcdb, d : C, \mathcal{S}', \Gamma \Rightarrow \Delta, b : B \circ C, d : C}{Rcdb, d : C, \mathcal{S}', \Gamma \Rightarrow \Delta, b : B \circ C, d : C}\,\mathrm{R1}}{Rcdb, c : B, d : C, R0ab, Rcda, \Gamma \Rightarrow \Delta, b : B \circ C}\,R\circ}{c : B, d : C, R0ab, Rcda, \Gamma \Rightarrow \Delta, b : B \circ C}\,\mathrm{R3'}}{R0ab, a : B \circ C, \Gamma \Rightarrow \Delta, b : B \circ C}\,L\circ \quad (c, d \text{ fresh})
$$

where $\mathcal{S} = d : C, R0ab, Rcda$ and $\mathcal{S}' = c : B, R0ab, Rcda$.

If $A = B + C$, then:

$$\dfrac{\dfrac{R0cc, Rcda, c : B, \mathcal{S}, \Gamma \Rightarrow \Delta, c : B, d : C}{Rcda, c : B, \mathcal{S}, \Gamma \Rightarrow \Delta, c : B, d : C}\text{R1} \quad \dfrac{R0dd, Rcda, d : C, \mathcal{S}, \Gamma \Rightarrow \Delta, c : B, d : C}{Rcda, d : C, \mathcal{S}, \Gamma \Rightarrow \Delta, c : B, d : C}\text{R1}}{\dfrac{Rcda, Rcdb, R0ab, a : B + C, \Gamma \Rightarrow \Delta, c : B, d : C}{(c, d \text{ fresh})\ \dfrac{Rcdb, R0ab, a : B + C, \Gamma \Rightarrow \Delta, c : B, d : C}{R0ab, a : B + C, \Gamma \Rightarrow \Delta, b : B + C}R{+}^1}\text{R3}''}L{+}^1$$

where $\mathcal{S} = Rcdb, R0ab, a : B + C$.

If $A = B \bullet C$, then:

$$\dfrac{\dfrac{R0cc, Scdb, c : B, \mathcal{S}, \Gamma \Rightarrow \Delta, b : B \bullet C, c : B}{Scdb, c : B, \mathcal{S}, \Gamma \Rightarrow \Delta, b : B \bullet C, c : B}\text{R1} \quad \dfrac{R0dd, Scdb, d : C, \mathcal{S}', \Gamma \Rightarrow \Delta, b : B \bullet C, d : C}{Scdb, d : C, \mathcal{S}', \Gamma \Rightarrow \Delta, b : B \bullet C, d : C}\text{R1}}{\dfrac{Scdb, c : B, d : C, R0ab, Scda, \Gamma \Rightarrow \Delta, b : B \bullet C}{(c, d \text{ fresh})\ \dfrac{c : B, d : C, R0ab, Scda, \Gamma \Rightarrow \Delta, b : B \bullet C}{R0ab, a : B \bullet C, \Gamma \Rightarrow \Delta, b : B \bullet C}L\bullet}\text{R6}}R\bullet$$

where $\mathcal{S}$ is $d : C, R0ab, Scda$, and $\mathcal{S}'$ abbreviates $c : B, R0ab, Scda$.

If $A = \mathsf{t}$, then:

$$\dfrac{\dfrac{\dfrac{}{R0cb, \mathsf{P}c, R0ca, R0ab, a : \mathsf{t}, \Gamma \Rightarrow \Delta, b : \mathsf{t}}R\mathsf{t}}{\mathsf{P}c, R0ca, R0ab, a : \mathsf{t}, \Gamma \Rightarrow \Delta, b : \mathsf{t}}\text{R5}}{R0ab, a : \mathsf{t}, \Gamma \Rightarrow \Delta, b : \mathsf{t}}L\mathsf{t}$$

In all cases the premises are derivable by the inductive hypothesis. Derivations for A being either $\sim B$ or $B \to C$ can be found in [13]. □

Let φ be standing for one of the relational atoms employed in some of the proof systems mentioned in this paper.

Lemma 5.2 *The rules of weakening:*

$$\dfrac{\Gamma \Rightarrow \Delta}{a : A, \Gamma \Rightarrow \Delta}\text{LW} \qquad \dfrac{\Gamma \Rightarrow \Delta}{\Gamma \Rightarrow \Delta, a : A}\text{RW} \qquad \dfrac{\Gamma \Rightarrow \Delta}{\varphi, \Gamma \Rightarrow \Delta}\text{LW}_L$$

are height-preserving admissible.

Proof By induction on the height n of the derivation of the premise(s).

If $n = 0$, the cases are trivial. Let $n > 0$. For derivations concluding with rules without variable condition, the result follows straightforwardly by applying the inductive hypothesis to the premise(s) of the rule.

In case the derivation terminates with one of the rules with the eigenvariable condition, we first apply the substitution lemma to the premise(s) of the rule in order to obtain a fresh eigenvariable not clashing with those in the weakening formula ($a : A$ or φ). The conclusion is then obtained by applying the inductive hypothesis and, finally, the rule. As an example, we consider $L\bullet$:

$$\dfrac{\vdash^{n} Sazx, a : B, z : C, \Gamma \Rightarrow \Delta}{\vdash^{n+1} x : B \bullet C, \Gamma \Rightarrow \Delta}\,L\bullet \quad \underset{Lem.5.1}{\overset{\textsc{sub}(y/a)}{\rightsquigarrow}} \quad \dfrac{\dfrac{\vdash^{n} Syzx, y : B, z : C, \Gamma \Rightarrow \Delta}{\vdash^{n} a : A, Syzx, y : B, z : C, \Gamma \Rightarrow \Delta}\,i.h.}{\vdash^{n+1} a : A, x : B \bullet C, \Gamma \Rightarrow \Delta}\,L\bullet$$

where the clash of variables is avoided by substituting the clashing variable a with a fresh one, namely y.

The same procedure applies if the derivation terminates with an application of a relational rule, and $a : A$ or φ contain some of its eigenvariables. □

Definition 5.4 A rule **r** is *height-preserving invertible* just in case: if there is a derivation of the conclusion of **r**, then there is a dedrivation of premise(s) of **r** (with the height at most n, where n is the maximal height of the derivation of the conclusion).

Lemma 5.3 *All logical and relational rules introduced in this article are height-preserving invertible.*

Proof For each rule **r**, we have to show that if there is a derivation δ of the conclusion, then there is a derivation δ' of the premise(s), of the same height. For all rules without eigenvariable condition, we use a standard induction on the height of the derivation. For rules subject to the eigenvariable condition as well, but we need to be sure that in the transformed derivation we make use of a fresh label by applying the substitution lemma inside δ', if needed. The same procedures apply to all relational rules as well.

As an interesting example, we show height-preserving invertibility of $L\bullet$. It is proved by induction on the height n of the derivation of $a : A \bullet B, \Gamma \Rightarrow \Delta$. We distinguish three main cases. (1) If $n = 0$, $a : A \bullet B, \Gamma \Rightarrow \Delta$ is an initial sequent, and then also $Sbca, b : A, c : B, \Gamma \Rightarrow \Delta$, is an initial sequent. Let $n > 0$. (2) If $\vdash^{n+1} a : A \bullet B, \Gamma \Rightarrow \Delta$ is concluded by any rule **r** other than $L\bullet$, we apply the inductive hypothesis to the premise(s) $a : A \bullet B, \Gamma' \Rightarrow \Delta'$ ($a : A \bullet B, \Gamma'' \Rightarrow \Delta''$) to obtain derivation(s) of height n of $Sbca, b : A, c : B, \Gamma' \Rightarrow \Delta'$ ($Sbca, b : A, c : B, \Gamma'' \Rightarrow \Delta''$). By applying **r** we obtain a derivation of height $n + 1$ of $Sbca, b : A, c : B, \Gamma \Rightarrow \Delta$, as desired. (3) If $\vdash^{n+1} a : A \bullet B, \Gamma \Rightarrow \Delta$ is concluded by $L\bullet$, then $Sbca, b : A, c : B, \Gamma \Rightarrow \Delta$ is the requested conclusion of height n, possibly with different eigenvariables, but the desired ones can be obtained by height-preserving substitutions (Lemma 5.1). An analogous reasoning can be applied to all logical and relational rules with variable condition. □

Lemma 5.4 *The rules of contraction:*

$$\frac{a : A, a : A, \Gamma \Rightarrow \Delta}{a : A, \Gamma \Rightarrow \Delta}\text{LC} \qquad \frac{\Gamma \Rightarrow \Delta, a : A, a : A}{\Gamma \Rightarrow \Delta, a : A}\text{RC} \qquad \frac{\varphi, \varphi, \Gamma \Rightarrow \Delta}{\varphi, \Gamma \Rightarrow \Delta}\text{LC}_L$$

are height-preserving admissible.

Proof By induction on the height n of the derivation. If $n = 0$, then the premise is an initial sequent and, therefore, also the contracted sequent is an initial one. (Similarly for conclusions of Rt in **G3rE**^{t}.)

Let $n > 0$. We consider the last rule applied to derive the premise of contraction.

If the contraction formula is not principal, both occurrences are to be found in the premise(s) of the rule application, which has (have) a decreased derivation height. By applying the inductive hypothesis, we contract those occurrences and apply the rule to obtain our desired sequent.

If the contraction formula is principal, we distinguish three cases:

1. **r** is a rule in which the principal formulas appear also in the premise(s);
2. **r** is a rule where the premises consist of proper subformulas of the conclusion;
3. **r** is a rule where both, labels and proper subformulas of the principal formula, are active formulas.

Ad 1 Contraction is applied by the inductive hypothesis to the premise(s) of the rule. Let $A = B \bullet C$ and $R\bullet$ be the last rule applied:

$$\frac{\vdash^n Sbca, \Gamma \Rightarrow \Delta, a : B \bullet C, a : B \bullet C, b : B \qquad \vdash^n Sbca, \Gamma \Rightarrow \Delta, a : B \bullet C, a : B \bullet C, c : C}{\vdash^{n+1} Sbca, \Gamma \Rightarrow \Delta, a : B \bullet C, a : B \bullet C} R\bullet$$

By applying the inductive hypothesis to the premises, we obtain the requested derivation:

$$\frac{\vdash^n Sbca, \Gamma \Rightarrow \Delta, a : B \bullet C, b : B \qquad \vdash^n Sbca, \Gamma \Rightarrow \Delta, a : B \bullet C, c : C}{\vdash^{n+1} Sbca, \Gamma \Rightarrow \Delta, a : B \bullet C} R\bullet$$

Analogously if the last rule applied is $L \rightarrow$.

Let $A = B \circ C$ and suppose that the last rule applied is $R\circ$:

$$\frac{\vdash^n Rbca, \Gamma \Rightarrow \Delta, a : B \circ C, a : B \circ C, b : B \qquad \vdash^n Rbca, \Gamma \Rightarrow \Delta, a : B \circ C, a : B \circ C, c : C}{\vdash^{n+1} Rbca, \Gamma \Rightarrow \Delta, a : B \circ C, a : B \circ C} R\circ$$

By applying the inductive hypothesis to the premises, we obtain the requested derivation:

$$\frac{\vdash^n Rbca, \Gamma \Rightarrow \Delta, a : B \circ C, b : B \qquad \vdash^n Rbca, \Gamma \Rightarrow \Delta, a : B \circ C, c : C}{\vdash^{n+1} Rbca, \Gamma \Rightarrow \Delta, a : B \circ C}\ R\circ$$

Finally, let $A = B + C$ and suppose that the last step is $L+^1$:

$$\frac{\vdash^n b : B, Rbca, a : B + C, a : B + C, \Gamma \Rightarrow \Delta \qquad \vdash^n c : C, Rbca, a : B + C, a : B + C, \Gamma \Rightarrow \Delta}{\vdash^{n+1} Rbca, a : B + C, a : B + C, \Gamma \Rightarrow \Delta}\ L+^1$$

By applying the inductive hypothesis to the premises, we obtain the desired derivation:

$$\frac{\vdash^n b : B, Rbca, a : B + C, \Gamma \Rightarrow \Delta \qquad \vdash^n c : C, Rbca, a : B + C, \Gamma \Rightarrow \Delta}{\vdash^{n+1} Rbca, a : B + C, \Gamma \Rightarrow \Delta}\ L+^1$$

Ad 2 The procedure is rather straightforward as applications of contraction are simply reduced to contractions on less complex formulas.

Ad 3 In this case, the inductive hypothesis is used on formulas of smaller complexity, previously obtained by applications of the inversion lemma. Let $A = B \bullet C$ and suppose that the derivation ends with an application of $L\bullet$:

$$\frac{\vdash^n Sbca, b : B, c : C, a : B \bullet C, \Gamma \Rightarrow \Delta}{\vdash^{n+1} a : B \bullet C, a : B \bullet C, \Gamma \Rightarrow \Delta}\ L\bullet \quad \underset{Lem.5.4}{\overset{\text{INV.}}{\rightsquigarrow}} \quad (\text{x3})\ \frac{\dfrac{\vdash^n Sbca, b : B, c : C, Sbca, b : B, c : C, \Gamma \Rightarrow \Delta}{\vdash^n Sbca, b : B, c : C, \Gamma \Rightarrow \Delta}\ i.h.}{\vdash^{n+1} a : B \bullet C, \Gamma \Rightarrow \Delta}\ L\bullet$$

An identical reasoning can be carried out for $R \rightarrow$.

If $A = B \circ C$, and it is derived by $L\circ$, then:

$$\frac{\vdash^n Rbca, b : B, c : C, a : B \circ C, \Gamma \Rightarrow \Delta}{\vdash^{n+1} a : B \circ C, a : B \circ C, \Gamma \Rightarrow \Delta}\ L\circ \quad \underset{Lem.5.4}{\overset{\text{INV.}}{\rightsquigarrow}} \quad (\text{x3})\ \frac{\dfrac{\vdash^n Rbca, b : B, c : C, Rbca, b : B, c : C, \Gamma \Rightarrow \Delta}{\vdash^n Rbca, b : B, c : C, \Gamma \Rightarrow \Delta}\ i.h.}{\vdash^{n+1} a : B \circ C, \Gamma \Rightarrow \Delta}\ L\circ$$

If $A = B + C$, and the derivation ends with an application of $R+^1$, then:

$$\frac{\vdash^n Rbca, \Gamma \Rightarrow \Delta, a : B + C, b : B, c : C}{\vdash^{n+1} \Gamma \Rightarrow \Delta, a : B + C, a : B + C}\ R+^1 \quad \underset{Lem.5.4}{\overset{\text{INV.}}{\rightsquigarrow}} \quad (\text{x3})\ \frac{\dfrac{\vdash^n Rbca, Rbca, \Gamma \Rightarrow \Delta, b : B, c : C, b : B, c : C}{\vdash^n Rbca, \Gamma \Rightarrow \Delta, b : B, c : C}\ i.h.}{\vdash^{n+1} \Gamma \Rightarrow \Delta, a : B + C}\ R+^1$$

As a final example, let $A = \mathsf{t}$ and consider a derivation ending with $L\mathsf{t}$:

$$\dfrac{\vdash^{n} a:\mathsf{t}, \mathsf{P}b, R0ba, a:\mathsf{t}, \Gamma \Rightarrow \Delta}{\vdash^{n+1} a:\mathsf{t}, a:\mathsf{t}, \Gamma \Rightarrow \Delta}\,L\mathsf{t} \quad \overset{i.h.}{\rightsquigarrow} \quad \dfrac{\vdash^{n} \mathsf{P}b, R0ba, a:\mathsf{t}, \Gamma \Rightarrow \Delta}{\vdash^{n+1} a:\mathsf{t}, \Gamma \Rightarrow \Delta}\,L\mathsf{t}$$

Notice that when both contraction formulas are principal, we apply the closure condition. □

Theorem 5.1 *The rule of* CUT*:*

$$\dfrac{\Gamma \Rightarrow \Delta, a:A \qquad a:A, \Gamma' \Rightarrow \Delta'}{\Gamma, \Gamma' \Rightarrow \Delta, \Delta'}\,\text{CUT}$$

is admissible.

Proof The proof is carried out by a lexicographic induction on the complexity of the CUT-formula $a : A$ and the sum of the heights $h(\delta_1) + h(\delta_2)$. We perform a case analysis on the last rule used in the derivation above the CUT, and consider whether it applies to the CUT-formula or not. We show that each application of CUT can either be eliminated, or be replaced by one or more applications of CUT of smaller complexity. The proof proceeds similarly to the CUT-elimination proofs for several labelled calculi, e.g., [13, 24, 29, 41]. Intuitively, we eliminate the left- and topmost CUT first, and proceed by repeating the procedure until we reach a CUT-free derivation. We start by showing that CUT can be eliminated if one of the CUT premises is an initial sequent (**case 1**). Then we show that the CUT-height can be reduced in all cases in which the CUT-formula is not principal in at least one of the CUT-premises (**case 2**). Finally, we show that if the CUT-formula is principal in both CUT-premises, then the CUT is reduced to one or more CUTs on less complex formulas or on shorter derivations (**case 3**). We present some interesting cases where the CUT-formula A is principal in both premises.

(**3.1**) We start by considering a derivation where the last rules applied to obtain the CUT-premises are $R\circ$ and $L\circ$, respectively. Let A be $B \circ C$:

$$\dfrac{\dfrac{Rbca, \Gamma \Rightarrow \Delta, a:B\circ C, b:B \qquad Rbca, \Gamma \Rightarrow \Delta, a:B\circ C, c:C}{Rbca, \Gamma \Rightarrow \Delta, a:B\circ C}\,R\circ \quad {\scriptstyle (x,y\text{ fresh})}\dfrac{Rxya, x:B, y:C, \Gamma' \Rightarrow \Delta'}{a:B\circ C, \Gamma' \Rightarrow \Delta'}\,L\circ}{Rbca, \Gamma, \Gamma' \Rightarrow, \Delta, \Delta'}\,\text{CUT}$$

It is transformed into the following one:

$$\dfrac{\dfrac{\dfrac{Rbca, \Gamma \Rightarrow \Delta, a:B\circ C, c:C \qquad a:B\circ C, \Gamma' \Rightarrow \Delta'}{Rbca, \Gamma, \Gamma' \Rightarrow \Delta, \Delta', c:C}\,\text{CUT} \qquad \begin{matrix}\vdots\,\delta_1\\ Rbca, Rbca, c:C, \Gamma, \Gamma' \Rightarrow \Delta, \Delta'\end{matrix}}{Rbca, Rbca, Rbca, \Gamma, \Gamma, \Gamma', \Gamma' \Rightarrow \Delta, \Delta, \Delta', \Delta'}\,\text{CUT}}{Rbca, \Gamma, \Gamma' \Rightarrow \Delta, \Delta'}\,{\scriptstyle(\text{Lem. 5.5})}\ \text{LC+ RC+ LC}_L$$

where δ_1 is concluded by:

$$\dfrac{\dfrac{Rbca,\Gamma\Rightarrow\Delta,a:B\circ C,b:B \qquad a:B\circ C,\Gamma'\Rightarrow\Delta'}{Rbca,\Gamma,\Gamma'\Rightarrow\Delta,\Delta',b:B}\;\text{CUT} \qquad \dfrac{\text{(Lem. 5.1)}\;\dfrac{\text{(Lem. 5.1)}\;\dfrac{Rxya,x:B,y:C,\Gamma'\Rightarrow\Delta'}{Rbya,b:B,y:C,\Gamma'\Rightarrow\Delta'}\;\text{SUB}(b/x)}{Rbca,b:B,c:C,\Gamma'\Rightarrow\Delta'}\;\text{SUB}(c/y)}{}}{Rbca,Rbca,c:C,\Gamma,\Gamma'\Rightarrow,\Delta,\Delta'}\;\text{CUT}$$

(**3.2**) Assume that the premises of CUT are derived by $R+^1$ and $L+^1$, respectively. Let $A = B + C$:

$$\dfrac{(b,c\text{ fresh})\;\dfrac{Rbca,\Gamma\Rightarrow\Delta,b:B,c:C}{\Gamma\Rightarrow\Delta,a:B+C}\;R+^1 \qquad \dfrac{d:B,Rdea,a:B+C,\Gamma'\Rightarrow\Delta' \qquad e:C,Rade,a:B+C,\Gamma'\Rightarrow\Delta'}{Rdea,a:B+C,\Gamma'\Rightarrow\Delta'}\;L+^1}{Rdea,\Gamma,\Gamma'\Rightarrow\Delta,\Delta'}\;\text{CUT}$$

It is transformed into the following derivation:

$$\text{(Lem. 5.5)}\;\dfrac{\dfrac{\begin{matrix}\vdots\;\delta_1\\ Rdea,\Gamma,\Gamma'\Rightarrow\Delta,\Delta',e:C\end{matrix} \qquad \dfrac{\Gamma\Rightarrow\Delta,a:B+C \qquad e:C,Rdea,a:B+C,\Gamma'\Rightarrow\Delta'}{e:C,Rdea,\Gamma,\Gamma'\Rightarrow\Delta,\Delta'}\;\text{CUT}}{Rdea,Rdea,\Gamma,\Gamma,\Gamma',\Gamma'\Rightarrow\Delta,\Delta,\Delta',\Delta'}\;\text{CUT}}{Rdea,\Gamma,\Gamma'\Rightarrow\Delta,\Delta'}\;\text{LC+ RC+ LC}_L$$

where the conclusion of δ_1 is derived by:

$$\text{(Lem. 5.5)}\;\dfrac{\dfrac{\text{(Lem. 5.1)}\;\dfrac{\text{(Lem. 5.1)}\;\dfrac{Rbca,\Gamma\Rightarrow\Delta,b:B,c:C}{Rbea,\Gamma\Rightarrow\Delta,b:B,e:C}\;\text{SUB}(e/c)}{Rdea,\Gamma\Rightarrow\Delta,d:B,e:C}\;\text{SUB}(d/b) \qquad \dfrac{\Gamma\Rightarrow\Delta,a:B+C \qquad d:B,Rdea,a:B+C,\Gamma'\Rightarrow\Delta'}{d:B,Rdea,\Gamma,\Gamma'\Rightarrow\Delta,\Delta'}\;\text{CUT}}{Rdea,Rdea,\Gamma,\Gamma,\Gamma'\Rightarrow\Delta,\Delta,\Delta',e:C}\;\text{CUT}}{Rade,\Gamma,\Gamma'\Rightarrow\Delta,\Delta',e:C}\;\text{LC+ RC+LC}_L$$

(**3.3**) Assume that the premises of CUT are derived by $R\bullet$ and $L\bullet$, respectively. Let $A = B \bullet C$:

$$\dfrac{\dfrac{Sbca,\Gamma\Rightarrow\Delta,a:B\bullet C,b:B \qquad Sbca,\Gamma\Rightarrow\Delta,a:B\bullet C,c:C}{Sbca,\Gamma\Rightarrow\Delta,a:B\bullet C}\;R\bullet \qquad (x,y\text{ fresh})\;\dfrac{Sxya,x:B,y:C,\Gamma'\Rightarrow\Delta'}{a:B\bullet C,\Gamma'\Rightarrow\Delta'}\;L\bullet}{Sbca,\Gamma,\Gamma'\Rightarrow,\Delta,\Delta'}\;\text{CUT}$$

The desired transformed derivation is obtained as follows:

$$\text{(Lem. 5.5)}\;\dfrac{\dfrac{\dfrac{Sbca,\Gamma\Rightarrow\Delta,a:B\bullet C,c:C \qquad a:B\bullet C,\Gamma'\Rightarrow\Delta'}{Sbca,\Gamma,\Gamma'\Rightarrow\Delta,\Delta',c:C}\;\text{CUT} \qquad \begin{matrix}\vdots\;\delta_1\\ Sbca,Sbca,c:C,\Gamma,\Gamma'\Rightarrow\Delta,\Delta'\end{matrix}}{Sbca,Sbca,Sbca,\Gamma,\Gamma,\Gamma',\Gamma'\Rightarrow\Delta,\Delta,\Delta',\Delta'}\;\text{CUT}}{Sbca,\Gamma,\Gamma'\Rightarrow\Delta,\Delta'}\;\text{LC+ RC+ LC}_L \tag{1}$$

where δ_1 is concluded by:

$$\dfrac{\dfrac{Sbca, \Gamma \Rightarrow \Delta, a : B \bullet C, b : B \quad a : B \bullet C, \Gamma' \Rightarrow \Delta'}{Sbca, \Gamma, \Gamma' \Rightarrow \Delta, \Delta', b : B}\,\text{CUT} \qquad \dfrac{\dfrac{Sxya, x : B, y : C, \Gamma' \Rightarrow \Delta'}{Sbya, b : B, y : C, \Gamma' \Rightarrow \Delta'}\,\text{SUB}(b/x)\ (\text{Lem. 5.1})}{Sbca, b : B, c : C, \Gamma' \Rightarrow \Delta'}\,\text{SUB}(c/y)\ (\text{Lem. 5.1})}{Sbca, Sbca, c : C, \Gamma, \Gamma' \Rightarrow, \Delta, \Delta'}\,\text{CUT}$$

(**3.4**) As a final example, assume that the premises of CUT are derived by $R\mathsf{t}$ and $L\mathsf{t}$, respectively. Let $A = \mathsf{t}$:

$$\dfrac{\dfrac{}{\mathsf{P}b, R0ba, \Gamma \Rightarrow \Delta, a : \mathsf{t}}\,R\mathsf{t} \qquad \dfrac{\mathsf{P}x, R0xa, a : \mathsf{t}, \Gamma' \Rightarrow \Delta'}{a : \mathsf{t}, \Gamma' \Rightarrow \Delta'}\,L\mathsf{t}\ (x \text{ fresh})}{\mathsf{P}b, R0ba, \Gamma, \Gamma' \Rightarrow \Delta, \Delta'}\,\text{CUT}$$

It is transformed into the following derivation:

$$\dfrac{\dfrac{\mathsf{P}b, R0ba, \Gamma \Rightarrow \Delta, a : \mathsf{t} \qquad \dfrac{\mathsf{P}x, R0xa, a : \mathsf{t}, \Gamma' \Rightarrow \Delta'}{\mathsf{P}b, R0ba, a : \mathsf{t}, \Gamma' \Rightarrow \Delta'}\,\text{SUB}(b/x)\ (\text{Lem. 5.1})}{\mathsf{P}b, R0ba, \mathsf{P}b, R0ba, \Gamma, \Gamma' \Rightarrow \Delta, \Delta'}\,\text{CUT}}{\mathsf{P}b, R0ba, \Gamma, \Gamma' \Rightarrow \Delta, \Delta'}\,\text{LC+LC}_L\ (\text{Lem. 5.5})$$

The proofs for A being either $\sim B$ or $B \to C$ can be found in [13]. □

Proposition 5.2 *The following results can be shown:*

1. The heredity rules for atomic formulas are height-preserving admissible:

$$\dfrac{b : p, R0ab, a : p, \Gamma \Rightarrow \Delta}{R0ab, a : p, \Gamma \Rightarrow \Delta}\,\text{AtHer-L} \qquad \dfrac{R0ab, \Gamma \Rightarrow \Delta, b : p, a : p}{R0ab, \Gamma \Rightarrow \Delta, b : p}\,\text{AtHer-R}$$

2. The heredity rules for compound formulas are admissible:

$$\dfrac{b : A, R0ab, a : A, \Gamma \Rightarrow \Delta}{R0ab, a : A, \Gamma \Rightarrow \Delta}\,\text{GenHer-L} \qquad \dfrac{R0ab, \Gamma \Rightarrow \Delta, b : A, a : A}{R0ab, \Gamma \Rightarrow \Delta, b : A}\,\text{GenHer-R}$$

3. The following rules, relating the concepts of entailment and implication (i.e., the sequent-style version of the Verification Lemma, p. 270), are admissible:

$$\dfrac{\Rightarrow 0 : A \to B}{a : A \Rightarrow a : B}\,\text{VER1} \qquad \dfrac{a : A \Rightarrow a : B}{\Rightarrow 0 : A \to B}\,\text{VER2}$$

4. The following rules for the two defined fission connectives (see Sect. 2, p. 274) are admissible:

$$\frac{b^* : A, Rabc, a : A+B, \Gamma \Rightarrow \Delta \qquad c : B, Rabc, a : A+B, \Gamma \Rightarrow \Delta}{Rabc, a : A+B, \Gamma \Rightarrow \Delta}\ L{+}^2$$

$$(b, c \text{ fresh})\ \frac{Rabc, \Gamma \Rightarrow \Delta, b^* : A, c : B}{\Gamma \Rightarrow \Delta, a : A+B}\ R{+}^2$$

$$\frac{b^* : A, Rbca^*, a : A \oplus B, \Gamma \Rightarrow \Delta \qquad c^* : B, Rbca, a : A \oplus B, \Gamma \Rightarrow \Delta}{Rbca^*, a : A \oplus B, \Gamma \Rightarrow \Delta}\ L\oplus$$

$$(b, c \text{ fresh})\ \frac{Rbca^*, \Gamma \Rightarrow \Delta, b^* : A, c^* : B}{\Gamma \Rightarrow \Delta, a : A \oplus B}\ R\oplus$$

5. The "generation rule" (see Sect. 3, p. 279) is height-preserving admissible.

$$\frac{\mathsf{A}\mathsf{G}a, a : A, \Gamma \Rightarrow \Delta}{\mathsf{A}\mathsf{G}a, \Gamma \Rightarrow \Delta}\ L\mathsf{G}$$

6. The "necessitation rule" (i.e., Rt2*, Sect. 4, p. 285) is admissible.*

$$\frac{\Rightarrow 0 : A}{\Rightarrow 0 : \mathsf{t} \rightarrow A}\ \text{NEC}$$

7. The rules for ■ (see Sect. 4, pp. 288 and ff.) are admissible.

$$(x \text{ fresh})\ \frac{\mathsf{P}x, Raxb, \Gamma \Rightarrow \Delta, b : A}{\Gamma \Rightarrow \Delta, a : \blacksquare A}\ R\blacksquare \qquad \frac{b : A, a : \blacksquare A, \mathsf{P}x, Raxb, \Gamma \Rightarrow \Delta}{a : \blacksquare A, \mathsf{P}x, Raxb, \Gamma \Rightarrow \Delta}\ L\blacksquare$$

Proof The proofs of *1.*, *2.* and *3.* can all be found in [13].

Ad 4 We proceed by root-first derivation. Let's start by considering the rules for $A + B =_{\mathsf{df}} {\sim}A \rightarrow B$. For $L{+}^2$:

$$\dfrac{\dfrac{\dfrac{b^* : A, Rabc, a : A+B, \Gamma \Rightarrow \Delta}{b^* : A, Rabc, a : {\sim}A \rightarrow B, \Gamma \Rightarrow \Delta}\ \mathsf{df}{+}}{Rabc, a : {\sim}A \rightarrow B, \Gamma \Rightarrow \Delta, b : {\sim}A}\ R{\sim} \qquad \dfrac{c : B, Rabc, a : A+B, \Gamma \Rightarrow \Delta}{c : B, Rabc, a : {\sim}A \rightarrow B, \Gamma \Rightarrow \Delta}\ \mathsf{df}{+}}{\dfrac{Rabc, a : {\sim}A \rightarrow B, \Gamma \Rightarrow \Delta}{Rabc, a : A+B, \Gamma \Rightarrow \Delta}\ \mathsf{df}{+}}\ L\rightarrow$$

For $R{+}^2$:

$$\dfrac{(b,c\text{ fresh})\;\dfrac{\dfrac{Rabc,\Gamma\Rightarrow\Delta,b^*:A,c:B}{Rabc,b:{\sim}A,\Gamma\Rightarrow\Delta,c:B}\,L{\sim}}{\Gamma\Rightarrow\Delta,a:{\sim}A\to B}\,R{\to}}{\Gamma\Rightarrow\Delta,a:A+B}\,\mathsf{df}{+}$$

Finally, we consider $A\oplus B=_{\mathsf{df}}{\sim}({\sim}A\circ{\sim}B)$. For $L\oplus$:

$$\dfrac{(\text{Lem. }5.5)\;\dfrac{\dfrac{\dfrac{(\text{Lem. }5.3)\;\dfrac{\dfrac{b^*:A,Rbca^*,a:A\oplus B,\Gamma\Rightarrow\Delta}{b^*:A,Rbca^*,a:{\sim}({\sim}A\circ{\sim}B),\Gamma\Rightarrow\Delta}\,\mathsf{df}\oplus}{\dfrac{b^*:A,Rbca^*,a:{\sim}({\sim}A\circ{\sim}B),\Gamma\Rightarrow\Delta,a^*:{\sim}A\circ{\sim}B}{Rbca^*,a:{\sim}({\sim}A\circ{\sim}B),\Gamma\Rightarrow\Delta,a^*:{\sim}A\circ{\sim}B,b:{\sim}A}\,R{\sim}}\,\text{RW} \qquad (\text{Lem. }5.3)\;\dfrac{\dfrac{c^*:B,Rbca^*,a:A\oplus B,\Gamma\Rightarrow\Delta}{c^*:B,Rbca^*,a:{\sim}({\sim}A\circ{\sim}B),\Gamma\Rightarrow\Delta}\,\mathsf{df}\oplus}{\dfrac{c^*:B,Rbca^*,a:{\sim}({\sim}A\circ{\sim}B),\Gamma\Rightarrow\Delta,a^*:{\sim}A\circ{\sim}B}{Rbca^*,a:{\sim}({\sim}A\circ{\sim}B),\Gamma\Rightarrow\Delta,a^*:{\sim}A\circ{\sim}B,c:{\sim}B}\,R{\sim}}\,\text{RW}}{Rbca^*,a:{\sim}({\sim}A\circ{\sim}B),\Gamma\Rightarrow\Delta,a^*:{\sim}A\circ{\sim}B}\,R\circ}{Rbca^*,a:{\sim}({\sim}A\circ{\sim}B),a:{\sim}({\sim}A\circ{\sim}B),\Gamma\Rightarrow\Delta}\,L{\sim}}{Rbca^*,a:{\sim}({\sim}A\circ{\sim}B),\Gamma\Rightarrow\Delta}\,\text{LC}}{Rbca^*,a:A\oplus B,\Gamma\Rightarrow\Delta}\,\mathsf{df}\oplus$$

For $R\oplus$:

$$\dfrac{\dfrac{(b,c\text{ fresh})\;\dfrac{\dfrac{\dfrac{Rbca^*,\Gamma\Rightarrow\Delta,b^*:A,c^*:B}{Rbca^*,c:{\sim}B,\Gamma\Rightarrow\Delta,b^*:A}\,L{\sim}}{Rbca^*,b:{\sim}A,c:{\sim}B,\Gamma\Rightarrow\Delta}\,L{\sim}}{a^*:{\sim}A\circ{\sim}B,\Gamma\Rightarrow\Delta}\,L\circ}{\Gamma\Rightarrow\Delta,a:{\sim}({\sim}A\circ{\sim}B)}\,R{\sim}}{\Gamma\Rightarrow\Delta,a:A\oplus B}\,\mathsf{df}\oplus$$

Ad 5 We consider **G3rB**$^{\bullet}$ (and extensions thereof) and prove the claim by induction on the height n of the derivation. If $n=0$, then the sequent is an initial one, and we obtain our desired result as follows:

$$A\mathsf{G}a,a:A,R0ab,a:p,\Gamma\Rightarrow\Delta,b:p\overset{i.h.}{\rightsquigarrow}A\mathsf{G}a,R0ab,a:p,\Gamma\Rightarrow\Delta,b:p$$

where the rightmost sequent is obtained by applying the induction hypothesis to the leftmost one.

Let $n>0$. If the derivation ends with an application of rules without eigenvariable condition, then we simply apply the inductive hypothesis to the premises and finally the rule. We display the cases for derivations ending with $L\bullet$ and $R\to$ (subject to the eigenvariable condition). If the formula A is not principal, we proceed as follows:

$$\dfrac{\vdash^n Sbca,b:B,c:C,A\mathsf{G}a,a:A,\Gamma\Rightarrow\Delta}{\vdash^{n+1}a:B\bullet C,A\mathsf{G}a,a:A,\Gamma\Rightarrow\Delta}\,L\bullet\quad\overset{i.h.}{\rightsquigarrow}\quad\dfrac{\vdash^n Sbca,b:B,c:C,A\mathsf{G}a,\Gamma\Rightarrow\Delta}{\vdash^{n+1}a:B\bullet C,A\mathsf{G}a,\Gamma\Rightarrow\Delta}\,L\bullet$$

And, finally:

$$\dfrac{\vdash^{n} Rabc, b : B, A\mathsf{G}a, a : A, \Gamma \Rightarrow \Delta, c : C}{\vdash^{n+1} A\mathsf{G}a, a : A, \Gamma \Rightarrow \Delta, a : B \to C}\ R{\to} \quad \overset{i.h.}{\rightsquigarrow} \quad \dfrac{\vdash^{n} Rabc, b : B, A\mathsf{G}a, \Gamma \Rightarrow \Delta, c : C}{\vdash^{n+1} A\mathsf{G}a, \Gamma \Rightarrow \Delta, a : B \to C}\ R{\to}$$

If the last rule applied is a relational rule, the procedure is identical.

Let's consider a rule application with A principal. Let $A = B \bullet C$:

$$\dfrac{\vdash^{n} Sbca, b : B, c : C, (B \bullet C)\mathsf{G}a, \Gamma \Rightarrow \Delta}{\vdash^{n+1} a : B \bullet C, (B \bullet C)\mathsf{G}a, \Gamma \Rightarrow \Delta}\ L\bullet \quad \underset{Lem.5.3}{\overset{\text{LW}}{\rightsquigarrow}} \quad \begin{array}{c} \text{(Lem. 5.4)}\ \dfrac{\vdash^{n} a : B \bullet C, Sbca, b : B, c : C, (B \bullet C)\mathsf{G}a, \Gamma \Rightarrow \Delta}{\text{(Lem. 5.5)}\ \dfrac{\vdash^{n} a : B \bullet C, a : B \bullet C, (B \bullet C)\mathsf{G}a, \Gamma \Rightarrow \Delta}{\vdash^{n} a : B \bullet C, (B \bullet C)\mathsf{G}a, \Gamma \Rightarrow \Delta}\ \text{LC}}\ \text{INV} \\ \hdashline \vdash^{n} (B \bullet C)\mathsf{G}a, \Gamma \Rightarrow \Delta \quad i.h. \end{array}$$

Finally, let $A = B \to C$:

$$\dfrac{\vdash^{n} Rabc, a : B \to C, (B \to C)\mathsf{G}a, \Gamma \Rightarrow \Delta, b : B \qquad \vdash^{n} c : C, Rabc, a : B \to C, (B \to C)\mathsf{G}a, \Gamma \Rightarrow \Delta}{\vdash^{n+1} Rabc, a : B \to C, (B \to C)\mathsf{G}a, \Gamma \Rightarrow \Delta}\ L\to$$

By applying LW (Lemma 5.2) to the premises we construct the following derivation:

$$\begin{array}{c} \text{(Lem. 5.4)}\ \dfrac{\vdash^{n} a : B \to C, Rabc, a : B \to C, (B \to C)\mathsf{G}a, \Gamma \Rightarrow \Delta, b : B \qquad \vdash^{n} a : B \to C, c : C, Rabc, a : B \to C, (B \to C)\mathsf{G}a, \Gamma \Rightarrow \Delta}{\text{(Lem. 5.5)}\ \dfrac{\vdash^{n} a : B \to C, Rabc, a : B \to C, (B \to C)\mathsf{G}a, \Gamma \Rightarrow \Delta}{\vdash^{n} Rabc, a : B \to C, (B \to C)\mathsf{G}a, \Gamma \Rightarrow \Delta}\ \text{LC}}\ \text{INV} \\ \hdashline \vdash^{n} Rabc, (B \to C)\mathsf{G}a, \Gamma \Rightarrow \Delta \quad i.h. \end{array}$$

Ad 6 Consider the system $\mathbf{G3rE^{t}}$. We show that, if $\Rightarrow 0 : A$, then $\Rightarrow 0 : \mathsf{t} \to A$.

$$\text{(Lem. 5.1)}\ \dfrac{\Rightarrow 0 : A}{\Rightarrow b : A}\ \text{SUB}(b/0)$$
$$\text{(Lem. 5.3)}\ \dfrac{\Rightarrow b : A}{R0ab, a : \mathsf{t} \Rightarrow b : A}\ \text{LW+LW}_{L} \ (\text{double line})$$
$$(a, b\ \text{fresh})\ \dfrac{R0ab, a : \mathsf{t} \Rightarrow b : A}{\Rightarrow 0 : \mathsf{t} \to A}\ R\to$$

Ad 7 Consider the rules of $\mathbf{G3rE^{t}}$. We proceed by root-first derivation. For $R\blacksquare$:

$$\text{(Lem 5.3)}\ \dfrac{\mathsf{P}x, Raxb, \Gamma \Rightarrow \Delta, b : A}{\mathsf{P}x, Raxb, R0xy, Rayb, y : \mathsf{t}, \Gamma \Rightarrow \Delta, b : A}\ \text{LW+LW}_{L} \ (\text{double line})$$
$$\dfrac{\mathsf{P}x, Raxb, R0xy, Rayb, y : \mathsf{t}, \Gamma \Rightarrow \Delta, b : A}{\mathsf{P}x, R0xy, Rayb, y : \mathsf{t}, \Gamma \Rightarrow \Delta, b : A}\ \text{T3}$$
$$(x\ \text{fresh})\ \dfrac{\mathsf{P}x, R0xy, Rayb, y : \mathsf{t}, \Gamma \Rightarrow \Delta, b : A}{Rayb, y : \mathsf{t}, \Gamma \Rightarrow \Delta, b : A}\ L\mathsf{t}$$
$$(y, b\ \text{fresh})\ \dfrac{Rayb, y : \mathsf{t}, \Gamma \Rightarrow \Delta, b : A}{\Gamma \Rightarrow \Delta, a : \mathsf{t} \to A}\ R\to$$
$$\begin{array}{c} \Gamma \Rightarrow \Delta, a : \mathsf{t} \to A \\ \hdashline \Gamma \Rightarrow \Delta, a : \blacksquare A \end{array}\ \mathsf{df}\blacksquare$$

For $L\blacksquare$:

$$
\dfrac{
\text{(Lem 5.3)}\ \dfrac{\dfrac{\dfrac{}{\mathsf{P}x, R0xx, \Gamma \Rightarrow \Delta, x:\mathsf{t}}\,Rt}{\mathsf{P}x, \Gamma \Rightarrow \Delta, x:\mathsf{t}}\,\mathrm{R1}}{a:\mathsf{t}\rightarrow A, \mathsf{P}x, Raxb, \Gamma \Rightarrow \Delta, x:\mathsf{t}}\,\text{LW+LW}_L
\qquad
\dfrac{b:A, a:\blacksquare A, \mathsf{P}x, Raxb, \Gamma \Rightarrow \Delta}{b:A, a:\mathsf{t}\rightarrow A, \mathsf{P}x, Raxb, \Gamma \Rightarrow \Delta}\,\mathsf{df}\blacksquare
}{
\dfrac{a:\mathsf{t}\rightarrow A, \mathsf{P}x, Raxb, \Gamma \Rightarrow \Delta}{a:\blacksquare A, \mathsf{P}x, Raxb, \Gamma \Rightarrow \Delta}\,\mathsf{df}\blacksquare
}\,L\rightarrow
$$

□

6 Soundness and Completeness

We conclude the paper by showing that our labelled calculi are sound and complete. Roughly, we prove that:

$\Gamma \Rightarrow \Delta$ is provable in **G3rX** $\Longleftrightarrow$ $\Gamma \Rightarrow \Delta$ is valid in every frame for **X**

where **X** is any of the relevant logics discussed in this paper.

Soundness ($\Longrightarrow$) The proof consists in showing that the rules of each labelled calculus presented in this article preserve validity over Routley-Meyer frames. Let $\mathscr{F}$ and $\mathscr{M}$ be, respectively, any of the Routley-Meyer frames and models discussed in this paper. We start by extending semantic notions to sequents as follows:

Definition 6.1 Let **s** be any labelled sequent of the form $\Gamma \Rightarrow \Delta$. An **s**-*interpretation in* $\mathscr{M}$ is a mapping $[\![\cdot]\!]$ from the labels in **s** to the set K of states in $\mathscr{M}$, such that (i) $0 = [\![0]\!]$ and (ii) if $Rabc$ (or $Sbca$, or $A\mathsf{G}a$, or $\mathsf{P}a$) is in Γ, then $R[\![a]\!][\![b]\!][\![c]\!]$ (or $S[\![b]\!][\![c]\!][\![a]\!]$, or $A\mathsf{G}[\![a]\!]$, or $P[\![a]\!]$) holds in $\mathscr{M}$. Now we can define:

$$\mathscr{M}, [\![\cdot]\!] \Vdash \mathbf{s} \text{ iff if for all } a:A \in \Gamma, \text{ we have } \mathscr{M}, [\![a]\!] \Vdash A, \text{ then there exists } b:B \in \Delta, \text{ such that } \mathscr{M}, [\![b]\!] \Vdash B.$$

Definition 6.2 A sequent **s** is *satisfied* in $\mathscr{M}$ if for all **s**-interpretations $[\![\cdot]\!]$ we have $\mathscr{M}, [\![\cdot]\!] \Vdash$ **s**. A sequent **s** is valid in a frame $\mathscr{F}$, if for all valuations v, the sequent **s** is satisfied in $\mathscr{M}$.

Finally, we can prove the requested soundness result:

Theorem 6.1 *If a sequent* **s** *is provable in* **G3rX**, *then it is valid in every Routley-Meyer frame for* **X**.

Proof We proceed by induction on the height of the derivation of **s**. We show that for each rule **r** of the form $\mathcal{P}_1, \ldots, \mathcal{P}_n/\mathcal{C}$, if the premises $\mathcal{P}_1, \ldots, \mathcal{P}_n$ are valid in all Routley-Meyer frames, then so is $\mathcal{C}$. It follows from a case analysis on **r**:

In. By way of contradiction, assume that $R0ab, a : p, \Gamma \Rightarrow \Delta, b : p$ is not valid in all Routley-Meyer frames. This means that there is a model $\mathscr{M}$ and an interpretation $[\![\cdot]\!]$, such that $\mathscr{M}, [\![\cdot]\!] \not\Vdash R0ab, a : p, \Gamma \Rightarrow \Delta, b : p$, i.e., $R[\![0]\!][\![a]\!][\![b]\!]$ and $\mathscr{M}, a \Vdash p$, but $\mathscr{M}, b \not\Vdash p$. However, this is not possible given heredity (see p. 270).

$R+^1$. By way of contradiction, assume that $Rbca, \Gamma \Rightarrow \Delta, b : A, c : B$ is valid in all Routley-Meyer frames, but $\Gamma \Rightarrow \Delta, a : A + B$ is not, where $b, c \notin \Gamma, \Delta$. The latter means that there is a model $\mathscr{M}$ and an interpretation $[\![\cdot]\!]$, such that $\mathscr{M}, [\![\cdot]\!] \not\Vdash \Gamma \Rightarrow \Delta, a : A + B$. In particular, we know that there are worlds b' and c' such that $Rb'c'[\![a]\!]$ and $\mathscr{M}, b' \not\Vdash A$, and $\mathscr{M}, c' \not\Vdash B$. Now we define an extension $[\![\cdot]\!]'$ of $[\![\cdot]\!]$ such that $[\![b]\!]' = b'$, $[\![c]\!]' = c'$ and $[\![\cdot]\!]' = [\![\cdot]\!]$. Then, $\mathscr{M}, [\![\cdot]\!]' \not\Vdash Rbca, \Gamma \Rightarrow \Delta, b : A, c : B$. Contradiction.

$L+^1$. By way of contradiction, assume that $b : A, Rbca, \Gamma \Rightarrow \Delta$ and $c : B, Rbca, \Gamma \Rightarrow \Delta$ are valid in all Routley-Meyer frames, but $Rbca, a : A + B, \Gamma \Rightarrow \Delta$ is not. The latter means that there is a model $\mathscr{M}$ and an interpretation $[\![\cdot]\!]$, such that $\mathscr{M}, [\![\cdot]\!] \not\Vdash Rbca, a : A + B, \Gamma \Rightarrow \Delta$, i.e., $R[\![b]\!][\![c]\!][\![a]\!]$ and $\mathscr{M}, a \Vdash A + B$, but $\mathscr{M}, x \not\Vdash D$ for all $x : D \in \Delta$. However, by the clause for $+$ (p. 271) we also have $\mathscr{M}, b \Vdash A$ or $\mathscr{M}, c \Vdash B$. Consequently, $\mathscr{M}, [\![\cdot]\!] \not\Vdash b : A, Rbca, \Gamma \Rightarrow \Delta$ or $\mathscr{M}, [\![\cdot]\!] \not\Vdash c : B, Rbca, \Gamma \Rightarrow \Delta$. Contradiction.

$R\bullet$. By way of contradiction, assume that $Sbca, \Gamma \Rightarrow \Delta, b : A$ and $Sbca, \Gamma \Rightarrow \Delta, c : B$ are valid in all Routley-Meyer frames, but $Sbca, \Gamma \Rightarrow \Delta, a : A \bullet B$ is not. The latter means that there is a model $\mathscr{M}$ and an interpretation $[\![\cdot]\!]$, such that $\mathscr{M}, [\![\cdot]\!] \not\Vdash Sbca, \Gamma \Rightarrow \Delta, a : A \bullet B$, i.e., $S[\![b]\!][\![c]\!][\![a]\!]$ but $\mathscr{M}, a \not\Vdash A \bullet B$. By the forcing clause for $\bullet$ (Sect. 3), then we have $\mathscr{M}, b \not\Vdash A$ or $\mathscr{M}, c \not\Vdash B$. As a consequence, we obtain $\mathscr{M}, [\![\cdot]\!] \not\Vdash Sbca, \Gamma \Rightarrow \Delta, b : A$ or $\mathscr{M}, [\![\cdot]\!] \not\Vdash Sbca, \Gamma \Rightarrow \Delta, c : B$. Contradiction.

$L\bullet$. By way of contradiction, assume that $Sbca, b : A, c : B, \Gamma \Rightarrow \Delta$ is valid, but $a : A \bullet B, \Gamma \Rightarrow \Delta$ is not, where $b, c \notin \Gamma, \Delta$. The latter means that there is a model $\mathscr{M}$ and an interpretation $[\![\cdot]\!]$, such that $\mathscr{M}, [\![\cdot]\!] \not\Vdash a : A \bullet B, \Gamma \Rightarrow \Delta$, i.e., $\mathscr{M}, a \not\Vdash A \bullet B$, but $\mathscr{M}, x \not\Vdash D$, for $x : D \in \Delta$. By the forcing clause for $\bullet$, then we have that there are worlds b', c' such that $Sb'c'[\![a]\!]$, $\mathscr{M}, b' \Vdash A$ and $\mathscr{M}, c' \Vdash B$. Let $[\![\cdot]\!]'$ be an extension of $[\![\cdot]\!]$ such that $[\![b]\!]' = b'$, $[\![c]\!]' = c'$ and $[\![\cdot]\!]' = [\![\cdot]\!]$. As a consequence, we obtain $\mathscr{M}, [\![\cdot]\!]' \not\Vdash Sbca, b : A, c : B, \Gamma \Rightarrow \Delta$. Contradiction.

An analogous reasoning can be carried out with rules $R\circ$ and $L\circ$ of Sect. 2. As a final example, we consider the rules for t of $\mathbf{G3rE^t}$ (Sect. 4):

$R\mathsf{t}$. By way of contradiction, assume that $\mathsf{P}b, R0ba, \Gamma \Rightarrow \Delta, a : \mathsf{t}$ is not valid in all Routley-Meyer frames. This means that there is a model $\mathscr{M}$ and an interpretation $[\![\cdot]\!]$, such that $\mathscr{M}, [\![\cdot]\!] \not\Vdash \mathsf{P}b, R0ba, \Gamma \Rightarrow \Delta, a : \mathsf{t}$, i.e., $\mathsf{P}[\![b]\!]$ and $R[\![0]\!][\![b]\!][\![a]\!]$, but $\mathscr{M}, a \not\Vdash \mathsf{t}$. However, by the forcing condition for t, from $\mathsf{P}[\![b]\!]$ and $R[\![0]\!][\![b]\!][\![a]\!]$, it follows that $\mathscr{M}, a \Vdash \mathsf{t}$. Contradiction.

$L\mathsf{t}$. By way of contradiction, assume that $\mathsf{P}b, R0ba, a : \mathsf{t}, \Gamma \Rightarrow \Delta$ is valid in all Routley-Meyer frames, but $a : \mathsf{t}, \Gamma \Rightarrow \Delta$ is not, where $b \notin \Gamma, \Delta$. The latter means that there is a model $\mathscr{M}$ and an interpretation $[\![\cdot]\!]$, such that $\mathscr{M}, [\![\cdot]\!] \not\Vdash a : \mathsf{t}, \Gamma \Rightarrow \Delta$. In particular, we know that there are worlds b' such that

$\mathsf{P}b'$, $R[\![0]\!]b'[\![a]\!]$ and $\mathscr{M}, b' \Vdash \mathsf{t}$, but $\mathscr{M}, x \not\Vdash D$, for all $x : D \in \Delta$. Now we define an extension $[\![\cdot]\!]'$ of $[\![\cdot]\!]$ such that $[\![b]\!]' = b'$ and $[\![\cdot]\!]' = [\![\cdot]\!]$. Then, $\mathscr{M}, [\![\cdot]\!]' \not\Vdash \mathsf{P}b, R0ba, a : \mathsf{t}, \Gamma \Rightarrow \Delta$. Contradiction.

The remaining cases, especially those concerning the rules for $\sim$ and $\rightarrow$, can be recovered from [13]. □

Completeness ($\Longleftarrow$) In what follows, we prove that, for every sequent **s**, the proof search either terminates in a proof or fails, and the failed proof tree is used to obtain a countermodel for **s**. Intuitively, to see whether A is derivable, we check if it is valid at the actual world $0 \in W$, i.e., $0 \Vdash A$. This, indeed, will correspond to have the sequent $\Rightarrow 0 : A$ in our calculus.[17] Again, let **X** be any of the relevant logics dealt with in this article.

Theorem 6.2 *Let* **s** *be a sequent. Then either* **s** *is derivable in* **G3rX** *or* **s** *has a countermodel with the frame properties peculiar for* **X**.

Proof We proceed with the construction of a derivation tree for $\Gamma \Rightarrow \Delta$ by applying the rules of **G3rX** root-first. (For the complete development of the argument see [13, Appendix 1]). If the reduction tree is finite, i.e., all leaves are initial sequents, we have a proof in **G3rX**. Assume that the derivation tree is infinite. By König's lemma, it has an infinite branch that is used to build the needed counterexample. Suppose that $\Gamma \Rightarrow \Delta \equiv \Gamma_0 \Rightarrow \Delta_0, \Gamma_1 \Rightarrow \Delta_1, \ldots, \Gamma_i \Rightarrow \Delta_i \ldots$ is one of such branches. Consider the sets $\mathbf{\Gamma} \equiv \bigcup \Gamma_i$ and $\mathbf{\Delta} \equiv \bigcup \Delta_i$, for $i \geq 0$. We now construct a countermodel, i.e. a model that makes all labelled formulas and relational atoms in $\mathbf{\Gamma}$ true and all labelled formulas in $\mathbf{\Delta}$ false. We show that, A is forced in the model at 0 if $0 : A$ is in $\mathbf{\Gamma}$ and A is not forced at 0 if $0 : A$ is in $\mathbf{\Delta}$. We will end up with a countermodel to the endsequent. We display some relevant cases:

- If p is atomic, the claim holds by definition of the model.
- If $0 : A + B$ is in $\mathbf{\Delta}$, at the successive step of the reduction tree we find that $Rab0$ and that both $a : A, b : B$ are in $\mathbf{\Delta}$. By the inductive hypothesis we obtain $Rab0$ and $a \not\Vdash A$ and $b \not\Vdash B$, that is, $0 \not\Vdash A + B$ in the model.
- If $0 : A + B$ is in $\mathbf{\Gamma}$, we consider all the relational atoms $Rab0$ that occur in $\mathbf{\Gamma}$. If there's no relational atom, the accessibility condition is vacuously satisfied and, therefore, $0 \Vdash A + B$ is in the model. For any occurrence of $Rba0$ in $\mathbf{\Gamma}$, by construction of the tree $a : A$ is in $\mathbf{\Gamma}$ or $b : A$ is in $\mathbf{\Gamma}$. By the inductive hypothesis $a \Vdash A$ or $b \Vdash B$, and since $Rab0$, we obtain $0 \Vdash A + B$ in the model.
- If $0 : A \bullet B$ is in $\mathbf{\Delta}$, we consider all the relational atoms $Sab0$ that occur in $\mathbf{\Gamma}$. If there's no relational atom, the accessibility condition is vacuously satisfied and, therefore, $0 \not\Vdash A \bullet B$ is in the model. For any occurrence of $Sab0$ in $\mathbf{\Gamma}$, by construction of the tree $a : A$ is in $\mathbf{\Delta}$ or $b : A$ is in $\mathbf{\Delta}$. By the inductive hypothesis $a \not\Vdash A$ or $b \not\Vdash B$, and since $Sab0$, we obtain $0 \not\Vdash A \bullet B$ in the model. (The case for $0 : A \circ B$ in $\mathbf{\Delta}$ is analogous.)

[17] As remarked in [13], this correspond to reflect, at the calculus level, the *actualistic* notion of validity employed in reduced Routley-Meyer models. See also [46, 276].

- If $0 : A \bullet B$ is in $\mathbf{\Gamma}$, at the successive step of the reduction tree we find that $Sab0$, $a : A$ and $b : B$ are in $\mathbf{\Gamma}$. By the inductive hypothesis we obtain $Sab0$, $a \Vdash A$ and $b \Vdash B$, that is, $0 \Vdash A \bullet B$ in the model. (The case for $0 : A \circ B$ in $\mathbf{\Gamma}$ is analogous.)
  As a final example, we consider the rules for t of Sect. 4:
- If $0 : \mathsf{t}$ is in $\mathbf{\Delta}$ and $\mathsf{P}b$, $R0b0$ is in $\mathbf{\Gamma}$, then by the inductive hypothesis we obtain $\mathsf{P}b$, $R0b0$, but $0 \nVdash \mathsf{t}$ in the model.
- If $0 : \mathsf{t}$ is in $\mathbf{\Gamma}$, then at the successive step of the reduction tree we find that $\mathsf{P}b$, $R0b0$ and $0 : \mathsf{t}$ are in $\mathbf{\Gamma}$. By the inductive hypothesis we obtain $\mathsf{P}b$, $R0b0$, and $0 \Vdash \mathsf{t}$ in the model.

The remaining cases can be checked in [13]. □

This proof directly implies our desired result:

Corollary 6.1 *If a sequent* **s** *is valid in every Routley-Meyer frame for* **X**, *then* **s** *is derivable in the system* **G3rX**.

7 Alternative Proof Systems and Final Remarks

Despite their complexities and subtleties, proof-theoretic studies of relevant logics have received extensive attention. In this paper, we have introduced labelled sequent calculi for a broad range of relevant logics including fusion, fission, and Ackermann's truth constant. Our approach involves incorporating semantic information from reduced Routley-Meyer models at the syntactic level. We have established soundness and completeness of these calculi. Additionally, we have demonstrated height-preserving invertibility of the rules, height-preserving admissibility of structural rules, and CUT-admissibility. In this final section, we will contextualize our current work within the broader landscape of proof systems for relevant logics by exploring various frameworks. Additionally, we will conclude this section by discussing both the advantages and limitations of the approach taken in this paper, while also suggesting potential directions for future research.

Firstly, by employing standard Gentzen sequents, as outlined, for instance, in [49, 50–51], one can start with a sequent calculus for linear propositional logic and incorporate the left and right contraction rules to define the *distribution-free* version of **R**. Indeed, in such a sequent system one cannot deduce the distribution axiom $A \wedge (B \vee C) \rightarrow (A \wedge B) \vee (A \wedge C)$, meaning it lacks the strength required to establish completeness with respect to the full Hilbert-style presentation of **R**. Other systems, such as $\mathbf{R}^{\rightarrow}$ and $\mathbf{E}^{\rightarrow}$, which represent the implicational fragments of **R** and **E** respectively, can be characterized using Gentzen's sequent calculus for intuitionistic logic **LJ** as base framework. Specifically, these systems are defined by introducing specific constraints, particularly on the right rule for $\rightarrow$, if $\mathbf{E}^{\rightarrow}$ is considered. This additional constraint requires that in each root-first application of the right rule for $\rightarrow$, every formula in the antecedent must take the form of $A \rightarrow B$ (see [25, 264]). In addition to the previously mentioned fragments of both **R** and **E**, Bimbó [9, 103–122] also explores ordinary sequent systems for the $\{\sim, \rightarrow\}$-fragments of **R** and **E**, their reformulation with the truth constant t, as well as systems for some of **R**'s and **E**'s modal expansions. However, the success of these strategies is mostly limited to the mentioned fragments of **R** and **E**, and to capture a broader spectrum of relevant logics, more finely-grained proof-theoretic techniques are needed.

To begin with, Anderson and Belnap's early work [5] provides well-known examples of proof systems, encompassing both sequent and natural deduction calculi for relevant logics. Concerning natural deduction systems, they proposed rules with labels, specifically natural numbers, to track the *actual uses* of assumptions in derivations. The rules, presented in Jaśkowski-Fitch style, for introducing and eliminating implications take the following form:

$$(\rightarrow E) \quad \left| \begin{array}{l} A_a \\ A \rightarrow B_b \\ B_{a \cup B} \end{array} \right. \qquad\qquad (\rightarrow I) \quad \left| \begin{array}{l} \left| \begin{array}{l} \underline{A_{\{k\}}} \\ \vdots \\ B_a \end{array} \right. \\ A \rightarrow B_{a-\{k\}} \end{array} \right.$$

Alternatively, these rules can be presented in a Gentzen-Prawitz style as follows:

$$\frac{A_a \qquad A \rightarrow B_b}{B_{a \cup B}}\ (\rightarrow E) \qquad\qquad \frac{\begin{array}{c} \overline{A_{\{k\}}} \\ \vdots \\ B_a \end{array}}{A \rightarrow B_{a-\{k\}}}\ (\rightarrow I)$$

Importantly, in $(I \rightarrow)$, we require that $k \in a$. Intuitively, these rule represent an easy way of handling with numerals to express the relevance conditions by ensuring that for a formula B to follow from a formula A, the assumption A must be *actually used* in proving the conclusion B. More precisely, the rules for $\rightarrow$ displayed above allow for: (1) the inference of $B_{a \cup b}$ from A_a and $A \rightarrow B_b$; and (2) the deduction of $A \rightarrow B_{a-\{k\}}$ from a proof of B_a based on the hypothesis $A_{\{k\}}$, provided that $k \in a$ (with $a, b, \ldots$ ranging over numerals). The resulting natural deduction systems, extensively studied in [5], have subsequently been adapted to encompass a wide range of quantified relevant logics in Brady's work [10]. Brady's research in [12] also focuses on normalization procedures, using the Fitch-style natural deduction system for propositional **DW** as a case study. Additionally, Standefer in [64] devised Fitch-style natural deduction systems for propositional **E**, while Anderson, in his earlier publication [3], introduced Fitch-style calculi also for quantified **E**. Interestingly, Standefer's work in [66] also explores translations between linear and tree natural deduction systems for relevant logics.

As for sequent-style frameworks, it's worth mentioning that besides the use of ordinary sequents, Anderson and Belnap in [5, 57–69] introduced the systems they labelled *merge calculi* (see also [9, 135–137]). The idea is that in the formulation of rules, $\Gamma, \Delta, \ldots$ represent sequences of formulas subject to a special operation called *merge*, which combines sequences of formulas. Specifically, the elements of the sequence represented by Γ may be distributed in the "interstices" of Δ as long as the elements of Γ retain their internal ordering, formally denoted as $\mu(\Gamma, \Delta)$.[18] For example, if $\Gamma = A_1, A_2, A_3$ and $\Delta = B_1, B_2, B_3$, then the sequence $\Pi = A_1, B_1, B_2, A_2, A_3, B_3$ is one of the merges of Γ and Δ, that is $\Pi \in \mu(\Gamma, \Delta)$. These systems were used to capture relevant logics $\mathbf{T}^{\rightarrow}$, $\mathbf{E}^{\rightarrow}$, and $\mathbf{R}^{\rightarrow}$. The following rules for $\rightarrow$ are used:

$$\frac{\Gamma, \Delta \Rightarrow A \qquad \Pi, B, \Sigma \Rightarrow C}{\mu(\Gamma, \Pi, A \rightarrow B), \Delta, \Sigma \Rightarrow C}\ L\rightarrow_\mu \qquad\qquad \frac{\Gamma, A \Rightarrow B}{\Gamma \Rightarrow A \rightarrow B}\ R\rightarrow_\mu$$

In addition to the single-conclusion restriction on the right-hand side of sequents, the following restrictions must be imposed on $L \rightarrow_\mu$: for $\mathbf{T}^{\rightarrow}$, we impose that $\Delta \neq \varnothing$; for $\mathbf{E}^{\rightarrow}$, it is required that either $\Delta \neq \varnothing$ or, if $\Delta = \varnothing$, then A must be of the form $B \rightarrow C$, for some B, C. For $\mathbf{R}^{\rightarrow}$, no restriction on $L \rightarrow_\mu$ is needed.

To explore more complex languages, including those with conjunction and disjunction connectives, and hence to reintroduce the distribution axiom at the level of Hilbert-style systems, it becomes necessary to introduce those frameworks known as *Dunn-Mints* (or *consecution*) *calculi*. These systems were independently developed by Mints in his 1972 paper [39] and by Dunn in his 1973 contribution [14]. In this framework, there are two distinct operations for composing data on the left-hand side of sequents. In addition to the standard comma ",", which is used in Gentzen's sequents to represent extensional conjunction at the meta-level, we introduce the semicolon ";" to denote intensional conjunction (see [49, 127–130], [9, pp. 123–129], [25, 267–268]). Usually, the elements that in consecution calculi are grouped together by the separators "," and ";" are referred to as *structures* and are denoted by X, Y, etc. Formally, (X, Y) and $(X; Y)$ represent extensional and intensional structures, respectively, and the notation $X[Y]$ is used to denote that Y is a substructure of X. For example, the consecution calculus for the $\{\wedge, \vee, \rightarrow, \circ, \mathsf{t}\}$-fragment of $\mathbf{R}$ comprises the following rules for $\rightarrow$, $\circ$ and t employing ";" rather than ",":

$$\frac{X \Rightarrow A \qquad Y[B] \Rightarrow C}{Y[X; A \rightarrow B] \Rightarrow C}\ L\rightarrow; \qquad\qquad \frac{X; A \Rightarrow B}{X \Rightarrow A \rightarrow B}\ R\rightarrow;$$

$$\frac{X[A; B] \Rightarrow C}{X[A \circ B] \Rightarrow C}\ L\circ; \qquad \frac{X \Rightarrow A \qquad Y \Rightarrow B}{X; Y \Rightarrow A \circ B}\ R\circ; \qquad \frac{X[A] \Rightarrow C}{X[A; \mathsf{t}] \Rightarrow C}\ L\mathsf{t};$$

[18] "Metaphorically speaking, a merge of two sequences is like closing a zipper with unevenly spaced juts: the juts on each strip retain their place with respect to each other, but a group of juts on one strip can jostle between a pair of juts on the other strip" [9, 135].

Furthermore, the precise distinction between ";" and "," enables one to preserve a version of the rule of weakening while simultaneously preventing the derivation of irrelevant formulas. Such a weakening rule takes the following form:

$$\frac{X[Y] \Rightarrow C}{X[Y,Z] \Rightarrow C}\ LW,$$

where $Y \neq \varnothing$, and any root-first application of the rule can be performed only on extensional structures.

Consecution calculi represent a cornerstone in the development of proof theory for relevant logics. However, their application becomes more complex when negation is among the connectives. A successful approach to dealing with negation is provided by the display calculus, a framework first discussed by Belnap in [8]. In this framework, not only are ";" and "," introduced, but a variety of *structural connectives* is also employed, making the structural reasoning used in derivations *visually* explicit. The fundamental feature of display calculi is that given a formula and a sequent containing it, it is always possible to transform that sequent into an equivalent one where the given formula is either the entire antecedent or the entire succedent (such a feature is called *display property*, hence the name display calculus). Structures denoted by X, Y, etc., are used, multiple-conclusion sequents are allowed, and as a consequence structures are combined not only on the left-hand side but also on the right-hand side of sequents. Belnap introduced display logic as a proof-theoretic tool for the investigation of a wide range of logics, not limited to relevant logics alone. His most significant contribution was the proof of a general CUT-elimination theorem applicable to all display calculi whose rules adhere to a small number of easily verified formal conditions (see [8, 389–392], [9, 199–208]). This result demonstrates that display logic is a powerful framework for the unified treatment of the proof theory of various logics. For example, we can accommodate $\sim$ by introducing an additional structural connective denoted by "$*$" and provide a calculus for the full language of **R**. The rules, taken from Restall's emendation of Belnap's original calculus, can be defined as follows (see [57], [49, 130–135]):

$$\frac{X \Rightarrow \varnothing}{X \Rightarrow \mathsf{f}}\ R\mathsf{f};\qquad \frac{\varnothing \Rightarrow X}{\mathsf{t} \Rightarrow X}\ L\mathsf{t};\qquad \frac{X \Rightarrow *A}{X \Rightarrow {\sim}A}\ R{\sim}*\qquad \frac{*A \Rightarrow X}{{\sim}A \Rightarrow X}\ L{\sim}*$$

$$\frac{X;A \Rightarrow B}{X \Rightarrow A \to B}\ R\to;\qquad \frac{X \Rightarrow A \quad B \Rightarrow Y}{A \to B \Rightarrow *X;Y}\ L\to;*\qquad \frac{X \Rightarrow A \quad Y \Rightarrow B}{X;Y \Rightarrow A \circ B}\ R\circ;$$

$$\frac{A;B \Rightarrow X}{A \circ B \Rightarrow X}\ L\circ;\qquad \frac{X \Rightarrow A;B}{X \Rightarrow A+B}\ R+;\qquad \frac{A \Rightarrow X \quad B \Rightarrow Y}{A+B \Rightarrow X;Y}\ L+;$$

Similar to consecution calculi, the rules governing $\wedge$ and $\vee$ rely on the use of the extensional separator "," instead of ";". Moreover, just as the comma on the left-(resp. right-)hand side of sequents denotes extensional conjunction (resp. disjunction), the semicolon on the left (resp. right) of $\Rightarrow$ denotes intensional conjunction (resp. disjunction). The empty structure, $\varnothing$, is employed to introduce t (resp. f) on the left (resp. right) of $\Rightarrow$. As illustrated by the rule $L \rightarrow; *$, different structural connectives can be combined. For further information on display calculi one can refer to Wansing's book [71].

Another approach, aimed at preserving the distinctions between extensional and intensional combinations of premises, was introduced by Read [56]. This approach is sometimes referred to as the *Scottish Plan* in relevant logics (see [56, 131–143]). The calculus consists of sequents of the following form: "$X : A$", where X is referred to as a *bunch* of premises, and A is a formula. In this setting, a bunch of premises X can be conjoined both extensionally and intensionally, and the formula A is the conclusion inferred based on those premises. Bunches are represented using two pairing operations symbolized by the comma and the semicolon: the comma denotes extensional conjunction, while the semicolon denotes intensional conjunction [56, 58]. Within this framework, Read introduces two general schemas, one for introduction and one for elimination rules. Based on these schemas, he defines several bunch proof systems for a variety of relevant logics, with particular attention to **DW**, **E**, and especially **R** [56, pp. 40–41, 51–77].

All the approaches we have discussed so far modify the sequent calculus in a manner that enhances its expressive power, enabling it to capture more nuanced distinctions in the intensional and extensional combination of data. These distinctions are not only important for the study of relevant logics but also extend to several non-classical systems. Another noteworthy approach involves the introduction of *hypersequent* calculi, as presented in Avron's work [7]. In essence, a hypersequent takes the form $\Gamma_1 \Rightarrow \Delta_1 \mid \cdots \mid \Gamma_n \Rightarrow \Delta_n$, where each $\Gamma \Rightarrow \Delta$ is an ordinary sequent, with Γ and Δ representing sequences of formulas. The hypersequent is essentially a sequence of sequents, where the separator "|" is interpreted as a meta-level (extensional) disjunction. Avron initially introduced the hypersequent formulation for **RM**, and, for the sake of simplicity, we present it in a simplified manner, relying on the hypersequent calculus for linear logic as a foundational framework (see [25, 266]). The hypersequent structure is used in formulating the additional rules necessary for transitioning from the linear logic base framework to **RM**. These additional rules are presented as follows:

$$\frac{G \mid \Gamma, \Pi \Rightarrow \Delta, \Sigma}{G \mid \Gamma \Rightarrow \Delta \mid \Pi \Rightarrow \Sigma}\ \textit{Split} \qquad \frac{G \mid \Gamma \Rightarrow \Delta \qquad H \mid \Pi \Rightarrow \Sigma}{G \mid H \mid \Gamma, \Pi \Rightarrow \Delta, \Sigma}\ \textit{Mingle}$$

where, G and H represent side hypersequents. This hypersequent formulation of **RM** is deductively equivalent with respect to the axiomatization of **RM**, i.e., **R** with the mingle axiom, $A \rightarrow (A \rightarrow A)$. That is, the hypersequent calculus allows the derivation of all axioms of **RM** and, conversely, that each hypersequent, suitably translated into an object language formula, is derivable in the axiomatic system for **RM**. The translation, denoted by τ, can be displayed as follows: $\tau(\Gamma_1 \Rightarrow \Delta_1 \mid \cdots \mid \Gamma_n \Rightarrow \Delta_n) = (\circ\Gamma_1 \rightarrow +\Delta_1) \vee \cdots \vee (\circ\Gamma_n \rightarrow +\Delta_n)$, where $\circ\Gamma$ and $+\Delta$ denote the intensional conjunction and disjunction of formulas in Γ and Δ, respectively.

On a different, albeit related, note, besides the natural deduction systems labelling formulas with numerals to express relevance between premises and conclusions in derivations, a variety of labelling techniques have been used in the development of the proof theory for relevant logics. For example, important instances of labelled calculi for relevant logics include the contribution made by E. Orlowska. In [48], she introduced a methodology aimed at constructing natural deduction-style systems for various propositional relevant logics. This method shares some resemblance with Negri's approach, as it involves characterizing relevant logics using ternary relational frames and subsequently formulating deduction rules based on these relational frames. Similarly, Priest [54, pp. 188–220, 535–563] introduced labelled tableau systems for various propositional and quantified relevant logics, referring to the unreduced Routley-Meyer relational semantics.

An important aspect of the development of labelled proof theory has encompassed the study of semilattice relevant logics (referenced in works such as [23, 27, 28, 69, 74]). The semantics incorporated into the proof-theoretic structure does not involve three different states a, b, c related by a ternary relation of the form $Rabc$; instead, rather than introducing a third state c, the approach considers the union of a and b. Urquhart, intuitively, motivates his semantic choice by arguing that for a piece of information b to determine a formula $A \to B$ is to say that, whenever we can conclude A based on a piece of information a, we can also conclude B based on both pieces of information a and b taken together, formally $a \cup b$ [69, 160]. The resulting rules for $\to$ can be expressed as follows:

$$\frac{\Gamma, A_a \Rightarrow \Delta, A \to B_b, B_{b \cup a}}{\Gamma \Rightarrow \Delta, A \to B_b} R \to' \qquad \frac{\Gamma, A \to B_b \Rightarrow \Delta, A_a \quad \Gamma, A \to B_b, B_{b \cup a} \Rightarrow \Delta}{\Gamma, A \to B_b \Rightarrow \Delta} L \to'$$

Another important example of labelled deductive systems for relevant logics was presented by L. Viganò in [70]. As noted in [42, 125], the main difference between our approach and Viganò's lies in the construction of sequent calculi with explicit structural rules, along with the use of single-conclusion rules for the accessibility relation (Horn clauses) confined within Harrop theories, i.e., theories lacking disjunctions in the right-hand side of axioms (see also [70, 61–70]). In addition to sequent systems, Viganò also introduced labelled natural deduction calculi. It's worth noting that the proof systems presented in this paper can also be represented as natural deduction systems, with significant differences between Orlowska's and Viganò's approaches. To elaborate, Negri and von Plato [46, 268–277] describe a methodology and its corresponding philosophical advantages for establishing a set of introduction and elimination rules, which mirror the left and right rules of a certain labelled sequent system. This method is based on starting with the semantic clause of the particular connective (such as $\sim$, $\to$, $\circ$, $+$, $\bullet$, t) and following the instructions outlined in Negri and von Plato's adaptation of the *inversion principle*, originally formulated by Gentzen and Prawitz. In essence, Negri and von Plato's strategy can be summarized as follows: guided by their inversion principle, which stipulates that *whatever can be deduced from the direct reasons for asserting a proposition must also be deducible from that proposition*, one examines the introduction rule of a given connective and deduces the corresponding well-formed labelled elimination rule.[19]

In line with the mentioned labelled techniques, our approach has enabled us to *modularly capture* various systems within the same family of logics, alongside the already formal results, through the incorporation of simple additions to a base system. In our case, instead of altering the logical rules of a given labelled system, it amounts to making adjustments within the group of relational rules. Additionally, it is essential to emphasize that many of the techniques outlined in this section often enable us to characterize only small fragments of relatively strong relevant logics. However, labelled techniques broaden our horizons, enabling the consideration of more extensive sets of connectives. This expanded expressive power stems from the incorporation of labels into the syntax. Such an observation is of utmost importance, as labelled methods facilitate the development of calculi for weaker systems of relevant logics possibly including a variety of both intensional and extensional connectives. Achieving such results in non-labelled frameworks would be considerably more challenging. Secondly, as noted on several occasions, labelled calculi constructed using the methodology we employed enable the straightforward extrapolation of a countermodel for a sequent from a failed proof-search procedure (see, e.g., [43, 44]). However, in relation to specific limitations of labelled sequents, we present two important remarks.

Firstly, a common weakness associated with labelled calculi is the absence of a *formula interpretation*, i.e., a formal procedure for translating sequents into object-language formulas. One approach to address this challenge stems from the recognition that the proliferation of non-classical logics has spurred the exploration of various generalizations of sequent systems, prompting inquiries into their interrelationships. Therefore, an avenue to mitigate the necessity for independent proofs could involve establishing translations between labelled calculi, such as those introduced in this paper, and other characterizations. To obtain alternative versions of the formula interpretation, one might investigate translations into non-labelled frameworks, such as the previously mentioned hypersequents (e.g., [7]) and display sequents (e.g., [57]).

Secondly, certain concerns arise when investigating the *subformula property* and its implications within labelled frameworks. Without delving into the intricacies of (un)decidability issues, let us outline potential avenues for future research in this area.[20] CUT-elimination plays a crucial role, yielding the subformula property in sequent systems, ensuring that all formulas in a derivation are subformulas of those in the endsequent. However, labelled sequent calculi lack a full subformula property due to geometrical rules, wherein relational atoms vanish from premises

[19] For the sake of brevity, I won't delve into the details of this approach. However, one can find the details in [46]. See also [45, 17–23].

[20] It is important to acknowledge that relevant logics encounter challenges in establishing decidability, with many of them being undecidable.

to conclusions. Nevertheless, they exhibit a *weaker* version of the property, where all formulas in a derivation are either subformulas of formulas in the endsequent or relational atoms such as $Rabc$ (see, e.g., [16, 41]). Yet, this property alone is insufficient to prove syntactic decidability. To address this, one needs to avoid non-terminating proof-search procedures. Roughly, this involves putting bounds on the number of eigenvariables in derivations, along with additional bounds on the number of applications of rules with endformulas in the premises.

Lastly, a concluding observation. In our paper, we have extensively explored labelled rules that encompass various intensional operators. However, it is worth mentioning that relevant logics might include additional connectives, such as modal operators, or quantifiers as well.[21] Despite these possibilities, we have made a deliberate choice to focus solely on the specific set of connectives we considered, leaving the exploration of broader sets of connectives for future research. Nevertheless, it is essential to note that the methodology we have employed thus far can, in principle, be extended to accommodate the mentioned extensions.

References

1. Ackermann, W. (1956). Begründung einer strengen implikation. *The Journal of Symbolic Logic, 21*(2), 113–128.
2. Aloni, M. (2023). Disjunction. In E. N. Zalta & U. Nodelman (Eds.), *The Stanford encyclopedia of philosophy*, Spring 2023 edition. Stanford: Metaphysics Research Lab, Stanford University.
3. Anderson, A. R. (1960). Completeness theorems for the systems **E** of entailment and **EQ** of entailment with quantification. *Zeitschrift für mathematische Logik und Grundlagen der Mathematik, 6*(7–14), 201–216.
4. Anderson, A. R., & Belnap, N. D. (1959). Modalities in Ackermann's "rigorous implication". *The Journal of Symbolic Logic, 24*(2), 107–111.
5. Anderson, A. R., & Belnap, N. D. (1975). *Entailment. The logic of relevance and necessity* (Vol. I). Princeton: Princeton University Press.
6. Angell, R. B. (1962). A propositional logic with subjunctive conditionals. *The Journal of Symbolic Logic, 27*(3), 327–343.
7. Avron, A. (1987). A constructive analysis of **RM**. *The Journal of Symbolic Logic, 52*(4), 939–951.
8. Belnap, N. D. (1982). Display logic. *Journal of Philosophical Logic, 11*, 375–417.
9. Bimbó, K. (2014). *Proof theory: Sequent calculi and related formalisms*. Boca Raton: CRC Press.
10. Brady, R. T. (1984). Natural deduction systems for some quantified relevant logics. *Logique et Analyse, 27*(108), 355–377.
11. Brady, R. T. (1989). A Routley-Meyer affixing style semantics for logics containing Aristotle's Thesis. *Studia Logica, 48*, 235–241.
12. Brady, R. T. (2006). Normalized natural deduction systems for some relevant logics I: The logic **DW**. *The Journal of Symbolic Logic, 71*(1), 35–66.

[21] Recently, there has been notable progress in the field of quantified and modal relevant logics. One can explore e.g., [18, 65, 67, 68].

13. De Martin Polo, F. (2023). Modular labelled calculi for relevant logics. *Australasian Journal of Logic, 20*(1), 47–87.
14. Dunn, J. M. (1973). A 'Gentzen system' for positive relevant implication. *The Journal of Symbolic Logic, 38*, 356–357 (abstract).
15. Dunn, J. M., & Restall, G. (2002). Relevance logic. In D. Gabbay & F. Guenthner (Eds.), *Handbook of philosophical logic* (pp. 1–128). Dordrecht: Kluwer.
16. Dyckhoff, R., & Negri, S. (2012). Proof analysis in intermediate logics. *Archive for Mathematical Logic, 51*(1), 71–92.
17. Estrada-González, L. (2020). The Bochum Plan and the foundations of contra-classical logics. *CLE e-Prints, 19*(1), 1–22.
18. Ferenz, N. (2023). Quantified modal relevant logics. *The Review of Symbolic Logic, 16*(1), 210–240.
19. Francez, N. (2019). Relevant connexive logic. *Logic and Logical Philosophy, 28*(3), 409–425.
20. Francez, N. (2020). A poly-connexive logic. *Logic and Logical Philosophy, 29*(1), 143–157.
21. Francez, N. (2021). *A view of connexive logic*. London: College Publications.
22. Giambrone, S. (1992). Real reduced models for relevant logics without WI. *Notre Dame Journal of Formal Logic, 33*(3), 442–449.
23. Giambrone, S., & Urquhart, A. (1987). Proof theories for semilattice logics. *Zeitschrift Für Mathematische Logik Und Grundlagen Der Mathematik, 33*(5), 433–439.
24. Hóu, Z., Goré, R., & Tiu, A. (2018). A labelled sequent calculus for BBI: Proof theory and proof search. *Journal of Logic and Computation, 28*(4), 809–872.
25. Indrzejczak, A. (2021). *Sequents and trees. An introduction to the theory and applications of propositional sequent calculi*. Cham: Birkhäuser.
26. Kamide, N., & Wansing, H. (2015). *Proof theory of N4-paraconsistent logics* (Vol. 54). London: College Publications.
27. Kashima, R. (2001). Proof theoretical studies on semilattice relevant logics. *Research Reports on Mathematical and Computing Sciences, TIT*. Published online: Link: https://citeseerx.ist.psu.edu/document?repid=rep1&type=pdf&doi=40a1ddb3e829188c787012139c4004ab28eee360
28. Kashima, R. (2003). On semilattice relevant logics. *Mathematical Logic Quarterly: Mathematical Logic Quarterly, 49*(4), 401–414.
29. Kurokawa, H., & Negri, S. (2020). Relevant logics: From semantics to proof systems. In A. Rezus (Ed.), *Contemporary logic and computing* (pp. 259–306). London: College Publications.
30. Lenzen, W. (2020). A critical examination of the historical origins of connexive logic. *History and Philosophy of Logic, 41*(1), 16–35.
31. Lenzen, W. (2022). Rewriting the history of connexive logic. *Journal of Philosophical Logic, 51*(3), 525–553.
32. Lukasiewicz, J. (1957). *Aristotle's syllogistic from the standpoint of modern formal logic*. Oxford: Clarendon Press.
33. Mares, E. (2006). Relevance logic. In: D. Jacquette (Ed.), *A companion to philosophical logic* (pp. 607–627). London: Blackwell.
34. Mares, E., & Standefer, S. (2017). The relevant logic **E** and some close neighbours. A reinterpretation. *IfCoLog Journal of Logics and Their Applications, 4*(3), 695–730.
35. McCall, S. (1966). Connexive implication. *The Journal of Symbolic Logic, 31*(3), 415–433.
36. McCall, S. (1967). Connexive implication and the syllogism. *Mind, 76*(303), 346–356.
37. Méndez, J. M. (2009). A Routley-Meyer semantics for Ackermann's logics of "strenge Implikation". *Logic and Logical Philosophy, 18*(3–4), 191–219.
38. Meyer, R. K. (1970). **E** and **S4**. *Notre Dame Journal of Formal Logic, 11*(2), 181–199.
39. Mints, G. E. (1976). Cut-elimination theorem for relevant logics. *Journal of Soviet Mathematics, 6*(4), 422–428 (Translated from the original Russian article printed in "Notes of the scientific seminars" of the Leningrad Department of the Mathematical Institute V. A. Steklov of the USSR Academy of Sciences, Vol. 32, 1972).

40. Mortensen, C. (1984). Aristotle's thesis in consistent and inconsistent logics. *Studia Logica, 43*, 107–116.
41. Negri, S. (2005). Proof analysis in modal logic. *Journal of Philosophical Logic, 34*(5), 507–544.
42. Negri, S. (2007). Proof analysis in non-classical logics. In C. Dimitracopoulos, L. Newelski, D. Normann, & J. Steel (Eds.), *ASL lecture notes in logic* (Vol. 28, pp. 107–128). Cambridge: Cambridge University Press.
43. Negri, S. (2009). Kripke completeness revisited. In G. Primiero (Ed.), *Acts of knowledge: History, philosophy and logic: Essays dedicated to Göran Sundholm* (pp. 247–282). London: College Publications.
44. Negri, S. (2014). Proofs and countermodels in non-classical logics. *Logica Universalis, 8*(1), 25–60.
45. Negri, S., & von Plato, J. (2011). *Proof analysis: A contribution to Hilbert's last problem.* Cambridge: Cambridge University Press.
46. Negri, S., & von Plato, J. (2015). Meaning in use. In H. Wansing (Ed.), *Dag Prawitz on proofs and meaning*. Outstanding contributions to logic (pp. 239–257). Berlin: Springer.
47. Omori, H., & Wansing, H. (2019). Connexive logics. An overview and current trends. *Logic and Logical Philosophy, 28*(3), 371–387.
48. Orlowska, E. (1992). Relational proof system for relevant logics. *The Journal of Symbolic Logic, 57*(4), 1425–1440.
49. Paoli, F. (2002). *Substructural logics: A primer*. Berlin: Springer.
50. Pizzi, C. (1991). Decision procedures for logics of consequential implication. *Notre Dame Journal of Formal Logic, 32*(4), 618–636.
51. Pizzi, C. (1993). Consequential implication: A correction. *Notre Dame Journal of Formal Logic, 34*(4), 621–624.
52. Pizzi, C. (2005). Aristotle's thesis between paraconsistency and modalization. *Journal of Applied Logic, 3*(1), 119–131.
53. Pizzi, C. (2008). Aristotle's cubes and consequential implication. *Logica Universalis, 2*, 143–153.
54. Priest, G. (2008). *An introduction to non-classical logic: From if to is*. Cambridge: Cambridge University Press.
55. Read, S. (1982). Disjunction. *Journal of Semantics, 1*(3–4), 275–285.
56. Read, S. (1988). *Relevant logic. A philosophical examination of inference*. Oxford: Basil Blackwell Publisher (Corrected edition, 2012).
57. Restall, G. (1998). Displaying and deciding substructural logics 1: Logics with contraposition. *Journal of Philosophical Logic, 27*(2), 179–216.
58. Restall, G. (2000). *An introduction to substructural logics*. London: Routledge.
59. Routley, R. (1978). Semantics for connexive logics I. *Studia Logica, 37*(4), 393–412.
60. Routley, R., & Meyer, R. K. (1973). The semantics of entailment. In H. Leblanc (Ed.), *Truth, syntax, and modality (Proceedings of The Temple University conference on alternative semantics)* (pp. 199–243). Amsterdam: North-Holland Publishing Company.
61. Routley, R., Plumwood, V., Meyer, R. K., & Brady, R. T. (1982). *Relevant logics and their rivals* (Vol. 1). Atascadero: Ridgeview Publishing Company.
62. Ruge, F. (2023). Connexivity in Aristotle's Logic. *History and Philosophy of Logic, 44*(4), 353–372.
63. Slaney, J. K. (1987). Reduced models for relevant logics without WI. *Notre Dame Journal of Formal Logic, 28*(3), 395–407.
64. Standefer, S. (2018). Trees for **E**. *Logic Journal of the IGPL, 26*(3), 300–315.
65. Standefer, S. (2020). Actual issues for relevant logics. *Ergo. An Open Access Journal of Philosophy, 7*.
66. Standefer, S. (2021). Translations between linear and tree natural deduction systems for relevant logics. *The Review of Symbolic Logic, 14*(2), 285–306.
67. Standefer, S. (2023). Varieties of relevant **S5**. *Logic and Logical Philosophy, 32*(1), 53–80.

68. Tedder, A., & Ferenz, N. (2022). Neighbourhood semantics for quantified relevant logics. *Journal of Philosophical Logic, 51*(3), 457–484.
69. Urquhart, A. (1972). Semantics for relevant logics. *The Journal of Symbolic Logic, 37*(1), 159–169.
70. Viganò, L. (2000). *Labelled non-classical logics*. Berlin: Springer.
71. Wansing, H. (1998). *Displaying modal logic*. Dordrecht: Springer.
72. Wansing, H. (2005). Connexive modal logic. In R. Schmidt, I. Pratt-Hartmann, M. Reynolds, & H. Wansing (Eds.), *Advances in modal logic* (Vol. 5, pp. 387–399). London: College Publications.
73. Wansing, H. (2023). Connexive logic. In E. N. Zalta & U. Nodelman (Eds.), *The Stanford encyclopedia of philosophy* (summer 2023 edition). Stanford: Metaphysics Research Lab, Stanford University.
74. Weiss, Y. (2021). A characteristic frame for positive intuitionistic and relevance logic. *Studia Logica, 109*(4), 687–699.
75. Weiss, Y. (2022). Did Aristotle endorse Aristotle's thesis? A case study in Aristotle's metalogic. *Notre Dame Journal of Formal Logic, 63*(4), 551–579.

Entailment Generalized

Tore Fjetland Øgaard

Abstract Two Hilbert calculi in the use-restricted relevance tradition of Anderson and Belnap are presented. One yields Anderson and Belnap's conjunctive version of the so-called "entailment theorem," whereas the other yields the implication-iterated version. In contrast to previous such use-restricted calculi, those presented in this paper apply to arbitrary strengthenings of the weak contractionless logic **B** without need for modifying the use-conditions. As such they help to identify the core idea of the "use-account" of relevance.

1 Introduction

Anderson and Belnap claimed that the standard account of logical consequence fails to take account of a substantial notion of premise use and so devised a use-restricted Hilbert calculus in order to better account for this notion. The first in-depth investigation into relevant logics came with the first volume of Anderson and Belnap's *Entailment: The Logic of Relevance and Necessity*. In it, they state the so-called *Entailment theorem* as follows: "If there is a proof in **E** that $A_1, \ldots, A_n$ entail(s) B, then $(A_1 \& \ldots \& A_n) \rightarrow B$ is provable in **E**" [p. 278 3]. The *entail(s)* on the left-hand side denotes a use-restricted notion of Hilbert derivability, whereas "provable in" on the right-hand side denotes standard Hilbert derivability. The converse version of the entailment theorem also holds, and so the theorem may be viewed as a definition of Anderson and Belnap's notion of entailment in terms of a logically true conditional statement.

I would like to extend my heartfelt thanks to the referees for their engaging comments and helpful suggestions; this paper is significantly better because of these. I would also like to thank the audience at the conference *New Directions in Relevant Logic* for helpful comments and suggestion.

T. F. Øgaard (✉)
Department of Philosophy, University of Bergen, Bergen, Norway
e-mail: Tore.Ogaard@uib.no

I. Sedlár et al. (eds.), *New Directions in Relevant Logic*, Trends in Logic 63,
https://doi.org/10.1007/978-3-031-69940-5_13

Anderson and Belnap's notion of entailment only applies to logics with only adjunction and modus ponens as primitive rules. It was shown in [16] that the Entailment theorem, using Anderson and Belnap's notion of entailment, not only holds for Anderson and Belnap's favorite logic **E**, but in fact that it also applies to logics arbitrarily stronger, provided, that is, that the logic in question has no more primitive rules than adjunction and modus ponens. [7] generalized Anderson and Belnap's notion of entailment so as to apply to a range of logics weaker than **E**. However, only logics extending the relevant logic **B** augmented *with* contraction—**B**-extensions which validate the rule allowing one to infer $A \to B$ from $A \to (A \to B)$, or, equivalently, the *pseudo modus ponens* axiom $(A \wedge (A \to B)) \to B$—are covered by Brady's result. This paper fills this lacuna by showing that the notion of entailment can be improved upon so as to yield an entailment theorem for both contractive and non-contractive logics extending a weak sublogic of **B**. Furthermore, whereas Brady's notion of entailment varies across logics due to the variety of use-conditions for the primitive rules, the notion of entailment used in this paper is the same for all logics.

Anderson and Belnap's entailment theorem yields that a non-empty set of formulas Δ entails B if and only if $A \to B$ is a logical theorem of the corresponding standard Hilbert calculus, where A, then, is some conjunction of formulas from Δ. I will in the following therefore call their notion of entailment, as well as the generalizations of it, *C-entailment*, 'C' being mnemonic for *conjunctive*. There is a related notion of entailment that will be called *I-entailment* in the following—'I' being mnemonic for *implicative*—with a corresponding I-entailment theorem. Whereas C-entailment is a fully structural (and Tarskian) consequence relation, I-entailment is defined as a consequence relation taking sequences and a single formula as input. For many logics, this yields a substructural consequence relation. This is a topic that will be commented upon in Sect. 5. The I-entailment theorem that the notion of I-entailment begets is that for any non-empty sequence $\langle A_1 \ldots, A_n\rangle$ of formulas, $\langle A_1 \ldots, A_n\rangle$ I-entails B if and only if $A_1 \to (\ldots \to (A_n \to B)\ldots)$ is a logical truth of the corresponding standard Hilbert calculus. The related claim that $\langle A_1 \ldots, A_n, A_{n+1}\rangle$ I-entails B if and only if $\langle A_1 \ldots, A_n\rangle$ I-entails $A_{n+1} \to B$, will also be shown to hold for arbitrary strengthenings of the weak relevant logic **B**.

This kind of result in the case of the implication fragment of the relevant logic **R** goes back to [9] and [23], and was generalized so as to apply to Anderson and Belnap's three favorite logics—**T**, **E**, and **R**—in [10]. Kron's result was further expanded upon in [7] so as to apply to a range of logics, including the weak contractionless logic **B**. Both Kron's and Brady's results hold by varying the use-condition of modus ponens. The I-entailment theorems, then, hold due to *different* use conditions. It is shown in this paper, however, that it is possible to devise a use-calculus which works for all logics extending **B**, thus showing that the notion of use may be held constant without giving up the I-entailment theorem.

The paper is structured as follows: Sect. 2 is the main part of the paper. A weak logic called '**ML**' is therein defined which has the minimal amount of "logical juice" to allow the C-entailment theorem to hold. Sect. 3 explains why the proof of the C-entailment theorem generalizes to more or less arbitrarily stronger logics—the relevant logics **B**, classical logic and classical modal logics are given as examples. Sect. 4 gives a brief comparison of the current notion of entailment with Anderson and Belnap's and Brady's notions of entailment. The already mentioned Sect. 5 delves into the notion of I-entailment before Sect. 6 gives a short discussion of the notion of entailment vs. that of logical theoremhood, before Sect. 7 provides a short summary. The proof of the I-entailment theorem has been relegated to the Appendix.

This is a technical paper. I would like to note, however, that some of the results found in this paper are used in chapter 12 of this volume [18] to mount a more philosophical critique of the Anderson-Belnap account of relevance.

2 Entailment and the Entailment Theorem

In this section I will show how the standard Hilbert consequence relation for any given logic is to be defined as well as how to define the concept of conjunctive entailment, *C-entailment* for short. The main goal is to prove a version of Anderson and Belnap's "Entailment theorem" for a minimal logic **ML**. After having given the proof, I will comment on how it generalizes to stronger logics.

Definition 1 (Parentheses Conventions) $\vee$, $\wedge$ and $\circ$ are to bind tighter than $\rightarrow$, and so I'll usually drop parentheses enclosing conjunctions and disjunctions. Association is otherwise to the left so that $A \wedge B \wedge C$ is simply shorthand for $(A \wedge B) \wedge C$.

A logic will in this paper be identified using a list of axioms and rules. All such will be schemas, although I will simply use the shorter terms *axiom* and *rules*. Three different ways of defining a consequence relation using such a list will be show. The "standard" such will be called *H-entailment* in this paper and will prove central to the current enterprise as it will be used to define both the other two consequence relations, but also because it is in this paper the relation used to determine whether one logic extends another.[1]

Definition 2 (H-entailment) A HILBERT PROOF (H-proof) of a formula A from a set of formulas Γ in the logic **L** is defined to be a finite list $C_1, \dots, C_n$ such that $C_n = A$ and every $C_{i \leq n}$ is either a member of Γ, a logical axiom of **L**, or there is a set $\Delta \subseteq \{C_j \mid j < i\}$ such that $\Delta \Vdash C_i$ is an instance of a rule of **L**. The existential claim that there is such a proof is written $\Gamma \vdash^H_{\mathbf{L}} A$, and expressed as Γ H-ENTAILS A.

[1] Certain logics require several different types of rules. For instance, the necessitation rule of modal logic should only apply to logical theorems, and not to mere premises. The following definition of H-entailment could be molded so as to take this into account, but for present purposes it ought to suffice to only consider logics with one type of rule.

Note, then, that there are three possible justifications that a formula occurring in a Hilbert proof may have—either *premise*, or *axiom*, or *obtained by a rule*. The rules of modus ponens and adjunction—the rules $\{A, A \rightarrow B\} \Vdash B$ and $\{A, B\} \Vdash A \wedge B$, respectively—will receive special focus in defining the other two notions of consequence—*C-entailment* and *I-entailment*. For that reason it will be convenient to make the notion of a justification for a H-proof a bit more explicit:

Definition 3 A JUSTIFICATION for a H-proof $C_1, \ldots, C_n$ to the effect that $\Gamma \vdash^H_{\mathbf{L}} A$, where **L** is a logic having modus ponens and adjunction among its primitive rules, is a list $\mathcal{J}_1, \ldots, \mathcal{J}_n$ such that $\mathcal{J}_{i \leq n}$ satisfies one of the following conditions:

(1) $\mathcal{J}_i =$ 'premise' & $C_i \in \Gamma$
(2) $\mathcal{J}_i =$ 'axiom' & C_i is an instance of an axiom of **L**
(3) $\mathcal{J}_i =$ 'MP: (j, k)' & $C_k = C_j \rightarrow C_i$ & $j, k < i$
(4) $\mathcal{J}_i =$ 'adj: (j, k)' & $C_k = C_j \wedge C_i$ & $j, k < i$
(5) $\mathcal{J}_i =$ 'other: $(i_1, \ldots, i_j)'$ & $i_1, \ldots, i_j < i$ & $\{C_{i_1}, \ldots, C_{i_j}\} \Vdash C_i$ is an instance of a rule of **L** different from modus ponens and adjunction.

I trust that the idea here is clear enough, and so will not explicitly mention any such justification-list going forward.

The weakest logic dealt with in this paper is the logic **ML** which can be determined using the following list of axioms and rules:

Definition 4 (The Logic ML)

(A1)	$A \rightarrow A$	
(A2)	$A \wedge B \rightarrow A$	
(A3)	$A \wedge B \rightarrow B$	
(R1)	$\{A, A \rightarrow B\} \Vdash B$	modus ponens
(R2)	$\{A, B\} \Vdash A \wedge B$	adjunction
(R3)	$\{A \rightarrow B, A \rightarrow C\} \Vdash A \rightarrow B \wedge C$	lattice conjunction
(R4)	$\{A \rightarrow B, B \rightarrow C\} \Vdash A \rightarrow C$	transitivity

Before we proceed we need to specify what it is to mean that one logic *extends* another.[2]

[2] Sometimes the notion of one logic extending another is defined only with regards to logical theorems. Although fine for some applications, doing so allows for logics to be identified even though they differ with regards to H-entailment. A well-known example where this is the case is Anderson and Belnap's logic **E** vs. the logic Π' first set forth in [1]. Since the latter logic can be axiomatized as **E** augmented by the γ-rule $\{A, {\sim}A \vee B\} \Vdash B$, and it was shown in [15] that this rule is admissible in **E**, it follows that **E** and Π' are identical with regards to logical theorems. However, whereas $\vdash^H_{\Pi'}$ is explosive—meaning that $\{A, {\sim}A\} \vdash^H_{\Pi'} B$ for every formula A and B—$\vdash^H_{\mathbf{E}}$ is well known to be *paraconsistent*, that is not explosive.

Definition 5

- If $\mathbf{L}_1$ and $\mathbf{L}_2$ are two lists of axioms and rules such that the set of connectives of $\mathbf{L}_1$ is included in that of $\mathbf{L}_2$, then $\mathbf{L}_2$ EXTENDS $\mathbf{L}_1$ just in case $\vdash^H_{\mathbf{L}_1} \subseteq \vdash^H_{\mathbf{L}_2}$.
- $\mathbf{L}_1$ is a SUBLOGIC of $\mathbf{L}_2$ just in case $\mathbf{L}_2$ EXTENDS $\mathbf{L}_1$.
- Two lists $\mathbf{L}_1$ and $\mathbf{L}_2$ determine the same logic just in case $\vdash^H_{\mathbf{L}_1} = \vdash^H_{\mathbf{L}_2}$.

By a phrase such as *A is an axiom of the logic* $\mathbf{L}$, then, I shall simply mean that A is an instance of an axiom of a list of axioms and rules used to determine $\vdash^H_{\mathbf{L}}$. Also, to avoid confusion, let's agree to call $A \to B$ the *major rule premise* of modus ponens, and A its *minor* rule premise.

Lemma 1 $\varnothing \vdash^H_{\mathbf{ML}} (A_1 \wedge \ldots \wedge A_i \wedge \ldots \wedge A_n) \to A_i$

Proof If $n = 1$, then it is simply an instance of (A1). If $i = n$, then it is simply an instance of (A3). Assume, then, that $1 \leq i < n$. $A_1 \wedge \ldots \wedge A_i \to A_i$ is an instance of (A3). Furthermore, all of the following are instance of (A2):

$$(A_1 \wedge \ldots \wedge A_n) \to (A_1 \wedge \ldots \wedge A_{n-1})$$
$$(A_1 \wedge \ldots \wedge A_{n-1}) \to (A_1 \wedge \ldots \wedge A_{n-2})$$
$$\vdots$$
$$(A_1 \wedge \ldots \wedge A_{i+1}) \to (A_1 \wedge \ldots \wedge A_i)$$

Using (R4) possibly several times will yield that $\varnothing \vdash^H_{\mathbf{ML}} (A_1 \wedge \ldots \wedge A_n) \to A_i$. □

It is easy to see that the concept of Hilbert-derivability yields that any logical theorem follows from any sentence, that is if $\varnothing \vdash^H_{\mathbf{L}} A$, then $\Gamma \vdash^H_{\mathbf{L}} A$ for any set of formulas Γ, even though, then, no formula in Γ need be used in deriving A. Anderson and Belnap initiated the relevant research program in the late fifties. One of its main targets was that both classical and intuitionistic logic, as well as the classical modal logic **S4**, were all incorrect theories of logical consequence because all fail to make sense of logical consequence as a *following from* relation: "the defect lies in the definition, which fails to take seriously the word 'from' in 'proof *from* hypothesis.' " [3, p. 18]. Their suggestion for how to fix the "Official definition of a deduction" was to restrict the use of the rules in such a way as to make sure that at least some of the premises must be used in deriving the conclusion. To facilitate this, a marking strategy was introduced. In this paper I'll use '$\star$.' The idea, then, is that all premises start off as marked, and then it is demanded that the conclusion be marked as a consequence of the premises being used in rules. If, or so was Anderson and Belnap's idea, one then chance upon a suitable restriction on the use of the rules, one will ensure that at least some logical content will be shared between the premises and the conclusion. The following is my modification of Brady's version of Anderson and Belnap's definition of what *a proof that* $A_1, \ldots, A_n$ *entail(s)* B is to mean.[3]

Definition 6 (C-entailment) A C-PROOF of a formula A from a set of formulas Γ in the logic $\mathbf{L}$ is a Hilbert-proof $C_1, \ldots, C_n$ of A from Γ, along with the justification for each $C_{i \leq n}$, such that it is possible to $\star$-mark the $C_{i \leq n}$'s according to the following rules:

(1) If $C_i \in \Gamma$ and the justification for C_i is indeed "premise," then C_i is marked.
(2) R1 is only used if the major rule premise is unmarked.
(3) If C_i is obtained using R1, then it is marked if its minor premise is marked.
(4) R2 is only used on premises which are either both marked or both unmarked.
(5) If C_i is obtained from C_j and C_k using R2 and both of C_j and C_k are marked, then C_i is marked.
(6) Any rule different from R1&R2 is only used on unmarked formulas.
(7) No other formulas are marked.
(8) As a consequence of (1–7), C_n is marked.

The existential claim that there is such a proof is written $\Gamma \vdash^C_{\mathbf{L}} A$, and expressed as Γ C-ENTAILS A.

There are two features that are worth noticing right away. First of all, at least one premise must be used—the $\star$ can only be introduced by way of a premise, and since the conclusion must be starred, at least one premise must occur in any C-proof. It follows, then, that $\varnothing \nvdash^C_{\mathbf{L}} A$ for every formula A and every logic $\mathbf{L}$. Secondly, an instance of a formula which is both a premise and an instance of an axiom may occur in a derivation as starred while another instance of the same formula may occur unmarked. The following proof, then, is an acceptable proof that $A \to A$ C-entails $A \to A$:

$$\begin{array}{ll} (1) & A \to A \text{ axiom} \\ (2) & \star\, A \to A \text{ premise} \end{array}$$

Lemma 2 *Let* $\mathbf{L}$ *be any logic extending* $\mathbf{ML}$*. Assume that* $\Gamma \vdash^C_{\mathbf{L}} A$ *and let* $C_1, \ldots, C_n$ *be a proof of this with* $\star$*-markings according to Definition 6. Claim: If* $C_{i \leq n}$ *is unmarked, then* $\varnothing \vdash^H_{\mathbf{L}} C_i$*.*

Proof By strong induction: Assume for induction hypothesis that all unmarked $C_{j<i}$ are logical theorems of $\mathbf{L}$. We must prove the claim that if C_i is unmarked, then $\varnothing \vdash^H_{\mathbf{L}} C_i$, so assume that C_i is indeed unmarked. Since it is unmarked it must have been obtained either with the justification of being an instance of an axiom, in

[3] A comparison between this notion and that set forth by Anderson and Belnap, and that of Brady is presented in Sect. 4.

which case it is obviously the case that $\varnothing \vdash^H_{\mathbf{L}} C_i$, or it could have been obtained by way of one of the rules of **L**. If so, then since it itself is unmarked it must have been obtained from unmarked rule premises only. From the induction hypothesis, then, it follows that these premises are logical theorems, and since every primitive rule of any logic is theorem-preserving, it follows that C_i is a logical theorem as well. By induction, then, it follows that every unmarked $C_{i \leq n}$ is a logical theorem. □

Theorem 1 (C-entailment Theorem) $\Gamma \vdash^C_{\mathbf{ML}} B$ *if and only if for some conjunction* $A := A_1 \wedge \ldots \wedge A_n$, *where* $\{A_1, \ldots, A_n\} \subseteq \Gamma$, $\varnothing \vdash^H_{\mathbf{ML}} A \to B$.

Proof ($\Leftarrow$) Assume that $\varnothing \vdash^H_{\mathbf{ML}} A \to B$, where A is a formula as described. Assume that $B_1, \ldots, B_m$ is a Hilbert-proof of $A \to B$ from $\varnothing$. Claim:

$$B_1, \ldots, B_m, A_1^\star, A_2^\star, \ldots, A_n^\star, (A_1 \wedge A_2)^\star, \ldots, (A_1 \wedge \ldots \wedge A_n)^\star, B^\star$$

is a proof that Γ C-entails B (with markings shown). That this is so is seen from the following considerations:

- The first $B_1, \ldots, B_m$-part is acceptable since all the rules of **ML** may be applied without restriction to unmarked formulas, and any such rule-use will result in an unmarked formula.
- The A_i's are all members of Γ, and so the $B_1, \ldots, B_m, A_1^\star, A_2^\star, \ldots, A_n^\star$-part is acceptable.
- $(A_1 \wedge A_2)^\star$ up to $(A_1 \wedge \ldots \wedge A_n)^\star$ are obtained from previous $\star$-marked formulas using R2 and so are all acceptable.
- B is finally obtained from the unmarked B_m which is the formula $A \to B$ and the marked formula A which is the formula $A_1 \wedge \ldots \wedge A_n$, and so B follows using R1 without violating $\star$-rules (2) and (3).

($\Rightarrow$) Assume that $\Gamma \vdash^C_{\mathbf{ML}} B$, and let $B_1, \ldots, B_m$ be a C-proof to this effect. Let $A = \bigwedge(\Gamma \cap \{B_1, \ldots, B_m\})$—the conjunction of every formula occurring both in the set of premises and on the C-proof.[4] The rest of the proof is by way of a strong induction to establish that $\varnothing \vdash^H_{\mathbf{ML}} A \to B_i$ for every $B_{i \leq m}$ marked by $\star$. Since $B = B_m$ is marked according to $\star$-rule (8), this will then be sufficient for proving that $\varnothing \vdash^H_{\mathbf{ML}} A \to B$.

Assume, then, for inductive hypothesis that B_i is marked and that for all $j < i$ such that B_j is marked that $\varnothing \vdash^H_{\mathbf{ML}} A \to B_j$. The goal is to show that $\varnothing \vdash^H_{\mathbf{ML}} A \to B_i$.

[4] Notice that $\Gamma \cap \{B_1, \ldots, B_m\}$ cannot be the empty set: Since $B_1, \ldots, B_m$ is a C-proof, B_m must be $\star$-marked and only premises can introduce stars. It follows, then, that Γ and $\{B_1, \ldots, B_m\}$ have at least one common member.

Since B_i is marked, it follows from Definition 6 that it could have been obtained in one of three different ways, namely with the justification of being a member of Γ, or derived using either (R1) or (R2).

(Γ) If the justification for B_i was that $B_i \in \Gamma$, then B_i is in fact one of the conjuncts of A, and so $\varnothing \vdash^H_{\mathbf{ML}} A \to B_i$ by Lemma 1.

(R1) If B_i was obtained from B_j and $B_j \to B_i$ using modus ponens, then since B_i is assumed to be marked it follows from $\star$-rules of Definition 6 that only B_j is marked. From Lemma 2 we then can infer that $\varnothing \vdash^H_{\mathbf{ML}} B_j \to B_i$. Since $j < i$ it follows from the inductive hypothesis that $\varnothing \vdash^H_{\mathbf{ML}} A \to B_j$. We can now use (R4) to infer that $\varnothing \vdash^H_{\mathbf{ML}} A \to B_i$.

(R2) If B_i is $B_j \wedge B_k$ and was obtained using adjunction, then according to the $\star$-rules of Definition 6, both B_j and B_k are marked since B_i is assumed to be marked. Since $j, k < i$ it follows from the induction hypothesis that both $\varnothing \vdash^H_{\mathbf{ML}} A \to B_j$ and $\varnothing \vdash^H_{\mathbf{ML}} A \to B_k$. Using (R3) we can then infer that $\varnothing \vdash^H_{\mathbf{ML}} A \to B_i$.

In all cases it follows that $\varnothing \vdash^H_{\mathbf{ML}} A \to B_i$, and so standard induction yields that $\varnothing \vdash^H_{\mathbf{ML}} A \to B_i$ for every $B_{i \leq m}$ marked by $\star$, and therefore also that $\varnothing \vdash^H_{\mathbf{ML}} A \to B$ since $B = B_m$ and B_m is $\star$-marked. □

3 Generalizations

ML is the *minimal* logic of this paper in the sense that every axiom and rule of it are needed for the entailment theorem to go through. It could, however, be weakened slightly: the rules (R3) and (R4) are only needed as *admissible rules*. What is more interesting, however, is how it can be extended. The answer is quite simply that it can be extended more or less *ad libitum*. First of all, note that the proof of the C-entailment theorem only requires axioms (A1)–(A3); further axioms, either involving $\to$ or $\wedge$, or other different connectives altogether, may be added, without altering the above proof. In terms of adding rules: The only requirement is that the rule be *theorem-preserving*. Being theorem-preserving is a rather low bar on the requirement for rule-hood, and so the claim that any extension of **ML** is covered seems quite legitimate. As an example, let's consider the relevant logics **B** with both the Ackermann constant **t** and the intensional conjunction $\circ$ called *fusion*:

Definition 7 (The Logic B)

(BA1) $A \to A$

(BA2) $A \to A \vee B$ and $B \to A \vee B$

(BA3) $A \wedge B \to A$ and $A \wedge B \to B$

(BA4) $A \wedge (B \vee C) \to (A \wedge B) \vee (A \wedge C)$

(BA5) $(A \to B) \wedge (A \to C) \to (A \to B \wedge C)$

(BA6) $(A \to C) \wedge (B \to C) \to (A \vee B \to C)$

(BA7) $\sim\sim A \to A$

(BA8) $A \to (B \to A \circ B)$

(BA9) $\mathbf{t}$

(BR1) $\{A, A \to B\} \Vdash B$

(BR2) $\{A, B\} \Vdash A \wedge B$

(BR3) $\{A \to B\} \Vdash (C \to A) \to (C \to B)$

(BR4) $\{A \to B\} \Vdash (B \to C) \to (A \to C)$

(BR5) $\{A \to \sim B\} \Vdash B \to \sim A$

(BR6) $\{A \to (B \to C)\} \Vdash A \circ B \to C$

(BR7) $\{A\} \Vdash \mathbf{t} \to A$

B evidently extends **ML** in the sense of Definition 5. Furthermore, by using the standard definition of conjunction using implication and negation, so does even classical logic, here stated using Łukasiewicz's famous axiomatization:[5]

Definition 8 (The Logic CL)

(CA1) $(A \to B) \to ((B \to C) \to (A \to C))$

(CA2) $(\neg A \to A) \to A$

(CA3) $A \to (\neg A \to B)$

(CR1) $\{A, A \to B\} \Vdash B$

[5] See [13, ch. 2] for this axiomatization. Łukasiewicz's axiomatization allows for $\wedge$ being defined as $\neg(A \to \neg B)$. It is with this definition in mind that Łukasiewicz's axiomatization of classical logic can be said to extend **ML**.

The same goes for intuitionistic logic which also extends **ML**. Furthermore, even classical *modal* logics are covered as these are simply obtained by adding modal axioms to classical logic, along with the theorem-preserving rule of necessitation. Thus all of the three logics Anderson and Belnap thought of as rival accounts of logical consequence—**S4**, intuitionistic, and classical logic—satisfy the C-entailment theorem using the same use-restricted Hilbert calculus as Anderson and Belnap's three favorite logics **T**, **E**, and **R**.

B, like **ML**, **T**, **E**, and **R**, satisfies the variable sharing property [cf. 5]. Although they regarded this as a necessary property, Anderson and Belnap seem to have thought of the notion of premise use as necessary and sufficient for relevance, somehow.[6] From the above result, however, it is evident that neither the C-entailment theorem nor the use-restrictions encoded into the notion of a C-proof can be viewed as differentiating features in the same way that the variable sharing property separates, say, classical logic from **B**.[7]

The next section gives a quick comparison of the current notion of entailment with the corresponding notion set forth by Anderson and Belnap and that of Brady.

4 Comparing C-entailment to Anderson and Belnap's Notion of Entailment, and to Brady's Notion of Entailment

Anderson and Belnap's use-restricted Hilbert calculus for their notion of entailment was set forth in [3, § 23.6]. The same notion of Hilbert derivability is therein used, but the ⋆-rules are slightly different than the ones used in this paper. As already mentioned, they only considered adjunction and modus ponens as primitive rules, and so they have no rule corresponding to ⋆-rule (6). The significant difference, however, is with regards to modus ponens. The current one restricts the use of modus ponens to unmarked major premises. Anderson and Belnap's, on the other hand, allows for unrestricted usage, and only demands that if A_i is obtained from A_j and $A_j \rightarrow A_i$ using modus ponens, then A_i is marked if *either* A_j *or* $A_j \rightarrow A_i$ is marked.

The trick of restricting any rule different from adjunction and modus ponens to unmarked formulas is inspired from [7, thm. 3] which thus restricts the **B**-rules

[6] See [18] for a discussion.

[7] That the entailment theorem also holds for classical logic was shown in [16, sect. 5] using Anderson and Belnap's original notion of entailment. That this is the case was first indicated by [6] in which Bennett reviewed [2]. Other authors have picked up on Bennett's claim, but beyond Brady's paper detailed in the next section there have not been any published attempts at trying to make the claim rigorous and to generalize it as the present paper does. See [18] for a critical and historical account of Anderson and Belnap's "use criterion" of relevance which is the philosophical motivation behind their "Entailment theorem."

(BR3)–(BR5).[8] Brady's restriction on modus ponens, however, is quite different. It reads:

> *(v) For Modus Ponens,*
> *$\mathcal{B}$ is starred iff either $\mathcal{A}$ or $\mathcal{A} \to \mathcal{B}$ is starred,* and
> *if $\mathcal{A} \to \mathcal{B}$ is starred then $\mathcal{A}$ is starred.*
> [7, p. 274]

Notice, then, that both Brady's and the current definition of C-entailment are simply different twists on Anderson and Belnap's notion of entailment. What Brady's notion of entailment rules out, but which Anderson and Belnap's notion allows, is the case in which the major premise $B_j \to B_i$ is marked, but the minor premise B_j is not. Brady's notion of entailment was designed to fit logics between **B** augmented by the pseudo modus ponens axiom $A \wedge (A \to B) \to B$ and the Anderson-Belnap logic **T**. One of the theorems of **T** is known as *conjunctive syllogism*:

$$(A \to B) \wedge (B \to C) \to (A \to C),$$

an instance of which is

$$((A \to A) \to B) \wedge (B \to C) \to ((A \to A) \to C).$$

However, $((A \to A) \to B) \wedge (B \to C) \to C$ is not a logical theorem of **T**.[9] This contrasts to the stronger logic **E** which has $((A \to A) \to C) \to C$ as one of its characteristic axioms.[10] Thus **E** does have $((A \to A) \to B) \wedge (B \to C) \to C$ as a logic theorem. Anderson and Belnap's notion of entailment allows for the following entailment proof of C from $\{(A \to A) \to B, B \to C\}$ where we for simplicity assume that conjunctive syllogism is an axiom of the logic and the stars are displayed in a separate column:

(1)		$((A \to A) \to B) \wedge (B \to C) \to ((A \to A) \to C)$	axiom
(2)	$\star$	$(A \to A) \to B$	premise
(3)	$\star$	$B \to C$	premise
(4)	$\star$	$((A \to A) \to B) \wedge (B \to C)$	R2: 2, 3
(5)	$\star$	$(A \to A) \to C$	R1: 1, 4
(6)		$A \to A$	axiom
(7)	$\star$	C	R1: 5, 6

[8] [7] does not consider fusion or the Ackermann constant.

[9] A counter-model is easily obtained using John Slaney's computer program MaGIC [24].

[10] See [3, § 27.1] for the axiomatization of **E**.

The above $\star$-proof uses modus ponens twice: first in line 5 in which only the minor premise is marked, and then in line 7 with only the major premise marked. Brady's marking rules, then, allows only the latter of the two uses of modus ponens. However, his rules do allow the use of modus ponens in the case where *both* premises are marked. As a consequence it forces B to be C-entailed by $\{A, A \to B\}$. Thus Brady's marking rules can only be fit for logics which validates *Pseudo Modus Ponens*: $A \wedge (A \to B) \to B$. This axiom is interderivable with the *contraction* rule,

$$\{A \to (A \to B)\} \Vdash A \to B,$$

which many relevant logics—including some of Brady's own favorite logics [cf. 8]—fail to validate. This is where the current definition of entailment improves upon Brady's definition. It should also be remarked that the improvement is without loss as the current notion with the accompanying C-entailment theorem applies equally to, say, **T** and **E**, as it does **ML**.

Anderson and Belnap's marking technique is a way of ensuring that the conclusion must have some logical connection with at least one of the premises; the $\star$-mark is introduced via a premises and must be passed along to the conclusion using either modus ponens or adjunction. Anderson and Belnap wanted to read the conditional as expressing entailment: they regarded $\to$ as an object-language representative of the relation of entailment. The current account of the marking rules are such as to even highlight this intended reading as the major premise $A \to B$ must be unmarked which was shown in Lemma 2 to imply that $A \to B$ must be a logical theorem, and therefore to, according to Anderson and Belnap's reading, represent a fact of entailment. Thus restricted, modus ponens can be used in two scenarios: The first is towards establishing further facts of entailment, the other in passing along a connection between the premises and the conclusion. If, then, such a major premise is involved in passing along a connection between the premises and conclusion—when its minor rule premise is marked—then that is because the premises suffice for the antecedent of a logically true entailment-conditional, which, given the C-entailment theorem, represents, given the logic at hand, a true fact of, if not entailment as such, then at least C-entailment.

There is a last aspect which differentiates Brady's from the current marking technique. Brady demands that every premise be marked, but that all logical theorems not be marked [cf. 7, p. 574]. For premises which are also logical theorems, this makes for an inconsistent demand, however. Although not inconsistent, Anderson and Belnap's $\star$-rules regarding premises vs. axioms also have an unfortunate feature. These rules read as follows:

(a) if S_i is a premise, then S_i is starred;
(b) if S_i is an axiom which is not a premiss, then S_i is not starred;
[3, p. 277]

These rules, then, leave it impermissible, it would seem, to let a formula which happens to be both a premise and a logical axiom, occur in a derivation once as marked on account of it being a premise, but to also occur unmarked. Without going further into this, I end this section by simply noting that the explicit mention of having *premise* as its justification in Definition 6(1) is precisely to make it clear that it is indeed permissible in a proof of C-entailment for one instance of a logical axiom to occur marked yet another instance of the very same formula to be unmarked.

5 Implicative Entailment

C-entailment is a notion of consequence for which premises are aggregated to mirror the nature of extensional conjunction. The C-entailment theorem shows, then, that this extensional aggregation is even mirrored in a particularly strong way, namely as conjunctive antecedents of logically true conditionals. Note, then, that $A \wedge B \rightarrow C$ is not equivalent to $A \rightarrow (B \rightarrow C)$ for any relevant logic: $B \wedge A \rightarrow A$ is a logical truth of any extension of **ML**, but $B \rightarrow (A \rightarrow A)$ cannot be if the logic is to satisfy the variable sharing property. This opens up the possibility of a more intensional take on the nature of premise-aggregation. Instead of having premises be aggregated to mirror the extensional conjunction, it is possible to have premise-aggregation mirror the nature of the *intensional* conjunction $\circ$ that we saw in connection with the relevant logic **B**. One corresponding entailment theorem for such an intensional take on premise aggregation yields that claims of entailment can be represented as logically true conditionals on the form $(A_1 \circ \ldots) \circ A_n \rightarrow B$. Using the $\circ$-axiom (BA8) together with the prefixing rule (BR3), it is easy to verify that such a sentence suffices for deriving $A_1 \rightarrow (\ldots \rightarrow (A_n \rightarrow B)\ldots)$. Similarly, $A_1 \rightarrow (\ldots \rightarrow (A_n \rightarrow B)\ldots)$ suffices for deriving $(A_1 \circ \ldots) \circ A_n \rightarrow B$ using (BR6). Thus the recourse to the intensional conjunction $\circ$ is not needed in order to represent intensional premise-aggregation as antecedents of logically true conditionals. However, without such a conjunction, the fact that $A_1 \rightarrow (\ldots \rightarrow (A_n \rightarrow B)\ldots)$ may happen to be a logical truth in some logic, need not correspond to a fact about C-entailment for that logic. For instance, $(A \rightarrow B) \wedge (A \rightarrow C) \rightarrow (A \rightarrow B \wedge C)$ is a logical theorem of **B**. However, being a sublogic of the three-valued Łukasiewicz logic, $((A \rightarrow B) \wedge (A \rightarrow C) \wedge A) \rightarrow (B \wedge C)$ fails to be a logical theorem of **B**.[11]

In the already mentioned paper by Brady—[7]—it was shown that there is a use-restricted notion of derivability for the logic **B** which captures the notion of implicative entailment. The relation can be defined as follows:

Definition 9 (I-entailment) An I-PROOF of a formula B from a non-empty sequence $\mathbb{S} = \langle A_1, \ldots, A_m \rangle$ in a logic **L** extending $\mathbf{B}^{\wedge\rightarrow}$ is a Hilbert-proof

[11] The standard three-valued algebra for $Ł_3$ in which 1 is the only designated (true) value, and conjunction and disjunction are interpreted as, respectively, minimum and maximum over the three values $\{0, 1/2, 1\}$, and negation and the conditional according to the displayed matrices, is provided in Fig. 1 in which it is also displayed one way to invalidate $((A \rightarrow B) \wedge (A \rightarrow C) \wedge A) \rightarrow (B \wedge C)$. See [12, 20] for more on $Ł_3$.

$\rightarrow$	0	1/2	1	$\sim$
0	1	1	1	1
1/2	1/2	1	1	1/2
1	0	1/2	1	0

$[\![((1/2 \rightarrow 0) \wedge (1/2 \rightarrow 0) \wedge 1/2) \rightarrow (0 \wedge 0)]\!] = 1/2$

Fig. 1 The $Ł_3$ algebra

$C_1, \ldots, C_n$ of B from $\{A_1, \ldots, A_m\}$, along with the justification for each $C_{i \leq n}$, such that for each $i \leq m$, $C_i = A_i$, and it is possible to mark each $C_{i \leq n}$ with $\tau_i \subseteq \{1, \ldots, m\}$ according to the following rules:

(1) If $i \leq m$, then $\tau_i = \{i\}$.
(2) (BR1) is only used on premises C_j and $C_k = C_j \rightarrow C_i$ provided that if $\tau_k \neq \varnothing$, then for some $o > max(\tau_k)$, $\tau_j = \{o\}$. If C_i is thus obtained, then $\tau_i = \tau_j \cup \tau_k$.
(3) (BR2) is only used on premises C_j and C_k such that $\tau_j = \tau_k$. If C_i is thus obtained, then $\tau_i = \tau_k$.
(4) Any rule different from (BR1)&(BR2) is only used on $\varnothing$-marked formulas.
(5) All other formulas are $\varnothing$-marked.
(6) As a consequence of (1–5), $\tau_n = \{1, \ldots, m\}$.

The existential claim that there is such a proof is written $\mathbb{S} \vdash^{I}_{\mathbf{L}} B$, and expressed as $\mathbb{S}$ I-ENTAILS B.

The first thing to notice here is that I-proofs require that *all* the premises be listed first. Listing all the premises first would be cumbersome in practice. Since, however, any H-proof is finite, only finitely many premises will ever occur on such a proof. It is, furthermore, evident that one could demand that any H-proof start off by listing every premise, and so the "all premises first" requirement on I-proofs isn't really restrictive. The reason for the requirement is simply to ensure that the markings will be correct since the fact that an instance of a formula occurs in an I-proof with the justification of being a premise should not rule out that another instance of the same formula can occur with the justification of being an axiom or the justification of being derive from previous formulas.

The purpose of the I-entailment calculus is to have a calculus which lines up the premises correctly so that they yield the conclusion just in case the corresponding iterated implication formula is logically true. Now, mingle—the formula $A \rightarrow (A \rightarrow A)$—isn't a logical theorem for most relevant logics, and so for such logics it should not generally be the case that $\langle A, A\rangle$ I-entails A. Furthermore, in order to derive B from $A \rightarrow (A \rightarrow B)$ and A, we need in general to use A twice—first in order to derive $A \rightarrow B$, and then to finally derive B. For I-entailment, this corresponds to the fact that $\langle A \rightarrow (A \rightarrow B), A, A\rangle$ I-entails B in every logic. However, if the logic at hand allows one to derive $A \rightarrow B$ from $A \rightarrow (A \rightarrow B)$ without using A, then one may "contract" the two instances of A into one, yielding that $\langle A \rightarrow (A \rightarrow B), A\rangle$ I-entails B. For I-entailment, then, we need to be able to keep track of how many times a formula occurs on the premise list $\mathbb{S}$. The same goes for the *order* of the premises: $\langle A \rightarrow B, A\rangle$ I-entails B in every logic since this simply corresponds to the logical truth $(A \to B) \to (A \to B)$. Strong logics such as **R**, however, allow one to permute the order of any premises, and so it will be the case that $\langle A_1, \ldots, A_i, \ldots, A_j, \ldots, A_m \rangle$ I-entails B if and only $\langle A_1, \ldots, A_j, \ldots, A_i, \ldots, A_m \rangle$ I-entails B. Weaker logics like **B**, however, do not allow such unrestricted permutation.

Let's look at a simple I-proof before moving on. The following list of formulas is easily verified to be an I-proof to the effect that the sequence $\langle B \to B, B \to B \rangle$ I-entails $B \to B$ in the logic **B** augmented by the mingle axiom:

(1)	$\{1\}$	$B \to B$	1. premise
(2)	$\{2\}$	$B \to B$	2. premise
(3)	$\varnothing$	$(B \to B) \to ((B \to B) \to (B \to B))$	axiom (mingle)
(4)	$\varnothing$	$B \to B$	axiom (BA1)
(5)	$\{1\}$	$(B \to B) \to (B \to B)$	1, 3, BR1
(6)	$\{1, 2\}$	$B \to B$	2, 5, BR1

The current marking strategy is to demand that any I-proof simply starts off by copying the premise list $\mathbb{S}$ and marking the i-th premise by the singleton set $\{i\}$. Thus if a formula occurs several times on $\mathbb{S}$, than each instance will be indexed by a different singleton set. This is the case with the first two occurrences of $B \to B$ which, then, are indexed using the singleton sets $\{1\}$ and $\{2\}$, respectively. By Definition 9(5), axioms may occur $\varnothing$-marked, as indeed is the case in lines 3–4 of the above I-proof. Note also that line 4 isn't utilized in any way. This is generally OK for instances of logical axioms. That the formulas occurring in lines 1, 2, 4 and 6 are all instances of the same formula is also OK, despite the fact that they are all marked differently. The conclusion must be marked by the set $\{1, \ldots, m\}$ to ensure that every premise on the list $\mathbb{S} = \langle A_1, \ldots, A_m \rangle$ is used in deriving it. The current proof, then, has $B \to B$ also as its last formula where it is marked by $\{1, 2\}$ in accordance with this.[12]

Brady, following [10, 11], tailor-made the use-conditions for modus ponens for various sublogics of **R**. Brady's notion of I-entailment fit for the logic **RW**,[13] for instance, is obtained from the above by replacing the use-restriction on modus ponens (3) by the following clause:

[Brady's **RW** Clause (BRWC)]

[12] The idea of listing all the premises first is new to this paper. It is, however, simply a way of making sure that the idea behind the marking strategy in [7] works as intended. Unfortunately, Brady's formulation of it makes it inconsistent, due to the same reason as explained in Sect. 4.

[13] **RW** is obtained from **B** by replacing the rules (BR3)–(BR6) with their corresponding conditional axioms. Without **t** and $\circ$, however, the assertion axiom $A \to ((A \to B) \to B)$ must be added, which is how **RW** is presented in [7].

(BR1) is only used on premises C_j and $C_k = C_j \to C_i$ provided if $\tau_k \neq \varnothing$, then $\tau_k \cap \tau_j = \varnothing$. If C_i is obtained using (BR1), then $\tau_i = \tau_j \cup \tau_k$

Let's look at how this allows the sequence $\langle A, A \to B \rangle$ to entail B. The most straightforward proof looks like this:

(1)	$\{1\}$	A	1. premise
(2)	$\{2\}$	$A \to B$	2. premise
(3)	$\{1, 2\}$	B	1, 2, BR1

The use of modus ponens here evidently satisfies (BRWC), but violates the use-condition for modus ponens stated above for I-entailment since $\tau_1 = \{1\}$ and it is not the case that $1 > max(\tau_2) = 2$. However, that does not mean that it is not the case that $\langle A, A \to B \rangle$ I-entails B in **RW**. The following proof shows that it is indeed the case that $\langle A, A \to B \rangle \vdash^{I}_{\mathbf{RW}} B$, using, then, the current definition of an I-proof:

(1)	$\{1\}$	A	1. premise
(2)	$\{2\}$	$A \to B$	2. premise
(3)	$\varnothing$	$A \to ((A \to B) \to B)$	axiom of **RW**
(4)	$\{1\}$	$(A \to B) \to B$	1, 3, BR1
(5)	$\{1, 2\}$	B	2, 4, BR1

By minor modifications of Brady's proof, both of the following statements are provable:[14]

Theorem 2 (I-entailment Theorem) *Let* **L** *be any logic extending* $\mathbf{B}^{\wedge\to}$*—the conjunction-implicational fragment* **B**.[15] *For any non-empty sequence* $\mathbb{S} = \langle A_1, \ldots, A_m \rangle$,

$$\mathbb{S} \vdash^{I}_{\mathbf{L}} B \Longleftrightarrow \varnothing \vdash^{H}_{\mathbf{L}} A_1 \to (\ldots \to (A_m \to B)\ldots).$$

Theorem 3 (Deductive Equivalence Theorem) *Let* **L** *be any logic extending* $\mathbf{B}^{\wedge\to}$*: For any non-empty sequence* $\mathbb{S} = \langle A_1, \ldots, A_m \rangle$,

$$\langle A_1, \ldots, A_m, A_{m+1} \rangle \vdash^{I}_{\mathbf{L}} B \Longleftrightarrow \mathbb{S} \vdash^{I}_{\mathbf{L}} A_{m+1} \to B.$$

[14] Proofs are provided in the Appendix.

[15] The result obviously also holds for any logic extending $\mathbf{B}^{\to}$ without, then, having $\wedge$ as a logical connective. If, however, **L** is a logic extending $\mathbf{B}^{\to}$ with a binary connective $\sqcap$ for which adjunction is a primitive rule, then it must satisfy the $\sqcap$-version of the axiom (BA5) if the use-conditions are to be those corresponding to Definition 9(3).

Some relevantists have attributed a special significance to the above I-entailment theorem, or the related Deductive equivalence theorem. Read, for instance, calls it "the correct statement of the Deduction Equivalence" [21, § 3.8], whereas Mares claims that it tell us something about inferences in relevant logic [cf. 14, p. 164]. Note, then, that if this is indeed the case, then that is so for independent reasons: The notion of I-entailment and the I-entailment and Deductive equivalence theorems hold for irrelevant and relevant logics alike, which we have seen was also the case with Anderson and Belnap's preferred notion of entailment, namely C-entailment.

Whereas the focus in previous investigations have been on finding the correct notion of use for some logic, what this and Sects. 2–3 have shown is that there is a single such notion which yields corresponding C- and I-entailment theorems for very weak logics, but also for more or less any way of strengthening these. Of course, one might find philosophical support for some stronger notion of premise use than what is encoded in the current notions of C- and I-entailment. What I have tried to highlight, however, is that unless the use-conditions are stated in some fundamentally different way than previous accounts have, then such notions of entailment will still yield corresponding C- and I-entailment theorems as well as deductive equivalence theorems, for logics which violate the variable sharing property. By inspecting the proofs of these theorems, it is evident that if such a theorem holds for a logic **L**, then it also holds for that logic augmented by implicational paradoxes such as $A \rightarrow (B \rightarrow B)$ or $A \wedge \sim A \rightarrow B$.[16]

6 Entailment and Logical Theorems

We have seen that the notion of entailment brought out using the use-restricted Hilbert calculi can be framed both as a conjunctive-antecedent notion, as well as a notion for which premises take the role of conditional-nested antecedents. The latter notion—I-entailment—is a *substructural* notion of logical consequence for many logics. For logics such as **R** for which

$$(A \rightarrow (B \rightarrow C)) \rightarrow (B \rightarrow (A \rightarrow C))$$

is a logical theorem, for instance, it is easily seen that the order of the premise-sequence is insignificant. For weaker logics like **B**, however, it is. **R** does not have $A \rightarrow (A \rightarrow A)$ as a logical theorem, and so even though $\langle A \rangle$ I-entails A for every A, some A's fail to be I-entailed by $\langle A, A \rangle$. For logics like classical and intuitionistic

[16] For a critical examination of the more interpretational side of the notion of use which some relevantists have attributed to aspects of use-restricted calculi as presented in this paper, see [18].

logic, however, both

$$(A \to (B \to C)) \to (A \wedge B \to C)$$

$$(A \wedge B \to C) \to (A \to (B \to C))$$

are logical truths, and so it is easily seen that I- and C-entailment for these logics are the same notion.[17] Note, however, that both C- and I-entailment are in fact *not* identical to standard Hilbert consequence even for classical logic. The reason is simply that both C- and I-entailment are stated in such a way as to require the premise-set or premise-sequence be non-empty. As a consequence, these notions of logical consequence admit of no logical theorems in the sense of being a consequence from the empty collection of premises. Although this may seem in line with at least Anderson and Belnap's critique of the hallow notion of *from* encoded in standard Hilbert consequence, many relevantists that do argue for related notions of entailment in fact seem to have a more liberal notion of logical consequence in mind in that logical theorems are permitted. Both [14] and [21], for instance, argue that an iterated notion of entailment is correct, but seem both to accept logical theorems as following from, at least in some sense, nothing.

Note, then, that the notion of I-entailment can easily be modified so as to allow for logical theorems: By redefining the notion of an I-proof by simply allowing proofs from empty sequences of formulas, it is easy to verify that indeed $\langle\rangle \vdash_{\mathbf{L}}^{I^\dagger} A$ if and only if $\varnothing \vdash_{\mathbf{L}}^{H} A$. The current choice, however, has been to obtain a notion of implicative entailment most alike Anderson and Belnap's notion of conjunctive entailment. There is an option for those agreeing with the Anderson and Belnap's sentiment that following from nothing—be it the empty set or the empty sequence—is offensive to the meaning of *from*, but for whom giving up the category of logical theorems is too high a price. One can redefine the notion of a logical theorem as one following from, in the sense of I-entailment, the sequence $\langle \mathbf{t} \rangle$—$\mathbf{t}$ being the Ackermann constant. This is in essence how logical theorems are presented in [21]. One can also retain such a notion of logical theoremhood in the case of C-entailment by defining, then, A to be a logical C-theorem in the logic just in case $\{\mathbf{t}\} \vdash_{\mathbf{L}}^{C} A$.[18]

There is, I should like to note, a rather different approach to making room for logical theorems that goes back to Church's original formulation of a deduction theorem for the implicational fragment of **R**. Church states the deduction theorem

[17] I-entailment is only defined for *finite* lists of formulas, whereas C-entailment allows for infinite premise sets. The sameness, then, holds with this restriction in mind.

[18] A much neater proof theory than that used here is the *consecution calculus* set forth in [22]. All of the three consequence relations investigated in this paper can be defined using it, although it takes some work to show that the resultant relations are indeed the very same [cf. 17]. Note, then, that Restall's proof theory allows one to define logical theoremhood without the recourse the object-language $\mathbf{t}$ as it allows for a 0-ary premise combinator—'0'—which is a sort of proof-theoretic interpretation of $\mathbf{t}$ in the same way as the comma which separates premises in a set of formulas (list of formulas) can be said to be the proof-theoretic interpretation of the object-language connective '$\wedge$' ('$\circ$').

as the claim that if B is derivable from the list of formulas $A_1, \ldots, A_n$, then either B or $A_n \rightarrow B$ are derivable from $A_1, \ldots, A_{n-1}$ [cf. 9, p. 24]. This type of disjunctive deductive theorem is taken up in [4]—therein called the *relevant deduction property*—as central to the very claim that $\rightarrow$ is a properly relevant connective.[19] Going further into how the notion of a logical theorem ought to be understood is, regrettably, beyond the scope of this paper.

7 Conclusion

This paper has shown how to generalize Anderson and Belnap's notion of entailment so as to apply to **ML**—a rather weak sublogic of the contractionless logics relevant logic **B**. The use-restricted Hilbert calculus was shown to yield Anderson and Belnap's "Entailment theorem" for **ML** and extends to arbitrarily stronger logics without needing to modify the use-restrictions. Anderson and Belnap's notion of entailment has it that a non-empty set of formulas entails B just in case a conditional having B as its consequent and which has some of the premises conjuncted together into a single antecedent is logically true. A related notion of entailment for which premises must be nested implication-antecedents was shown to be definable in the case of the relevant logic **B**, and applying to arbitrarily stronger logics, and which was such as to yield a corresponding version of the entailment theorem, namely that a non-empty *sequence* of formula $\langle A_1, \ldots, A_n \rangle$ implicative-entails B just in case $A_1 \rightarrow (\ldots \rightarrow (A_n \rightarrow B)\ldots)$ is logically true. The calculus for implication-entailment also applies to any extension of **B** without needing to modify the use-restrictions.

The calculi used to define C- and I-entailment both restrict the use of modus ponens. In the first case the restriction is rather severe in that the major rule premise must itself be a logical theorem. In the case of I-entailment, the major rule premise must be such as to not depend on what the minor premises does. In both cases, then, the minor rule premise must either itself be a logical theorem, or contribute something new from the collection of premises from which the conclusion is to follow so as to assure that the premises—at least some of them in the case of C-entailment, and all of them in the case of I-entailment—be used in deriving the conclusion.

Anderson and Belnap thought there to be something fundamentally wrong with the accounts of logical consequence associated with what Anderson and Belnap took to be the main rival accounts of entailment, namely classical logic, intuitionistic logic, and classical modal logics such as **S4**. Their notion of entailment was intended as an account of logical consequence which would truly require that the conclusion follow *from* the premises in a substantial way not accounted for by other theories of

[19] For a discussion of Avron's take on relevance, see [19]. For a discussion of what a relevant connective is, see [25].

logical consequence. Although different from the standard Hilbertian consequence relation, this paper has shown that both C-entailment and I-entailment along with the corresponding entailment theorems apply equally to relevant logics such as **B** and **E** as it does to "irrelevant" logics such as intuitionistic logic, classical logic, and classical modal logics such as **S4**. In fact, the theorems hold true of any extension of the **ML** in the case of C-entailment, and of **B** in the case of I-entailment.

Appendix: Proof of the I-entailment Theorems

Lemma 3 *Let* $\mathbf{L}$ *be any logic extending* $\mathbf{B}^{\wedge\rightarrow}$*—the conjunction-implicational fragment of* **B**. *Assume that* $\mathbb{S} \vdash^{I}_{\mathbf{L}} A$ *and let* $C_1, \ldots, C_n$ *be a proof according to the rules in Definition 9. Claim: If* $A_{i \leq n}$ *is unmarked, then* $\varnothing \vdash^{H}_{\mathbf{L}} A_i$.

Proof Similar to the proof of Lemma 2. □

Lemma 4 *Let* $\mathbf{L}$ *be any logic extending* $\mathbf{B}^{\wedge\rightarrow}$. *For any 1-element sequence* $\mathbb{S} = \langle A \rangle$,

$$\mathbb{S} \vdash^{I}_{\mathbf{L}} B \Longleftrightarrow \varnothing \vdash^{H}_{\mathbf{L}} A \rightarrow B.$$

Proof ($\Leftarrow$) Assume that $\varnothing \vdash^{H}_{\mathbf{L}} A \rightarrow B$, and let $B_1, \ldots, B_n$ be a Hilbert-proof of $A \rightarrow B$ from $\varnothing$. Then

$$A^{\{1\}}, B_1^{\varnothing}, \ldots, B_n^{\varnothing}, B^{\{1\}}$$

is a proof that $\mathbb{S}$ I-entails B (with markings shown) and in which the last formula is obtained from B_n and A using modus ponens.

($\Rightarrow$) Assume that $\mathbb{S} \vdash^{I}_{\mathbf{L}} B$ and let $B_1, \ldots, B_n$ be a proof of this. The proof is by strong induction over natural numbers showing that for all $i \leq n$, if $B_{i \leq n}$ is marked by $\{1\}$, then $\varnothing \vdash^{H}_{\mathbf{L}} A \rightarrow B_i$. Since $B_n = B$ is marked, this will suffice for showing that $\varnothing \vdash^{H}_{\mathbf{L}} A \rightarrow B$.

Inductive hypothesis (IH): Let $1 \leq i \leq n$ be any number and assume that for all $j < i$ such that B_j is $\{1\}$-marked, that $\varnothing \vdash^{H}_{\mathbf{L}} A \rightarrow B_j$.

We must show that if B_i is $\{1\}$-marked, then $\varnothing \vdash^{H}_{\mathbf{L}} A \rightarrow B_i$. Assume, then, that B_i is $\{1\}$-marked. There are two justifications which can have brought about B_i' marking: either if it was obtained as the first premise occurring in $\mathbb{S}$, or if it was obtained using a rule. If it was the first premise, then $B_i = A$ and so it is obviously the case that $\varnothing \vdash^{H}_{\mathbf{L}} A \rightarrow B_i$. If it was obtained using a rule, then since it is only (BR1) and (BR2) which can beget marked formulas, it suffices to consider these two rules.

(BR1) If B_i was obtained using (BR1) from B_j and $B_k = B_j \to B_i$, then since B_i is $\{1\}$-marked it must be the case that $\tau_k = \varnothing$ and $\tau_j = \{1\}$. From (IH) it then follows that $\varnothing \vdash^H_{\mathbf{L}} A \to B_j$ and from Lemma 3 that $\varnothing \vdash^H_{\mathbf{L}} B_j \to B_i$, and so it follows using the derivable rule (R4) that $\varnothing \vdash^H_{\mathbf{L}} A \to B_i$.

(BR2) If $B_i = B_j \wedge B_k$ was obtained using (BR2), then since B_i is $\{1\}$-marked it must be the case that both $\tau_k = \tau_j = \{1\}$. From (IH) it then follows that both $\varnothing \vdash^H_{\mathbf{L}} A \to B_j$ and $\varnothing \vdash^H_{\mathbf{L}} A \to B_k$, and so it follows that $\varnothing \vdash^H_{\mathbf{L}} A \to B_i$ using the derivable rule (R3).

□

Definition 10 If $B_1, \ldots, B_k, \ldots$ is a list of formulas with accompanying index sets, such that at least $B_1, \ldots, B_k$ is a correct initial segment of a proof that $\mathbb{S} \vdash^I_{\mathbf{L}} B$, then $B_1, \ldots, B_k$ will be called an ACCEPTABLE PROOF-TAIL FOR THE CLAIM THAT $\mathbb{S} \vdash^I_{\mathbf{L}} B$.

Definition 11 If L is the list of formulas $B_1, \ldots, B_k$, then $L + \langle A_1, \ldots, A_m \rangle$ is the list of formulas $B_1, \ldots, B_k, A_1, \ldots, A_m$.

Theorem 4 *Let* $\mathbf{L}$ *be any logic extending* $\mathbf{B}^{\wedge\to}$*. For any non-empty sequence* $\mathbb{S} = \langle A_1, \ldots, A_m \rangle$,

$$\langle A_1, \ldots, A_m, A_{m+1} \rangle \vdash^I_{\mathbf{L}} B \Longleftrightarrow \mathbb{S} \vdash^I_{\mathbf{L}} A_{m+1} \to B.$$

Proof ($\Leftarrow$) Assume that $\mathbb{S} \vdash^I_{\mathbf{L}} A_{m+1} \to B$, and let $A_1^{\{1\}}, \ldots, A_m^{\{m\}}, B_1^{\tau_{B_1}}, \ldots, B_n^{\{1,\ldots,m\}}$ be a proof to this effect with markings as indicated. Then

$$A_1^{\{1\}}, \ldots, A_m^{\{m\}}, A_{m+1}^{\{m+1\}}, B_1^{\tau_{B_1}}, \ldots, B_n^{\{1,\ldots,m\}}, B^{\{1,\ldots,m+1\}}$$

where the last formula is obtained using (BR1) from A_{m+1} and $B_n = A_{m+1} \to B$, is obviously an I-proof of B from $\langle A_1, \ldots, A_m, A_{m+1} \rangle$.

($\Rightarrow$) Assume that $\langle A_1, \ldots, A_m, A_{m+1} \rangle \vdash^I_{\mathbf{L}} B$ and let

$$A_1, \ldots, A_m, A_{m+1}, B_1, \ldots, B_n$$

be a proof of this. The rest of the proof, like that indicated in [7], is an induction to show that for each $0 \leq i \leq n$, there is a list L_i, all starting with the formulas $A_1, \ldots, A_m$ which is an acceptable proof-tail for the claim that $\mathbb{S} \vdash^I_{\mathbf{L}} A_{m+1} \to B$. Since L_n will end with the formula $A_{m+1} \to B$ which will be indexed by the set $\{1, \ldots, m\}$, L_n will be a correct I-proof to the effect that $\mathbb{S} \vdash^I_{\mathbf{L}} A_{m+1} \to B$.

For $1 \leq i \leq n$, let τ_i be the index of the formula B_i, and let a sequence of modified indices and formulas be defined as follows:

$$
\begin{aligned}
\sigma_0 &= \varnothing \\
\sigma_i &= \begin{cases} \tau_i \setminus \{m+1\} & \text{if } m+1 \in \tau_i \\ \tau_i & \text{else} \end{cases} \\
C_0 &= A_{m+1} \rightarrow A_{m+1} \\
C_i &= \begin{cases} A_{m+1} \rightarrow B_i & \text{if } m+1 \in \tau_i \\ B_i & \text{else} \end{cases}
\end{aligned}
$$

The proof-tails are now defined recursively as follows:

$$
L_0 = A_1^{\{1\}}, \ldots, A_m^{\{m\}}, C_0^{\sigma_0}
$$

$$
L_{1 \leq i \leq n} = \begin{cases}
L_{i-1} + \langle P_{i_1}^{\varnothing}, C_i^{\sigma_i} \rangle & \begin{cases} \text{if } B_i \text{ was obtained from } B_j \text{ and } B_k = B_j \rightarrow B_i \text{ using (BR1), and } \tau_k = \varnothing \text{ and } m+1 \in \tau_i. \\ P_{i_1} = (A_{m+1} \rightarrow B_j) \rightarrow (A_{m+1} \rightarrow B_i) \end{cases} \\
L_{i-1} + \langle P_{i_2}^{\varnothing}, C_i^{\sigma_i} \rangle & \begin{cases} \text{if } B_i \text{ was obtained from } B_j \text{ and } B_k = B_j \rightarrow B_i \text{ using (BR1), and } \tau_k \neq \varnothing \text{ and } m+1 \in \tau_i. \\ P_{i_2} = (B_j \rightarrow B_i) \rightarrow (A_{m+1} \rightarrow B_i) \end{cases} \\
L_{i-1} + \langle Q_{i_1}^{\sigma_i}, Q_{i_2}^{\varnothing}, C_i^{\sigma_i} \rangle & \begin{cases} \text{if } B_i \text{ was obtained from } B_j \text{ and } B_k \text{ using (BR2), and } m+1 \in \tau_i. \\ Q_{i_1} = (A_{m+1} \rightarrow B_j) \wedge (A_{m+1} \rightarrow B_k) \\ Q_{i_2} = Q_{i_1} \rightarrow (A_{m+1} \rightarrow B_j \wedge B_k) \end{cases} \\
L_{i-1} + \langle C_i^{\sigma_i} \rangle & \text{else}
\end{cases}
$$

Tail Claim: All $L_{i \leq n}$ are acceptable proof-tails for the claim that $\mathbb{S} \vdash_{\mathbf{L}}^{I} A_{m+1} \rightarrow B$. □

Proof of Tail Claim By standard induction on the numbers.

Base case: L_0 is an acceptable proof-tail for the claim that $\mathbb{S} \vdash_{\mathbf{L}}^{I} A_{m+1} \rightarrow B$ since it is simply the list $\mathbb{S}$ indexed as the premises ought to be in any I-proof from $\mathbb{S}$, and ends with the formula $A_{m+1} \rightarrow A_{m+1}$ indexed by $\sigma_{m+1} = \varnothing$.

(IH): Assume that L_{i-1} is an acceptable proof-tail for the claim that $\mathbb{S} \vdash_{\mathbf{L}}^{I} A_{m+1} \rightarrow B$.

(1) $L_i = L_{i-1} + \langle P_i^{\varnothing}, C_i^{\sigma_i} \rangle$. If B_i was obtained using (BR1) with $\tau_k = \varnothing$ and $m+1 \in \tau_i$, then since $C_k = B_k = B_j \rightarrow B_i$ is indexed by $\varnothing$ and $k \leq i-1$ with L_{i-1} an acceptable proof-tail, it follows that adding $P_{i_1} = (A_{m+1} \rightarrow B_j) \rightarrow (A_{m+1} \rightarrow B_i)$ indexed by $\varnothing$ is acceptable, where P_{i_1}, then, is obtained from the $\varnothing$-indexed

formula C_k using the prefixing rule (BR3).[20] Since $m+1 \in \tau_i$, $m+1 \in \tau_j$, and so $C_j = A_{m+1} \to B_j$. But then applying (BR1) to C_j and P_{i_1} yields the formula $C_i = A_{m+1} \to B_i$ with index

$$\sigma_i = \sigma_j \cup \varnothing = (\tau_j \setminus \{m+1\}) \cup \varnothing = \sigma_j.$$

Thus L_i is an acceptable proof-tail for the claim that $\mathbb{S} \vdash^I_{\mathbf{L}} A_{m+1} \to B$.

(2) $L_i = L_{i-1} + \langle P_i^{\varnothing}, C_i^{\sigma_i} \rangle$. If B_i was obtained using (BR1) with $\tau_k \neq \varnothing$ and $m+1 \in \tau_i$, then it follows from Definition 9 that $\tau_j = \{m+1\}$ and $m+1 \notin \tau_k$. But then $C_j = A_{m+1} \to B_j$ and is indexed by $\varnothing$. Since L_{i-1} is an acceptable proof-tail and $j \leq i-1$, it follows that adding $P_{i_2} = (B_j \to B_i) \to (A_{m+1} \to B_i)$ indexed by $\varnothing$ is acceptable, where P_{i_2}, then, is obtained from C_j using the suffixing rule (BR4).[21] Since $m+1 \notin \tau_k$, $C_k = B_k = B_j \to B_i$, and so applying (BR1) to C_k and P_{i_2} yields the formula $C_i = A_{m+1} \to B_i$ with index

$$\sigma_i = \sigma_k \cup \varnothing = \sigma_k.$$

Thus L_i is an acceptable proof-tail for the claim that $\mathbb{S} \vdash^I_{\mathbf{L}} A_{m+1} \to B$.

(3) $L_i = L_{i-1} + \langle Q_{i_1}^{\sigma_i}, Q_{i_2}^{\varnothing}, C_i^{\sigma_i} \rangle$. If B_i was obtained using (BR2) with $m+1 \in \tau_i$, then it follows from Definition 9 that $\tau_i = \tau_j = \tau_k$, and that $B_i = B_j \wedge B_k$. By the definitions of the σ's and the C's, it furthermore follows that

$$C_j = A_{m+1} \to B_j \big| C_k = A_{m+1} \to B_k \big| C_i = A_{m+1} \to B_j \wedge B_k$$
$$\sigma_j = \sigma_k = \sigma_i = \tau_i \setminus \{m+1\}$$

By (IH), L_{i-1} is an acceptable proof-tail for the claim that $\mathbb{S} \vdash^I_{\mathbf{L}} A_{m+1} \to B$ and since both $j \leq i-1$ and $k \leq i-1$, it follows that we can use (BR2) and expand L_{i-1} with $Q_{i_1} = C_j \wedge C_k$ which, then, becomes correctly indexed by $\sigma_j = \sigma_k$. Q_{i_2} is an instance of the logical axiom (BA5), and so L_{i-1} can also be expanded with this formula indexed by $\varnothing$. C_i is then obtained from Q_{i_1} and Q_{i_2} using (BR1) with index $\sigma_j \cup \varnothing$ as it should be.

(4) $L_i = L_{i-1} + \langle C_i^{\sigma_i} \rangle$. In the "else" case, it is not the case that both $m+1 \in \tau_i$ and that B_i was obtained by either (BR1) or (BR2). If it was, then, obtained using either of these two rules, then it follows from Definition 9 that both premises B_j and B_k which it was obtained from have indices such that $m+1 \notin \tau_j$ and $m+1 \notin \tau_k$. But then C_i follows from $C_j = B_j$ and $C_k = B_k$ using the very same rule. If B_i was rather obtained on account of being an instance of an axiom, then B_i was indexed by $\varnothing$ and so $C_i = B_i$. Lastly, if B_i was obtained by means of some rule different from

[20] In case $\mathbf{L}$ does not have (BR3) as a primitive rule, one will need to rather insert the whole derivation of P_{i_1}.

[21] As above, if (BR4) isn't a primitive rule, one will need to rather insert the whole derivation of P_{i_2}.

(BR1) and (BR2), then B_i along with all the premises $B_{i_1}, \ldots, B_{i_l}$ from which it was obtained must have been $\varnothing$-indexed, and so $C_{i_1} = B_{i_1}, \ldots, C_{i_l} = B_{i_l}, C_i = B_i$. In all "else" cases, then, it follows that L_i will be an acceptable proof-tail for the claim that $\mathbb{S} \vdash^I_{\mathbf{L}} A_{m+1} \to B$ since L_{i-1} is by (IH) assumed to be.

It follows, then, by standard induction that all $L_{i \leq n}$ are acceptable proof-tails for the claim that $\mathbb{S} \vdash^I_{\mathbf{L}} A_{m+1} \to B$. ■

Since L_n, then, is an acceptable proof-tail for the claim that $\mathbb{S} \vdash^I_{\mathbf{L}} A_{m+1} \to B$, with the last formula $C_n = A_{m+1} \to B$ indexed by $\sigma_n = \{1, \ldots, m\}$, it follows that L_n is a proof to the effect that $\mathbb{S} \vdash^I_{\mathbf{L}} A_{m+1} \to B$ □

Theorem 5 *Let* **L** *be any logic extending* $\mathbf{B}^{\wedge\to}$. *For any sequence* $\mathbb{S} = \langle A_1, \ldots, A_m \rangle$,

$$\mathbb{S} \vdash^I_{\mathbf{L}} B \Longleftrightarrow \varnothing \vdash^H_{\mathbf{L}} A_1 \to (\ldots \to (A_m \to B)\ldots).$$

Proof This follows from Lemma 4 and Theorem 4. □

References

1. Ackermann, W. (1956). Begründung einer strengen Implikation. *Journal of Symbolic Logic, 21*(2), 113–128. https://doi.org/10.2307/2268750
2. Anderson, A. R., & Belnap, N. D. (1962). The pure calculus of entailment. *Journal of Symbolic Logic, 27*(1), 19–52. https://doi.org/10.2307/2963676
3. Anderson, A. R., & Belnap, N. D. (1975). *Entailment: The logic of relevance and necessity* (Vol. 1). Princeton: Princeton University Press.
4. Avron, A. (2014). What is relevance logic? *Annals of Pure and Applied Logic, 165*(1), 26–48. https://doi.org/10.1016/j.apal.2013.07.004
5. Belnap, N. D. (1960). Entailment and relevance. *Journal of Symbolic Logic, 25*(2), 144–146. https://doi.org/10.2307/2964210
6. Bennett, J. (1965). Review of *Entailment and deducibility* by T. J. Smiley, and *The pure calculus of entailment* by Alan Ross Anderson and Nuel D. Belnap, Jr. *The Journal of Symbolic Logic, 30*(2), 240–241. https://doi.org/10.2307/2270141
7. Brady, R. T. (1993). Rules in relevant logic—II: Formula representation. *Studia Logica, 52*(4), 565–585. https://doi.org/10.1007/BF01053260
8. Brady, R. T. (2006). *Universal logic*. Stanford: CSLI Publication.
9. Church, A. (1951). The weak theory of implication. In A. Menne, A. Wilhelmy, & H. Angsil (Eds.), *Kontrolliertes Denken: Untersuchungen zum Logikkalkül und zur Logik der Einzelwissenschaften* (pp. 22–37). Munich: Kommissions-Verlag Karl Alber.
10. Kron, A. (1973). Deduction theorems for relevant logics. *Mathematical Logic Quarterly, 19*(3–6), 85–92. https://doi.org/10.1002/malq.19730190306
11. Kron, A. (1976). Deduction theorems for *T*, *E* and *R* reconsidered. *Zeitschrift für Mathematische Logik und Grundlagen der Mathematik, 22*, 261–264. https://doi.org/10.1002/malq.19760220135
12. Łukasiewicz, J. (1920). O logice trówartościowej. *Ruch Filozoficzny, 5*, 169–171. English translation: On Three-Valued Logic. In: Jan Łukasiewicz: Selected Works (Ed.) North Holland, 87–88, 1990.

13. Łukasiewicz, J. (1966). *Elements of mathematical logic*. Oxford: Pergamon Press. Translated from Polish by O. Wojtasiewicz.
14. Mares, E. D. (2004). *Relevant logic: A philosophical interpretation*. Cambridge: Cambridge University Press. https://doi.org/10.1017/CBO9780511520006
15. Meyer, R. K., & Dunn, J. M. (1969). E, R and γ. *Journal of Symbolic Logic, 34*, 460–474. https://doi.org/10.2307/2270909
16. Øgaard, T. F. (2021). Non-Boolean classical relevant logics I. *Synthese, 198*, 6993–7024. https://doi.org/10.1007/s11229-019-02507-z
17. Øgaard, T. F. (2021). From Hilbert proofs to consecutions and back. *Australasian Journal of Logic, 18*(2). https://doi.org/10.26686/ajl.v18i2.6770
18. Øgaard, T. F. (2024). Withered relevance: Evaluating the Anderson-Belnap account of relevant logics. In S. Standefer, I. Sedlár, & A. Tedder (Eds.), *New directions in relevant logics*, chap. 3. Berlin: Springer.
19. Øgaard, T. F. (forthcoming). Suppressing irrelevance: Notes on Arnon Avron's view of relevance. *Logique et Analyse*.
20. Priest, G. (2008). *An introduction to non-classical logic. From if to is* (2nd ed.). Cambridge: Cambridge University Press. https://doi.org/10.1017/CBO9780511801174
21. Read, S. (1988). *Relevant logic: A philosophical examination of inference*. Oxford: Basil Blackwell.
22. Restall , G. (2000). *An introduction to substructural logics*. London: Routledge. https://doi.org/10.4324/9780203016244
23. Shaw-Kwei, M. (1950). The deduction theorems and two new logical systems. *Methodos, 2*(5), 56–75.
24. Slaney, J. (1995). MaGIC, matrix generator for implication connectives: Release 2.1 notes and guide. Technical Report TR-ARP-11/95, Automated Reasoning Project, Australian National University. http://users.cecs.anu.edu.au/~jks/magic.html
25. Standefer, S. (2022). What is a relevant connective? *Journal of Philosophical Logic, 51*(4), 919–350. https://doi.org/10.1007/s10992-022-09655-7

Proofs with Star and Perp

Greg Restall

Abstract In this paper, I show how to incorporate insights from the model-theoretic semantics for negation (insights due the late J. Michael Dunn among others), into a *proof-first* understanding of the semantics of negation. I then discuss the different ways a logical pluralist may understand the underlying accounts of proofs and their significance.

J. Michael Dunn's pioneering work in non-classical logic is profoundly important, covering many different topics. He has given us new tools and techniques, inspired new questions, and opened up us fresh perspectives on what we thought was familiar. In this paper I will transplant some of his important insights about the semantics of negation [7]—insights grown in model-theoretic soil—to a new field, that of proof-theoretic semantics.

The key idea to be applied can be simply described. The behaviour of negation in a range of logical systems (classical logic, intuitionistic logic, orthologic, relevant logics, linear logic, etc.) can be understood in terms of an underlying relation of *incompatibility* on a model. Given some collection P of points, and a relation $\Vdash$ evaluating formulas at those points, then *whatever* we think about the details of the particular behaviour of negation, we can agree that if we have two points a and b where $a \Vdash \neg A$ (according to a, $\neg A$ holds) and $b \Vdash A$ (according to b, A holds), then in some sense, a and b are *incompatible*. If $\perp$ is the incompatibility relation,

Thanks to Franz Berto, Aaron Cotnoir and colleagues in the Metaphysics and Logic research seminar in Arché at the University of St Andrews, for discussion on material that lead to these results. Thanks also to the European Network for the Philosophy of Logic, and the Philosophy Department at the University of Bergen opportunities to present on these results, for useful questions and feedback. Finally, thanks an anonymous referees of this chapter for suggestions that helped improve the formulation of these results.

G. Restall (✉)
Philosophy Department, The University of St Andrews, St Andrews, UK
e-mail: gr69@st-andrews.ac.uk

I. Sedlár et al. (eds.), *New Directions in Relevant Logic*, Trends in Logic 63,
https://doi.org/10.1007/978-3-031-69940-5_14

we have $a \perp b$. An incompatibility model (or a 'perp'-model, for short) takes this 'only if' to be an 'if and only if', and for it to be the defining feature of negation.

- $a \Vdash \neg A$ if and only if for each b, if $b \Vdash A$ then $a \perp b$.

This semantics for negation certainly sounds plausible enough [4], but it might look unfamiliar.[1] Despite its unfamiliarity, it is ubiquitous. If we impose different conditions on the incompatibility relation $\perp$, familiar logics arise. The truth-functional classical semantics of negation is recovered if $\perp$ is simply defined to be the nonidentity relation $\neq$. A point is incompatible with everything else, it is compatible with itself only. On a Kripke model for intuitionistic logic, we have $a \perp b$ iff a and b have no common descendants: that is, there is no c where $a \sqsubseteq c$ and $b \sqsubseteq c$.

However, if we allow incompatibility to range more widely, we can model paraconsistent logics. If a point a supports both A and its negation $\neg A$, then we have $a \perp a$. Contradictory points clash *with themselves*. Such points may clearly be defective in some sense or other, but that does not mean that a semantics should rule them out *tout court*. Ringing the changes with different kinds of incompatibility relations gives a wide range of logics. So wide, in fact, that if we allow incompatibility to fail to be *symmetric*, then we have models that provide counterexamples to the inference from A to $\neg\neg A$,[2] and we can independently add double negation *introduction* (in the presence of an order relation $\sqsubseteq$ on points, at least) given another more complex condition on incompatibility [23].

One of Dunn's many insights is that the Routley star semantics [31] for a de Morgan negation, with the familiar clause

- $a \Vdash \neg A$ if and only if $a^* \nVdash A$

can be naturally understood as a kind of incompatibility semantics. If we understand a^* as the *maximal* point compatible with a (so $a \not\perp b$ if and only if $b \sqsubseteq a^*$) then indeed the Routley semantics is a species of the broader genus of incompatibility models. Incompatibility models with a Routley star are particularly well-suited to modelling a non-classical *de Morgan* negation. In these models, the constructively

[1] In the incompatibility models we will discuss, conjunction and disjunction have the standard 'local' truth conditions: $a \Vdash A \wedge B$ iff $a \Vdash A$ and $a \Vdash B$; $a \Vdash A \vee B$ iff $a \Vdash A$ or $a \Vdash B$, and so, the resulting logics extend distributive lattice logic with an order inverting 'negation' operator. To model logics without distribution we must vary the semantics for either conjunction or disjunction. This is not difficult to do, and in fact, the initial treatment of negation by means of an incompatibility relation, due to Rob Goldblatt [9] in 1974, was used to model orthologic, in which distribution fails. In these models, Goldblatt uses the local truth conditions for conjunction, and $A \vee B$ is a definitional abbreviation for $\neg(\neg A \vee \neg B)$.

[2] Here is a simple model that does the job. Take two points a, b where $b \perp a$, but no other incompatibilities hold. Let $a \Vdash p$ and $b \nVdash p$. Then we have $b \Vdash \neg p$, since a is the only p-point, and $b \perp a$. However, in this model we do not have $a \Vdash \neg\neg p$, since $b \Vdash \neg p$, but we do *not* have $a \perp b$.

invalid de Morgan law (from $\neg(A \wedge B)$ to $\neg A \vee \neg B$) is valid, and if our model satisfies the condition that $a^{**} = a$, then if $a \Vdash \neg\neg A$, we have $a \Vdash A$ too. So, in the rest of this paper we will consider 'perp' models, of the form $\langle P, \perp, \Vdash \rangle$ and 'star' models, the form $\langle P, ^*, \Vdash \rangle$ as two kinds of *incompatibility* models for negation. As is usual with models of this general type, the underlying structure, *sans* evaluation relation $\Vdash$, is called a *frame*, so $\langle P, \perp \rangle$ is a perp frame, while $\langle P, ^* \rangle$ is a star frame.

Incompatibility models provide a unifying framework for understanding the semantics of negation. We now have a map of the terrain, that takes in a great deal of the logical scenery indeed, including the placid and well-understood downtowns of classical and intuitionistic negation, and the rather less familiar wilds of paraconsistency.

One question that lurks in the background when we use this semantics for different logical systems is—what do such models *model*? What *are* these points that stand in a relation of incompatibility with each other? Do they have any independent status, apart from their use in our semantic theorising? If so, where do we find them? And *how* do we come in contact with them? If they have no independent status, other than in our models, how do they do such a good job of modelling the behaviour of negation? How is it that negation could have the properties that are appropriately modelled by models like these?[3]

I will not answer these questions in this essay, but I will provide for the friend of the incompatibility semantics a new perspective on the meaning of negation and how we might grasp it, in such a way as to help explain how the incompatibility model semantics turns out to be an appropriate class of models for the concept as we use it. For that, we don't start with models. We start with *proofs*.

1 Proofs and Contexts

However we grasp concepts like conjunction, disjunction and negation, that grasp is intimately bound up with the way we use these concepts in our everyday reasoning practice. There are many things we could mean by terms like 'reasoning' and 'inference.'[4] Here, though, let us focus on a very simple connection between our

---

[3] The literature in models of non-classical logic describes points in such models as 'pieces of information' [35, p. 159], 'set-ups' [30, p. 135], 'worlds' [20, §5.2], 'states' [22, p. 60], and different analyses are offered concerning what these entities might be.

[4] See Harman's influential *Change in View* [10] for a convincing defence of the idea that inference as understood as reasoned change in view is *not* to be identified with deductive validity, and see Bob Brandom's *Articulating Reasons* [5] for an introduction to *inferentialism*, according to which a primitive notion of *material inference* (not to be confused with formal deductive logical validity) lies at the root of our conceptual capacities. Then, for good measure, read Catarina Dutilh Novaes' superb *The Dialogical Roots of Deduction* [8] for an account of how the roots of our reasoning practice take nourishment in fertile soil of our dialogical norms for dispute resolution.

$$A\ (\textit{Assumption}) \qquad \frac{A \quad B}{A \wedge B}\wedge I \qquad \frac{A \wedge B}{A}\wedge E \qquad \frac{A \wedge B}{B}\wedge E$$

$$\frac{A}{A \vee B}\vee I \qquad \frac{B}{A \vee B}\vee I \qquad \frac{A \vee B \quad \begin{matrix}[A]^j \\ \Pi \\ C\end{matrix} \quad \begin{matrix}[B]^k \\ \Pi' \\ C\end{matrix}}{C}\vee E^{j,k}$$

$$\frac{\begin{matrix}[A]^i \\ \Pi \\ \bot\end{matrix}}{\neg A}\neg I^i \qquad \frac{\neg A \quad A}{\bot}\neg E \qquad \frac{\bot}{A}\bot E$$

Fig. 1 Natural deduction proof rules

practices of *proof*, and our grasp of logical concepts such as conjunction and negation.

If someone were to grant A and grant B and still take it (after this had been pointed out) that it remained an open question, in need of further justification, as to whether $A \wedge B$ were the case, this would seem to be good reason to doubt that this person understood that '$\wedge$' means *and*, in the usual logicians' sense of conjunction. Similarly, if you had just granted a conjunction, such as $A \wedge B$, a competent user of '$\wedge$' could see that if the issue of whether or not A is the case were to arise, the answer would again be 'yes,' and no further justification would be needed than to point back to $A \wedge B$, or whatever it was that justified $A \wedge B$. We can understand the inference rules of a natural deduction system (such as the rules in Fig. 1) as encoding basic competencies with the logical concepts, at least under a specific idealisation.

One key feature of natural deduction rules like these is that they do not merely tell us how to grind out indubitable consequences of things that have been granted as true. The rules allowing for *discharging* of assumptions (here, $\vee E$ and $\neg I$), and the negation elimination rule $\neg E$ rely on our ability to engage in the question-and-answer practice of proving from assumptions that we don't necessarily take to be true. The rule, $\neg E$ tells us that if we have ever proved $\neg A$ and also proved A, then we have landed in an inconsistency (marked here by writing $\bot$) in the conclusion point of the proof tree. The negation introduction rule gives us something to *do* when we land in such an inconsistency. We can take one of the assumptions we used to lead to the inconsistency, 'blame' it, and conclude on the basis of the *other* assumptions, that it is not true—that is, its negation is true. One further feature of the rules as presented here is that the smallest proof is a single formula, A, which serves as both the premise and the conclusion of this atomic proof. Each other proof rule serves as a means to generate new proofs from old, by extending a pre-existing proof with the addition of a new conclusion.[5]

Here is a simple example of these rules in use in a proof from the assumption $\neg p \vee \neg q$ to the conclusion $\neg(p \wedge q)$.

$$\dfrac{\dfrac{\neg p \vee \neg q \qquad \dfrac{[\neg p]^1 \qquad \dfrac{[p \wedge q]^3}{p}\,\wedge E}{\bot}\,\neg E \qquad \dfrac{[\neg q]^2 \qquad \dfrac{[p \wedge q]^3}{q}\,\wedge E}{\bot}\,\neg E}{\bot}\,\vee E^{1,2}}{\neg(p \wedge q)}\,\neg I^3$$

This proof can be understood as not only reassuring us that the argument from $\neg p \vee \neg q$ to $\neg(p \wedge q)$ is valid, but that it gives us a way to *justify* the claim of $\neg(p \wedge q)$, appealing only to the assumption $\neg p \vee \neg q$ and using only the primitive inference rules. These rules, as we have indicated, are to be understood as giving us basic constraints on competent use of the logical concepts in a context where we are asking yes/no questions against a background in which certain assumptions may be taken for granted. The resulting justification is straightforward. Working backwards, we can conclude $\neg(p \wedge q)$ because the assumption of $p \wedge q$ together with $\neg p \vee \neg q$ is inconsistent. Why is that inconsistent? Because that assumption is inconsistent with either disjunct of our disjunctive assumption. Why is *that* the case? Because $p \wedge q$ leads to p, which is immediately inconsistent with the first disjunct $\neg p$, and $p \wedge q$ also leads to q and this is immediately inconsistent with the second disjunct. Each of the transitions in this piece of reasoning corresponds to one of the primitive inference rules, and these are the kinds of things we can *follow*.

This seems a very long way away from incompatibility models for non-classical logics of negation. At most, this gives us intuitionistic logic.[6] However, the gap is not as wide we might worry, once we note that the norms governing assertion and proof constrain not only assertions made under the scope of *additive* suppositions (where we add extra assumptions to draw out their consequences, at least locally), but also *alternative* suppositions. To illustrate what I have in mind, consider the following stretch of reasoning.

> Let's grant that either p is necessary or q is necessary. Let's show that the disjunction $p \vee q$ is necessary. To do that, let's suppose things go some way or other, totally arbitrarily. On the option that it was p that was necessary, that means that in this case, we've still got p, and so, $p \vee q$, too. Alternatively, in the option that it was q was necessary, similarly, in that

[5] A consequence of this definition is that, if we declare that the contradiction marker $\bot$ is not a formula, but a structural constant, then $\bot$ can enter into a proof only through a $\neg E$ inference. It never appears as a leaf in a proof tree.

[6] Or minimal logic, if we are prepared to do without $\bot E$ which *does* look rather out of place, it must be said [13]. But for more on $\bot E$, see the discussion in Sect. 3.

case we've got q, and so, $p \lor q$, too. So, on either option, we have $p \lor q$. So, it follows from all this that $p \lor q$ is necessary.

The wording here is clumsy, but one way to articulate the logical structure of this stretch of reasoning is following proof, in which the formulas are annotated with *context* markers.

$$\dfrac{\dfrac{\Box p \lor \Box q \cdot a \qquad \dfrac{\dfrac{[\Box p \cdot a]^1}{p \cdot b}\,\Box E}{p \lor q \cdot b}\,\lor I \qquad \dfrac{\dfrac{[\Box q \cdot a]^2}{q \cdot b}\,\Box E}{p \lor q \cdot b}\,\lor I}{p \lor q \cdot b}\,\lor E^{1,2}}{\Box(p \lor q) \cdot a}\,\Box I$$

Here, the formulas tagged with 'a' are those asserted (or supposed) in our home context, not under the scope of the "suppose things go some way or other" shift in the dialogue. That shift opens up another *context* in which assertions or inferences can go. We tag these formulas with 'b.' The $\Box E$ inference is a context-shifting inference. Granting $\Box A$ (in some context of the discussion) can be used to justify A in other contexts. That is the force of a claim of necessity: to grant $\Box A$ is to agree that even had things gone otherwise, we still would have had A.[7] How, then, do we show that something is necessary? The natural way to do so is to prove it *in an arbitrary context*. There is nothing special about a context that makes it arbitrary, other than the fact that at the step of the reasoning at which the inference us made (here, it is a $\Box I$ step), none of the active commitments are invoked in that context. In our proof, we proved $p \lor q$ (in context b) from an assumption featuring context a. So, at least if we understand deductive practice with necessity in this way, our inference is legitimate.

It is worth reflecting on the two different ways that *supposition* functions in this reasoning. As with unlabelled natural deduction, there is the supposition involved in the making of assumptions to later be discharged (in $\lor E$ or $\neg I$ inferences). This kind of supposition does not involve the addition of a new context. Rather, to prove something from the disjunction $A \lor B$ in context a, the supposition involved when we grant the antecedent A, without any context shift, involves temporarily adding A to our commitments and treating it as if it were asserted (in this very context), without any of the usual requirements of evidence or warrant, using it on the way to prove some desired conclusion. If we can also reach that conclusion when we suppose the other disjunct, B (in the very same context), then we can discharge

[7] This is not to say that all forms of "going otherwise" are the same. There is more than one notion of necessity, and more than one way that we can suppose that things go otherwise [6, 14, 27]. The details of these distinctions, as interesting and fruitful as they are, are not important for the present discussion.

both suppositions, retaining the conclusion. The supposition involved in making this assumption and then discharging it does not involve a change of context, but is a temporary reprieve from the norms of warrant and evidence involved in assertion. It is appropriate for you to query the grounds for my assertion of p, and for me to withdraw my assertion if the answer I give is found wanting. If I ask you to *suppose* p, it is no problem at all if there is no evidence at all for p.

The 'indicative' supposition involved in $\vee E$ and $\neg I$ inferences are to be distinguished from the counterfactual or *subjunctive* supposition involved in our modal reasoning. In this reasoning, when we "suppose things go some way or other", this marks the introduction of a new context, which in our labelled natural deduction proof is represented by way of a new context label, b. In both cases, supposition involves the introduction of new material without any need for justification. Indicative supposition introduces a claim, treated like an assertion, but without the usual quality control measure of calling out for a justification. Subjunctive supposition introduces a new context. In a $\Box E$ inference, we use the claim $\Box A$ granted in some given context a to justify the claim A, given in another context b, which may have been introduced to the discourse by subjunctive supposition.

Notice that in the signed modal proof given above it was natural to allow the disjunction rules to operate on formulas tagged with contexts, even when the contexts shift in the course of our reasoning. If the aim was to extend our propositional logic with modal operators, the natural choice for rules for the propositional connectives is relatively straightforward. Since we ordinarily do not think of context shifts as changing the meaning of the connectives, the rules remain almost unchanged, except for the addition of context labels. The results are compiled in Fig. 2.

With the addition of context labels, the grammar of our proofs has changed. To be precise, given some set $Atom$ of atomic formulas, the set $Form$ of formulas is defined by the following grammar:

$$A\cdot x\ (\mathit{Assumption}) \qquad \frac{A\cdot x \quad B\cdot x}{A\wedge B\cdot x}\,\wedge I \qquad \frac{A\wedge B\cdot x}{A\cdot x}\,\wedge E \qquad \frac{A\wedge B\cdot x}{B\cdot x}\,\wedge E$$

$$\frac{A\cdot x}{A\vee B\cdot x}\,\vee I \qquad \frac{B\cdot x}{A\vee B\cdot x}\,\vee I \qquad \frac{A\vee B\cdot x \quad \begin{matrix}[A\cdot x]^j\\ \Lambda\\ C\end{matrix} \quad \begin{matrix}[B\cdot x]^k\\ \Lambda'\\ C\end{matrix}}{C}\,\vee E^{j,k}$$

$$\frac{\begin{matrix}[A\cdot x]^i\\ \Lambda\\ \bot\end{matrix}}{\neg A\cdot x}\,\neg I^i \qquad \frac{\neg A\cdot x \quad A\cdot x}{\bot}\,\neg E \qquad \frac{\bot}{A\cdot x}\,\bot E$$

Fig. 2 Natural deduction proof rules with *contexts*

$$Form \quad := \quad Atom \mid \neg Form \mid (Form \wedge Form) \mid (Form \vee Form)$$

and the constituents of proofs are now *judgements*, consisting of a formula together with a context label (selected from some set Ctx), or the contradiction marker, $\bot$.

$$Judg \quad := \quad \bot \mid Form \cdot Ctx$$

In the statement of the rules given in Fig. 2, the A and B are arbitrary formulas, x is a context label, and C is a judgement (either a signed formula or a contradiction marker).

In most of these rules, the contexts are fixed from premise to conclusion. From A in a given context, I can conclude $A \vee B$ in *that* context. If I have A and B in the same context, I can conclude $A \wedge B$ in that context. And in particular, given our particular interest in negation, we reach a contradiction from A and $\neg A$ *in the same context*.

Some details deserve comment. The first is that the contradiction marker serves only as a sign that our assumptions, taken together, clash. As we have seen, the assumptions leading up to those contradictory conclusions may well have been granted in some other contexts, so there is no sense that we need locate the blame for that clash in any particular context.

The second detail is the special role of the conclusion C in the $\vee E$ rule. Here, C is anything that can be the conclusion of a proof: either a tagged formula or the contradiction marker. In the case of a tagged formula, there is no requirement that this tagged formula be tagged with the same context marker as the disjunctive assumption leading to that conclusion. We saw this in our proof from $\Box p \vee \Box q$ to $\Box(p \vee q)$. The premise was granted in context a, and so, we split into two cases in that context. In both of those two cases, we conclude $p \vee q$ in the fresh context b, and so we hold onto that conclusion on the basis of the disjunctive assumption. What drives the case-based reasoning of a disjunction elimination is that the same conclusion is derived in each case, and each case in question is opened up in the *same* context as the disjunctive conclusion being eliminated. The conclusion of each inference must be the same, but it need not be another claim in that original context. It could also be a contradiction marker, as seen in the first (untagged) natural deduction proof on page 347.

The final thing to note about natural deduction proofs with context labels is that they at least open the way to understand how a possible world semantics might truly *model* the logic of necessity, without having to take some antecedently given notion of possible worlds as the starting point of our semantic theorising. While it is understandably tempting to take the context label in a proof as naming a *world*, there is no need to do so. There is no need to take the world labels to be referring expressions of any kind, any more than the superscript in the second variable in the ordered pair expression $\langle x, x' \rangle$ should be taken as referring to anything. The role of the context labels in proof exhausted by their identity and difference. $A \cdot a$ and $\neg A \cdot a$ clash, while $A \cdot a$ and $\neg A \cdot b$ need not, since there is some context shift between the two judgements. It is that shift in context that is represented by the different labels. Of course we can—if we like—treat labels as referring expressions, and evaluate tagged formulas in a possible worlds model. We can think of assertions in some fixed context as evaluated with respect to some *possible world*. Perhaps this interpretation can do some explanatory work. However, it is a further question as to which direction the explanation goes. Do we justify the proofs on the terms that they fit with possible worlds models, or do we justify the models on the basis that they provide models that fit with the antecedently grasped rules for reasoning with modal expressions and the context shifts they involve.[8] The question of the direction of explanation (whether the model theory or the proof theory comes first when doing semantics) is related to the issue of 'semantic pollution' in proof theory. Labelled proof systems are rightly called into question as importing notions from model theory into proof calculi [18, 21]. There is no space to enter into a full discussion of the extent to which labelled proof systems like those under consideration here are semantically polluted. However, insofar as there *are* context shifts at work in natural reasoning, and that in the one dialogue we might grant a context in which A holds, and another in which $\neg A$ holds, then a natural deduction representation of such reasoning which keeps track of different contexts by tagging assertions, suppositions and the like with different labels need not immediately fall prey to the criticism that it is semantically polluted in the sense of importing notions from some antecedently given model-theoretic semantics. If the labels are used to represent contexts in a discourse, then they have some claim to be as semantically innocent as other devices used to render the inferential structure of reasoning.

Since any questions concerning the nature of contexts and worlds and the pollution or purity of proof rules with context labels remain live when we talk of *inconsistent* or *incomplete* contexts (or 'worlds'), and since new questions are added to them, it seems worthwhile to see if a story like this one can be told for incompatibility models for negations, so let's turn our attention there.

2 Proofs with Perp

If we start with the idea that proofs can combine claims made in different contexts, the issue immediately arises: what, if anything, follows if we conclude $\neg A$ one context and A in another? The rules discussed in the previous section (see Fig. 2) give us no advice on this issue. However, having learned the lessons Dunn taught us, we see that there *is* a way to describe what follows from a negation $\neg A$ in one context a and its negand A, in another context b, and that is that the contexts a and b are incompatible, since what is claimed of b is ruled out by what is claimed of the other. So, let's mark that in our reasoning writing '$a \perp b$' for this incompatibility claim.

[8] It is possible to develop the explanatory connection between proofs and models in this second direction, justifying models in terms of an antecedently motivated system of proof rules [26]. Although the proof system discussed in that paper is a hypersequent calculus and not a labelled natural deduction system, the broad lesson is the same.

$$\frac{\neg A \cdot a \quad A \cdot b}{a \perp b}\,\neg E \qquad \frac{\begin{array}{c}[A \cdot b]^i \\ \Lambda \\ a \perp b\end{array}}{\neg A \cdot a}\,\neg I^{i\dagger}$$

†(In $\neg I$, b appears nowhere else in the undischarged assumptions of the proof Λ.)

Fig. 3 Negation rules with perp

With our candidate for a negation *elimination* rule, what would be a matching *introduction* rule? Harmony considerations dictate that we are in position to *derive* $\neg A$ (in a given context a) whenever we can vouchsafe the transition from A in some context to the conclusion that a is incompatible with *that* context—whichever context that turns out to be. So, the natural negation introduction rule is no real surprise. The *incompatibility* rules for negation are given in Fig. 3. If we have some labelled proof from $A \cdot b$ to the conclusion $a \perp b$, then provided we have made no other assumption in context b, we can conclude $\neg A$ in context a. That context, indeed, has what it takes to rule A out.

With candidate introduction and elimination rules using the notion of incompatibility between contexts, our rules no longer feature the context neutral contradiction marker, and so we revise the grammar of judgements used in our proofs as follows:

$$Judg \quad := \quad Ctx \perp Ctx \mid Form \cdot Ctx$$

When we replace the negation rules from Fig. 2 those in Fig. 3, we have a simple, well-behaved natural deduction system for the logic of incompatibility models, and it is one can motivate the logic by general considerations of reasoning with negation *in contexts*.[9] To illustrate, here is a proof, from $\neg p \vee \neg q$ to $\neg(p \wedge q)$, in the same context, a.

$$\dfrac{\dfrac{\neg p \vee \neg q \cdot a \qquad \dfrac{[\neg p \cdot a]^1 \quad \dfrac{[p \wedge q \cdot b]^3}{p \cdot b}\,\wedge E}{a \perp b}\,\neg E \qquad \dfrac{[\neg q \cdot a]^2 \quad \dfrac{[p \wedge q \cdot b]^3}{q \cdot b}\,\wedge E}{a \perp b}\,\neg E}{a \perp b}\,\vee E^{1,2}}{\neg(p \wedge q) \cdot a}\,\neg I^3$$

[9] While this chapter was in final review, a reader let me know that incompatibility rules for negation of these form were independently formulated by Edwin Mares, in his 2024 book *The Logic of Entailment and its History* [15, p. 183].

Notice that this proof has the same general shape as the unlabelled proof of $\neg(p \wedge q)$ from the premise $\neg p \vee \neg q$. The difference is the presence of the labels. We need to check that the negation introduction step at the conclusion of this proof satisfies the side-condition of that rule, that the context label b, occurring in the conclusion and in the assumptions discharged at that step (the two occurrences of $p \wedge q \cdot b$) occur nowhere else among the undischarged assumptions, and so, the b in this part of the proof is perfectly *general*. This is indeed the case, so the proof is well-formed.

There are a few notable features of labelled proofs: first, incompatibility claims $a \perp b$ arise only *inside* proofs, and never as assumptions occurring in the leaves. (The smallest proof is a labelled formula $A \cdot a$ on its own, in which the formula appears as both premise and conclusion.) Second, notice that the proof rules each have the same shape as the unlabelled proof rules in Fig. 1. If we take *any* labelled proof and simply erase all the labels, the result is a proof in intuitionistic natural deduction. Third, as we had hoped as we introduced these proofs, they precisely match Dunn's incompatibility models for negation. We start with *soundness*: and to state this fact we need to be precise with how the premises and conclusions of a proof are to be evaluated on a model. We will use the following, natural, definition:

Definition 1 (Holding, Validity) The judgement $A \cdot a$ HOLDS in $\langle P, \perp, \Vdash \rangle$, relative to the assignment p of points to labels when $p(a) \Vdash A$. The the judgement $b \perp c$ HOLDS in the model when $p(b) \perp p(c)$. An argument from the premises X to some judgement C is said to be VALID ON THE MODEL if and only if for any assignment p, if the premises hold in the model, so does the conclusion.

This definition makes explicit in the *models* what is already explicit in the labelled proof theory. There are two notions of consequence at play in our logical systems. First, there is *internal* notion of consequence, relative to a fixed choice of context (A entails B where we have a proof from $A \cdot a$ to $B \cdot a$, or correspondingly, in any model, and any point x in the model, if $x \Vdash A$, then $x \Vdash B$). For internal consequence, the premises and conclusions are formulas, and each is evaluated at the same context. Our definition here utilises a more general *external* notion of consequence, which allows contexts to vary among premises and conclusions. The premises and conclusions are taken from the class of judgements, which carry along their contexts with them. Since labelled natural deduction proofs operate on judgements, our soundness and completeness proofs show that the more general *external* notion of consequence definable by proofs and by models agree. Since internal consequence is definable in terms of external consequence (by restricting our attention to judgements with a single given context), soundness and completeness for internal consequence follows immediately.

With this definition in place, we have the following fact:

Fact 1 (Soundness (For Perp Proofs)) If Λ is a labelled proof with premises X and conclusion $B \cdot c$ [or $b \perp c$], then the argument from X to $B \cdot c$ [or $b \perp c$] is valid on every incompatibility model.

Proof This is shown by a straightforward induction on the construction of the proof Λ, appealing to the match between the proof rules and the evaluation conditions for the connectives in incompatibility models. The base case consists of a proof with the single premise $A \cdot a$, with the very same conclusion, so the induction hypothesis holds trivially. The cases for the distributive lattice connectives are straightforward and need no extensive discussion, but let's consider $\vee E$ since it allows for context shifting. Let's fix on some incompatibility model $\langle P, \perp, \Vdash \rangle$ and some assignment p sending labels to points, and we suppose the hypothesis applies to the labelled proofs Λ, Λ', Λ'' leading up to this application of $\vee E$:

$$\dfrac{\begin{matrix} X \\ \Lambda \\ A \vee B \cdot a \end{matrix} \qquad \begin{matrix} Y,\ [A \cdot a]^j \\ \Lambda' \\ C \end{matrix} \qquad \begin{matrix} Z,\ [B \cdot a]^k \\ \Lambda'' \\ C \end{matrix}}{C} \vee E^{j,k}$$

So, the arguments from X to $A \vee B \cdot a$, and from $Y, A \cdot a$ to C and $Z, B \cdot a$ to C are valid on our incompatibility model $\langle P, \perp, \Vdash \rangle$. Consider now our assignment p. We know that if each of the members of X holds in the model, so does $A \vee B \cdot a$, and that if the members of $Y, A \cdot a$ hold in the model, so does C, and if the members of $Z, B \cdot a$ hold in the model, C does as well. To show that the entire proof is valid in the model, suppose that each of the members of X, Y, Z hold in the model. Since the members of X hold, so does $A \vee B \cdot a$, and by the disjunction clause ($a \Vdash A \vee B$ iff $a \Vdash A$ or $a \Vdash B$), we know that either $A \cdot a$ holds in the model or $B \cdot a$ holds in the model. In the first case, the validity of the argument from $Y, A \cdot a$ to C assures us that C holds in the model, and in the second case, it is the validity of the second argument that does the job, and so, in either case, the desired conclusion indeed holds in the model. (Notice that this reasoning applies whether the conclusion C is a labelled formula, or an incompatibility claim. The only condition requirement is that it be derived in either case.)

For the negation rules, consider $\neg E$:

$$\dfrac{\begin{matrix} X \\ \Lambda \\ \neg A \cdot a \end{matrix} \qquad \begin{matrix} Y \\ \Lambda' \\ A \cdot b \end{matrix}}{a \perp b} \neg E$$

By hypothesis, the arguments from X to $\neg A \cdot a$ and from Y to $A \cdot b$ are valid on our model. So, to show that the whole argument is valid, suppose that both X and Y hold in our model under assignment p. It follows that $p(a) \Vdash \neg A$, and $p(b) \Vdash A$. Given the evaluation clause for negation, this means that $p(a) \perp p(b)$, and hence, $a \perp b$ holds in our model, as desired.

Finally, consider $\neg I$:

$$\dfrac{\begin{array}{c} X,\ [A \cdot b]^i \\ \Lambda \\ a \perp b \end{array}}{\neg A \cdot a}\ \neg I$$

The induction hypothesis is that the argument from $X, A \cdot b$ to $a \perp b$ is valid on our model. We wish to show that the argument from X to $\neg A \cdot a$ is *also* valid on our model. The one extra piece of information is that the label b is absent from X. So, let's suppose that p is some assignment for which X holds in our model. We wish to show that $\neg A \cdot a$ holds in our model, too. For that, we need to show that whenever $c \Vdash A$, then $p(a) \perp c$. So, suppose b is a point in the frame where $c \Vdash A$. Select a new assignment p' of points to labels such that $p'(b) = c$ while $p'(d) = p(d)$ for every *other* label d. (This includes the label a.) This includes the labels occurring in the premises X, since b is absent from X. The induction hypothesis applied to *this* assignment p' tells us that since every member of $X, A \cdot y$ hold under *this* assignment, so does the conclusion $a \perp b$. Since $p'(a) = p(a)$ and $p'(b) = c$, we have $p(a) \perp c$ as desired, and so, since c was arbitrary, we have shown that $\neg A \cdot a$ holds in our model. ■

Now, since nothing in this reasoning assumed *any* features of the incompatibility relation $\perp$, this soundness fact tells us that the logic arising out of these proof rules is *very* weak. The non-symmetric model given above[10] provides a counterexample to the argument from $p \cdot a$ to $\neg\neg p \cdot a$. In other words, there is no way to assign labels to the unlabelled proof

$$\dfrac{\dfrac{[\neg p]^1 \quad p}{\perp}\ \neg E}{\neg\neg p}\ \neg I^1$$

—or to any other unlabelled proof from p to $\neg\neg p$—satisfying the conditions of our labelled calculus. Models where compatibility fails to be reflexive (and so, we allow $x \perp x$ for some points x) give us means to satisfy contradictory conjunctions $p \wedge \neg p$ at point, and so, to provide ways to fail to validate $\neg(p \wedge \neg p)$. This means that we also have no way to convert this

$$\dfrac{\dfrac{\dfrac{[p \wedge \neg p]^1}{\neg p}\ \wedge E \qquad \dfrac{[p \wedge \neg p]^1}{p}\ \wedge E}{\perp}\ \neg E}{\neg(p \wedge \neg p)}\ \neg I^1$$

[10] See footnote 2.

—or any other proof to this conclusion from no premises—into a properly labelled proof. Of course, there are simple ways to extend the calculus with symmetry or irreflexivity rules to allow for labellings of these proofs:

$$\dfrac{\dfrac{\dfrac{[\neg p\cdot b]^1 \qquad p\cdot a}{b\perp a}\,\neg E}{a\perp b}\,\mathit{Sym}}{\neg\neg p\cdot a}\,\neg I^1 \qquad\qquad \dfrac{\dfrac{\dfrac{\dfrac{\dfrac{[p\wedge\neg p\cdot b]^1}{\neg p\cdot b}\,\wedge E \qquad \dfrac{[p\wedge\neg p\cdot b]^1}{p\cdot b}\,\wedge E}{b\perp b}\,\neg E}{\perp}\,\mathit{Irrefl}}{a\perp b}\,\perp E}{\neg(p\wedge\neg p)\cdot a}\,\neg I^1$$

Rather than exploring the details of different rules such as *Sym* and *Irrefl* and the corresponding frame conditions (for space reasons only—the proving soundness for the extended proof systems with respect to the corresponding class of models is straightforward) let us end this section discussing the *Completeness* Theorem.

Fact 2 (Completeness (For Perp Proofs)) If an argument with premises X and conclusion $B\cdot c$ [or $b\perp c$] is valid on every incompatibility model, then there is some labelled proof with from X to $B\cdot c$ [or $b\perp c$].

There are a number of different ways to show that our labelled proof rules are complete for validity on incompatibility models. We will sketch a direct proof of completeness of an extended proof system in Sect. 4. Here I will take a different approach, favouring theft over toil.

Proof (Sketch) We borrow the completeness result for the *sequent* natural deduction system of for distributive lattice logic with split negation, given in *An Introduction to Substructural Logic* (ISL), Chapter 11 [24]. There, it is shown that incompatibility frames provide a semantics with respect to which a simple *unlabelled* sequent calculus for distributive lattice logic extended with two rules for *two* negations, is sound and complete:

$$\dfrac{A\vdash\neg B \qquad X\vdash B}{X\vdash{\sim}A}\,\neg E/{\sim}I \qquad\qquad \dfrac{A\vdash{\sim}B \qquad X\vdash B}{X\vdash\neg A}\,{\sim}E/\neg I$$

Incompatibility models are extended to interpret the converse negation $\sim$ in the obvious way, using the *converse* of incompatibility.[11] The two negations are modelled together, like this:

[11] To be precise, in ISL, the models are described in terms of *compatibility* rather than an *incompatibility* relation, but the easy translation (defining $\perp$ as the negation of compatibility) works as expected.

- $a \Vdash \neg A$ if and only if for each b, if $b \Vdash A$ then $a \perp b$,
- $a \Vdash {\sim}A$ if and only if for each b, if $b \Vdash A$ then $b \perp a$.

We then can show that unlabelled sequent proofs (which are complete for compatibility frames) can be translated into labelled proofs. A sequent consists of a multiset X of formulas on the left-hand side and a single formula on the right. For any label a, and any natural deduction proof π ending in a sequent $X \vdash A$, the a-translation of π will be some labelled proof π_a with undischarged assumptions selected from X, each labelled with a, and with conclusion $A \cdot a$. The axiomatic sequent $A \vdash A$ is a-translated into the atomic proof $A \cdot a$. The rules for conjunction and disjunction act exactly as one would expect with a homophonic translation from one system to the other. For sequent natural deduction proofs using the negation rules we translate as follows, using the expected converse *I* and *E* rules for the negation $\sim$.

$$\dfrac{\begin{array}{cc}\pi & \pi' \\ A \vdash \neg B & X \vdash B\end{array}}{X \vdash {\sim}A}\ \neg E/{\sim}I \qquad \Rightarrow \qquad \dfrac{\dfrac{\begin{array}{cc}[A \cdot b]^1 & X \cdot a \\ \pi_b & \pi'_a \\ \neg B \cdot b & B \cdot a\end{array}}{b \perp a}\ \neg E}{{\sim}A \cdot a}\ {\sim}I^1$$

$$\dfrac{\begin{array}{cc}\pi & \pi' \\ A \vdash {\sim}B & X \vdash B\end{array}}{X \vdash \neg A}\ {\sim}E/\neg I \qquad \Rightarrow \qquad \dfrac{\dfrac{\begin{array}{cc}[A \cdot b]^1 & X \cdot a \\ \pi_b & \pi'_a \\ {\sim}B \cdot b & B \cdot a\end{array}}{a \perp b}\ {\sim}E}{\neg A \cdot a}\ \neg I^1$$

The end of the translation process is some labelled proof from the premises to the conclusion, littered, alas, with the converse negation $\sim$ which may be foreign matter to the premises and conclusion under discussion. This is no problem. We have been delivered a labelled natural deduction proof. It can be normalised, as usual (remember, if we delete the labels, this is simply an intuitionist propositional proof, satisfying extra conditions). The normalisation process (as we will see in the final section) preserves the special properties of the negation rules, and so, we can transfer our proof into a thoroughly normal one, which will satisfy the subformula property. This will delete the extraneous converse negations from our proof and the result is a proof in the original calculus. ■

With soundness and completeness proved, we see that the external consequence relation definable by proofs agrees with that definable by way of models, and so the *internal* consequence relations given by proofs and given by models also agrees. Since the internal logic of general incompatibility frames (in the language of conjunction, disjunction and negation) is a proper sublogic of first-degree entailment, it is clearly a *relevant* logic, and so, the internal entailment relation given by way of proofs is also relevant. This raises the interesting question of whether the *external* consequence relation, defined over arbitrary judgements, is also, in any sense a *relevant* consequence relation. The entailment from $p \cdot a$ and $\neg p \cdot b$ to $a \perp b$ may give some pause, since when we erase context labels we get the inference from $p, \neg p$ to $\perp$ which is problematic on at least *some* relevance grounds [34]. However, it is clear that there is some shared subject matter in the inference from $p \cdot a$ and $\neg p \cdot b$ to $a \perp b$: the context labels are playing an important role. It is in virtue of the claim of p at a and $\neg p$ at b that a and b are taken to be incompatible. There seems to be no irrelevance here, but what relevance there is seems to be at the level of *context* rather than an identifiable *content* asserted *in* one context or other. Space forbids an extended account of relevance for external consequence, so we must leave this matter for another occasion.

What, then, of the de Morgan negation represented by Routley star models? On these models, the argument from $\neg\neg A$ to A comes out as *valid*, but no straightforward rules governing $\perp$ will give us a means to bridge that gap—at least, none will do so if the added rules remain intuitionistically acceptable when we erase labels. We'll have to do something else to bridge the gap. For that, we turn to natural deduction for *classical logic*.

3 Assertion and Denial

There are many ways to modify the rules for Prawitz-style natural deduction to allow for properly classical proofs. The simplest thing to do is to add extra principles to some or other connective, beyond the harmonious introduction and elimination rules. This will *not* be our strategy here, since it is possible keep the connective rules fixed, and vary what Belnap calls the 'antecedently given context of deducibility' [3], the underlying framework in which proofs are constructed. To do this, we will pay attention to the kinds of speech acts we use in dialogue. Not only do we assert and suppose in different contexts. We also *deny*. Taking assertion and denial to be on equal footing has become an important theoretical perspective on proof-theoretical semantics for classical logic [11, 12, 19, 25, 29, 32, 33].

However, instead of complicating our natural deduction system to incorporate positively and negatively signed judgements at every step (which results in a *fully bilateral* natural deduction system), it suffices for our purposes to be only very *mildly* bilateral, by allowing for negative *premises* in our proofs, and leaving the rest of the proof system unchanged.[12] So, for our original unlabelled natural deduction system, we allow not only formulas to occur in the leaves of a proof, we also allow *slashed* formulas, $\not A$, indicating the (primitive) *denial* of the claim A. If to assert A is to answer *yes* to the polar question A?, then to deny A is to answer the question A? with a *no*.

[12] For a more extensive treatment of this mildly bilateralist natural deduction, its connection to Michel Parigot's $\lambda\mu$-calculus for classical logic [16, 17], and some of its many pleasant properties, see "Structural Rules in Natural Deduction with Alternatives" [28].

$$\frac{\cancel{A}\quad A}{\bot}\uparrow(\mathit{Store})\qquad \begin{array}{c}[\cancel{A}]^i\\ \Pi\\ \dfrac{\bot}{A}\end{array}\downarrow^i(\mathit{Retrieve})\qquad\qquad \frac{\cancel{A\cdot a}\quad A\cdot a}{\bot}\uparrow\qquad \begin{array}{c}[\cancel{A\cdot a}]^j\\ \Lambda\\ \dfrac{\bot}{A\cdot a}\end{array}\downarrow^j$$

Fig. 4 *Denial* rules, unlabelled and labelled

The rules governing this primitive denial are straightforward, and are presented in Fig. 4 (focus on the first pair, of *unlabelled* rules, for the moment). Asserting and denying the same thing is, of course, inconsistent. On the other hand, if we reach an inconsistency, one way out of that dead end is to say 'yes' to something we had previously ruled out. (These rules are named '*Store*' and '*Retrieve*' because we can think of the negatively tagged formulas as temporarily stored conclusions, which can be retrieved from storage when needed.)[13]

With the addition of these purely structural rules, the existing logical rules for our unlabelled system now suffice for all of classical logic. Here is a proof from $\neg\neg p$ to p:

$$\dfrac{\dfrac{\neg\neg p \qquad \dfrac{\dfrac{[\cancel{p}]^2 \quad [p]^1}{\bot}\uparrow}{\neg p}\neg I^1}{\bot}\neg E}{p}\downarrow^2$$

At the first step we have assumed p and denied it at the same time, to reach a contradiction. We immediately blame that contradiction on the assumption of p, discharging that to deduce $\neg p$. So, at this stage of the proof we have inferred $\neg p$ in a context where p is denied. This, of course, clashes with the assumption $\neg\neg p$, and so, in the presence of *this* contradiction we retrieve the denied p, to conclude p, as desired. We recover classical reasoning by paying the coin of two structural rules involving denials.

[13] In the labelled calculus, our family of judgements is expanded to allow for *denial* judgements in addition to *assertions*, and to return the unadorned contradiction marker to the fold:

$$Judg \quad := \quad \bot \mid Ctx \perp Ctx \mid Form \cdot Ctx \mid \cancel{Form \cdot Ctx}$$

Notice that in our proofs, *denial* judgements occur only ever as leaves in proof trees, and never as conclusions.

Adding these two rules to the unlabelled natural deduction system has an added benefit. The awkwardly justified $\bot E$ rule is now redundant, and can be replaced by the *Retrieve* rule, in the special (*explicitly* irrelevant) case where we retrieve *zero* instances of a stored A.

$$\dfrac{p}{q \to p}\,{\to}I \qquad\qquad \dfrac{\dfrac{\neg p \quad p}{\bot}\,\neg E}{q}\,\downarrow$$

In this case, in just the same way that the classically valid inference from p to $q \to p$ arises from assuming p, and discharging *zero* occurrences of the assumption of q to conclude $q \to p$, the classically valid inference from $p, \neg p$ to q arises in the same way. The p and $\neg p$ lead to a contradiction, and we retrieve *zero* instances of the denied q to conclude q. The two kinds of irrelevance present in classical logic come down to the same underlying phenomenon, vacuous discharge of positive or negative assumptions, and if we wish to be properly *relevant*, one way to do this is to ban vacuous discharge, whether positive or negative.

Instead of exploring the finer details of this classical natural deduction system, let's immediately see how we can use ideas from it in the setting of a *labelled* natural deduction system. The generalisation we use, as seen in the second pair of rules in Fig. 4, is the simplest possible. Just as in the unlabelled system we allowed for leaves of the tree to be slashed to represent denial, we do the same thing here, so the denial of A in the context x is represented by slashing the entire judgement, like so: $\cancel{A\ x}$. The sense of such a judgement is straightforward: just as the assertion, or supposition, of some content can have a different significance in different contexts (recall the modal proof from $\Box p \vee \Box q$ to $\Box(p \vee q)$, in which $p \vee q$ was proved in an 'arbitrary' context), so, too, can a *denial*. As with other kinds of bilateral proof systems, we take it that the denial of a claim A might in some sense be conceptually prior to the assertion (or supposition) of its negation, and that denials, like assertions, are sensitive to shifts in context. This is marked in this proof system with the *denial* judgement $\cancel{A\ x}$, in which A is denied in context x.

However we are to understand the *significance* of such speech acts, this one addition to the formal system will allow us to construct natural deduction proofs for the Routley star semantics, as we will see in the next section.

4 Proofs with Star

Recall the truth conditions for negation in Routley star models:

- $x \Vdash \neg A$ if and only if $x^* \nVdash A$.

$$\frac{\neg A\cdot a \qquad A\cdot a^*}{\bot}\,\neg E \qquad\qquad \frac{\begin{matrix}[A\cdot a^*]^i\\ \Lambda\\ \bot\end{matrix}}{\neg A\cdot a}\,\neg I^i \qquad\qquad \frac{A\cdot a}{A\cdot a^{**}}\,{}^{**}I \qquad\qquad \frac{A\cdot a^{**}}{A\cdot a}\,{}^{**}E$$

Fig. 5 Negation rules with star

If we use this modelling condition in a frame with a map $^* : P \to P$ of period two (so $x^{**} = x$), then the appropriate proof rules for negation to match these models will be as given in Fig. 5.[14]

Despite the presence of the Routley star with period two, the $\neg I/\neg E$ rules, together with the label-manipulating rules $^{**}I$ and $^{**}E$ do not suffice to derive the constructively invalid inference from $\neg\neg p$ to p, or the constructively invalid de Morgan's law. These rules, as usual, remain constructively valid if we erase the labels, so on their own, they would not be enough to generate the non-constructive de Morgan validities. In the presence of the denial rules, however, they suffice. For example, we have this proof for double negation elimination:

$$\dfrac{\dfrac{\dfrac{\neg\neg p\cdot a \qquad \dfrac{\dfrac{[\cancel{p\cdot a^{**}}]^2 \qquad [p\cdot a^{**}]^1}{\bot}\uparrow}{\neg p\cdot a^*}\,\neg I^1}{\bot}\,\neg E}{p\cdot a^{**}}\downarrow^2}{p\cdot a}\,{}^{**}E$$

[14] The class of judgements used in Routley star proofs is again, different. Now, we generate new general contexts by closing our original context labels under the star operator.

$$GCtx \;:=\; Ctx \mid GCtx^* \qquad Judg \;:=\; \bot \mid Form\cdot GCtx \mid \cancel{Form\cdot GCtx}$$

This longer proof suffices for the constructively invalid de Morgan's Law:

$$\dfrac{\dfrac{[\cancel{\neg p \vee \neg q \cdot a}]^4 \qquad \dfrac{\dfrac{\dfrac{[\cancel{\neg p \vee \neg q \cdot a}]^4 \qquad \dfrac{\dfrac{\dfrac{[\cancel{\neg p \cdot a}]^3 \qquad \dfrac{\dfrac{\neg(p \wedge q) \cdot a \qquad \dfrac{[p \cdot a^*]^1 \quad [q \cdot a^*]^2}{p \wedge q \cdot a^*}\,\wedge I}{\bot}\,\neg E}{\neg p \cdot a}\,\neg I^1}{\bot}\,\uparrow}{\neg q \cdot a}\,\neg I^2}{\neg p \vee \neg q \cdot a}\,\vee I}{\bot}\,\uparrow}{\neg p \cdot a}\,\downarrow^3}{\neg p \vee \neg q \cdot a}\,\vee I}{\bot}\,\uparrow}{\neg p \vee \neg q \cdot a}\,\downarrow^4$$

We can combine the *denial* rules with the incompatibility rules for negation, too. With either the star rules or the incompatibility rules in place for negation, the resulting proofs are still sound for the logic modelled by our star or perp models. To show this, we need to expand our definition of validity and holding, to allow for slashed premises in our arguments.

Definition 2 (Holding and Validity, Again) For any model with an evaluation $\Vdash$, the slashed labelled formula $\cancel{A \cdot a}$ HOLDS in the model when $p(a) \not\Vdash A$. The inconsistency sign $\bot$ never holds in a model.

In any *star* model, a map p from labels to points must respect stars on labels: that is, $p(a^*) = p(a)^*$ for each label a. Then, as before, an argument from X to C is VALID ON A MODEL, if and only if, under any assignment p, if the members of X hold in the model, so does C.

With this definition, soundness is straightforward to state and prove.

Fact 3 (Soundness For Labelled Proofs with Denial) If Λ is a labelled incompatibility [star] proof with premises X and conclusion $B \cdot c$ [or $b \perp c$, or $\bot$], then corresponding argument is valid on every incompatibility [star] model.

Proof The proof takes the same shape as our previous soundness proof, an induction on the construction of the labelled proof. The new cases to consider are the *Store* and *Retrieve* rules, as well as the star rules. Consider the *Store* and *Retrieve* rules first. Suppose the argument from X to $A \cdot a$ is valid on our model, and we extend some proof from X to $A \cdot a$ with the *Store* rule, adding the premise $\cancel{A \cdot a}$, leading to the conclusion $\bot$. Since on any choice of values for the labels in which the members of X hold, $A \cdot a$ holds, there is no assignment of values in which the members of X and $\cancel{A \cdot a}$ holds, so the argument from X, $\cancel{A \cdot a}$ to $\bot$ is indeed valid on our model.

Conversely, if we have a proof Λ from X, ~~$A \cdot a$~~ to $\perp$ and this argument is valid on our model, this means there is no assignment of points to labels that such that every member of X, ~~$A \cdot a$~~ holds. So, for any assignment of points to labels where every member of X holds, $A \cdot a$ must hold, too, and the argument from X to $A \cdot a$ is valid on our model, too.

Soundness is straightforward for the $^{**}I$ and $^{**}E$ rules, since any assignment of points to labels respects star, and in any Routley star model, $a^{**} = a$ for every point a. Soundness for $\neg I$ and $\neg E$ is also straightforward, given the truth conditions for negations on Routley star models. There is *no* model in which $\neg A \cdot x$ and $A \cdot x^*$ can both hold, on the same assignment of values to points. On the other hand, if the argument from X and $A \cdot x^*$ to $\perp$ is valid on a model, that means that there is no assignment of points to labels where X and $A \cdot x^*$ holds. That means on any assignment where which each member of X holds is one where $A \cdot x^*$ doesn't, but that means $\neg A \cdot x$ holds, and so, the argument from X to $\neg A \cdot x$ is valid, as desired. ∎

The *completeness* fact for proofs with denial is straightforward to state and prove.

Fact 4 (Completeness for Labelled Proofs with Denial) An argument from X and C valid in every incompatibility [star] model, also has some labelled incompatibility [star] proof.

Proof (Sketch) This proof is a standard canonical model construction, and as usual, we prove the contrapositive. If there is no proof from X to the conclusion C, we construct a model and assignment of points to labels in which each member of X holds, under that assignment and C fails to hold, again, under that assignment. Instead of constructing the model out of maximal sets of formulas (or maximal prime/ideal pairs, as is appropriate for distributive lattice logics [24]), the construction is rather straightforward. We start with our the pair $\langle X, \{C\}\rangle$, which is *available*, in the sense that there is no proof from any subset of elements on the left side to any member of the right side. Our aim is to fill this out, constructing a partition $\langle Yes, No\rangle$ of the whole language, where the statements on the left side are ruled *in* while those on the right are ruled *out*. The constraint, as we fill out our pair of sets is that the pair *remains* available as we fill it out. We must be careful, since the constraint of availability is stricter than the condition that is no proof from members of *Yes* to any members of *No*, since the pair $\langle\{p \vee q \cdot a\}, \{p \cdot a, q \cdot a\}\rangle$ is indeed available in that weak sense (there is no proof from $p \vee q \cdot a$ to $p \cdot a$, and nor is there a proof to $q \cdot a$) but there will be no model in which $p \vee q \cdot a$ holds and both $p \cdot a$ and $q \cdot a$ doesn't. No, this pair shouldn't be available, and the appropriate definition of availability of a pair $\langle Y, N\rangle$ is that there is no proof from members of Y together with *denied* (slashed) members of N to a member of N. There is a proof from $p \vee q \cdot a$, ~~$p \cdot a$~~ to $q \cdot a$, and so the pair $\langle\{p \vee q \cdot a\}, \{p \cdot a, q \cdot a\}\rangle$ is not available, as desired.

So, the construction of our partition $\langle Yes, No\rangle$ can proceed in the way familiar from Henkin completeness proofs. We take for ourselves a countable supply of labels, and enumerate our collection of labelled formulas and compatibility claims.

Then, given any finite available pair $\langle Y, N \rangle$, we select the next object in our enumeration C (either a labelled formula or an incompatibility statement). We consider whether we have a proof from $Y, \not{N}$ to C, where we allow ourselves to appeal not only to formulas in Y, slashes of formulas occuring in N.[15] If we do, then there is no proof from $Y, \not{N}, C$ to $\bot$. If we *did* have such a proof, we could compose this with the proof from $Y, \not{N}$ to C, to give us a proof which leads us from $Y, \not{N}$ to $\bot$, in which case $\langle Y, N \rangle$ would not be available. So, either $\langle Y \cup \{C\}, N \rangle$ is available or $\langle Y, \{C\} \cup N \rangle$ is. Choose one to be $\langle Y', N' \rangle$, and continue the process until the limit, where every labelled formula and compatibility statement has been chosen. The result is a partition of the language, and it is available.

There is one complication to the process. If the labelled formula C is a *negation* $\neg A \cdot a$, we add it to the *right* of the partition, we must add a *witness*, too. In this case, we know that $\langle Y, \{\neg A \cdot a\} \cup N \rangle$ is available. So, there is no proof from $Y, \not{N}$ to $\neg A \cdot a$. So, choose a fresh label b. We know that there is no proof $Y, \not{N}, \not{\neg A \cdot a}, A \cdot b$ to $a \perp b$, since if there *were* such a proof, we could extend it by $\neg E$ (and some store/retrieve steps) to give a proof of $\neg A \cdot a$ from $Y, \not{N}$, and we know there is no such proof. So, it follows that $\langle Y \cup \{A \cdot b\}, \{\neg A \cdot a, a \perp b\} \cup N \rangle$ is also available, so whenever our construction asks us to make a negation *not* true, we add a fresh witness label and witnessing facts in this way.

Then, the limit $\langle Yes, No \rangle$ of this process will be a description of a compatibility model. The class of points is the set of all labels used in the process, the incompatibility relation is read off the set of $a \perp b$ claims added to the No side (these are the incompatibility claims that *don't* hold in the canonical model), in the sense that we set $a \perp b$ to hold in our model when the corresponding statement $a \perp b \notin No$. Then we set $a \Vdash A$ when $A \cdot a \in Yes$ and $a \nVdash A$ when $A \cdot a \in No$. The standard truth conditions can be seen to hold by an easy inspection. For example, if $a \Vdash A$ and $a \Vdash B$ then since $A \cdot a, B \cdot a \in Yes$, we must have $A \wedge B \cdot a \in Yes$ since $\langle Yes, No \rangle$ is a partition, and we cannot have $A \wedge B \cdot a \in No$, since there is a one-step proof from $A \cdot a, B \cdot a$ to $A \wedge B \cdot a$, so $a \Vdash A \wedge B$ as desired. For the converse, if $a \Vdash A \wedge B$, then $A \wedge B \cdot a \in Yes$, and since there are one-step proofs from the conjunction to each conjunct, we must have $A \cdot a$ and $B \cdot a \in Yes$ too.

The most difficult cases to check are negation. If $a \Vdash \neg A$, then we wish to show that for any b where $b \Vdash A$, we have $a \perp b$ in the model. That is, we want $a \perp b \notin No$. This is immediate, since if $\neg A \cdot a \in Yes$ and $A \cdot b \in Yes$, we cannot have $a \perp b \notin No$, since we can infer $a \perp b$ by $\neg E$. For the converse, we show that if $a \nVdash \neg A$ then there is some b where $b \Vdash A$ and $a \perp b$ *fails* in our model. But this is *exactly* the condition on partitions that is ensured by the special condition on adding negations to the No side of the partition, covered above.

That sketches the completeness construction for *incompatibility* frames. The case for *star* frames is much simpler, since we do not need to throw in fresh labels for

[15] Recall the grammar of proofs: incompatibility statements $a \perp b$ do not occur slashed in proofs, so any incompatibility facts in N do not occur in $\not{N}$.

witnesses for negations in the construction. The details are simpler, and are left to the reader as an exercise. ■

Let's leave the formal reflections on this proof system here. There are many more questions to be answered, of course, and there is much more to explore, but I hope that I have at least sketched a way forward for developing proof-first understanding of how we might acquire the concept of negation, and *why* such a concept might appropriately be modelled with an incompatibility semantics. If we use the concepts of conjunction, disjunction, negation like *this*, by making suppositions, assertions, denials, and inferences *in context*; if context shifts allow for us to at least potentially take *some* suppositional contexts as self-undermining (in the sense of being incompatible with themselves), then we have given some sense to a paraconsistent negation. And we have done so without having to start with a worlds semantics and its ontology. But the result is not moving away from the logic of incompatibility frames. On the contrary, we have shown how just such a logic can arise.

5 Classical and Non-classical Proof

We could stop there, but this raises one further question which I will begin to address. We have a plurality of logics of negation. There is the logic of the *unlabelled* proof system, and the logic with the labelled proof system.[16] Must we be forced to choose, or can we find a way for those to live alongside one another, as two parts of a larger picture, as two kinds of logical consequence, as two different kinds of *proof*, each corresponding to a different criterion for deductive validity?

In this last section of the paper, I will sketch an approach to harmonising a classical and a non-classical account of *proof*, which can be seen as extending the approach to harmonising classical and relevant consequence in "Negation in Relevant Logics" [22]. That paper uses the incompatibility semantics to defend *both* a relevant paraconsistent consequence relation, and a classical consequence relation. In any incompatibility model $\langle P, \perp, \Vdash \rangle$ we can isolate a *subset* of the points, consisting of all those points that act like *worlds*. These 'worlds' are points that are consistent and complete with respect to negation. If w is such a world, then we do not have $w \perp w$ (it is *consistent*), and furthermore, if $w \nVdash A$ and we have $y \Vdash A$, then we must have $w \perp y$, since y verifies A and w, *per force* must somehow rule A out, since it is *complete*. If the model is a *star* model, the condition for w being a world is simpler to state: it is simply the condition that $w^* = w$.

Once we have worlds at hand, we can recover classical logic by attending to whether the model furnishes a *world* that renders the premises true and the conclusion false. Our model has both a coarse-grained consequence relation (quantifying

[16] There is also the question of whether to use denial or not, but I will set that aside and help myself to denial in what follows.

$$\dfrac{Cons\,x \qquad A\cdot x}{A\cdot x^*}\,ConsE \qquad\qquad \dfrac{Comp\,x \qquad A\cdot x^*}{A\cdot x}\,CompE$$

Fig. 6 Consistency and completeness rules with star

only over worlds in the specification of validity) and a finer-grained one (casting its eyes over all points, whether worlds or not). We have two validity notions for the price of one, the kind of pluralism about validity defended elsewhere by Beall and Restall [1, 2, 22].

How can we take such a perspective we start with proofs, rather than with models? If we start with *star* proofs, we can split the consistency and completeness conditions on points in two. A point is x *consistent* if whenever $x \Vdash A$, we also have $x^* \Vdash A$. It is *complete* if the converse holds. We can capture this notion in a pair of proof rules, in Fig. 6.

These rules use a *flag*, entered into the proof tree as an assumption, which records that we are treating the context x as *consistent*, in the first case, and *complete*, in the second. (These need not be taken to be formulas, and I will not do so here.) Once we add the consistency and completeness rules to the proof system, we can recover classical principles. Here is a proof of the law of the excluded middle, showing why $p \vee \neg p$ holds in a complete context.

$$\dfrac{\dfrac{\dfrac{}{[\cancel{p\vee\neg p\cdot x}]^2} \qquad \dfrac{\dfrac{\dfrac{\dfrac{[\cancel{p\vee\neg p\cdot x}]^2 \qquad \dfrac{\dfrac{Comp\,x \qquad [p\cdot x]^1}{\neg p\cdot x}\,CompE}{p\vee\neg p\cdot x}\,\vee I}{\bot}\,\uparrow}{\neg p\cdot x}\,\neg I^1}{p\vee\neg p\cdot x}\,\vee I}{}}{\bot}\,\uparrow}{p\vee\neg p\cdot x}\,\downarrow^2$$

Proving a small number of results like of this form (showing how to prove the principles that suffice to lift us from the weak paraconsistent logic to classical logic) would be enough to ensure that what is *provable* classically can be *proved* in the paraconsistent proof system, whether with *star* or with *perp*. This is enough if all we are concerned with is *validity*. However, we could ask for more. We could wonder whether any classical (unlabelled) *proof* can be rewritten as a paraconsistent (labelled) proof, enhanced only with judicious applications of the assumption that

$$\frac{\mathit{Cons}\,x \qquad x \perp x}{\bot}\,\mathit{ConsE} \qquad\qquad \frac{\mathit{Comp}\,x \qquad \begin{matrix}[A\cdot x]^i\\ \Lambda\\ \bot\end{matrix} \qquad A\cdot y}{x \perp y}\,\mathit{CompE}^i$$

Fig. 7 Consistency and completeness rules with perp

the context is consistent and complete. In this section I will explain why this is indeed the case, there is a simple embedding of classical proof into labelled proofs, an embedding that is so intimate that the steps we use for *normalising* classical proofs correspond to steps used for normalising labelled proofs. This is strong evidence for the view that classical unlabelled proofs can be further analysed into labelled incompatibility proofs, which make explicit where the assumptions of consistency and completeness play a role in the reasoning, assumptions that are left completely *implicit* in classical proof.

To start presenting this translation from classical proofs into labelled proofs, let's see the consistency and completeness rules appropriate for *incompatibility* proofs. The rules are presented in Fig. 7.

The *completeness* rule is a little more complex than the completeness rule for star proofs, but it encodes the idea that if a context x fails to include A whereas a context y includes A, then, provided that x is complete, then x is indeed incompatible with y. In the presence of the denial rules, this condition is not too strong, though it might be thought that the second premise of the rule, which requires a reduction of A (at x) to absurdity, which seems stronger than the idea that A simply fails to hold at x. This worry is unfounded, because the labelled proof Λ may have other premises, including simply the premise that A fails at x.

$$\frac{\mathit{Comp}\,x \qquad \dfrac{\cancel{A\cdot x} \qquad [A\cdot x]^i}{\bot}\uparrow \qquad A\cdot y}{x \perp y}\,\mathit{CompE}^i$$

With the consistency and completeness rules at hand, we can encode the classical unlabelled proof rules in to the incompatibility proofs with labels. The translation will systematically send an unlabelled proof Π from X to C [or to $\bot$] to a *labelled* proof Π_w from the premises in X, all labelled with the single context label w, (we abbreviate this set $X \cdot w$, for obvious reasons), to the conclusion $C \cdot w$ [or to $\bot$], with the addition of some additional premises $\mathit{Comp}\ w$ and $\mathit{Cons}\ w$ when necessary.

Most of the rules in the labelled and unlabelled proof system differ merely in the presence or absence of labels, and the conjunction, disjunction, and store and retrieve rules may be homophonically translated by adding w labels everywhere. The work must be done with the *negation* rules, which differ substantially from one system to the other. In particular, the $\neg I$ rule in the labelled system has the side condition involving the freshness of the context label used in the discharged assumption. There is no such condition in the classical system, and the translation of classical proofs to labelled proofs annotates all the premises in the proof with the same context label, so we must find a way to insert a fresh label *somewhere*. But first, let's see the translation for $\neg E$.

$$\dfrac{\begin{matrix}\Pi \\ \neg A\end{matrix} \quad \begin{matrix}\Pi' \\ A\end{matrix}}{\bot}\,\neg E \quad \Rightarrow \quad \dfrac{\dfrac{\begin{matrix}\Pi_w \\ \neg A \cdot w\end{matrix} \quad \begin{matrix}\Pi'_w \\ A \cdot w\end{matrix}}{w \perp w}\,\neg E \qquad Cons\,w}{\bot}\,ConsE$$

Given a proof that ends in a $\neg E$ step, we translate it by translating the subproofs first, and then the work of the unlabelled $\neg E$ step is achieved first by a labelled $\neg E$ step, which gives us the conclusion that the world context w is inconsistent. We then appeal to consistency to reduce that conclusion to absurdity.

The translation of $\neg I$ has the same form, except in reverse. The subproof of our $\neg I$ proof is a proof of absurdity. To transform this into a proof of *incompatibility*, we appeal to the *completeness* of our context w, and this assumption provides just what we need. We take y to be a fresh label, which can be immediately discharged in a $\neg I$ step, which gives us the negation at w as desired.

$$\dfrac{\begin{matrix}[A]^i \\ \Pi \\ \bot\end{matrix}}{\neg A}\,\neg I^i \quad \Rightarrow \quad \dfrac{\dfrac{Comp\,w \qquad \begin{matrix}[A \cdot w]^i \\ \Pi_w \\ \bot\end{matrix} \qquad [A \cdot y]^j}{w \perp y}\,CompE^i}{\neg A \cdot w}\,\neg I^j$$

With these translations, and homophonic translations of all the other inference rules, we have a systematic translation of classical unlabelled proofs into labelled incompatibility proofs. We can see the distinctions that classical *proofs* draw, faithfully represented inside the wilder and more varied world of labelled incompatibility proofs. We have two logics of *proofs*, for the price of one.[17]

Let me end this section (and this paper) by showing how this representation is so faithful that normalisation steps operating on classical proofs correspond to normalisation steps operating on the underlying labelled proofs. The key normalisation steps for our purposes, again, are the negation rules, since these differ most, so I will explain these and leave the rest to the reader. Recall, the principle case of normalisation in a proof is the elimination of a *detour* where a formula is *introduced* and then *eliminated.* In the unlabelled proof system, a negation introduction followed by an elimination goes like the proof on the left, and the detour formula $\neg A$ is boxed.

$$\dfrac{\dfrac{\begin{matrix}[A]^i\\ \Pi\\ \bot\end{matrix}}{\boxed{\neg A}}\,\neg I^i \qquad \begin{matrix}\Pi'\\ A\end{matrix}}{\bot}\,\neg E \qquad\rightsquigarrow\qquad \begin{matrix}\Pi'\\ A\\ \Pi\\ \bot\end{matrix}$$

We can eliminate this detour by replacing this proof section by the proof on the right, which stacks the subproof Π' some number of times into the proof Π. (How many times? Exactly as many times as the discharged assumption A was used in Π. In general, this is any finite number, including zero.)

In the case of a *labelled* proof, we can normalise in exactly the same way, though with the wrinkle of incorporating labels:

[17] It is worth noting (and thanks to an anonymous referee for prompting me on this point) that the symmetry of the compatibility relation is not used in this result: it applies generally to any incompatibility frame. For example, even though double negation introduction (the inference from $p \cdot x$ to $\neg\neg p \cdot x$) is not, in general valid, in the absence of symmetry, it *is* valid for *worldly* consequence. The standard natural deduction proof from p to $\neg\neg p$ is transformed into a proof using $\neg I$, $\neg E$, and the completeness and consistency conditions, without any appeal to symmetry of incompatibility.

$$\dfrac{\dfrac{\mathit{Comp}\,w \qquad \dfrac{\dfrac{[\neg p \cdot w]^1 \quad p \cdot w}{w \perp w}\,\neg E \qquad \mathit{Cons}\,w}{\bot}\,\mathit{ConsE} \qquad [\neg p \cdot y]^2}{w \perp y}\,\mathit{CompE}^1}{\neg\neg p \cdot w}\,\neg I^2$$

$$\dfrac{\dfrac{\begin{array}{c}[A\cdot y]^i\\ \Lambda\\ x\perp y\end{array}}{\boxed{\neg A\cdot x}}\neg I^i \qquad \begin{array}{c}\Lambda'\\ A\cdot z\end{array}}{x\perp z}\neg E \qquad\rightsquigarrow\qquad \begin{array}{c}\Lambda'\\ A\cdot z\\ \Lambda[y/z]\\ x\perp z\end{array}$$

Here, we take the subproof Λ' of $A \cdot z$ and use this in the place of the discharged assumptions $A \cdot y$. To do this, we replace each y in Λ by z, reassured that this does not change any *other* premise in Λ, and the result replaces the conclusion of the proof by $x \perp z$, the required conclusion.

Another normalising pair to consider in our labelled proof system involves the consistency and completeness rules. These could be iterated. If I stack a $CompE$ before a $ConsE$ step in some proof, we have a subproof which ends in this pair of steps:

$$\dfrac{Cons\,x \qquad \dfrac{Comp\,x \qquad \begin{array}{c}[A\cdot x]^i\\ \Lambda\\ \bot\end{array} \qquad \begin{array}{c}\Lambda'\\ A\cdot x\end{array}}{\boxed{x\perp x}}CompE^i}{\bot}ConsE$$

It seems quite natural to see the $x \perp x$ as a detour formula, since the detour through this formula (and the concomitant appeals to completeness and consistency) is thoroughly redundant. We could replace this subproof by this subproof

$$\begin{array}{c}\Lambda'\\ A\cdot x\\ \Lambda\\ \bot\end{array}$$

which makes no such appeal to consistency and completeness, and arrives at the same conclusion from the same remaining premises. With this pair of normalisation steps at hand, consider the labelled *translation* of an unlabelled proof which involves a $\neg I/\neg E$ detour. The translation of such a classical detour has this shape:

$$\cfrac{\cfrac{\cfrac{\cfrac{Comp\,w \quad \begin{matrix}[A\cdot w]^i\\ \Pi_w\\ \bot\end{matrix} \quad [A\cdot y]^j}{w\perp y}\,CompE^i}{\boxed{\neg A\cdot w}}\,\neg I^j \qquad \begin{matrix}\Pi'_w\\ A\cdot w\end{matrix}}{w\perp w}\,\neg E \qquad Cons\,w}{\bot}\,ConsE$$

In this labelled proof, the $\neg A \cdot w$ is indeed a detour formula, and in this translated proof, it is introduced by a $\neg I$ and immediately eliminated by a $\neg E$. So, we can normalise it away. Once we do, we get exactly the *Completeness*/*Consistency* pair we saw

$$\cfrac{\cfrac{Comp\,w \quad \begin{matrix}[A\cdot w]^i\\ \Pi_w\\ \bot\end{matrix} \quad \begin{matrix}\Pi'_w\\ A\cdot w\end{matrix}}{\boxed{w\perp w}}\,CompE^i \qquad Cons\,w}{\bot}\,ConsE \qquad \rightsquigarrow \qquad \begin{matrix}\Pi'_w\\ A\cdot w\\ \Pi_w\\ \bot\end{matrix}$$

and so, it normalises away to *exactly* what we would expect, no more and no less than the translation of the classical proof that eliminated the unlabelled detour. I take it that this is *some* evidence that we have a faithful representation of classical proof inside the more flexible, more general and more detailed world of labelled incompatibility proofs, and we understand both these worlds better by seeing one reflected inside the other.

References

1. Beall, J. C., & Restall, G. (2000). Logical pluralism. *Australasian Journal of Philosophy, 78*, 475–493.
2. Beall, J. C., & Restall, G. (2006). Logical pluralism. Oxford: Oxford University Press.
3. Belnap, N. D. (1962). Tonk, Plonk and Plink. *Analysis, 22*, 130–134.
4. Berto, F., & Restall, G. (2019). Negation on the Australian plan. *Journal of Philosophical Logic, 48*(6), 1119–1144.
5. Brandom, R. B. (2000). *Articulating reasons: An introduction to inferentialism*. Cambridge: Harvard University Press.
6. Davies, M., & Humberstone, L. (1980). Two notions of necessity. *Philosophical Studies, 38*(1), 1–30.
7. Dunn, J. M. (1994). Star and perp: Two treatments of negation. In J. E. Tomberlin (Ed.), *Philosophical perspectives* (Vol. 7, pp. 331–357). Atascadero: Ridgeview Publishing Company.

8. Dutilh Novaes, C. (2021). *The dialogical roots of deduction*. Cambridge: Cambridge University Press.
9. Goldblatt, R. (1974). Semantic analysis of orthologic. *Journal of Philosophical Logic, 3*(1–2), 19–35.
10. Harman, G. (1986). *Change in view: Principles of reasoning*. Bradford Books. Cambridge: MIT Press.
11. Incurvati, L., & Schlöder, J. J. (2017). Weak rejection. *Australasian Journal of Philosophy, 95*(4), 741–760.
12. Incurvati, L., & Smith, P. (2010). Rejection and valuations. *Analysis, 70*(1), 3–10.
13. Kürbis, N. (2019). *Proof and falsity*. Cambridge: Cambridge University Press.
14. Lance, M., & Heath White, W. (2007). Stereoscopic vision: Persons, freedom, and two spaces of material inference. *Philosophers– Imprint, 7*(4), 1–21.
15. Mares, E. (2024). *The logic of entailment and its history*. Cambridge: Cambridge University Press.
16. Parigot, M. (1992). $\lambda\mu$-Calculus: An algorithmic interpretation of classical natural deduction. In A. Voronkov (Ed.), *International Conference on Logic for Programming Artificial Intelligence and Reasoning*. Lecture notes in artificial intelligence (Vol. 624, pp. 190–201). Berlin: Springer.
17. Parigot, M. (1993). Classical proofs as programs. In G. Gottlob, A. Leitsch, & D. Mundici (Eds.), *Computational logic and proof theory*. Lecture notes in computer science (Vol. 713, pp. 263–276). Berlin: Springer.
18. Poggiolesi, F., & Restall, G. (2012). Interpreting and applying proof theories for modal logic. In G. Restall & G. Russell (Eds.), *New waves in philosophical logic* (pp. 39–62). London: Palgrave Macmillan.
19. Price, H. (1990). Why 'not'?. *Mind, 99*(394), 222–238.
20. Read, S. (1988). *Relevant logic: A philosophical examination of inference*. Oxford: Basil Blackwell.
21. Read, S. (2015). Semantic pollution and syntactic purity. *The Review of Symbolic Logic, 8*(4), 649–661.
22. Restall, G. (1999). Negation in relevant logics: How I stopped worrying and learned to love the routley star. In D. Gabbay & H. Wansing (Eds.), *What is negation?*. Applied logic series (Vol. 13, pp. 53–76). Dordrecht: Kluwer Academic Publishers.
23. Restall, G. (2000). Defining double negation elimination. *The Logic Journal of the IGPL, 8*(6), 853–860.
24. Restall, G. (2000). *An introduction to substructural logics*. Milton Park: Routledge.
25. Restall, G. (2005). Multiple conclusions. In P. HàJek, L. Valdès-Villanueva, & D. Westerståhl (Eds.), *Logic, Methodology and Philosophy of Science: Proceedings of the Twelfth International Congress* (pp. 189–205). London: KCL Publications.
26. Restall, G. (2009). Truth values and proof theory. *Studia Logica, 92*(2), 241–264.
27. Restall, G. (2012). A cut-free sequent system for two-dimensional modal logic, and why it matters. *Annals of Pure and Applied Logic, 163*(11), 1611–1623.
28. Restall, G. (2023). Structural rules in natural deduction with alternatives. *Bulletin of the Section of Logic, 52*(2), 109–143.
29. Ripley, D. (2017). Bilateralism, coherence, warrant. In F. Moltmann & M. Textor (Eds.), *Act-based conceptions of propositional content* (pp. 307–324). Oxford: Oxford University Press.
30. Routley, R., Plumwood, V., Meyer, R. K., & Brady, R. T. (1982). *Relevant logics and their rivals*. Atascadero: Ridgeview.
31. Routley, R., & Routley, V. (1972). Semantics of first degree entailment. *Noûs, 6*(4), 335–359.
32. Rumfitt, I. (2000). "Yes" and "no". *Mind, 109*(436), 781–823.
33. Smiley, T. (1996). Rejection. *Analysis, 56*, 1–9.
34. Standefer, S. (2022). What is a relevant connective?. *Journal of Philosophical Logic, 51*(4), 919–950.
35. Urquhart, A. (1972). Semantics for relevant logics. *The Journal of Symbolic Logic, 37*, 159–169.

Morphing Rules of Evaluation into Rules of Deduction: Preserving Relevance and Epistemic Gain

Neil Tennant

Abstract This study seeks to reveal the proper source of the (*correct*) rules of natural deduction (and their associated rules of the sequent calculus). Perhaps surprisingly, this source consists of just the familiar *truth tables* (deriving from Frege). These tables can be construed inferentially. The primitive steps of value-computation correspond to primitive steps of 'inference'. We shall call them, however, primitive steps (or rules) of *evaluation*. These can be steps of *verification* or of *falsification*. The rules of evaluation constitute the inductive clauses in a metalinguistic co-inductive definition of model-relative verifications and falsifications.

We then show how the rules of evaluation can be 'morphed' into rules of natural deduction. Rules of verification thereby become introduction rules, and rules of falsification become elimination rules. The morphing produces model-*invariant* rules in the simplest way possible. It preserves, for natural deduction, the feature of *relevance* that is involved in truth-tabular computation. This makes for a system of natural deduction (and a directly corresponding sequent calculus) that is relevant.

1 Introduction

This study is inferentialist in spirit. We aim to show how to pass from the rules of evaluation (verification and falsification) that are enshrined in the truth tables, to rules of natural deduction (introduction and elimination). Rules of evaluation are *model relative*. Rules of natural deduction, however, are *model invariant*. We show how to morph the former rules into the latter rules. Our method of morphing

This paper is based on the invited presentation given to the Online Workshop 'New Directions in Relevant Logic' organized by Igor Sedlár, Shawn Standefer, and Andrew Tedder, November 10, 2022.

N. Tennant (✉)
Department of Philosophy, The Ohio State University, Columbus, OH, USA
e-mail: tennant.9@osu.edu

I. Sedlár et al. (eds.), *New Directions in Relevant Logic*, Trends in Logic 63,
https://doi.org/10.1007/978-3-031-69940-5_15

preserves the connections of relevance that are intrinsic to evaluations (which are computations, hence relevant); and produces rules of inference so formulated as to ensure the relevance of premises to conclusions of the proofs that they generate. The deductive rules that result from this investigation are those of Core Logic $\mathbb{C}$. We shall also have occasion to invoke its classicized extension, $\mathbb{C}^+$. See Sect. 3 for the rules of these systems.

2 Notation

Λ	A set of literals (encoding a model)
φ, ψ, θ	Sentences
Δ, Γ	Sets of sentences
$\Delta : \varphi$	A sequent (in single-conclusion logic)

For any logical operator @:

(@-$\mathcal{V}$)	the Verification rule for @
(@-$\mathcal{F}$)	the Falsification rule for @
(@-I)	the Introduction rule for @ in Natural Deduction
(@-E)	the Elimination rule for @ in Natural Deduction
(@-R)	the Right rule for @ in Sequent Calculus
(@-L)	the Left rule for @ in Sequent Calculus

Evaluation

$\mathcal{V}(V, \varphi, \Lambda)$	V is a verification of φ relative to Λ
$\mathcal{F}(F, \varphi, \Lambda)$	F is a falsification of φ relative to Λ

$\Lambda \Vdash \varphi$	$\exists V\ \mathcal{V}(V, \varphi, \Lambda)$	'Λ makes φ true'
$\Lambda, \varphi \Vdash \bot$	$\exists F\ \mathcal{F}(F, \varphi, \Lambda)$	'Λ makes φ false'

Deduction

$\mathcal{P}_{\mathcal{S}}(\Pi, \varphi, \Delta)$ Π is an $\mathcal{S}$-proof of φ, with the set Δ of premises
($\mathcal{S}$ can be the system of natural deduction, or the sequent calculus.)

$\Delta \vdash_{\mathcal{S}} \varphi$	$\exists \Gamma \subseteq \Delta\ \exists \Pi\ \mathcal{P}_{\mathcal{S}}(\Pi, \varphi, \Gamma)$	'Δ $\mathcal{S}$-proves φ', or 'φ is $\mathcal{S}$-deducible from Δ'

Note that $\mathcal{P}_{\mathcal{S}}(\Pi, \varphi, \Delta)$ implies that Δ is the exact set of premises (i.e., undischarged assumptions) of the proof Π. This set is finite, because proofs are finite.

3 The Rules of Core Logic $\mathbb{C}$ and Classical Core Logic $\mathbb{C}^+$

For the reader not familiar with the Core systems their rules will be set out here. The rules are stated rather more fastidiously than is usually the case in traditional presentations of natural deduction *à la* [1] and [2].

In the graphic statements of natural-deduction rules below, a box appended to the discharge-stroke over an 'assumption for the sake of argument' indicates that one must have made use of that assumption in deriving the subordinate conclusion; whereas a diamond indicates that 'vacuous discharge' is permitted.

Note that vertically descending dots in the following graphic rules indicate that non-trivial proof-work may be involved in descending from the assumptions indicated to the subordinate conclusion in question. The *absence* of such descending dots above a premise (which will be a major premise for an elimination) indicates that the premise in question *stands proud, with no non-trivial proof-work above it.* This ensures that all proof are in normal form. It also ensures that natural deductions and their corresponding sequent proofs are macro-structurally isomorphic.

In the list of rules below, the Introduction and Elimination rules of natural deduction are stated on the left; the corresponding Right and Left logical rules of the sequent calculus are stated on the right.[1] Note that all elimination rules in natural deduction are in *parallelized* form.

(¬I)
$$\dfrac{\begin{array}{c}\overset{\Box\!\_\_\,(i)}{\varphi}\\ \vdots\\ \bot\end{array}}{\neg\varphi}\,(i)$$

(¬$_R$)
$$\dfrac{\varphi\,,\Delta:\bot}{\Delta:\neg\varphi}$$

(¬E)
$$\dfrac{\neg\varphi \quad \begin{array}{c}\vdots\\ \varphi\end{array}}{\bot}$$

(¬$_L$)
$$\dfrac{\Delta:\varphi}{\neg\varphi,\Delta:\bot}$$

(∧I)
$$\dfrac{\begin{array}{c}\vdots\\ \varphi\end{array} \quad \begin{array}{c}\vdots\\ \psi\end{array}}{\varphi\wedge\psi}$$

(∧$_R$)
$$\dfrac{\Delta:\varphi \qquad \Gamma:\psi}{\Delta,\Gamma:\varphi\wedge\psi}$$

[1] The reader will find these rules in [9, pp. 161–164].

(∧E)
$$\dfrac{\varphi \wedge \psi \qquad \begin{array}{c} (i)\underline{\quad}\square\underline{\quad}(i) \\ \underbrace{\varphi\ ,\ \psi} \\ \vdots \\ \theta \end{array}}{\theta}\,{\scriptstyle (i)}$$

where □ means at least one of φ, ψ must have been used as an assumption

$(\wedge_L)$
$$\dfrac{\Delta : \theta}{\varphi \wedge \psi,\ \Delta \backslash \{\varphi, \psi\} : \theta}$$

where $\Delta \cap \{\varphi, \psi\} \neq \emptyset$

(∨I)
$$\dfrac{\begin{array}{c}\vdots \\ \varphi\end{array}}{\varphi \vee \psi} \qquad \dfrac{\begin{array}{c}\vdots \\ \psi\end{array}}{\varphi \vee \psi}$$

$(\vee_R)$
$$\dfrac{\Delta : \varphi}{\Delta : \varphi \vee \psi} \qquad \dfrac{\Delta : \psi}{\Delta : \varphi \vee \psi}$$

(∨E)
$$\dfrac{\varphi \vee \psi \qquad \begin{array}{c} \square\underline{\quad}(i) \\ \varphi \\ \vdots \\ \theta/\bot \end{array} \qquad \begin{array}{c} \square\underline{\quad}(i) \\ \psi \\ \vdots \\ \theta/\bot \end{array}}{\theta/\bot}\,{\scriptstyle (i)}$$

$(\vee_L)$
$$\dfrac{\varphi, \Delta : \theta/\bot \qquad \psi, \Gamma : \theta/\bot}{\varphi \vee \psi,\ \Delta,\ \Gamma : \theta/\bot}$$

(→I)(a)
$$\dfrac{\begin{array}{c} \square\underline{\quad}(i) \\ \varphi \\ \vdots \\ \bot \end{array}}{\varphi \rightarrow \psi}\,{\scriptstyle (i)}$$

(→I)(b)
$$\dfrac{\begin{array}{c} \Diamond\underline{\quad}(i) \\ \varphi \\ \vdots \\ \psi \end{array}}{\varphi \rightarrow \psi}\,{\scriptstyle (i)}$$

$(\rightarrow_R)$(a)
$$\dfrac{\Delta, \varphi : \bot}{\Delta : \varphi \rightarrow \psi}$$

$(\rightarrow_R)$(b)
$$\dfrac{\Delta : \psi}{\Delta \backslash \{\varphi\} : \varphi \rightarrow \psi}$$

(∃I)
$$\dfrac{\begin{array}{c}\vdots\\ \varphi^x_t\end{array}}{\exists x\varphi}$$

($\exists_R$)
$$\dfrac{\Delta:\varphi^x_t}{\Delta:\exists x\varphi}$$

(∃E)
$$\dfrac{\exists x\varphi^{@} \qquad \begin{array}{c}\square\text{——}(i)\\ \underbrace{@\ldots\varphi^x_a\ldots @}\\ \vdots\\ \psi^{@}\end{array}}{\psi}\,(i)$$

($\exists_L$)
$$\dfrac{\varphi^x_a,\Delta:\psi}{\exists x\varphi,\Delta:\psi}\ @$$

(∀I)
$$\dfrac{\begin{array}{c}@\\ \vdots\\ \varphi\end{array}}{\forall x\varphi^a_x}$$

($\forall_R$)
$$\dfrac{\Delta:\varphi}{\Delta:\forall x\varphi^a_x}\ @$$

(∀E)
$$\dfrac{\forall x\varphi \qquad \begin{array}{c}(i)\text{——}\cdots\square\cdots\text{——}(i)\\ \underbrace{\varphi^x_{t_1}\ ,\ \ldots\ ,\ \varphi^x_{t_n}}\\ \vdots\\ \theta\end{array}}{\theta}\,(i)$$

($\forall_L$)
$$\dfrac{\varphi^x_{t_1}\ ,\ \ldots\ ,\ \varphi^x_{t_n},\Delta:\theta}{\forall x\varphi,\Delta:\theta}$$

The foregoing rules form the system $\mathbb{C}$ of *Core Logic*. We obtain the system $\mathbb{C}^+$ of *Classical Core Logic* by adding the following Classical Rule of Dilemma (Natural Deduction form on the left, Sequent Calculus form on the right).

(*Dil*)
$$\dfrac{\begin{array}{c}\square\text{——}(i)\\ \varphi\\ \vdots\\ \psi\end{array} \qquad \begin{array}{c}\square\text{——}(i)\\ \neg\varphi\\ \vdots\\ \psi/\bot\end{array}}{\psi}\,(i)$$

$$\dfrac{\Delta,\varphi:\psi \qquad \Gamma,\neg\varphi:\psi/\bot}{\Delta,\Gamma:\psi}$$

4 The Method of Framing Rules of Verification and of Falsification

We turn now to a new and completely different kind of rules from the rules of formal deduction that have just been stated in Sect. 3. The new rules are those of verification and of falsification, which we describe collectively as rules of *evaluation*. These are rules for determining the truth value of a sentence in a model; so, in the broadest possible terms, they are semantic rules. Since they are applied to evaluate sentences in specific models, these rules are quite unlike the rules of formal deduction. Formal deductions (constructed in accordance with rules of inference) need to be *model invariant*; whereas evaluations (constructed in accordance with rules of verification and of falsification) are *model-relative*.

The rules of verification and of falsification (collectively: 'evaluation rules') are therefore, themselves, 'model-relative'—especially, as the reader will see, the rule of existential falsification and the rule of universal verification. These rules were put forward in [7] and [10]. They were also dealt with in Chapter 3 of [9].

Although the current expository order is to state the rules of inference first (in Sect. 3) and the rules of evaluation only thereafter (in this section), this is only because the former are likely to be more familiar as a *kind* of rule (prescinding from certain tweaking deviations from the norm), while the rules of evaluation are a relatively new kind of offering. It is important to stress, though, that the ideal presentational ordering would be to deal first with the raw materials of evaluations—i.e., rules of verification and of falsification—and only thereafter advance to the more highly processed materials of deduction—i.e., respectively, rules of introduction and of elimination (in a system of natural deduction). This is the main innovation on offer here, as the study's title suggests: we are going to morph the rules of evaluation into the rules of deduction. The means of logical reasoning (from complex premises to complex conclusions) are going to be shown to emerge in a smooth and subtle fashion from the means whereby we work out the truth values of sentences in particular models. And in showing how this is so, we shall gain a deeper understanding of how best to ensure *relevance* of one's premises to one's conclusions in formal deductions. It has its provenance in the relevance of inputs to outputs when one engages in truth-evaluative computations.

The value T (for true) and the value F (for false) are conceptually coequal. Each is just as important as the other, as far as the familiar truth tables are concerned. We remind the reader of a point made in [9], at p. 53:

> The truth tables should really be called *truth-value* tables, because they involve two truth values, T and F, with F figuring just as importantly within them as T. We need to bear that overriding consideration in mind, and not let it be eclipsed by our preference, when forming beliefs or making assertions, for true propositions over false ones.

The rules of verification and of falsification, likewise, are coequal. Verifications and falsifications are coinductively defined. The rules of verification and of falsification for the connectives can be read off from their familiar truth tables. Reciprocally, the familiar truth tables can be constructed directly from the rules of verification and of falsification themselves. This means that we can be inferentialists *au fond.*

If, however, one wishes to take the truth tables as one's starting point, one can advance to the rules of verification and of falsification in the way we shall now describe. The reader is invited to read each row within each of the following truth tables from left to right. We have supplied, in the final column, the valid sequents expressing the value-determining contribution of the pertinent row. Assignment of the value T to an immediate constituent is registered by having that constituent stand as a premise of the sequent, on the left. Assignment of the value F to an immediate constituent is registered by having the negation of that constituent stand as a premise of the sequent, on the left. When the compound sentence receives the value T, this is registered by having the compound sentence feature as the conclusion of the sequent, whose premises represent the assignments of values to the immediate constituents. When the compound sentence receives the value F, this is registered by having $\perp$ on the right as the conclusion of the sequent in question, and having the compound sentence featuring on the left as a premise.

φ	$\neg\varphi$	
T	F	$\neg\varphi, \varphi : \perp$
F	T	$\neg\varphi : \neg\varphi$ (trivial)

φ	ψ	$\varphi \wedge \psi$	
T	T	T	$\varphi, \psi : \varphi \wedge \psi$
T	F	F	$\varphi \wedge \psi, \neg\psi : \perp)$
F	T	F	$\varphi \wedge \psi, \neg\varphi : \perp)$
F	F	F	overkill

φ	ψ	$\varphi \vee \psi$	
T	T	T	overkill
T	F	T	$\varphi : \varphi \vee \psi$
F	T	T	$\psi : \varphi \vee \psi$
F	F	F	$\varphi \vee \psi, \neg\varphi, \neg\psi : \perp$

φ	ψ	$\varphi\rightarrow\psi$	
T	T	T	$\psi : \varphi\rightarrow\psi$
T	F	F	$\varphi\rightarrow\psi, \varphi, \neg\psi : \bot$
F	T	T	overkill
F	F	T	$\neg\varphi : \varphi\rightarrow\psi$

The rules of verification and of falsification for the quantifiers treat quantified sentences as generalized conjunctions or disjunctions of all instances of the predicate that is quantified. Such generalized conjunctions or disjunctions can of course be infinitary, when the domain of interpretation is infinite. That also leads to evaluations being infinitary—in particular, verifications of universal generalizations, and falsifications of existential generalizations. It is for this reason that we have to regard evaluations as very different indeed from proofs. For the latter are finite and effectively checkable for correctness; whereas evaluations need not be. Nevertheless, the seeds of proof—namely, rules of inference—are to be found in the rules of evaluation themselves. The present study aims to explain how this is so.

1. Our rules of evaluation determine whether a sentence is **verified** or **falsified** in a given interpretation.
2. An *interpretation*, or *model* (in the first-order case) consists of a *domain of individuals*, and a specification of *extensions of predicates* within that domain. In the propositional case, an interpretation is an assignment of truth values to the atomic sentences.
3. A **verification** of a sentence φ (relative to an interpretation M) establishes φ as its *conclusion*:

$$\begin{array}{c}\vdots\\ \varphi\end{array}$$

4. A **falsification** of a sentence φ (relative to an interpretation M) establishes absurdity ($\bot$) from φ as its *major premise*:

$$\begin{array}{c}\varphi\\ \vdots\\ \bot\end{array}$$

Note that major premises for steps of falsification always ‘stand proud’, with no ‘evaluation work’ above them. This important feature of major premises will be preserved when we make (in Sect. 7) the promised transition to Elimination rules in natural deduction. And we have already emphasized this point (how major premises for eliminations stand proud) in Sect. 3, when we set out the rules of Core Logic.

The simplest verifications (relative to a first-order interpretation M) are of the form

$$\frac{}{P(\alpha_1, \ldots, \alpha_n)}\,M$$

where in M the predicate P holds of the individuals $\alpha_1, \ldots, \alpha_n$.

This is a verification of $P(\alpha_1, \ldots, \alpha_n)$ relative to M.

The simplest falsifications (relative to an interpretation M) are of the form

$$\frac{P(\alpha_1, \ldots, \alpha_n)}{\bot}\,M$$

where in M the predicate P fails to hold of the individuals $\alpha_1, \ldots, \alpha_n$.

This is a falsification of $P(\alpha_1, \ldots, \alpha_n)$ relative to M.

Any interpretation M is *atomically complete*: every atomic sentence features either in a positive M-factoid, or in a negative M-factoid. Any interpretation M is also *coherent*: no atomic sentence features both in a positive M-factoid, and in a negative M-factoid.

More complex verifications and falsifications (relative to M) are built up in accordance with the rules (@-$\mathcal{V}$) and (@-$\mathcal{F}$) for logical operators @, which will be stated presently.

In the propositional case the simplest verifications and falsifications relative to a truth-value assignment τ take the respective forms

$$\frac{}{A}\,\tau \quad \text{and} \quad \frac{A}{\bot}\,\tau\,,$$

for atomic sentences A.

Note that we are dealing here, in the first-order case, with *saturated* formulae. These are created by taking a formula $\varphi(x_1, \ldots, x_n)$ (with the indicated variables free) and replacing each free occurrence of a variable x_i $(1 \leq i \leq n)$ with an individual α_i:

$$\varphi^{x_1}_{\alpha_1} \cdots {}^{x_n}_{\alpha_n} = \varphi(\alpha_1, \ldots, \alpha_n).$$

Saturated formulae do not have any free variables. That is why they are able to be true, or false, in an interpretation. They can be thought of as *sentences*, liable to come out true, or come out false, in any model that interprets their extra-logical vocabulary.

A *literal* is an atomic sentence, or the negation of an atomic sentence. Thus atomic saturated formulae and their negations are literals. Construed inferentially, literals are our simplest possible evaluations. This constitutes the *basis clause* in our co-inductive definition of (the tree-like arrays that we are calling) verifications and falsifications.

In the propositional case, the simplest possible verifications and falsifications are, respectively,

$$\frac{}{A}\,\tau \quad \text{and} \quad \frac{A}{\bot}\,\tau \,.$$

In the first-order case, they are

$$\frac{}{P(\alpha_1,\ldots,\alpha_n)}\,M \quad \text{and} \quad \frac{P(\alpha_1,\ldots,\alpha_n)}{\bot}\,M \,.$$

So: the simplest possible verifications and falsifications express, respectively, *positive* or *negative* factoids.

We shall use Λ as a placeholder for sets of literals of the interpretation in question. Every such Λ is a coherent set of factoids. We remind the reader that these sets Λ can in general contain both positive and negative factoids.

Such sets Λ feature as the sets of 'undischarged side assumptions' in the following graphic statements of our rules of evaluation. These rules are of the respective forms @-$\mathcal{V}$ and @-$\mathcal{F}$ for each logical operator @.

Each such rule can be construed as an inductive clause in our co-inductive definition of the notions

$$\mathcal{V}(\Pi, \varphi, M, D) \text{ and } \mathcal{F}(\Pi, \varphi, M, D).$$

These metalinguistic predications can be read, respectively, as
'Π is a verification of φ relative to model M with domain D'
and
'Π is a falsification of φ relative to model M with domain D'.

4.1 Verification and Falsification Rules for ¬

$$(\neg\text{-}\mathcal{V}) \qquad \frac{\begin{array}{c}\Box\overline{}^{(i)}\\ \underbrace{\varphi\ ,\ \Lambda}\\ \vdots\\ \bot\end{array}}{\neg\varphi}\,(i) \qquad \text{Note the box } \Box\,.$$

To verify a negation $\neg\varphi$ one must falsify φ. That is, one must actually *use* φ as an 'evaluative assumption' and show it to be false (*modulo* Λ). This requirement of 'non-vacuous discharge' is what is represented by the box that annotates the discharge stroke. The same remark applies to subsequent boxes in other evaluation rules.

$$(\neg\text{-}\mathcal{F}) \qquad \dfrac{\neg\varphi \quad \begin{array}{c}\Lambda\\ \vdots\\ \varphi\end{array}}{\bot}$$

To falsify a negation $\neg\varphi$ one must verify φ.
Note the classicism that has immediately crept in by virtue of this dualizing.

4.2 Verification and Falsification Rules for $\wedge$

$$(\wedge\text{-}\mathcal{V}) \qquad \dfrac{\begin{array}{c}\Lambda_1\\ \vdots\\ \varphi\end{array} \quad \begin{array}{c}\Lambda_2\\ \vdots\\ \psi\end{array}}{\varphi\wedge\psi}$$

To verify a conjunction, one must verify both conjuncts.

$$(\wedge\text{-}\mathcal{F}) \qquad \dfrac{\varphi\wedge\psi \quad \begin{array}{c}\Box\!\!-\!(i)\\ \underbrace{\varphi\ ,\ \Lambda}\\ \vdots\\ \bot\end{array}}{\bot}{\scriptstyle(i)} \quad \dfrac{\varphi\wedge\psi \quad \begin{array}{c}\Box\!\!-\!(i)\\ \underbrace{\psi\ ,\ \Lambda}\\ \vdots\\ \bot\end{array}}{\bot}{\scriptstyle(i)} \qquad \text{Note the boxes.}$$

To falsify a conjunction, one has to falsify one of the conjuncts.

4.3 Verification and Falsification Rules for ∨

$$(\vee\text{-}\mathcal{V})\qquad \dfrac{\begin{array}{c}\Lambda\\ \vdots\\ \varphi\end{array}}{\varphi\vee\psi}\qquad \dfrac{\begin{array}{c}\Lambda\\ \vdots\\ \psi\end{array}}{\varphi\vee\psi}$$

To verify a disjunction, one has to verify one of the disjuncts.

$$(\vee\text{-}\mathcal{F})\qquad \dfrac{\varphi\vee\psi\qquad \begin{array}{c}\overset{\Box\text{---}(i)}{}\\ \underbrace{\varphi\ ,\ \Lambda_1}\\ \vdots\\ \bot\end{array}\qquad \begin{array}{c}\overset{\Box\text{---}(i)}{}\\ \underbrace{\psi\ ,\ \Lambda_2}\\ \vdots\\ \bot\end{array}}{\bot}(i)$$

Note the boxes.

To falsify a disjunction, one must falsify both disjuncts.

4.4 Verification and Falsification Rules for →

$$(\rightarrow\text{-}\mathcal{V})\qquad \dfrac{\begin{array}{c}\overset{\Box\text{---}(i)}{}\\ \underbrace{\varphi\ ,\ \Lambda}\\ \vdots\\ \bot\end{array}}{\varphi\rightarrow\psi}(i)\qquad \dfrac{\begin{array}{c}\Lambda\\ \vdots\\ \psi\end{array}}{\varphi\rightarrow\psi}$$

Note the box.

To verify a conditional, one must either falsify its antecedent or verify its consequent.

$$(\rightarrow\text{-}\mathcal{F})\qquad \dfrac{\varphi\rightarrow\psi\qquad \begin{array}{c}\Lambda_1\\ \vdots\\ \varphi\end{array}\qquad \begin{array}{c}\overset{\Box\text{---}(i)}{}\\ \underbrace{\psi\ ,\ \Lambda_2}\\ \vdots\\ \bot\end{array}}{\bot}(i)$$

Note the box.

To falsify a conditional, one must both verify its antecedent and falsify its consequent.

4.5 Verification and Falsification Rules for ∃

$$(\exists\text{-}\mathcal{V})\qquad \frac{\begin{array}{c}\Lambda\\ \vdots\\ \varphi^x_{\boldsymbol{\alpha}}\end{array}}{\exists x\varphi}\qquad \text{where } \boldsymbol{\alpha} \text{ is any individual in the domain}$$

To verify an existential, one must verify some instance of it.

$$(\exists\text{-}\mathcal{F})\qquad \frac{\exists x\varphi \qquad \begin{array}{c}\underbrace{\overset{\square\text{---}(i)}{\varphi^x_{\boldsymbol{\alpha_1}}},\ \Lambda_1}\\ \vdots\\ \bot\end{array}\ \cdots\ \begin{array}{c}\underbrace{\overset{\square\text{---}(i)}{\varphi^x_{\boldsymbol{\alpha_n}}},\ \Lambda_n}\\ \vdots\\ \bot\end{array}\ \cdots}{\bot}(i)\qquad \text{Note the boxes.}$$

where $\boldsymbol{\alpha_1},\ldots,\boldsymbol{\alpha_n},\ldots$ are all the individuals in the domain

To falsify an existential, one must falsify every instance of it.

4.6 Verification and Falsification Rules for ∀

$$(\forall\text{-}\mathcal{V})\qquad \frac{\begin{array}{c}\Lambda_1\\ \vdots\\ \psi(\boldsymbol{\alpha_1})\end{array}\quad \cdots \quad \begin{array}{c}\Lambda_n\\ \vdots\\ \psi(\boldsymbol{\alpha_n})\end{array}\quad \cdots}{\forall x\psi(x)}$$

where $\boldsymbol{\alpha_1},\ldots,\boldsymbol{\alpha_n},\ldots$ are all the individuals in the domain

To verify a universal, one must verify every instance of it.

$$
(\forall\text{-}\mathcal{F}) \qquad \dfrac{\forall x\psi(x) \qquad \begin{array}{c} \square\!\!\overline{\qquad}^{(i)} \\ \underbrace{\psi(\alpha)\ ,\ \Lambda} \\ \vdots \\ \bot \end{array}}{\bot}\,{\scriptstyle (i)} \qquad \text{Note the box.}
$$

where α is any individual in the domain

To falsify a universal, one must falsify some instance of it.

5 Possession of a Verification Is Equivalent to Tarskian Truth

Theorem 1 Modulo *a metatheory which contains the mathematics of* $\overline{\overline{D}}$*-furcating trees of finite depth, we have, for all models M with domain D,*

$$\exists\Pi\ \mathcal{V}(\Pi, \varphi, M, D) \quad \Leftrightarrow \quad M \Vdash \varphi,\ \textit{i.e.},\ \varphi \textit{ is true in } M$$

where the right-hand side is in the sense of Tarski.

We are explaining here what Tarskian truth (in a model M) *consists in*. It consists in the *existence of an M-relative verification* of the sentence in question.

Proof The proof is by the obvious induction on the complexity of the sentence (or saturated formula) φ. Note that the proof is intuitionistic, provided only that the M-relative truthmakers for universals and the M-relative falsitymakers for existentials, in the case where D is infinite, can be assumed to exist, courtesy of the background metamathematics. This observation affects the right-to-left direction in the relevant cases of the inductive step.

6 Bivalence and Non-contradiction

We can see from the rules of evaluation for negation that Bivalence will hold for evaluations with respect to any interpretation M: every sentence being evaluated will either come out true (by having an M-relative verification), or come out false (by having an M-relative falsification).

Moreover, the Law of Non-Contradiction will hold at the metalevel: no sentence will have both an M-relative verification and an M-relative falsification. This can be proved by induction on the complexity of sentences, by appeal to the basis fact that the interpretation M is both *atomically complete* and *coherent.*

7 Framing Rules of Natural Deduction by Morphing Rules of Evaluation

The rules of *verification* and of *falsification*, to repeat, are *model-relative*. But rules of inference for *deduction* have to be *model-invariant*.

Our rules of inference for natural deduction are nevertheless going to be sourced in the rules of evaluation.

The rules of verification and of falsification will be respectively 'morphed into' *model-invariant* rules of introduction and elimination in natural deductions. The morphing will preserve as much of the former rules' overall features as possible. One such feature has already been mentioned: major premises for eliminations (like major premises earlier, for falsifications) must stand proud, with no proof-work above them. Another important feature that will be preserved is the *relevance* of constituents to compounds—now in the deductive setting, as earlier in the evaluational or computational setting. This means that in the deductive setting we continue to *respect the boxes* that annotated discharges in the evaluational setting.

The morphings also yield the Right and Left rules for the logical operators in sequent calculus. These *are* the Introduction and Elimination rules; they are just set out in a slightly different format.

This is how we shall pass from rules of evaluation to rules of deduction.

1. Replace sets Λ of literals with sets Δ of arbitrary sentences.
2. In the $\mathcal{F}$-rules for $\wedge$, $\vee$, $\rightarrow$, $\forall$ and $\exists$, allow for an arbitrary sentence θ as the conclusion, as an alternative to $\bot$. One can still, of course, have $\bot$ as a conclusion.
3. In step (2) above, when formulating $\vee$-E from $\vee$-$\mathcal{F}$, allow for θ to be the overall conclusion even when it is the conclusion of just one of the case-proofs, the other case-proof concluding with $\bot$. This is called 'liberalized' $\vee$-E.
4. In step (2) above, when formulating $\forall$-I from $\forall$-$\mathcal{V}$, and $\exists$-E from $\exists$-$\mathcal{F}$, replace the domain-many subproofs for instances with a single subproof template involving a parameter a. This respects the fact that the aim of deduction is to preserve truth over all models, and to do so by means of finitary proofs.

The following clarificatory remarks should help one understand the morphings.

1. We shall use V (with or without numerical subscripts) as a placeholder for verifications.
2. Reminder: A verification has a sentence as its conclusion.
3. We shall use F (with or without numerical subscripts) as a placeholder for falsifications.
4. Reminder: A falsification has the absurdity symbol $\bot$ as its conclusion.

5. We shall use Λ (with or without numerical subscripts) as a placeholder for sets of literals of the model, relative to which a sentence is being evaluated.
6. We shall use Π (with or without numerical subscripts) as a placeholder for proofs (natural deductions).
7. A proof can have either a sentence or the absurdity symbol $\bot$ as its conclusion.
8. We shall use Δ (with or without numerical subscripts) as a placeholder for sets of sentences (the set of premises of a proof)
9. The changes wrought by the morphings will be indicated by the squiggly arrow $\rightsquigarrow$.
10. We first take all the Verification rules, and morph them into Introduction rules:

$$(@\text{-}\mathcal{V}) \quad \rightsquigarrow \quad (@\text{-I})$$

11. We then take all the Falsification rules, and morph them into Elimination rules:

$$(@\text{-}\mathcal{F}) \quad \rightsquigarrow \quad (@\text{-E})$$

7.1 Morphing Verification Rules to Introduction Rules

7.1.1 Morphing $\neg$-$\mathcal{V}$ to $\neg$-I

$$(\neg\text{-}\mathcal{V}) \quad \dfrac{\begin{array}{c}\overset{\square\!\!-\!\!(i)}{\underbrace{\varphi\ ,\ \Lambda}} \\ F \\ \bot\end{array}}{\neg\varphi}{\scriptstyle (i)} \qquad \rightsquigarrow \qquad (\neg\text{-I}) \quad \dfrac{\begin{array}{c}\overset{\square\!\!-\!\!(i)}{\underbrace{\varphi\ ,\ \Delta}} \\ \Pi \\ \bot\end{array}}{\neg\varphi}{\scriptstyle (i)}$$

7.1.2 Morphing $\wedge$-$\mathcal{V}$ to $\wedge$-I

$$(\wedge\text{-}\mathcal{V}) \quad \dfrac{\begin{array}{c}\Lambda_1 \\ V_1 \\ \varphi_1\end{array} \qquad \begin{array}{c}\Lambda_2 \\ V_2 \\ \varphi_2\end{array}}{\varphi_1 \wedge \varphi_2} \qquad \rightsquigarrow \qquad (\wedge\text{-I}) \quad \dfrac{\begin{array}{c}\Delta_1 \\ \Pi_1 \\ \varphi_1\end{array} \qquad \begin{array}{c}\Delta_2 \\ \Pi_2 \\ \varphi_2\end{array}}{\varphi_1 \wedge \varphi_2}$$

7.1.3 Morphing ∨-$\mathcal{V}$ to ∨-I

$$(\vee\text{-}\mathcal{V})\quad \dfrac{\begin{matrix}\Lambda\\ V\\ \varphi\end{matrix}}{\varphi\vee\psi}\quad \dfrac{\begin{matrix}\Lambda\\ V\\ \psi\end{matrix}}{\varphi\vee\psi}\qquad\rightsquigarrow\qquad(\vee\text{-I})\quad \dfrac{\begin{matrix}\Delta\\ \Pi\\ \varphi\end{matrix}}{\varphi\vee\psi}\quad \dfrac{\begin{matrix}\Delta\\ \Pi\\ \psi\end{matrix}}{\varphi\vee\psi}$$

7.1.4 Morphing →-$\mathcal{V}$ to →-I

$$(\rightarrow\text{-}\mathcal{V})\quad \dfrac{\begin{matrix}\overset{\square\!\!-\!(i)}{\underbrace{\varphi\ ,\ \Lambda}}\\ F\\ \bot\end{matrix}}{\varphi\rightarrow\psi}{(i)}\quad \dfrac{\begin{matrix}\Lambda\\ V\\ \psi\end{matrix}}{\varphi\rightarrow\psi}\qquad\rightsquigarrow\qquad(\rightarrow\text{-I})\quad \dfrac{\begin{matrix}\overset{\square\!\!-\!(i)}{\underbrace{\varphi\ ,\ \Delta}}\\ \Pi\\ \bot\end{matrix}}{\varphi\rightarrow\psi}{(i)}\quad \dfrac{\begin{matrix}\Delta\\ \Pi\\ \psi\end{matrix}}{\varphi\rightarrow\psi}$$

Note that these two parts of (→-I) resulting from this direct morphing of the verification rule (→-$\mathcal{V}$) do not yet furnish that form of (→-I) that permits one to assume φ for the sake of argument, then to deduce ψ by using assumption φ, and finally to discharge φ when one infers the conclusion $\varphi \rightarrow \psi$. We shall see in due course how to get this 'missing part' of (→-I).

7.1.5 Morphing ∀-$\mathcal{V}$ to ∀-I

$$(\forall\text{-}\mathcal{V})\quad \dfrac{\left\{\begin{matrix}\Lambda_{\alpha}\\ V_{\alpha}\\ \psi(\alpha)\end{matrix}\right\}_{\alpha\in M}}{\forall x\psi(x)}\qquad\rightsquigarrow\qquad(\forall\text{-I})\quad \dfrac{\begin{matrix}\Delta^{@}\\ \Pi\\ \psi(a)\end{matrix}}{\forall x\psi(x)}$$

7.1.6 Morphing ∃-$\mathcal{V}$ to ∃-I

$$(\exists\text{-}\mathcal{V})\quad \dfrac{\begin{matrix}\Lambda\\ V\\ \varphi^{x}_{\alpha}\end{matrix}}{\exists x\varphi}\ \ \begin{matrix}\text{where } \alpha \text{ is any}\\ \text{individual in}\\ \text{the domain}\end{matrix}\qquad\rightsquigarrow\qquad(\exists\text{-I})\quad \dfrac{\begin{matrix}\Delta\\ \Pi\\ \varphi^{x}_{t}\end{matrix}}{\exists x\varphi}\ \ \begin{matrix}\text{where } t \text{ is any}\\ \text{closed term}\end{matrix}$$

7.2 *Morphing Falsification Rules to Elimination Rules*

7.2.1 Morphing ¬-$\mathcal{F}$ to ¬-E

$$(\neg\text{-}\mathcal{F})\qquad \dfrac{\neg\varphi \qquad \begin{matrix}\Lambda\\ V\\ \varphi\end{matrix}}{\bot} \qquad\rightsquigarrow\qquad (\neg\text{-E})\qquad \dfrac{\neg\varphi\qquad \begin{matrix}\Delta\\ \Pi\\ \varphi\end{matrix}}{\bot}$$

7.2.2 Morphing ∧-$\mathcal{F}$ to ∧-E

$$(\wedge\text{-}\mathcal{F})\quad \dfrac{\varphi\wedge\psi \quad \begin{matrix}\underbrace{\overset{\Box\text{—}(i)}{\varphi}\ ,\ \Lambda}_{F}\\ \bot\end{matrix}}{\bot}{\scriptstyle (i)} \quad \dfrac{\varphi\wedge\psi \quad \begin{matrix}\underbrace{\overset{\Box\text{—}(i)}{\psi}\ ,\ \Lambda}_{F}\\ \bot\end{matrix}}{\bot}{\scriptstyle (i)} \quad\rightsquigarrow\quad (\wedge\text{-E})\quad \dfrac{\varphi\wedge\psi \quad \begin{matrix}\underbrace{\overset{\Box\text{—}(i)}{\varphi}\ ,\ \Delta}_{\Pi}\\ \theta/\bot\end{matrix}}{\theta/\bot}{\scriptstyle (i)} \quad \dfrac{\varphi\wedge\psi \quad \begin{matrix}\underbrace{\overset{\Box\text{—}(i)}{\psi}\ ,\ \Delta}_{\Pi}\\ \theta/\bot\end{matrix}}{\theta/\bot}{\scriptstyle (i)}\ ;$$

hence, more efficiently:

$$\dfrac{\varphi\wedge\psi \quad \begin{matrix}\underbrace{\overset{(i)\text{—}\Box\text{—}(i)}{\varphi\ ,\ \psi}\ ,\ \Delta}_{\Pi}\\ \theta/\bot\end{matrix}}{\theta/\bot}{\scriptstyle (i)}\ ,$$

where the box between the two discharge strokes indicates that at least one of φ and ψ must feature as an undischarged assumption of the subproof Π.

7.2.3 Morphing ∨-$\mathcal{F}$ to ∨-E

$$(\vee\text{-}\mathcal{F})\quad \dfrac{\varphi_1\vee\varphi_2 \quad \begin{matrix}\underbrace{\overset{\Box\text{—}(i)}{\varphi_1}\ ,\ \Lambda_1}_{F_1}\\ \bot\end{matrix} \quad \begin{matrix}\underbrace{\overset{\Box\text{—}(i)}{\varphi_2}\ ,\ \Lambda_2}_{F_2}\\ \bot\end{matrix}}{\bot}{\scriptstyle (i)} \quad\rightsquigarrow\quad (\vee\text{-E})\quad \dfrac{\varphi_1\vee\varphi_2 \quad \begin{matrix}\underbrace{\overset{\Box\text{—}(i)}{\varphi_1}\ ,\ \Delta_1}_{\Pi_1}\\ \theta/\bot\end{matrix} \quad \begin{matrix}\underbrace{\overset{\Box\text{—}(i)}{\varphi_2}\ ,\ \Delta_2}_{\Pi_2}\\ \theta/\bot\end{matrix}}{\theta/\bot}{\scriptstyle (i)}$$

'If either of the two case-proofs Π_1, Π_2 has $\bot$ as its conclusion, bring down as the main conclusion the conclusion of the other case-proof.'

7.2.4 Morphing $\to$-$\mathcal{F}$ to $\to$-E

$$(\to\text{-}\mathcal{F})\quad \dfrac{\varphi\to\psi \qquad \begin{array}{c}\Lambda_1\\ V\\ \varphi\end{array} \qquad \begin{array}{c}\Box\text{---}(i)\\ \underbrace{\psi\,,\Lambda_2}\\ F\\ \bot\end{array}}{\bot}{\scriptstyle (i)} \qquad \rightsquigarrow \qquad (\to\text{-E})\quad \dfrac{\varphi\to\psi \qquad \begin{array}{c}\Delta_1\\ \Pi_1\\ \varphi\end{array} \qquad \begin{array}{c}\Box\text{---}(i)\\ \underbrace{\psi\,,\Delta_2}\\ \Pi_2\\ \theta/\bot\end{array}}{\theta/\bot}{\scriptstyle (i)}$$

7.2.5 Morphing $\forall$-$\mathcal{F}$ to $\forall$-E

$$(\forall\text{-}\mathcal{F})\quad \dfrac{\forall x\psi(x) \qquad \begin{array}{c}\Box\text{———}(i)\\ \underbrace{\psi(\boldsymbol{\alpha})\;,\;\Lambda}\\ F\\ \bot\end{array}}{\bot}{\scriptstyle (i)} \qquad \rightsquigarrow \qquad (\forall\text{-E})\quad \dfrac{\forall x\psi(x) \qquad \begin{array}{c}\Box\text{———}(i)\\ \underbrace{\psi(t)\;,\;\Delta}\\ \Pi\\ \theta/\bot\end{array}}{\theta/\bot}{\scriptstyle (i)}\;;$$

where $\boldsymbol{\alpha}$ is any individual in the domain

where t is any closed term

hence

$$\dfrac{\forall x\psi(x) \qquad \begin{array}{c}(i)\text{———}\cdots\Box\cdots\text{———}(i)\\ \underbrace{\psi(t_1)\;,\;\ldots\;,\;\psi(t_n)\;,\;\Delta}\\ \Pi\\ \theta/\bot\end{array}}{\theta/\bot}{\scriptstyle (i)}$$

where $t_1,\ldots,t_n$ are closed terms

7.2.6 Morphing ∃-$\mathcal{F}$ to ∃-E

$$(\exists\text{-}\mathcal{F})\quad \dfrac{\exists x\varphi \quad \left\{\begin{matrix} \overset{\square\overline{}(i)}{\underbrace{\varphi^x_{\boldsymbol{\alpha}}\,,\,\Lambda_{\boldsymbol{\alpha}}}} \\ F_{\boldsymbol{\alpha}} \\ \bot \end{matrix}\right\}_{\boldsymbol{\alpha}\in M}}{\bot}(i) \quad\rightsquigarrow\quad (\exists\text{-E})\quad \dfrac{\exists x\varphi^{@} \quad \begin{matrix} \overset{\square\overline{}(i)}{\underbrace{\varphi^x_a\,,\,\Delta^{@}}} \\ \Pi \\ \theta^{@}/\bot \end{matrix}}{\theta^{@}/\bot}(i)$$

8 Allowing →-I to *Discharge* an Assumption

Note that we have thus far obtained an introduction rule for → in only the two parts that correspond to the respective considerations that it suffices, for the truth of a conditional, to have its antecedent false, or to have its consequent true. But this leaves out of the picture the most familiar form of →-Introduction, known as Conditional Proof, which allows for the discharge of the antecedent φ as an assumption if it has been used in the derivation of the consequent ψ as the conclusion of the subordinate proof:

$$\dfrac{\begin{matrix} \overline{}^{(i)} \\ \varphi \\ \vdots \\ \psi \end{matrix}}{\varphi \rightarrow \psi}(i)$$

The first two parts of →-Introduction immediately below were obtained by morphing the rule of →-Verification. The third part—the usual 'Rule of Conditional Proof' allowing discharge of the antecedent-as-assumption—can now be added to them, in light of the derivations that follow the three parts displayed here.

$$\dfrac{\begin{matrix} \overset{\square\overline{}(i)}{\underbrace{\varphi\,,\,\Delta}} \\ \Pi \\ \bot \end{matrix}}{\varphi \rightarrow \psi}(i) \qquad \dfrac{\begin{matrix} \Delta \\ \Pi \\ \psi \end{matrix}}{\varphi \rightarrow \psi} \qquad \dfrac{\begin{matrix} \overset{\overline{}(i)}{\underbrace{\varphi\,,\,\Delta}} \\ \Pi \\ \psi \end{matrix}}{\varphi \rightarrow \psi}(i)$$

The following derivations employ Bivalence—the first on the antecedent φ as the positive horn-assumption, the second on the consequent ψ as the same:

$$\dfrac{\dfrac{\begin{array}{c}\overline{\varphi}^{\,(2)}\\ \vdots \\ \psi\end{array}}{\varphi\to\psi} \qquad \dfrac{\dfrac{{}^{(2)}\overline{\neg\varphi} \quad \overline{\varphi}^{\,(1)}}{\bot}}{\varphi\to\psi}{\scriptstyle(1)}}{\varphi\to\psi}{\scriptstyle(2)} \qquad\qquad \dfrac{\dfrac{{}^{(2)}\overline{\psi}}{\varphi\to\psi} \qquad \dfrac{\dfrac{{}^{(2)}\overline{\neg\psi} \quad \begin{array}{c}\overline{\varphi}^{\,(1)}\\ \vdots \\ \psi\end{array}}{\bot}}{\varphi\to\psi}{\scriptstyle(1)}}{\varphi\to\psi}{\scriptstyle(2)}$$

9 Succinct Graphic Statement of $\to$-I

We now have the following three 'parts' of the rule of $\to$-Introduction:

$$\dfrac{\begin{array}{c}\Box\overline{\varphi}^{\,(i)}\;,\;\Delta\\ \underbrace{\qquad\qquad}\\ \Pi\\ \bot\end{array}}{\varphi\to\psi}{\scriptstyle(i)} \qquad \dfrac{\begin{array}{c}\Delta\\ \Pi\\ \psi\end{array}}{\varphi\to\psi} \qquad \dfrac{\begin{array}{c}\overline{\varphi}^{\,(i)}\;,\;\Delta\\ \underbrace{\qquad\qquad}\\ \Pi\\ \psi\end{array}}{\varphi\to\psi}{\scriptstyle(i)}$$

The second and third of these can be melded so that the display becomes

$$\dfrac{\begin{array}{c}\Box\overline{\varphi}^{\,(i)}\;,\;\Delta\\ \underbrace{\qquad\qquad}\\ \Pi\\ \bot\end{array}}{\varphi\to\psi}{\scriptstyle(i)} \qquad \dfrac{\begin{array}{c}\Diamond\overline{\varphi}^{\,(i)}\;,\;\Delta\\ \underbrace{\qquad\qquad}\\ \Pi\\ \psi\end{array}}{\varphi\to\psi}{\scriptstyle(i)}$$

The diamond in the graphic rule-part on the *right* says 'vacuous' discharge is permitted: φ need not be an undischarged assumption of the subordinate proof Π. This is exactly the rule of ($\to$-I) in the Gentzen–Prawitz tradition. The graphic rule-part on the *left* is a 'new part' of ($\to$-I) supplied by the Core logician *via* the foregoing method of morphing. It directly respects the third and fourth lines of the truth table for $\to$. It is derivable in the Gentzen–Prawitz system by appeal to their rule of EFQ, *which we, by contrast, eschew*.

It is easy to see how to re-write the foregoing rules of natural deduction as sequent rules. Introduction rules can be re-written as Right rules, and Elimination rules can be re-written as Left rules. A single illustration, say with the connective $\to$, should suffice to indicate the way this is done.

$$(\to\text{-I})\quad \begin{array}{cc} \overset{\square\!\!-\!\!-(i)}{\underbrace{\varphi\,,\,\Delta}} & \overset{\diamond\!\!-\!\!-(i)}{\underbrace{\varphi\,,\,\Delta}} \\ \vdots & \vdots \\ \dfrac{\bot}{\varphi\to\psi}{\scriptstyle(i)} & \dfrac{\psi}{\varphi\to\psi}{\scriptstyle(i)} \end{array} \quad \text{are re-written as} \quad (\to\text{-R})\ \dfrac{\Delta,\varphi:\bot}{\Delta:\varphi\to\psi}\quad \dfrac{\Delta:\psi}{\Delta\setminus\{\varphi\}:\varphi\to\psi}$$

$$(\to\text{-E})\quad \dfrac{\varphi\to\psi\quad \begin{array}{c}\Delta_1\\ \vdots\\ \varphi\end{array}\quad \begin{array}{c}\overset{\square\!\!-\!\!-(i)}{\underbrace{\psi\,,\,\Delta_2}}\\ \vdots\\ \theta\end{array}}{\theta}{\scriptstyle(i)} \quad \text{is re-written as} \quad (\to\text{-L})\ \dfrac{\Delta_1:\varphi\quad \Delta_2,\psi:\theta}{\Delta_1,\Delta_2,\varphi\to\psi:\theta}$$

The result of this re-writing of rules of natural deduction as their corresponding rules of the sequent calculus is that every core proof in natural deduction is 'isomorphic' to the corresponding core proof in the sequent calculus. Here is an illustration of this.

$$\dfrac{A\wedge(B\vee C)\qquad \dfrac{\overset{(2)}{\overline{B\vee C}}\quad \dfrac{\dfrac{\overset{(2)}{\overline{A}}\quad \overset{(1)}{\overline{B}}}{A\wedge B}}{(A\wedge B)\vee(A\wedge C)}\quad \dfrac{\dfrac{\overset{(2)}{\overline{A}}\quad \overset{(1)}{\overline{C}}}{A\wedge C}}{(A\wedge B)\vee(A\wedge C)}}{(A\wedge B)\vee(A\wedge C)}{\scriptstyle(1)}}{(A\wedge B)\vee(A\wedge C)}{\scriptstyle(2)}$$

$$\dfrac{\dfrac{\dfrac{\dfrac{A:A\quad B:B}{A,B:A\wedge B}}{A,B:(A\wedge B)\vee(A\wedge C)}\qquad \dfrac{\dfrac{A:A\quad C:C}{A,C:A\wedge C}}{A,C:(A\wedge B)\vee(A\wedge C)}}{A,B\vee C:(A\wedge B)\vee(A\wedge C)}}{A\wedge(B\vee C):(A\wedge B)\vee(A\wedge C)}$$

Next we illustrate the tree structure of a successful proof search for the foregoing result.

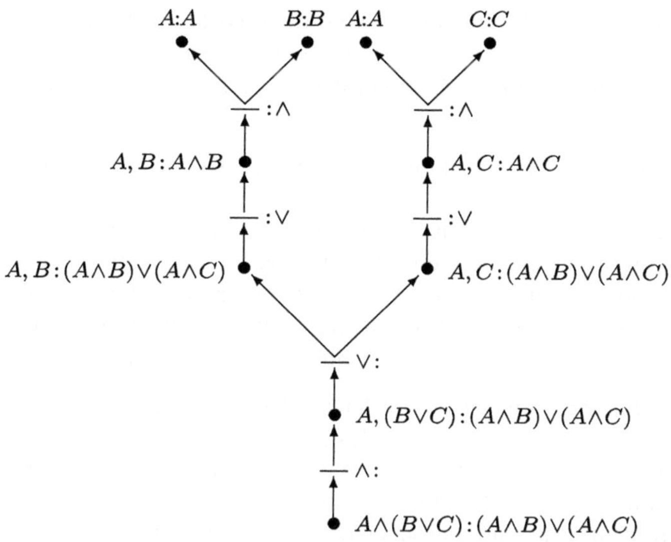

Core proofs can now be conceived of as the naturally arising objects that can be directly constructed by the method of proof search that proceeds by breaking down deductive problems into the naturally arising immediate subproblems resulting from either (i) choosing to focus on the dominant operator of the sought conclusion, so as to end up with a terminal Introduction (equivalently, application of a Right rule); or (ii) choosing to focus on the dominant operator in an available premise, so as to end up with a terminal Elimination (equivalently, application of a Left rule). Efficient proof-search is a matter of alternating insightfully between (i) and (ii) as the search proceeds. Normalization results in proof theory yield many such useful insights into how the syntactic form of a sequent to be proved can guarantee that if there is a proof at all to be had, then there is a proof of such-and-such a constrained form. (See [5].) To this end, similarly, a valuable constraint leading one to make successful choices during proof-search is the 'relevance filtration' afforded by the relevance property $\mathcal{R}(\Delta, \varphi)$ explicated in [8], and proved to hold between the premise-set Δ and the conclusion φ of any Classical Core proof.

10 Summary of Main Points

One starts with model-relative rules of evaluation, i.e. of verification and of falsification. One morphs those rules respectively into model-invariant introduction and elimination rules of natural deduction (or, equivalently, into Right rules and Left rules of sequent calculus).

The method of morphing, by its very design, produces no such rule as *Ex Falso Quodlibet* in natural deduction. Nor does it produce any rule of Cut or of Thinning in the sequent calculus. We therefore avoid the First Lewis Paradox ($A, \neg A : B$), which is the single most important aim of the relevantist.

Important features of the evaluation rules are preserved by the morphing. These are

- Major premises of falsifications stood proud; so too do major premises of elimination.
- Discharges were obligatory in evaluations; so too are they in natural deductions.
- Falsification rules were parallelized; so too are the elimination rules.
- Natural deductions become more 'general' by having arbitrary premise sets in place of sets of literals; and by having arbitrary sentences as conclusions of eliminations, in place of $\bot$ as the conclusion of a falsification.

Semantic *evaluations* are perforce *relevant*. This relevance is preserved by our morphing of rules of *evaluation* into rules of natural *deduction*.

Every rule of natural deduction (and in the sequent calculus) preserves fortuitous epistemic gains (i.e., gains in logical strength). If one succeeds fortuitously in finding a proof of a logically stronger result than the actual sequent that one set out to prove, that stronger result can be used at that very point in the construction of the desired proof. Put another way: logically stronger solutions to deductive subproblems generated during proof-search can always be 'inherited' to help deliver a stronger (and shorter) solution to the problem that generated those subproblems. The isomorphism between natural deductions and sequent proofs affords a conception of proofs as abstract objects of search, with their formal presentation being a matter of perspicuousness or pragmatic convenience.

11 Final Note

A reviewer helpfully inquired how the present study might be situated within the tradition of 'bilateralism' in proof theory. That tradition derives from the seminal paper [3], where the words 'bilateral', 'bilateralist', and 'bilateralism' were first introduced. The very first occurrence of any of these words was the occurrence of 'bilateral' on page 790, where Rumfitt wrote

> The standard introduction and elimination rules for "$\wedge$" and "$\vee$" and "$\rightarrow$" are harmonious in *this bilateral sense*. [Emphasis added]

The immediately preceding context supplied the sense in question. Rumfitt generously wrote

> For reasons that Neil Tennant [4, pp. 94–97] has explained, ... it makes sense to favour the strongest elimination rule that is a Prawitzian inverse of a given introduction rule.[fn] Tennant, indeed, shows how to detach Gentzen's underlying idea from the explanatory priority that he had accorded to introduction rules. Let us say that an introduction rule I is in harmony with an elimination rule E when (a) E's major premiss expresses the weakest proposition that can be eliminated using E, with I taken as given, and (b) I's conclusion expresses the strongest proposition that can be introduced using I, with E taken as given.

On page 97 of the passage that Rumfitt cited, the present author suggested that

> ... there is an intrinsic meaning to conjunction, for example, that is invariant across minimal, intuitionistic, and classical logic. I intend thereby to reveal as unjustifiable marginal excesses or excrescences the extra ingredients in the consequence relation of classical logic that have earned the generic labels (a) the fallacies of relevance, and (b) the classical laws of negation. The intrinsic meanings of the logical operators, characterized 'from below', as it were, provide no justification whatever for this fleshing out of the correct consequence relation.

The present study can be understood as underscoring further that very suggestion. Our methodology of morphing rules of evaluation into rules of deduction has avoided the fallacies of relevance, and has put us in a position to eschew the strictly classical rules of negation.

Rumfitt attributed another important facet of bilateralism as deriving from investigations by the present author. Just as in bilateralism the values T and F are coequal, along with the respective notions of assertion and rejection, so too in proof theory the notions of *proof* and of *disproof* are coequal. Rumfitt wrote (*loc. cit.*, p. 794

> ... as Tennant has put it, the expression [$\perp$] plays the role of a punctuation mark in the deduction (1999). It marks the point where the supposition ... has been shown to lead to a logical dead end, and is thus discharged, prior to an assertion of its negation.

The work [6] that Rumfitt cites presented the notions of proof and of disproof as coinductively defined. It is in the same spirit that we have here deployed the rules for generating the coinductively defined constructs that we called verifications and falsifications; and then morphed these rules into rules of natural deduction. In keeping with the metalogical prescription of [6], the latter rules can likewise be regarded as generating the coequal constructs that we called proofs and disproofs.

References

1. Gentzen, G. (1935). Untersuchungen über das logische Schliessen. *Mathematische Zeitschrift, 39*, 176–210, 405–431. https://doi.org/10.1007/BF01201353; https://doi.org/10.1007/BF01201363. Translated as 'Investigations into Logical Deduction'. In M. E. Szabo (Ed.), *The collected papers of Gerhard Gentzen* (pp. 68–131). Amsterdam: North-Holland (1969).
2. Prawitz, D. (1965). *Natural deduction: A proof-theoretical study*. Stockholm: Almqvist & Wiksell.
3. Rumfitt, I. (2000). "Yes" and "No". *Mind, 109*(436), 781–823.
4. Tennant, N. (1987, June). *Anti-realism and logic: Truth as eternal*. Clarendon Library of Logic and Philosophy. Oxford: Oxford University Press.
5. Tennant, N. (1992). *Autologic*. Edinburgh: Edinburgh University Press.
6. Tennant, N. (1999). Negation, absurdity and contrariety. In D. Gabbay & H. Wansing (Eds.), *What is negation?* (pp. 199–222). Dordrecht: Kluwer. https://doi.org/10.1007/978-94-015-9309-0_10
7. Tennant, N. (2010). Inferential semantics. In J. Lear & A. Oliver (Eds.), *The force of argument: Essays in honor of Timothy Smiley* (pp. 223–57). London: Routledge.

8. Tennant, N. (2015). The relevance of premises to conclusions of core proofs. *Review of Symbolic Logic, 8*(4), 743–784. https://doi.org/10.1017/S1755020315000040
9. Tennant, N. (2017). *Core logic*. Oxford: Oxford University Press. https://doi.org/10.1093/oso/9780198777892.001.0001
10. Tennant, N. (2018). A logical theory of truthmakers and falsitymakers. In M. Glanzberg (Ed.), *Handbook on truth* (pp. 355–393). Oxford: Oxford University Press.

Part IV
Applications

Frege Meets Belnap: Basic Law V in a Relevant Logic

Shay Allen Logan and Francesca Boccuni

Abstract Abstractionism in the philosophy of mathematics aims at deriving large fragments of mathematics by combining abstraction principles (i.e. the abstract objects $\S e_1$, $\S e_2$, are identical if, and only if, an equivalence relation $Eq_\S$ holds between the entities e_1, e_2) with logic. Still, as highlighted in work on the semantics for relevant logics, there are different ways theories might be combined. In exactly what ways must logic and abstraction be combined in order to get interesting mathematics? In this paper, we investigate the matter by deriving the axioms of second-order Peano Arithmetic from Frege's Basic Law V (the extension of F is identical with the extension of G if, and only if, F and G are extensionally equivalent) in the presence of a relevant higher-order logic. The results are interesting. Not only must we take on logic as true, and not only must we apply our logic to abstraction principles, but also we have to apply our theory of abstraction back to the logic in order to arrive at arithmetic. Thus, what Abstractionism gives us is not simply what we get from abstraction via logic, but also what we get from logic via abstraction.

1 Introduction

Abstractionism in the philosophy of mathematics is a varied logico-philosophical program, aiming at deriving large fragments of mathematics from (more often than not, classical higher-order) logic and abstraction principles. Informally, an abstraction principle states that the abstract objects $\S e_1$, $\S e_2$ are identical if, and only if, an equivalence relation $Eq_\S$ holds between e_1, e_2, where e_1, e_2 are entities of some kind. Syntactically, § is thus a functor that is defined on expressions of

S. A. Logan (✉)
Philosophy Department, Kansas State University, Manhattan, KS, USA
e-mail: salogan@ksu.edu

F. Boccuni
Philosophy Department, Vita-Salute San Raffaele University, Milan, Italy

I. Sedlár et al. (eds.), *New Directions in Relevant Logic*, Trends in Logic 63,
https://doi.org/10.1007/978-3-031-69940-5_16

whatever type the e's have, and which 'returns' as it were an expression of the type of singular terms. Semantically, § picks out a function that maps the set of equivalence classes induced by $Eq_{\S}$ to the first-order domain.

Some famous (or infamous) examples of abstraction principles trace back to Frege [7, 8]:

(**DIR**) the direction of line a is identical with the direction of line b if, and only if, a and b are parallel, for all a, b;

(**HP**) the number of the Fs is identical with the number of the Gs if, and only if, F and G are equinumerous, for all F, G;[1]

(**BLV**) the extension of F is identical with the extension of G if, and only if, F and G are extensionally equivalent, for all F, G.

Notice that DIR (the Direction Principle) is a first-order abstraction principle, since both lines and directions are first-order individuals. To the contrary, HP (Hume's Principle) and BLV (Basic Law V) are *second*-order abstraction principles, since they are formulated by quantifying over a second-order domain F, G belong to. Regardless of the interest first-order abstraction principles might have, in this article we will focus on second-order abstraction principles. We can then formulate a schematic second-order abstraction principle, of which HP and BLV are instances:

(**ABS**) for all F, G, the abstracts $\S F, \S G$ are identical if, and only if, an equivalence relation $Eq_{\S}$ holds between F, G.

There are several mathematical or philosophical aims an abstraction principle can be required to achieve. Especially the latter may vary from one abstractionist program to the other. Still, it is fair to say that all such programs:

- aim at deriving large fragments of mathematics by combining abstraction with logic;[2]
- argue that such a combination guarantees the existence of the objects abstraction governs, and provides identity and distinctness conditions for those objects.

This latter is crucial also as for the former, especially as regards those abstraction principles expected to generate denumerably many objects, e.g. natural numbers.

In order to examine ABS's commitments from above, it is helpful to factorize it in its two conjuncts:

($\mathbf{ABS}_{L2R}$) If $\S F$ and $\S G$ are identical, then $Eq_{\S}$ holds between F and G,

meaning that a partition of the second-order domain is necessary for the identity of abstracts;

($\mathbf{ABS}_{R2L}$) If $Eq_{\S}$ holds between F and G, then $\S F$ are identical $\S G$,

[1] Famously, HP is the non-logical principle that is central to so-called *Neologicism* of Hale and Wright—see [10, 22].

[2] See [4] for an effective survey.

i.e., a partition of the second-order domain is sufficient to deliver the existence of abstracts.

So far, we have formulated ABS, and some of its most famous instances such as HP and BLV, informally. This choice is dictated by our reluctance to take a stand on what kind of conditional "if... then" is yet. In what follows we will argue that, in case the conditional is relevant, the relation between the underlying logic and abstraction can be effectively brought to light in a very general fashion.

In order to appreciate how logic and abstraction are closely intertwined, and why such a relation is worth investigating under a respect as general as the one we are proposing, say that we read the conditional classically ($\supset$).

The right-to-left direction of ABS, $Eq_{\S}(F, G) \supset \S F = \S G$, states that, once we are committed to the existence of a certain partition of the second-order domain in a classical theory, we are thereby committed to the existence of the corresponding abstracts in that *very same theory*. On the other hand, the left-to-right direction, $\S F = \S G \supset Eq_{\S}(F, G)$, states that, only if we are committed to the existence of a certain partition of the second-order domain in a classical theory, we are thereby committed to the identity of abstracts in that *very same theory*.

There is nothing wrong with the material reading of ABS *per se*, but it can be questioned. As far as ABS_{R2L} is concerned, it might be objected that the existence of the individuals secured by a partition of the second-order domain is elicited by the underlying classical logic, which postulates the non-vacuity of the first-order domain.[3]

As for ABS_{L2R}, what does it mean that a certain partition be necessary for the identity of abstracts, since their identity is fixed in the very same theory in which the equivalence relation is required to hold? After all, abstracts have to be already identical or distinct, because they abide by classical logic.[4]

Still, in work that has come out on the semantics for relevant logics, there are different ways two theories might be combined. Disentagling the role logic and abstraction play both mathematically and philosophically is not only interesting *per se*, but leads to rather surprising results. More precisely, in this paper, we will tackle the following questions:

Q1. Can noncontractive *relevant* logics be used to derive a nontrivial amount of arithmetic from abstraction principles alone?

We will show that versions of the axioms of second-order Peano arithmetic PA^2 can be derived by relevant abstraction.[5] But there are interesting caveats. In particular, certain of the Peano axioms cannot be derived in their expected forms, but instead only in a form that makes explicit the various ways in which the abstractionist must

[3] Some authors complain about this feature of classical abstraction, which is crucial in e.g. Wright and Hale's Neologicism—see for instance Tennant [19, 20].

[4] Further authors complain about this feature of (extensionally) classical abstraction, which is also central in e.g. Wright and Hale's Neologicism—see e.g. Linnebo [11, 12].

[5] It is worth noticing that we will also get usual number-theoretic results from PA^2 axioms. This is nontrivial, modulo the well-known limitations of relevant arithmetic. See Sect. 5 below.

rely on the logic. Thus, in the course of answering Q1, we will raise and begin to answer:

Q2. In exactly what ways must logic and abstraction principles be combined, in order to get those mathematical results?

Q2 will sound nonsensical to classical logicians. But basement-dwelling relevance logicians (which is to say, those who work on systems at or below the logic E) will be quite familiar with the phenomenon in question: in weak relevance logics, one cannot conclude $A \to (B \to C)$ from $B \to (A \to C)$. Of particular importance in what follows is that even the so-called Ackermann constant t—a sentential constant that loosely interprets the conjunction of all the truths of the logic—is subject to this constraint: $\mathsf{t} \to (B \to C)$ does not follow from $B \to (\mathsf{t} \to C)$, nor do either of these follow from $(\mathsf{t} \wedge B) \to C$. Thus there is a rich diversity of different ways logic (in the form of t) might be combined with other information to arrive at a conclusion.

But also classical abstractionists (if not abstractionists in general) should be sensitive to this issue. Abstraction principles are often deemed non-logical (and also non-analytic, see e.g. [1]) because of their ontological consequences. And that's a fair enough accusation. But investigating how exactly the logic and abstraction have to be combined to that aim will shed light on the ontological responsibilities of the logic and abstraction respectively. As it will turn out, some of the patterns we see below suggest there is a connection between 'combinatorial' facts about how logic and abstraction combine to give results and facts about the existence and identity of abstract objects, and therefore on the relation between logic and abstraction. In particular, we will see how the existence of abstract objects is delivered by the application of the underlying logic to abstraction; whereas the identity of abstracts objects is obtained by the application of abstraction to logic. More in general: not only must we take on logic as true, and not only must we apply our logic to abstraction principles, but also we have to apply our theory of abstraction back to the logic in order to arrive at arithmetic. Thus, what Abstractionism gives us is not simply what we get from abstraction via logic, but also what we get from logic via abstraction.[6]

All of this will be said again—and said more precisely and more clearly—below. And on the note of what's below, here is a rough outline of what's to come. In Sect. 2, we will present the axiomatic system for a relevant logic and abstraction, and provide explanations of its main features. In Sect. 3, we will provide a derivation of (appropriate formulations of) the axioms of PA^2 in the system previously presented. In Sect. 4, we will comment on what such derivations reveal, i.e. what the order of application of the logic to abstraction, and of abstraction to the logic, is in order to derive PA^2, and what the ontological responsibilities of the logic and abstraction are respectively. Furthermore, we will provide a comparison with somewhat related

[6] See especially Sects. 2.4 and 4 below.

views in [15] and [21]. In Sect. 5, we will show how the recursiveness of addition is proved in our system. Finally, in Sect. 6, we provide closing remarks.

2 Explanation of the Apparatus and Formal Details

In this section, we will provide an axiomatization of the base second-order relevant logic we will use in the remainder. The particular choices we make will, at first, look a bit ad hoc, so we also take a moment to justify the design choices in play.

2.1 The Logic

We work in a second-order language with connectives $\neg$, $\wedge$, and $\rightarrow$ and quantifier $\forall$. With respect to these we adopt the usual definitions for $\vee$, $\leftrightarrow$, and $\exists$. Where A is a formula and F^n is an n-ary predicate variable, we write $A[F^n/B(y_1, \ldots, y_n)]$ for the formula that results from replacing each instance of a formula of the form $Ft_1 \ldots t_n$ in A with an instance of the formula $B(y_1/t_1, \ldots, y_n/t_n)$, where we assume that in the latter the substitution is done simultaneously via some device or other that allows one to avoid collision.

Language in place, the logic $\mathrm{DL2Q}^{\mathsf{t},fc}$ that we work in is axiomatized as follows:

A1. t
A2. $A \rightarrow A$
A3. $(A \wedge B) \rightarrow A$; $(A \wedge B) \rightarrow B$
A4. $((A \rightarrow B) \wedge (A \rightarrow C)) \rightarrow (A \rightarrow (B \wedge C))$
A5. $[A \wedge (B \vee C)] \rightarrow [(A \wedge B) \vee (A \wedge C)]$
A6. $\neg\neg A \rightarrow A$
A7. $(A \rightarrow \neg B) \rightarrow (B \rightarrow \neg A)$
A8. $((A \rightarrow B) \wedge (B \rightarrow C)) \rightarrow (A \rightarrow C)$
A9. $(A \rightarrow \neg A) \rightarrow \neg A$
A10. $\forall x A \rightarrow A(x/y)$ provided y is free for x in A.
A11. $\forall x(A \rightarrow B) \rightarrow (A \rightarrow \forall x B)$ provided x is not free in A.
A12. $\forall x(A \vee B) \rightarrow (A \vee \forall x B)$ provided x is not free in A.
A13. $\exists F^n \forall y_1 \ldots \forall y_n (A \leftrightarrow A(F^n/B(y_1, \ldots y_n)))$ where F doesn't occur free in B and B is free for F in A.

$$\text{R1.}\ \frac{A \qquad A \to B}{B}$$

$$\text{R2.}\ \frac{A \qquad B}{A \wedge B}$$

$$\text{R3.}\ \frac{A \to B \qquad C \to D}{(B \to C) \to (A \to D)}$$

$$\text{R4.}\ \frac{A}{\forall x A}$$

$$\text{R5.}\ \frac{A}{\mathsf{t} \to A}$$

As usual, we write '$\vdash A$' to mean that A is provable from the given axioms using the given rules. We note that in axioms A10-A12 and rule R4, the quantification can be either first- or second-order. The 'official' (but unweildy) name '$\text{DL2Q}^{\mathsf{t},fc}$' reflects the fact that the logic is a second-order quantified, includes the so-called Ackermann constant 't' and incorporates *f*ull *c*omprehension (axiom scheme A13).[7]

We'll now turn to answering some of the obvious questions one might ask about this logic, we'll note one further 'logical' abbreviation that will be helpful for our purposes: for $n > 0$ define $A \stackrel{n}{\to} B$ as follows:

- $A \stackrel{0}{\to} B := A \to B$
- $A \stackrel{n+1}{\longrightarrow} B := A \to (\mathsf{t} \stackrel{n}{\to} B)$

So, e.g. $A \stackrel{1}{\to} B$ abbreviates $A \to (\mathsf{t} \to B)$ and $A \stackrel{2}{\to} B$ abbreviates $A \to (\mathsf{t} \to (\mathsf{t} \to B))$, etc.

2.2 Answers to Some of the Obvious Questions

$\text{DL2Q}^{\mathsf{t},fc}$ certainly has a weird enough name. Readers of a suspicious bent are likely to find it suspicious on these grounds alone. The purpose of this sidebar is to allay such suspicions. The unsuspicious can thus safely skip it.

We allay the suspicions from left to right, as it were. First up, then, is DL. On its own, DL is a fairly well-known logic—it's axiomatized (see e.g. [3]) by A2–A8 together with R1–R3. DL *is* on the weaker end of the family of relevant logics. But it's also a dues-paying member in good standing of the relevance community. So those who found their suspicions aroused by the first two letters of our logic's name

[7] A referee informs us that the 'DL' part of the name originally stood for 'Dialectical Logic', though in the contexts in which that mattered, the logic often had explicit contradictions of a certain sort thrown in as well.

are invited to peruse e.g. [16] or other parts of the relevant literature to find ways to allay their worries.

Next up left-to-right is '2Q'. To extend from DL to DL2Q, we add A9–A11 and R4. That this is 'the right' way to extend is justified by the fact that it's the straightforward second-orderization of 'the right' way to extend from DL to first-order quantified DL, DLQ (see e.g. [6] or [13]).

The next bit—the so-called Ackermann constant t—is, proof-theoretically at least, straightforward. We've added t by adding axiom A1 and rule R5. These are the completely standard moves—see e.g. [16] again. What it means and why it's needed, on the other hand, requires a bit of explanation.

Intuitively, t represents the conjunction of all the truths of the logic, whatever it might be. As mentioned above, one main focus of the paper is determining exactly what ways logic must be applied to abstraction, and abstraction to logic in order to arrive at various bits of arithmetic. The use of t, whether in definitions or derivations of theorems, will make visible the order of both kinds of application. This should feel a bit loose at the moment, but when we find t playing this role below, we'll highlight it so that the reader can see exactly what we mean.

Finally is fc—full comprehension. This requires the most explanation. We've accomplished 'full' comprehension by adding axiom scheme A13. But the form of this axiom scheme might (should?) surprise the reader. Worth noting is that the instances of A13 where A is just $Fy_1 \dots y_n$, are exactly the instances of the 'usual' full comprehension axiom scheme:

$$\exists F \forall y_1 \dots \forall y_n (Fy_1 \dots y_n \leftrightarrow B)$$

So clearly by adopting all instances of the scheme we've given as axioms, we've also adopted all instances of the usual full (and fully impredicative) comprehension scheme. And as we leave to the reader to verify, in classical logic, once one has on hand all instances of the usual comprehension scheme, one can prove every instance of A13. Thus, in exactly the sense just spelled out, the two axiom schemes are classically equivalent. But in the setting at hand, this equivalence fails: the scheme we've added here is strictly stronger than the usual axiom scheme. One might wonder the reason for adopting the stronger option. Our reply is quite simple: it's necessary in order for universal elimination to work in the way one would hope; see Theorem 1 for the precise details.

We trust this to have allayed the suspicions of all those readers who held allayable suspicions, and thus move on.

2.3 A Few Important Facts

Our first task is to prove a few things about our system. We begin with a few straightforward results:

Lemma 1

- $\vdash A \to B$ *iff* $\vdash \neg B \to \neg A$. *(Contraposition)*
- *If* $\vdash \forall x(A \to B)$ *and* x *is not free in* A, *then* $\vdash A \to \forall x B$. *(Intensional Confinement)*
- *If* $\vdash A \to B$ *and* $\vdash B \to C$, *then* $\vdash A \to C$ *(Transitivity)*
- $\vdash \forall x(A \to B) \to (\exists x A \to B)$ *for all* B *in which* x *does not occur free. (Existential Elimination)*
- $\vdash \forall F(Fx \leftrightarrow Fx)$. *(Identity)*
- $\vdash A(c) \to \exists x A(x)$, *where here we write '$A(c)$' to mean* A *with one or more occurrences of 'c' highlighted in it and by '$A(x)$' we mean the corresponding formula in which the highlighted instances of 'c' have been replaced by 'x'. (Existential Introduction)*
- *If* $\vdash A \to (B_1 \to \cdots \to (B_n \to C)\ldots)$ *and* $\vdash C \to D$, *then* $\vdash A \to (B_1 \to \cdots \to (B_n \to D))$. *(Generalized Transitivity)*
- *If* $\vdash C_i \to B_i$ *for* $1 \le i \le n$, *then for all* A *and* D, $\vdash [A \to (B_1 \to (\cdots \to (B_n \to D)\ldots))] \to [A \to (C_1 \to (\cdots \to (C_n \to D)\ldots))]$ *(Embedded Prefixing)*

Proof The first five are proved in exactly the way one would expect, though a few come with an extra helping of tedium because of the second-order quantifiers. The last two follow by induction on n, though the second induction is a bit trickier than the first. □

In the remainder we will write '$a = b$' as shorthand for '$\forall F(Fa \leftrightarrow Fb)$'. We note that by Identity, $\vdash x = x$.

Corollary 1 (Generalized Cut) *If* $\vdash C_i \to B_i$ *for* $1 \le i \le n$ *and* $\vdash A \to (B_1 \to (\cdots \to (B_n \to D)\ldots))$, *then* $\vdash A \to (C_1 \to (\cdots \to (C_n \to D)\ldots))$.

Proof Immediate from Embedded Prefixing □

Corollary 2 *If* $\vdash B_i$ *for* $1 \le i \le n$ *and* $\vdash A \to (B_1 \to (\cdots \to (B_n \to D)\ldots))$, *then* $\vdash A \xrightarrow{n} D$.

Proof Since $\vdash B_i$, R5 gives $\vdash \mathsf{t} \to B_i$, so the result follows by Generalized Cut. □

And, because the paper would be incomplete without going through the details in at least one case, we'll prove the following important result in all its glory:

Theorem 1 (Instantiation) *If* $B(\overline{y}) = B(y_1, \ldots, y_n)$ *is free for* F^n *in* A *and does not contain* F^n *free, then* $\vdash \forall F^n A \to A(F^n / B(y_1, \ldots, y_n))$.

Proof To begin, note that by an instance of A3,

$$\vdash (A \leftrightarrow A[F^n/B(\overline{y})]) \to (A \to A[F^n/B(\overline{y})]) \tag{1}$$

Also where $\overline{z} = z_1, \ldots, z_k$ are the free variables in A,

$$\vdash \forall\overline{z}(A \leftrightarrow A[F^n/B(\overline{y})]) \rightarrow (A \leftrightarrow A[F^n/B(\overline{y})]) \tag{2}$$

Together (1) and (2) give that

$$\vdash \forall\overline{z}(A \leftrightarrow A[F^n/B(\overline{y})]) \rightarrow (A \rightarrow A[F^n/B(\overline{y})]) \tag{3}$$

Thus by Contraposition,

$$\vdash \neg(A \rightarrow A[F^n/B(\overline{y})]) \rightarrow \neg\forall\overline{z}(A \leftrightarrow A[F^n/B(\overline{y})]) \tag{4}$$

Next observe that by an instance of A9 we have

$$\vdash \forall F^n\neg(A \rightarrow A[F^n/B(\overline{y})]) \rightarrow \neg(A \rightarrow A[F^n/B(\overline{y})]) \tag{5}$$

Together (4) and (5) then give

$$\vdash \forall F^n\neg(A \rightarrow A[F^n/B(\overline{y})]) \rightarrow \neg\forall\overline{z}(A \leftrightarrow A[F^n/B(\overline{y})]) \tag{6}$$

Thus, by an application of R4 we have that

$$\vdash \forall F^n[\forall F^n\neg(A \rightarrow A[F^n/B(\overline{y})] \rightarrow \neg\forall\overline{z}(A \leftrightarrow A[F^n/B(\overline{y})])] \tag{7}$$

So since $\forall F$ does not occur free in $\forall F^n\neg(A \rightarrow A[F^n/B(\overline{y})]$, Intensional Confinement gives

$$\vdash \forall F^n\neg(A \rightarrow A[F^n/B(\overline{y})] \rightarrow \forall F^n\neg\forall\overline{z}(A \leftrightarrow A[F^n/B(\overline{y})]) \tag{8}$$

Thus, again by Contraposition,

$$\vdash \exists F^n\forall\overline{z}(A \leftrightarrow A[F^n/B(\overline{y})]) \rightarrow \exists F^n(A \rightarrow A[F^n/B(\overline{y})]) \tag{9}$$

But also by A13, $\vdash \exists F^n\forall\overline{z}(A \leftrightarrow A[F^n/B(\overline{y})])$. Thus

$$\vdash \exists F^n(A \rightarrow A[F^n/B(\overline{y})]) \tag{10}$$

Now observe that since by A9 $\forall F A \rightarrow A$, an instance of A2 and an application of R3 give that

$$\vdash (A \rightarrow A[F^n/B(\overline{y})]) \rightarrow (\forall F^n A \rightarrow A[F^n/B(\overline{y})]) \tag{11}$$

But then by R4,

$$\vdash \forall F^n[(A \rightarrow A[F^n/B(\overline{y})]) \rightarrow (\forall F^n A \rightarrow A[F^n/B(\overline{y})])] \tag{12}$$

But now observe that since X^n does not occur free in $\forall F^n A \rightarrow A[F^n/B(\overline{y})]$, it follows from (12) and Existential Elimination that

$$\vdash \exists F^n(A \rightarrow A[F^n/B(\overline{y})]) \rightarrow (\forall F^n A \rightarrow A[F^n/B(\overline{y})]) \tag{13}$$

Thus, finally, by (10) and (13) we have our desired result:

$$\vdash \forall F^n A \rightarrow A[F^n/B(\overline{y})] \tag{14}$$

□

Corollary 3 *If* $B(\overline{y}) = B(y_1, \ldots, y_n)$ *is free for* F^n *then* $\vdash \forall F^n A \rightarrow A(F^n/B(y_1, \ldots, y_n))$.

Proof If B does not contain F^n free, the result follows from the previous theorem. If it does, then consider $B' = B(F^n/G^n)$ where G^n does not occur in A. By the previous theorem $\vdash \forall F^n A \rightarrow A(F^n/B'(y_1, \ldots, y_n))$. Thus by R4, $\vdash \forall G^n(\forall F^n A \rightarrow A(F^n/B'(y_1, \ldots, y_n)))$. So by an instance of A9 and R1, we can instantiate at F^n and get $\vdash \forall F^n A \rightarrow A(F^n/B)$. □

2.4 *Adding Abstraction*

The final move is to augment $\mathrm{DL2Q}^{\mathrm{t},fc}$ by an abstraction principle. The abstraction principle we add should satisfy the following two desiderata:

- It should be strong enough to allow for a derivation of Peano axioms, but
- Not so strong that it interferes with the logic's ability to track 'combinatorial' details about how logic and abstraction need to be mixed to deliver these results.

Consider ABS again: $\forall F \forall G(\S F = \S G \leftrightarrow Eq_{\S}(F, G))$. We have two options here:

1. keep ABS as the sole non-logical axiom added to $\mathrm{DL2Q}^{\mathrm{t},fc}$, and derive significant second-order abstraction principles such as HP or BLV from it, and then derive PA^2 one way or another;
2. assume a particular abstraction principle, e.g. HP or BLV, directly as the sole non-logical axiom added to $\mathrm{DL2Q}^{\mathrm{t},fc}$, and derive PA^2 from it.

For the sake of simplicity, we will pursue the second option. In this respect, a further choice has to be made. Peano arithmetic can be recovered either by HP or BLV. The former would involve massive use of second-order definitional resources in order to recover so-called *Frege Arithmetic*, namely second-order logic plus HP which, via appropriate definitions of Fregean notions such as *predecessor*, *(weak) ancestral*, and *natural number*, interprets PA^2—this result is now known (see e.g. [23] and the references therein) as *Frege's theorem*.[8]

The problem is that, at least for agents like us, the complexity this adds ends up interfering with the second desideratum recorded above, as the formulas we end up at are so complex as to obscure rather than reveal the combinatorially interesting facts we're after. Thus, in the end, we settle on augmenting $\mathrm{DL2Q}^{\mathsf{t},fc}$ with one additional (and intuitively non-logical) axiom:

(BLV) $\forall F \forall G(\{F\} = \{G\} \leftrightarrow F \equiv G)$,

where $\equiv$ is defined as $\forall x(Fx \leftrightarrow Gx)$.[9]

An important initial observation is that BLV is formulated without any occurrences of t. More helpfully, contrast BLV in the form we've added it with axioms like the following:

- $\forall F \forall G((\{F\} = \{G\} \wedge \mathsf{t}) \leftrightarrow F \equiv G)$;
- $\forall F \forall G(\{F\} = \{G\} \leftrightarrow (F \equiv G \wedge \mathsf{t}))$;
- $\forall F \forall G(\{F\} = \{G\} \overset{n}{\leftrightarrow} F \equiv G)$; and so on.

In such axioms, the 'blame' (as it were) meted out for the presence of the abstracta in question (extensions in the case at hand) is being divided in various ways between logic and abstraction.

But it is often argued that abstraction principles are axioms of infinity that cannot be deemed logical, because of their ontological consequences. In fact, by axioms such as HP or BLV, it is possible to prove the existence of denumerably many first-order individuals, i.e. the *denotata* of abstract terms appearing on their left-hand sides. Still,

> [i]n logic we ban the empty domain as a concession to technical convenience but draw the line there: We firmly believe that the existence of even two objects, let alone infinitely many, cannot be guaranteed by logic alone. [1, 199]

In this respect, the lack of t in formulating BLV makes the non-logicality of the objects it guarantees the existence of apparent, in particular as regards its ontological consequences as we will see below. So, in a sense, BLV is *pure* abstraction—or, some might say, pure mathematics.

This latter remark also provides some historical justification for choosing BLV over HP. Famously, Frege argued, not without reservations, that BLV is a logical principle—see [8, *Foreword*, p. VII]. Later, he bitterly deplored BLV and its alleged logical status, because of Russell's paradox—[8, *Afterword*]; and eventually, he abandoned it altogether. In our framework, not only BLV requires no logic (i.e. no t)

---

8 A further path also worth investigating would be the derivation of HP from BLV as the sole non-logical axiom added to $\mathrm{DL2Q}^{\mathsf{t},fc}$, but that would involve at least the same amount of second-order resources that a derivation of PA^2 from HP as the sole non-logical axiom requires.

9 Clearly, the functional abstraction operator {}, (semantically) taking entities in the second-order domain as arguments and individuals from the first-order domain as values, is to be added to the underlying language.

to be formulated, but furthermore, in what follows, we will show where exactly BLV fails to be a logical principle: in differentiating among abstracts delivered by a given equivalence relation, and therefore, when appropriate, in delivering denumerably many abstracts.

In doing so, we will also show what the responsibility of the underlying logic is: it is the underlying logic that delivers (or has to deliver) the existence of at least an individual, starting from which abstraction works like an assembly line manufacturing denumerably many objects via identity and distinctness statements. Without the input of the logic, abstraction would run on empty.

By detaching the role BLV and the logic play in the derivation of Peano axioms (and the recursiveness of addition) from below, it will be visible that BLV is no logical principle, regardless of Russell's paradox and, most of all, *pace* Frege.

3 The Derivations of Peano Axioms

We'll now turn to deriving the Peano Axioms. But first, of course, we need numbers. Since they can be defined quite cleanly, we'll use (essentially) the von Neumann ordinals to do the job. Thus we adopt the following two definitions:

- $\overline{0} := \{x.\neg(x = x)\}$;
- $\{w\} := \{x.w = x\}$

We then define the (von Neumann) numbers as follows:

$$\mathbb{N}x := \forall F((\forall w(Fw \to F\{w\}) \wedge \mathsf{t}) \to (F\overline{0} \to Fx))$$

Note that classically, F is inductive along the von Neumann successor relation and holds of the initial von Neumann ordinal just if the following is true:

$$(F\overline{0} \wedge \forall w(Fw \to F\{w\}))$$

Thus, again speaking classically, one can say 'if F is inductive and holds of 0, then F holds of x' as follows:

$$(F\overline{0} \wedge \forall w(Fw \to F\{w\})) \to Fx$$

And of course, using classically-acceptable reasoning, the above is equivalent to

$$\forall w(Fw \to F\{w\}) \to (F\overline{0} \to Fx)$$

But the transformations this relies on are not permissible in DL2Q$^{\mathsf{t},fc}$. This raises two natural questions:

- Why use the latter formulation instead of the initial (conjunctive) formulation for capturing the numbers?
- Why the 't'?

The answer to the first question is this: since we lack conjunctive modus ponens (viz. the following: $(A \wedge (A \rightarrow B)) \rightarrow B$) as a theorem in our setting, the initial, conjunctive interpretation of 'inductive and holds of 0' is largely impotent.

For the second question, the answer is even easier: absent the t, $\overline{0}$ doesn't count as a number. This is fairly easy to see: what $\forall w(Fw \rightarrow F\{w\})$ guarantees is that, given anything F holds of, F also holds of the (von Neumann; we're going to stop saying that in the remainder) successor of F. But since 0 isn't a successor (see Lemma 2), this can't help us learn anything at all about what's going on with 0. Logic, however, *can* help us out on that front: by axiom A2, the following is a theorem of the logic: $F\overline{0} \rightarrow F\overline{0}$. Being a theorem, it's also implied by t. Thus, in order to include 0 among the numbers (which, as the proof of Peano1 below shows, crucially requires t to be carried out and will be commented on in Sect. 4), we have to add an explicit call to the logic (in the form of a '$\wedge\mathsf{t}$') to our definition of numbers. Arguably, it is because the logic is necessary to prove that indeed 0 is a number, namely that the set of natural numbers is nonempty because it contains at least an individual i.e. 0, that t is needed in the definition of $\mathbb{N}$. So, the logic is necessary to guarantee that there is at least a natural number, in particular, the one the ω-sequence starts from.

With definitions now expressed and discussed, we turn to deriving the Peano axioms, beginning with an explicit reconstruction of the argument given in the previous paragraph:

Theorem 2 (Peano1) $\vdash \mathbb{N}\overline{0}$

Proof $\vdash F\overline{0} \rightarrow F\overline{0}$ by A2 so by R5, $\vdash \mathsf{t} \rightarrow (F\overline{0} \rightarrow F\overline{0})$. Also by A3, $\vdash (\forall y(Fy \rightarrow F\{y\}) \wedge \mathsf{t}) \rightarrow \mathsf{t}$. So by Lemma 1 part (b), $\vdash (\forall y(Fy \rightarrow F\{y\}) \wedge \mathsf{t}) \rightarrow (F\overline{0} \rightarrow F\overline{0})$. Thus $\vdash \mathbb{N}\overline{0}$. □

Lemma 2 $\vdash \neg\{x\} = \overline{0}$.

Proof First note that by BLV, we have that

$$\vdash \{x\} = \overline{0} \rightarrow \forall z(z = x \rightarrow \neg z = z) \tag{15}$$

Next, note that by instances of A9 and A10 we have

$$\vdash \forall z(z = x \rightarrow \neg z = z) \rightarrow (x = x \rightarrow \neg x = x) \tag{16}$$

$$\vdash (x = x \rightarrow \neg x = x) \rightarrow \neg x = x \tag{17}$$

Together with transitivity (b), (15), (16), and (17) then give $\vdash \{x\} = \overline{0} \rightarrow \neg x = x$. So by A7 and R1, $\vdash x = x \rightarrow \neg\{x\} = \overline{0}$. But by Identity, $\vdash x = x$ as well. So $\vdash \neg\{x\} = \overline{0}$ □

Corollary 4 (Peano2) $\vdash (\mathbb{N}x \wedge \mathsf{t}) \rightarrow \neg\{x\} = \overline{0}$.

Theorem 3 $\vdash (\mathbb{N}x \wedge \mathsf{t}) \rightarrow \mathbb{N}\{x\}$.

Proof Note that by BLV and Instantiation, we have

$$\vdash \mathbb{N}x \rightarrow \underbrace{[(\forall y(Fy \rightarrow F\{y\}) \wedge \mathsf{t}) \rightarrow (F\overline{0} \rightarrow Fx)]}_{A} \tag{18}$$

Clearly we also have that

$$\vdash (\forall y(Fy \rightarrow F\{y\}) \wedge \mathsf{t}) \rightarrow (Fx \rightarrow F\{x\}) \tag{19}$$

Thus by R5, we also have

$$\vdash \mathsf{t} \rightarrow \underbrace{((\forall y(Fy \rightarrow F\{y\}) \wedge \mathsf{t}) \rightarrow (Fx \rightarrow F\{x\}))}_{B} \tag{20}$$

Thus we also have

$$\vdash (\mathbb{N}x \wedge \mathsf{t}) \rightarrow (A \wedge B) \tag{21}$$

But now observe that by A4 we get the following:

$$\vdash (A \wedge B) \rightarrow [(\forall y(Fy \rightarrow F\{y\}) \wedge \mathsf{t}) \rightarrow ([F\overline{0} \rightarrow Fx] \wedge [Fx \rightarrow F\{x\}])] \tag{22}$$

And by Generalized Transitivity we also have

$$\vdash ([F\overline{0} \rightarrow Fx] \wedge [Fx \rightarrow F\{x\}]) \rightarrow (F\overline{0} \rightarrow F\{x\}) \tag{23}$$

So, by two further instances of Generalized Transitivity, (first involving 22 and 23, then involving the resulting formula and 21) we have

$$\vdash (\mathbb{N}x \wedge \mathsf{t}) \rightarrow [(\forall y(Fy \rightarrow F\{y\}) \wedge \mathsf{t}) \rightarrow (F\overline{0} \rightarrow F\{x\})] \tag{24}$$

Thus by A4,

$$\vdash \forall F[(\mathbb{N}x \wedge \mathsf{t}) \rightarrow [(\forall y(Fy \rightarrow F\{y\}) \wedge \mathsf{t}) \rightarrow (F\overline{0} \rightarrow F\{x\})]] \tag{25}$$

And since F is not free in $\mathbb{N}x \wedge \mathsf{t}$, it follows that $\vdash (\mathbb{N}x \wedge \mathsf{t}) \rightarrow \mathbb{N}\{x\}$. □

It's worth noting that the above result does *not* seem to go through without the t. We suspect that a clever application of MaGIC should suffice to show that this is the case. But we couldn't quite be clever enough, so we'll instead offer the following heuristic reason to believe this: (18) and (19) are alone insufficient for showing that $\mathbb{N}x \rightarrow [(F\overline{0} \rightarrow Fx) \wedge (Fx \rightarrow F\{x\})]$ is a theorem. And absent that, there's no way to get to the desired result.

At any rate, with the above lemma in hand, we can quite easily prove the following:

Corollary 5 (Peano3) $\vdash \forall x((\mathbb{N}x \wedge \mathsf{t}) \rightarrow \exists y(\mathbb{N}y \wedge y = \{x\}))$

Lemma 3 $\vdash x = y \rightarrow \{x\} = \{y\}$

Proof It follows from the definition of '=' that $x = y \rightarrow \forall z(z = x \leftrightarrow z = y)$. But then by an instance of BLV, $x = y \rightarrow \{x\} = \{y\}$. □

Note that in all the results above, the needed 't's were all added *conjunctively*. In contrast, the next Peano axiom we derive requires an *intermediary* t—that is, rather than simply conjoining addition t's, we instead have to rely on the defined conditional $\overset{1}{\rightarrow}$.

Lemma 4 $\vdash \{x\} = \{y\} \overset{1}{\rightarrow} x = y$

Proof By BLV, Instantiation, and Transitivity, $\vdash \{x\} = \{y\} \rightarrow (x = x \rightarrow x = y)$. Thus since $\vdash \mathsf{t} \rightarrow x = x$ by Identity and R5, Corollary 2 gives that $\vdash \{x\} = \{y\} \rightarrow (\mathsf{t} \rightarrow x = y)$ as required. □

Corollary 6 (Peano4) $\vdash \forall x \forall y((\mathbb{N}x \wedge \mathbb{N}y \wedge \{x\} = \{y\}) \overset{1}{\rightarrow} x = y)$

This leaves us with just Peano 5. But the astute reader probably observed already that there's nothing to do here: our definition of '$\mathbb{N}x$' is essentially exactly 'x satisfies Peano 5'.

4 Philosophical Commentary

In this section, we will investigate how logic and abstraction interact in the derivation of Peano axioms from above, in order to shed light on the application both of logic to abstraction and of abstraction to logic in recovering them.

First, let us tackle the rather unequivocal cases: those in which only the t is involved in the proofs; and those in which only BLV is needed. Peano1 and Peano2 are the cases in point.

As anticipated in the previous section, in order to prove Peano1, the application of the background logic to abstraction is necessary. By BLV, the definition of $\overline{0}$ is provided, but BLV *per se* does not necessarily secure that there is an individual that $\overline{0}$ can be assigned to. Now consider two facts. First, $F\overline{0} \rightarrow F\overline{0}$ would be an instance of A2, even if $\overline{0}$ were an empty term. Secondly, it is via R5 that we eventually get to $\mathbb{N}\overline{0}$. So, in $\mathrm{DL2Q}^{\mathsf{t},fc}$ what the proof of Peano1 shows is that the non-vacuity of the set $\mathbb{N}$ of natural numbers, and in particular that $\mathbb{N}$ contains the individual 0, relies on the t. This presupposes that it is the background logic that guarantees (or has to guarantee) the existence of an individual which $\overline{0}$ is assigned to.[10]

Peano2, which captures the identity fact that 0 is no singleton, is proved as a corollary of Lemma 2, which states the functionality of the singleton. This latter result is proved on the basis on BLV—and by axioms and lemmas *not* concerning t. Consequently, BLV, but not the logic, is necessary to prove an identity fact concerning the abstract objects it governs.

The relation between the logic and abstraction is more nuanced as for Peano3, which states the existence of the successor, and Peano4, stating the injectivity of the singleton.

More specifically, Peano3 follows as a corollary from Theorem 3, stating that if x is a natural number (and the logic is around) so is its singleton. The proof of Theorem 3 requires both BLV and t, but in the proof BLV seems to provide just the raw material to which the logic has to be applied in order to get the desired result. In particular, by BLV if anything at all is F so will be its singleton, so that if 0 is F so will be any x, where x is a natural number—from which it also follows by Instantiation that, if anything at all is F so will be its singleton, so that if x is F so will be $\{x\}$. But *that* would be trivially true even if there were no numbers at all. As already mentioned, we suspect that without the logic we would not be able to move on from that to the proof that, if x is a natural number (and the logic is around), so will also be $\{x\}$. Consequently, without t it would not be possible to prove Peano3, which means that without the logic it would not be possible to prove the existence of the successor of any natural number.

Finally, let us come to Peano4. This latter is a corollary of Lemma 4, which states the injectivity of the singleton, *provided the logic is around.* At a closer look of the proof of Lemma 4, it is clear why this happens. BLV, by Transitivity, can only guarantee that $\{x\} = \{y\} \rightarrow (x = x \rightarrow x = y)$ is a theorem of $\mathrm{DL2Q}^{\mathsf{t},fc}$. But this theorem does *not* imply that $\vdash x = x \rightarrow (\{x\} = \{y\} \rightarrow x = y)$, since in $\mathrm{DL2Q}^{\mathsf{t},fc}$ $A \rightarrow (B \rightarrow C)$ does not follow from $B \rightarrow (A \rightarrow C)$. If this *were* the case, the injectivity of the singleton would follow by BLV and the definition of '=' (and Transitivity) alone. But this not being the case, we have to apply abstraction, i.e. $\{x\} = \{y\}$, to the logic, i.e. $\{x\} = \{y\} \rightarrow (\mathsf{t} \rightarrow \ldots)$, in order to deliver the desired result, i.e. $x = y$. This means that BLV is insufficient to prove the injectivity of the abstraction operator, at the very least as for singletons, which might sound rather surprising since BLV is the sole non-logical axiom governing the abstraction operator {}. Still, this makes sense as long as one considers that applying abstraction back to the logic is a completely different matter than applying the logic to abstraction: the logic *per se* does not deliver abstraction (sufficiently fine-grained identity conditions to distinguish between different individuals of a given kind),

[10] In classical Peano arithmetic, $\overline{0}$ is referential by assumption; in a setting with a free logic as the background logic, that $\overline{0}$ is indeed referential has to be proved, see e.g. [19, 20]. See also footnote 11 below.

whereas abstraction can only be added on top of a logic (which has to secure the existence of at least one individual) in order to be productive.

Given what's been argued so far, the logic, whether in isolation or as applied to abstraction, is needed in order to accomplish ontological results, (i.e. Peano1 and Peano3), whereas abstraction, either in isolation or as applied to the logic, is necessary when identity and distinctness facts have to be proved (i.e. Peano2 and Peano4). More precisely, once the existence of at least an individual is brought about by the logic, BLV starts proving (or at least is necessary to prove) the existence of denumerably many of them, via identity and, crucially, distinctness facts—as expected and as Lemmas 3 and 4 guarantee. But without the logic, BLV would run on empty.

Finally, our approach strikes some significant differences to [15] and [21]. First, there are some technical dissimilarities between our framework and the ones in [15] and [21] we deem important. For instance, both [15] and [21] are first-order systems with identity, so they suffer from the usual issues concerning identity in relevant settings. Being second-order, our framework can afford to define identity explicitly. This has consequences also as for the formulation of BLV, in particular in [21] where BLV has to be schematic. Since classical first-order logic plus schematic BLV is consistent, in order to bring about Russell's paradox [21] adds membership as a primitive—while we can define it explicitly *à la* Frege as $x \in y \leftrightarrow_{def} \exists F(y = \{F\} \wedge Fx)$. This highlights that Fregean extensions in our framework, unlike [21]'s, are delivered by a conception that is more faithful to Frege's (i.e. as objects strictly related to concepts) and, so, is remarkably different from the iterative (and even Cantorian) conception. Also, the system in [15] is modal, whereas our framework is plainly extensional.

Secondly, both in [15] and [21], the (respective) logics are meant to provide either *the* or at the very least *a* foundation for mathematics—or even all science as in [15]. This is immaterial in our perspective. Our use of the Ackermann constant t delivers a significantly different philosophical result. In any model, t is true at exactly those points that verify the logic of the model—that is, the set of sentences that all the theories in the model are closed under. Thus, those sentences that are a consequence of t in a given model are exactly the sentences that are contained in the logic of that model. *Theorems* involving t, then (such as the theorems we've proved above and will prove below) tell us something about what must be true in all logics—that is, what must be the case in all of our theory building endeavors whatsoever, whether they be scientific, mathematical or of any other nature.

Thus, what we are offering is a clear picture of the relation between the logic of a model (*whatever that might be*) and abstraction in deriving mathematical results. Our proposal embeds the bulk of logic and abstraction that is common to *any* logic and abstraction. What is common to any logic in our framework is that the logic has to provide at least an individual, so that abstraction does not 'run on empty' as it were.[11]

5 Addition

Before launching into the next section, it's worth noting why it's worth doing at all. The thing to observe is that it's one thing to show how to derive (versions of) the axioms of some system S in a novel logical environment and another thing to show how to arrive at (versions of) all the theorems of S in that environment. This is especially the case in derivations like those we're considering, since theories derived using classical logic are closed under an incredible array of rules that theories derived using relevant logics are not.

In the case at hand, what we need to now look at is the extent to which the above derivation of analogues of the Peano axioms also allow us to define and derive various bits of arithmetic. But there's rather a lot of arithmetic, so we'll only in fact be able to look one small but important piece of the picture.

To begin, we (roughly) follow the standard practice (see e.g. [5] and [23]) in defining the three-place addition relation $Sxyz$ as follows:

$$Sxyz := \forall F([\forall u \forall w(Fuw \rightarrow F\{u\}\{w\}) \wedge \mathsf{t}] \rightarrow [F\overline{0}x \rightarrow Fyz])$$

As with our definition of $\mathbb{N}$, our definition of S is classically equivalent to the usual definition, but modified to accommodate the vagaries of the system we're using here. And, as we trust the reader to pause and verify for themselves, $Sxyz$ intuitively holds of those triples where the third member is the sum of the first two.

Definition in hand, we can now state that our aim is to show that

$$\vdash \forall x \forall y(\mathbb{N}x \overset{2}{\rightarrow} (\mathbb{N}y \overset{2}{\rightarrow} \exists z Sxyz))$$

As an overview, here's roughly how we go about the task.

1. We begin by noting that the following is immediate from Theorem 1:

$$\vdash \mathbb{N}y \rightarrow ((\forall w((S\overline{0}ww \wedge \mathsf{t}) \rightarrow (S\overline{0}\{w\}\{w\} \wedge \mathsf{t})) \wedge \mathsf{t}) \rightarrow ((S\overline{000} \wedge \mathsf{t}) \rightarrow (S\overline{0}xx \wedge \mathsf{t})))$$

2. Thus, by Corollary 2, to show that $\vdash \mathbb{N}x \overset{2}{\rightarrow} S\overline{0}xx$, it suffices to show that $\vdash \forall w((S\overline{0}ww \wedge \mathsf{t}) \rightarrow (S\overline{0}\{w\}\{w\} \wedge \mathsf{t})) \wedge \mathsf{t}$ and $\vdash S\overline{000} \wedge \mathsf{t}$.

[11] This would rule out free logics, as pointed out in [17]. In our framework, it becomes clear why free logics don't work in this respect, unless the existence of at least an individual is proved in some way: the relation between the logic and abstraction is to be such that, for abstraction to work in the way we want it to, the underlying logic has to deliver the existence of at least an individual.

3. Since A1 gives $\vdash$ t, it suffices to show that $\vdash \forall w((S\bar{0}ww \wedge \mathsf{t}) \rightarrow (S\bar{0}\{w\}\{w\} \wedge \mathsf{t}))$ and $\vdash S\overline{000}$.
4. The second of these is essentially immediate from the fact that $F\overline{00} \rightarrow F\overline{00}$ is an axiom.
5. The first is demonstrated in Lemma 5 by an argument very like the argument we gave in Theorem 3.
6. We repeat the above procedure 'one level up' as it were. That is, let Fx abbreviate $\mathbb{N}y \overset{2}{\rightarrow} \exists zSxyz$. Then by Theorem 1, we have

$$\vdash \mathbb{N}x \rightarrow ((\forall w(Fw \rightarrow F\{w\}) \wedge \mathsf{t}) \rightarrow (F\bar{0} \rightarrow Fx))$$

7. Thus, by Corollary 2, to show that $\vdash \mathbb{N}x \overset{2}{\rightarrow} Fx$, it suffices to show that $\vdash \forall w((Fw \rightarrow F\{w\}) \wedge \mathsf{t}$ and $\vdash F\bar{0}$.
8. Since steps 1–5 will have already demonstrated (as recorded in Lemma 6) that $\vdash F\bar{0}$, it will suffice to show that $\vdash Fw \rightarrow F\{w\}$. And, having proved this in Lemma 8, two applications of R4 finish the job.

Plan in hand, we set to the job. The reader may find it useful to refer back to this summary to remind themselves why exactly we're doing the things we're doing.

Lemma 5 $\vdash \forall w((S\bar{0}ww \wedge \mathsf{t}) \rightarrow (S\bar{0}\{w\}\{w\} \wedge \mathsf{t}))$

Proof It will suffice to show $\vdash (S\bar{0}yy \wedge \mathsf{t}) \rightarrow S\bar{0}\{y\}\{y\}$. To see this, note first that Instantiation gives that

$$\vdash S\bar{0}yy \rightarrow [(\forall u\forall w(Fuw \rightarrow F\{u\}\{w\}) \wedge \mathsf{t}) \rightarrow (F\overline{00} \rightarrow Fyy)] \tag{26}$$

and also gives

$$\vdash (\forall u\forall w(Fuw \rightarrow F\{u\}\{w\}) \wedge \mathsf{t}) \rightarrow (Fyy \rightarrow F\{y\}\{y\}) \tag{27}$$

Thus, exactly as in Theorem 3 we get

$$\vdash (S\bar{0}yy \wedge \mathsf{t}) \rightarrow [(\forall u\forall w(Fuw \rightarrow F\{u\}\{w\}) \wedge \mathsf{t}) \rightarrow (F\overline{00} \rightarrow F\{y\}\{y\}) \tag{28}$$

From which our desired result follows exactly as it did in Theorem 3. □

Thus, after a helpful change of variables, we have

Lemma 6 $\vdash \mathbb{N}y \overset{2}{\rightarrow} S\bar{0}yy$

Corollary 7 $\vdash \mathbb{N}y \overset{2}{\rightarrow} \exists zS\bar{0}yz$

For this, we rely on the following Lemma:

Lemma 7 $\vdash Swyz \rightarrow S\{w\}y\{z\}$

Proof Note that $Swyz[F/Gx\{y\}(x, y)]$ is

$$(\forall u \forall w(Gu\{w\} \to G\{u\}\{\{w\}\}) \wedge \mathsf{t}) \to (G\bar{0}\{w\} \to Gy\{z\})$$

Thus, by Instantiation, we have that

$$\vdash Swyz \to [(\forall u \forall w(Gu\{w\} \to G\{u\}\{\{w\}\}) \wedge \mathsf{t}) \to (G\bar{0}\{w\} \to Gy\{z\})] \quad (29)$$

Also note that

$$\vdash (\forall u \forall w(Guw \to G\{u\}\{w\} \wedge \mathsf{t}) \to (\forall u \forall w(Gu\{w\} \to G\{u\}\{\{w\}\}) \quad (30)$$

Essentially by instantiation and then a few applications of Intensional Confinement. Thus by Generalized Cut and another application of confinement, we get our result. □

Corollary 8 $\vdash \forall w(Fw \to F\{w\})$

Proof Since $\vdash Swyz \to S\{w\}y\{z\}$, by Lemma 1, $\vdash \neg S\{w\}y\{z\} \to \neg Swyz$. Also by Theorem 1, $\vdash \forall z \neg S\{w\}yz \to \neg S\{w\}y\{z\}$. Thus by A2, R3, and R1, $\vdash \forall z \neg S\{w\}yz \to \neg Swyz$. So by R4, $\vdash \forall z[\forall z \neg S\{w\}yz \to \neg Swyz]$. And since z is not free in $\forall z \neg S\{w\}yz$, R1 and an instance of A10 then gives $\vdash \forall z \neg S\{w\}yz \to \forall z \neg Swyz$. So again by Lemma 1, $\exists z Swyz \to \exists z S\{w\}yz$.

From here we get that $[\mathbb{N}y \xrightarrow{2} \exists z Swyz] \to [\mathbb{N}y \xrightarrow{2} \exists z S\{w\}yz]$ by R3 applied to a few instances of A2. R4 then gives $\forall w(Fw \to F\{w\})$, as required. □

Theorem 4 $\vdash \forall x \forall y(\mathbb{N}x \xrightarrow{2} (\mathbb{N}y \xrightarrow{2} \exists z Sxyz))$

Proof By stringing together the pieces in exactly the way described in the plan. □

So there we have it: again, an analogue of a piece of arithmetic is available. But this time the 'combinatorial' complexity required to achieve it from logic and BLV is yet more. Such complexity notwithstanding and analogously to the proof of Peano3 above, though both BLV and the logic are involved in the proof of the existence of the sum, it is the (multiple!) application of the logic to BLV that guarantees such an existential result—in particular, initially by Corollary 7. Then, by providing denumerably many natural numbers, and identity and distinctness conditions for (distinct) pairs of those natural numbers, BLV delivers denumerably many natural numbers that are the sums of those (distinct) pairs. So, again, the logic guarantees that at least an individual is around for BLV to work with as an assembly line delivering denumerably many sums.

We'll end the section by noting that repeated investigations have strongly suggested to us that the corresponding result for multiplication will require more complexity yet. This is, in one sense not too surprising—after all, the complexity we see here is, in some sense, a straightforward reflection of the complexity present in Peano4, iterated across the induction required to get from Peano4 to here. Thus, since the usual recursive definition of multiplication involves both the successor and the addition operation, we expect that the complexity involved in expressing the appropriate analogue here will be at least the sum of these.

That, however, is at best a loose intuition. Fleshing it out in greater detail, and recovering larger fragments of arithmetic than this, will have to await future work.

6 Concluding Remarks

We started our investigation from two (related) questions:

Q1. Can noncontractive *relevant* logics be used to derive a nontrivial amount of arithmetic from abstraction principles alone?

The answer to Q1 seems to be in the positive. In particular, by a rather weak relevant second-order logic augmented by BLV, formulations of second-order Peano axioms (and the recursiveness of addition, which is a nice amount of arithmetic proper) are derivable. We are quite hopeful that the same can be accomplished for multiplication, though the combinatorial complexity might rise even more.

Of course, it does not, we want to emphasize, follow from our derivation of the second-order Peano axioms that we can in fact get all of second-order Peano arithmetic, since the logic has changed.[12] A result more in line with such a conclusion *is* available, however, in the case of first-order Peano arithmetic via the work in [9], which is nicely explained and expanded on in [18]. There's reason to suspect that perhaps the move to second-order logic would let us prove a good deal more, though this hope is cautious since we also have to work in a significantly weaker logic than Friedman and Meyer did.

Q2. In exactly what ways must logic and abstraction be combined, in order to get those mathematical results?

As shown above, the combination of logic and abstraction can be investigated, so that the order of application of logic to abstraction and abstraction to logic required to deliver the above mathematical results can finally surface. Interestingly, both as for the derivation of Peano axioms and the recursiveness of addition, it seems that ontological results are due to the application of the logic to abstraction; whereas identity results are brought about by the application of abstraction to logic. In a sense, the logic has to 'feed' at least an individual to abstraction, for this latter to start answering questions of identity and distinctness in a meaningful way such that, by abstraction, denumerably many individuals are made available.

Furthermore, as a consequence of answering Q2, we also anticipated we could answer a further question: Is BLV a logical principle? Well, unsurprisingly and *pace* Frege, BLV is not, but in a rather peculiar sense. It's not logical since it delivers the existence of denumerably many individuals by the identity and

---

[12] We owe thanks to an anonymous referee for pointing out that we should make this more clear.

distinctness conditions it incorporates. But, as mentioned above, it does so only if the background logic secures the existence of at least an individual, which, once is 'fed' to BLV, allows BLV to produce denumerably many more. So, BLV is not a logical principle, if the background logic 'feeds' it with at least an individual. But without the logic, BLV runs on empty: *per se* it guarantees the existence of no individual at all, so, if in logic the ban on the empty domain really just is a "concession to technical convenience", ironically, without the logic, BLV is trivially logical.

Acknowledgments We wish to sincerely thank: the San Raffaele visiting scholars program for hosting Shay Logan in 2021; the Philosophy Department of Kansas State University and the San Raffaele Center for Research in Experimental and Applied Epistemology (CRESA) for generously funding Francesca Boccuni to visit Kansas State University in 2022. We also wish to thank the audience and organizers of the *Perspectives on Logic and Philosophy* conference (Ruhr University Bochum, 2023), the audience at the CRESA Work in Progress seminar series 2021, Eileen Nutting, Andrew Tedder, Teresa Kouri, and an anonymous reviewer, for providing comments that dramatically improved the contents of the paper.

References

1. Boolos, G. (1998a). The consistency of Frege's foundations of arithmetic. In [2] (pp. 183–201).
2. Boolos, G. (1998b). Logic, logic, and logic. In J. Burgess & R. Jeffrey (Eds.). Cambridge: Harvard University Press.
3. Brady, R. T. (1984), Natural deduction systems for some quantified relevant logics. *Logique Et Analyse, 27*(8), 355–377.
4. Ebert, P. A., & Rossberg, M. (2016). *Abstractionism: Essays in philosophy of mathematics*. Oxford: Oxford University Press UK.
5. Ferreira, F. (2018). Zigzag and Fregean Arithmetic. In H. Tahiri (Ed.), *The philosophers and mathematics: Festschrift for Roshdi Rashed* (pp. 81–100). Berlin: Springer.
6. Fine, K. (1974). Models for entailment. *Journal of Philosophical Logic, 3*(4), 347–372.
7. Frege, G. (1884). *Die Grundlagen der Arithmetik*. Hildesheim: Georg Olms Verlagbuchhandlung. En. translation by J. L. Austin, *The foundations of arithmetic* (1950). Blackwell.
8. Frege, G. (1893–1903), *Grundgesetze der Arithmetik, begriffsschriftlick abgeleitet I-II*, Jena. En. translation by Philip A. Ebert & Marcus Rossberg (Eds.), *The basic laws of arithmetic* (2013). Oxford: Oxford University Press.
9. Friedman, H., & Meyer, R. K. (1992). Whither relevant arithmetic?' *The Journal of Symbolic Logic, 57*(3), 824–831.
10. Hale, B., & Wright, C. (2001). *The reason's proper study: Essays towards a Neo-Fregean philosophy of mathematics*. Oxford: Clarendon Press.
11. Linnebo, Ø. (2016). Impredicativity in the Neo-Fregean programme. In [4] (pp. 247–268).
12. Linnebo, Ø. (2018). *Thin objects: An abstractionist account*. Oxford: Oxford University Press.
13. Mares, E. D., & Goldblatt, R. (2006). An alternative semantics for quantified relevant logic. *Journal of Symbolic Logic, 71*(1), 163–187.
14. Ripley, D. (2015). Contraction and closure. *Thought, 4*(2), 131–138.
15. Routley, R. (2019). *Ultralogic as universal? The Sylvan jungle - Volume 4*. Cham: Springer.
16. Routley, R., Plumwood, V., Meyer, R. K., & Brady, R. T. (1982). *Relevant logics and their rivals*. Atascadero: Ridgeview.
17. Shapiro, S., & Weir, A. (2000). 'Neo-logicist' logic is not epistemically innocent. *Philosophia Mathematica, 8*(2), 160–189.

18. Slaney, J. (2022). Models of relevant arithmetic. *The Australasian Journal of Logic, 19*(1).
19. Tennant, N. (1987). *Anti-realism and logic: Truth as eternal*. Oxford: Oxford University Press.
20. Tennant, N. (2022). *The logic of number*. Oxford: Oxford University Press.
21. Weber, Z. (2021). *Paradoxes and inconsistent mathematics*. New York: Cambridge University Press.
22. Wright, C. (1983). *Frege's conception of numbers as objects*. Aberdeen: Aberdeen University Press.
23. Zalta, E. N. (2010). Frege's theorem and foundations for arithmetic. *The Stanford encyclopedia of philosophy* (Spring 2023 edition). In E. N. Zalta & U. Nodelman (Eds.), https://plato.stanford.edu/archives/spr2023/entries/frege-theorem/

Explicit and Implicit Belief in First Degree Entailment with Strict Implication

Igor Sedlár and Pietro Vigiani

Abstract We introduce $\mathsf{FDE}_c^{\prec}$, an extension of FDE with strict implication and a classicality constant, and we show that it formalizes the distinction between explicit and implicit belief. In the style of Levesque's formalization of these two concepts, explicit beliefs are modelled as sets of formulas closed under our extension of FDE, while implicit beliefs form, in a sense, the classical closure of explicit beliefs. We establish an embedding of Levesque's logic of explicit and implicit belief into $\mathsf{FDE}_c^{\prec}$. This result shows that $\mathsf{FDE}_c^{\prec}$ is a viable generalization of Levesque's logic lifting some of its limitations. Unlike a similar generalization introduced by Lakemeyer, $\mathsf{FDE}_c^{\prec}$ comes with an Australian-plan semantics and so it is an alternative potentially attractive to those who prefer the Australian plan over the American one.

1 Introduction

First Degree Entailment (FDE), the fragment of the relevant logic E of Anderson and Belnap [2] without nested implications, is a well-known model of reasoning with incomplete and inconsistent information [4, 5, 10]. Alternatively, FDE can be seen as a formula-formula sequent system where formulas appearing in sequents are implication-free. This perspective motivates extensions of FDE with various implication connectives; see the overview in [26], for instance. In this paper we introduce $\mathsf{FDE}_c^{\prec}$, an extension of FDE with *strict implication* $\prec$ that adds also a strict conjunction & (forming and adjoint pair with strict implication) and a classicality constant c. Our first technical result is a soundness and completeness theorem for $\mathsf{FDE}_c^{\prec}$. Next, we show that $\mathsf{FDE}_c^{\prec}$ provides a natural setting articulating

I. Sedlár (✉)
The Czech Academy of Sciences, Institute of Computer Science, Prague, The Czech Republic
e-mail: sedlar@cs.cas.cz

P. Vigiani
Department of Philosophy, Scuola Normale Superiore, Pisa, Italy
e-mail: pietro.vigiani@sns.it

I. Sedlár et al. (eds.), *New Directions in Relevant Logic*, Trends in Logic 63,
https://doi.org/10.1007/978-3-031-69940-5_17

the distinction between *explicit* and *implicit belief*, as famously modelled by Levesque [20]. Our second technical result is that Levesque's logic of explicit and implicit belief BL embeds into $\mathsf{FDE}_c^{\prec}$. At the same time, $\mathsf{FDE}_c^{\prec}$ avoids some of the restrictions of Levesque's logic. Most importantly, we allow nesting of epistemic operators and we do not assume some of the strong introspection properties implicit in Levesque's semantics (see Sect. 2).

Extensions of FDE with strict implication may be at odds with the relevant roots of FDE, but they are not unheard of—a prominent example is Priest's extension of FDE, which uses a version of strict implication as a basis of a logic of intentionality [27]. Priest's extension of FDE with strict implication builds on "American-plan" semantics for FDE with separate truth and falsity conditions and it is not closed under contraposition. Moreover, Priest's strict implication is "S5-style", meaning that the corresponding accessibility relation is universal. The novelty of our approach in comparison to Priest's logic is that we build on "Australian-plan" relational semantics and we do not assume any significant properties of the accessibility relation corresponding to strict implication. This variation in semantics is noteworthy in view of the recent debate on the respective merits of the Australian and American plans [7, 9, 25]. Our approach differs in a similar way from Lakemeyer's American-plan generalization of Levesque's logic [19].

The paper is structured as follows. Section 2 introduces $\mathsf{FDE}_c^{\prec}$; a relational Australian-plan semantics is defined, and soundness and completeness results for $\mathsf{FDE}_c^{\prec}$ and some of its extensions are established using the standard canonical model technique. In Sect. 3, we discuss the distinction between explicit and implicit belief, we briefly motivate our use of Australian-style semantics, and we show that $\mathsf{FDE}_c^{\prec}$ is able to express specific variants of the intuitive notions of explicit and implicit belief. It is shown that, in a precise sense, implicit belief of propositional formulas as modelled by $\mathsf{FDE}_c^{\prec}$ corresponds to the classical closure of explicit belief. Finally, Sect. 4 shows that Levesque's logic of explicit and implicit belief BL embeds into $\mathsf{FDE}_c^{\prec}$. The concluding section summarises the paper and discusses some interesting topics for future research.

2 FDE with Strict Implication and Classicality

In this section we introduce $\mathsf{FDE}_c^{\prec}$. Our semantics is relational and is obtained by adapting the standard relational semantics for relevant modal logics [16, 31, 33]. We obtain an Australian-plan relational semantic framework using the Routley star operation to interpret negation instead of separate truth and falsity conditions of the American plan.

Definition 1 (Languages) Fix a countable set Pr of propositional variables and let $\mathcal{L} = \{\wedge, \vee, \neg\}$, with sets of additional operators $I = \{\prec, \&\}$, where $\prec$, & are binary; $C = I \cup \{c\}$, where c is a constant; $BL = \{B, L\}$, where B, L are unary. For

$X \in \{\emptyset, I, C, BL\}$, let $\mathcal{L}_X = \mathcal{L} \cup X$ and Fm_X be the set of formulas of language $\mathcal{L}_X$, defined as expected. A Fm_X-sequent is an ordered pair of formulas in Fm_X, usually written as $\varphi \vdash \psi$.

The reason for adding strict conjunction & in addition to $\multimap$ is technical and has to do with the completeness proof (see Sect. 2.3). We note that the classicality constant c is related to da Costa's consistency constant [8] but it is stronger; the difference will be clear once we set up the semantics. The operators B and L will be used in Levesque's logic (see Sect. 4). We define $\top := p_0 \multimap p_0$, $\bot := \neg\top$, $\varphi \supset \psi := \neg\varphi \vee \psi$, $\varphi \equiv \psi := (\varphi \supset \psi) \wedge (\psi \supset \varphi)$, $\Box_e\varphi := \top \multimap \varphi$, $\Diamond_e\varphi := \varphi \,\&\, \top$ and $\Box_i\varphi := c \multimap \varphi$. In order to save parentheses, we stipulate that $\neg$ binds more strongly than $\wedge$ and $\vee$, which bind more strongly than $\multimap$, &, B and L. We define $Pr(\varphi) = \{p \in Pr \mid p \text{ occurs in } \varphi\}$.

2.1 Semantics

Definition 2 ($\mathsf{FDE}_c^{\multimap}$-models) A $\mathsf{FDE}_c^{\multimap}$-frame is a structure $(S, \leq, C, {}^*, R)$ where $(S, \leq)$ is a partially ordered set, * is a unary operation on S, R is a binary relation on S and C is a subset of S such that:

(F1) $x \leq y \Rightarrow y^* \leq x^*$;
(F2) $x^{**} = x$;
(F3) $Rxy, x' \leq x, y \leq y' \Rightarrow Rx'y'$;
(F4) if $x \in C$, then $x^* = x$.

A $\mathsf{FDE}_c^{\multimap}$-model M is a $\mathsf{FDE}_c^{\multimap}$-frame together with a function $V : Pr \rightarrow \mathscr{P}(S)$ such that $s \leq t$ and $s \in V(p)$ only if $t \in V(p)$, for all $p \in Pr$. $\mathbf{FDE}_c^{\multimap}$ is the class of all $\mathsf{FDE}_c^{\multimap}$-models.

Definition 3 ($\mathsf{FDE}_c^{\multimap}$-evaluation) Let $\models$ be a binary relation between states in $\mathsf{FDE}_c^{\multimap}$-models and formulas in Fm_C defined recursively as follows:

- $(M, s) \models p$ iff $s \in V(p)$;
- $(M, s) \models \neg\varphi$ iff $(M, s^*) \not\models \varphi$;
- $(M, s) \models \varphi \wedge \psi$ iff $(M, s) \models \varphi$ and $(M, s) \models \psi$;
- $(M, s) \models \varphi \vee \psi$ iff $(M, s) \models \varphi$ or $(M, s) \models \psi$;
- $(M, s) \models \varphi \multimap \psi$ iff for all t such that Rst, $(M, t) \models \varphi$ only if $(M, t) \models \psi$;
- $(M, s) \models \varphi \,\&\, \psi$ iff $(M, s) \models \psi$ and there is t such that Rts and $(M, t) \models \varphi$;
- $(M, s) \models c$ iff $s \in C$.

A sequent $\varphi \vdash \psi$ is valid in a $\mathsf{FDE}_c^{\multimap}$-model M ($\varphi \models_M \psi$) iff $(M, s) \models \varphi$ implies $(M, s) \models \psi$ for all $s \in S$ of the model; a sequent is valid in a $\mathsf{FDE}_c^{\multimap}$-frame iff it is valid in all $\mathsf{FDE}_c^{\multimap}$-models based on the frame; a sequent is valid in a class of $\mathsf{FDE}_c^{\multimap}$-models (frames) iff it is valid in each model (frame) in the class.

If X is a $\mathsf{FDE}_c^{\multimap}$-model, frame or a class of $\mathsf{FDE}_c^{\multimap}$-models or frames, then $Log(X)$ is the set of sequents valid in (each element of) X. A formula φ is valid in X ($\varphi \in Log(X)$) iff $\top \vdash \varphi$ is valid in X.

In what follows, we omit reference to the $\mathsf{FDE}_c^{\multimap}$-model M whenever it is clear from context. Note that strict implication $\multimap$ residuates over strict conjunction & from the right side:

$$\varphi \models_M \psi \multimap \chi \iff \varphi \,\&\, \psi \models_M \chi.$$

A "backward looking" strict implication would residuate from the left side, but we opted not to add another operator to the language. Note also that for every $\mathsf{FDE}_c^{\multimap}$-model M, if $\varphi \vdash \psi$ is in $Log(M)$, then $\varphi \multimap \psi$ is in $Log(M)$. The converse is not true. For example, $p \multimap (\top \,\&\, \top)$ is valid in all frames, but $p \vdash \top \,\&\, \top$ is not valid in frames where R is not backwards serial.

Our models for $\mathsf{FDE}_c^{\multimap}$ include a partially ordered set of information states $(S, \leq)$, the Routley star function $*$, a set C of classical states and the modal accessibility relation R, providing semantics to strict implication. As customary, we read $s \leq t$ as expressing that t contains at least as much information as s. Note that partially ordered models are not necessary. The first reason to include partial order, besides the fact that it is natural given the epistemic interpretation of our models, is that the partially ordered models introduced in this paper are ready for further extensions with other connectives that we want to pursue in the future (see Sect. 5 for a discussion of this point). The second reason is that the partially ordered framework aligns well with the related work of [35]; our first motivation for studying $\mathsf{FDE}_c^{\multimap}$ was the fact that it can be seen as a first step towards a formalization of implicit belief in the framework of [35].[1] We note that partially ordered frames for FDE with the Routley star were also used by Dunn [11].

The Routley star function maps each state s to its maximally compatible state s^*, where maximality is understood with respect to the information order $\leq$ and s^* does not verify anything that s falsifies (see e.g. [29]). Hence, by ($F4$) the set C of classical states are such that falsification and failure of verification coincide, i.e. $s \in C$ implies $s \models \neg\varphi$ iff $s \not\models \varphi$.

In Sect. 3, we will discuss some defined modal operators, $\Box_e$, $\Diamond_e$, $\Box_i$, from an epistemic perspective. Before we get there, the following comments should help the reader to associate some intuitions with our syntax. Strict implication $\multimap$ has the usual informal reading: $\varphi \multimap \psi$ means that, necessarily, if φ is true, then ψ is true as well. However, our reading of the relation R is epistemic: Rst iff all information available to some fixed agent according to state s is true in state t. Therefore, $\varphi \multimap \psi$ has a more specific meaning, expressing a notion close to *conditional belief:*

[1] [35] puts forward a framework in which epistemic modalities are closed under relevant consequence, while propositional logic is classical. It is worth noting that [36] develops a similar idea, introducing a version of justification logic where justification is closed under FDE-consequence but the propositional fragment of the logic remains classical.

$\varphi \dashv\!\!\!3\ \psi$ is true in s iff each state where the agent's information (according to s) is true, and where φ is true as well, is a state where ψ is true. This lends a particular epistemic reading to $\Box_e\varphi$ (i.e. $\top \dashv\!\!\!3\ \varphi$): $\Box_e\varphi$ is true in s iff φ is true in every state where the information that the agent possesses according to s is true. In line with Levesque's [20] use of the semantics for **FDE** to model explicit belief, we can say that $\Box_e\varphi$ means that φ is explicitly believed by the agent.[2] In contrast to Levesque's semantics, where explicit beliefs are modelled by a set of states B (see Sect. 4), we use an accessibility relation R. As such, Levesque's semantics implicitly commits to certain introspection properties: for instance, if φ is explicitly believed according to a state s in Levesque's model, then φ is explicitly believed in all states of the model and, similarly, if φ is not explicitly believed in s, then it is not explicitly believed according to all states. The adequacy of these semantic introspection properties of belief has been questioned in the epistemic logic literature (see e.g. [18, 22]), thus motivating the study of a more general semantic framework.

The classicality constant c is true in (denotes the set of) complete and consistent states; we may call such states "classical". Therefore, $\Box_i\varphi$ (i.e. $c \dashv\!\!\!3\ \varphi$) is satisfied in s iff φ is true in every *classical* state where the information that the agent possesses according to s is true. We will show later in Sect. 3.2 that, in a specific sense, this means that $\Box_i$ expresses "classical closure" of explicit beliefs held by the agent. In line with Levesque's terminology, we read $\Box_i\varphi$ as saying that φ is implicitly believed by the agent.[3] The operator $\ominus_e$ is auxiliary, but it has a relatively transparent epistemic reading: $\ominus_e\varphi$ holds in s iff s is epistemically accessible from a state where φ holds.

2.2 *Properties of* $\mathsf{FDE}_c^{\dashv\!3}$*-models*

Given our aim to set up a general framework, we study where stronger properties of the epistemic operators can be modularly adjusted according to the application at hand. Therefore, it will be of interest to examine more closely some properties of R in $\mathsf{FDE}_c^{\dashv\!3}$-frames. These frame properties correspond to sequents in the precise sense specified by Lemma 1 and the correspondences are summarised in Table 1.

[2] Levesque's approach has been criticised for not articulating a sensible notion of explicit belief since, in his semantics, "explicit beliefs" *are* closed under consequence, namely, consequence of FDE (see [15], for instance). A similar observation applies to our framework: $\Box_e\varphi$ has the flavour of implicit belief since it means that φ is the consequence, relatively to the given model, of the agent's information. However, we call this "explicit belief" for similar reasons Levesque does, namely, since $\Box_e$ avoids closure under classical logic. For example, explicit belief differs from implicit belief in that it is not closed under modus ponens, i.e. $\Box_e\varphi \wedge \Box_e(\varphi \supset \psi) \vdash \Box_e\psi$ is not valid in all $\mathsf{FDE}_c^{\dashv\!3}$-models.

[3] As clarified in Sect. 3.1, the definition of $\Box_i\varphi$ as $c \dashv\!\!\!3\ \varphi$ is naturally motivated by the conceptual understanding of implicit belief as what would be explicitly believed by the agent if the epistemically accessible states were of some ideal sort (classical possible worlds). Then, c represents the linguistic hallmark of classicality.

Table 1 Some frame correspondences for FDE_c^{-3}-frames

Label	Frame property		Sequents
($\Box_e 1$)	Reflexivity	Rss	$\Box_e\varphi \vdash \varphi$
($\Box_e 2$)	Transitivity	$Rst, Rtu \Rightarrow Rsu$	$\Box_e\varphi \vdash \Box_e\Box_e\varphi$
($\Box_e 3$)	*-euclideaness	$Rs^*t, Rsu \Rightarrow Ru^*t$	$\neg\Box_e\varphi \vdash \Box_e\neg\Box_e\varphi$
($\Box_e 4$)	C-reflexivity	$s \in C \Rightarrow Rss$	$c \vdash \Box_e\varphi \supset \varphi$
($\Box_e 5$)	C-transitivity	$s \in C, Rst, Rtu \Rightarrow Rsu$	$c \vdash \Box_e\varphi \supset \Box_e\Box_e\varphi$
($\Box_e 6$)	C-euclideaness	$s \in C, Rst, Rsu \Rightarrow Ru^*t$	$c \vdash \neg\Box_e\varphi \supset \Box_e\neg\Box_e\varphi$
($\Box_e\Box_i 1$)	ei-transitivity	$t \in C, Rst, Rtu \Rightarrow Rsu$	$\Box_e\varphi \vdash \Box_i\Box_e\varphi$
($\Box_e\Box_i 2$)	ei-euclideaness	$u \in C, Rs^*t, Rsu \Rightarrow Ru^*t$	$\neg\Box_e\varphi \vdash \Box_i\neg\Box_e\varphi$
($\Box_i 1$)	i-transitivity	$t, u \in C, Rst, Rtu \Rightarrow Rsu$	$\Box_i\varphi \vdash \Box_i\Box_i\varphi$
($\Box_i 2$)	i-euclideaness	$t, u \in C, Rs^*t, Rsu \Rightarrow Ru^*t$	$\neg\Box_i\varphi \vdash \Box_i\neg\Box_i\varphi$
(Lprec)	Left pre-classicality	$Rs^*t \Rightarrow Rst$	$(\varphi \multimap \psi) \wedge \neg(\varphi \multimap \psi) \vdash \bot$
(Rprec)	Right pre-classicality	$Rst \Rightarrow Rs^*t$	$\top \vdash (\varphi \multimap \psi) \vee \neg(\varphi \multimap \psi)$
(Lpostc)	Left post-classicality	$Rst^* \Rightarrow Rst$	$(\varphi \,\&\, \psi) \wedge \neg(\varphi \,\&\, \psi) \vdash \bot$
(Rpostc)	Right post-classicality	$Rst \Rightarrow Rst^*$	$\top \vdash (\varphi \,\&\, \psi) \vee \neg(\varphi \,\&\, \psi)$

The additional sequents can be used to define axiomatic extensions of FDE_c^{-3}, which we will show sound and complete with respect to classes of frames determined by the respective frame properties (see Theorem 2). With a slight abuse of notation, we will use the same label in Table 1 to denote the frame property and the corresponding sequent.

The sequents we consider in Table 1 underlie various epistemic properties one may wish to impose on explicit (conditional) belief. Most notably, ($\Box_e 1$)-($\Box_e 3$) express factivity, positive and negative introspection of explicit belief at information states; ($\Box_e 4$)-($\Box_e 6$) express factivity, positive and negative introspection of explicit belief at classical information states; and ($\Box_i 1$)-($\Box_i 2$) express positive and negative introspection of implicit belief. Next, ($\Box_e\Box_i 1$)-($\Box_e\Box_i 2$) capture the idea that the very fact that something is explicitly believed is implicitly believed. Note that a similar list, featuring some interaction axioms between explicit and implicit belief, appears in [19].[4] (Lprec) implies that s cannot support contradictory information about belief (either implicit or explicit), while (Rprec) implies that information about belief supported by s is complete (and similarly for (Lpost)-(Rpost)). Put differently, states satisfying pre-classicality (post-classicality) behave like possible worlds when it comes to formulas with $\multimap$ ($\&$).

Lemma 1 (**FDE_c^{-3}-definability**) *For all* FDE_c^{-3} *frames* F *and all frame properties* (X) *of Table 1,* R *and* C *satisfy* (X) *iff the corresponding sequent is valid in* F.

Proof For ($\Box_e 1$), assume $s \models \Box_e\varphi$. Hence, for all t such that Rst, $t \models \varphi$, from which we conclude by reflexivity that $s \models \varphi$. Conversely, assume F is not reflexive,

[4] Note that, while in [19] ($\Box_e\Box_i 1$) – ($\Box_i 2$) are valid in the basic model, here models can be fine-tuned, via frame correspondences, depending on specific applications.

i.e. for some s not Rss. It suffices to take a model based on F such that $u \in V(p)$ iff Rsu to obtain a counter-model to $\Box_e p \vdash p$. The case for ($\Box_e 4$) is similar. For ($\Box_e 2$), assume $s \models \Box_e \varphi$, Rst and Rtu. Hence, for all v such that Rsv, $v \models \varphi$ and by transitivity Rsu, by which we conclude $s \models \Box_e \Box_e \varphi$. Conversely, assume F is not transitive, i.e. for some s, t, u Rst, Rtu but not Rsu. It suffices to take a model based on F such that $v \in V(p)$ iff Rsv to obtain a counter-model to $\Box_e p \vdash \Box_e \Box_e p$. The cases for ($\Box_e 5$), ($\Box_e \Box_i 1$) and ($\Box_i 1$) are similar. For ($\Box_e 3$), assume $s \models \neg \Box_e \varphi$ and Rsu. Hence, there is t such that Rs^*t and $t \not\models \varphi$ and by *-euclideaness Ru^*t, by which $u \models \neg \Box_e \varphi$, by which we conclude that $s \models \Box_e \neg \Box_e \varphi$. Conversely, assume F is not *-euclidean, i.e. for some s, t, u Rs^*t, Rsu but not Ru^*t. It suffices to take any model based on F such that $V(p) = S \setminus \{t\}$ to obtain a counter-model to $\neg \Box_e p \vdash \Box_e \neg \Box_e p$. The cases for ($\Box_e 6$), ($\Box_e \Box_i 2$) and ($\Box_i 2$) are similar. For (Lprec), assume by contradiction that $s \models \varphi \multimap \psi$ and $s \models \neg(\varphi \multimap \psi)$. By the former, for all t such that Rst and $t \models \varphi$, $t \models \psi$. By the latter, there is u such that Rs^*u, $u \models \varphi$ and $u \not\models \psi$. Hence, by left pre-classicality Rsu, by which we conclude that $u \models \psi$, which is a contradiction. Conversely, assume F is not left pre-classical, i.e. there are s, t such that Rs^*t but not Rst. It suffices to take any model based on F such that $V(p) = S$ and $V(q) = S \setminus \{t\}$ to obtain a counter-model to $(p \multimap q) \wedge \neg(p \multimap q) \vdash \bot$. For (Rprec), we show that for all s, $s \models \varphi \multimap \psi$ or $s \models \neg(\varphi \multimap \psi)$. By contradiction, assume $s \not\models \varphi \multimap \psi$ and $s \not\models \neg(\varphi \multimap \psi)$. Hence, there is t such that Rst, $t \models \varphi$ and $t \not\models \psi$ and, for all u such that Rs^*u and $u \models \varphi$, $u \models \psi$. Hence, by right pre-classicality $t \models \varphi$, which is a contradiction. Conversely, assume F is not right pre-classical, i.e. there are s, t such that Rst but not Rs^*t. It suffices to take any model based on F such that $V(p) = S$ and $V(q) = S \setminus \{t\}$ to obtain a counter-model to $\top \vdash (p \multimap q) \vee \neg(p \multimap q)$. The cases for (Lpostc) and (Rpostc) are similar to those for (Lprec) and (Rprec). □

Lemma 2 (Heredity) *For all $\varphi \in Fm_I$, all $\mathsf{FDE}_c^{\multimap}$-models M and all $s, t \in S$, if $(M, s) \models \varphi$ and $s \leq t$, then $(M, t) \models \varphi$.*

Proof By induction on the complexity of φ. The base case and induction steps involving $\wedge$, $\vee$ are trivial, so we focus on the cases involving $\multimap$, $\&$. If $\varphi := \psi \multimap \chi$, assume $s \leq t$, $s \models \psi \multimap \chi$, Rtu and $u \models \psi$. By (F3), we have Rsu, hence by $s \models \psi \multimap \chi$ and $u \models \psi$ we have $u \models \chi$, by which we conclude that $t \models \psi \multimap \chi$. If $\varphi := \psi \,\&\, \chi$, assume $s \leq t$ and $s \models \psi \,\&\, \chi$. Hence, $s \models \chi$ and there is u such that Rus, $u \models \psi$. By (F3) we have Rut and using the induction hypothesis we conclude that $t \models \psi \,\&\, \chi$. □

2.3 Soundness and Completeness

In this section we establish one of the two main results of the paper, namely, a sound and complete axiomatization of various classes of $\mathsf{FDE}_c^{\multimap}$-frames. The basic axiom system we consider is quite natural. On top of a standard axiomatisation of FDE (see e.g. [26]), we consider sequents (I1)-(I3) that characterise the adjoint pair

$\multimap$, & syntactically[5] and sequents (C1)-(C3) that characterise c as a classicality constant. In addition to the basic axioms system, sound and complete with respect to the logic of all $\mathsf{FDE}_c^{\multimap}$-frames, we will consider axiomatic extensions given by the sequents in Table 1.

Definition 4 (Axiom Systems, Provability) The axiom system FDE consists of the following sequents and rules of inference, where $\dashv\vdash$ indicates a two-way sequent and $\Longleftrightarrow$ indicates a two-way rule of inference:

(FDE1)	$\varphi \wedge \psi \vdash \varphi$	(FDE6)	$\psi \vdash \varphi \vee \psi$
(FDE2)	$\varphi \wedge \psi \vdash \psi$	(FDE7)	$\varphi \dashv\vdash \neg\neg\varphi$
(FDE3)	$\varphi \vdash \psi,\ \varphi \vdash \chi \Rightarrow \varphi \vdash \psi \wedge \chi$	(FDE8)	$\varphi \vdash \psi \Rightarrow \neg\psi \vdash \neg\varphi$
(FDE4)	$\varphi \vdash \chi,\ \psi \vdash \chi \Rightarrow \varphi \vee \psi \vdash \chi$	(FDE9)	$\varphi \wedge (\psi \vee \chi) \vdash (\varphi \wedge \psi) \vee (\varphi \wedge \chi)$
(FDE5)	$\varphi \vdash \varphi \vee \psi$	(FDE10)	$\varphi \vdash \psi,\ \psi \vdash \chi \Rightarrow \varphi \vdash \chi$

The axiom system $\mathsf{FDE}_c^{\multimap}$ adds to FDE the following sequents and rules:

(I1)	$\varphi \vdash \psi \multimap \chi \Longleftrightarrow \varphi \,\&\, \psi \vdash \chi$	(C2)	$c \wedge \varphi \wedge \neg\varphi \vdash \bot$
(I2)	$\varphi \,\&\, (\psi \wedge \chi) \dashv\vdash (\varphi \,\&\, \psi) \wedge \chi$	(C3)	$\top \vdash c \vee \neg c$
(I3)	$\varphi \vdash \psi \Rightarrow \psi \multimap \chi \vdash \varphi \multimap \chi$	(C1)	$c \vdash \varphi \vee \neg\varphi$

The sequent $\varphi \vdash \psi$ is derivable in $\mathsf{FDE}_c^{\multimap}$ from a set of sequents Γ ($\varphi \vdash^{\Gamma}_{\mathsf{FDE}_c^{\multimap}} \psi$) iff there is a tree with root $\varphi \vdash \psi$, where each node is either a sequent of $\mathsf{FDE}_c^{\multimap}$, a sequent in Γ or the consequence of an instance of a rule of inference of $\mathsf{FDE}_c^{\multimap}$. A rule of inference $\alpha_1, \ldots, \alpha_n \Rightarrow \beta$ is derivable in $\mathsf{FDE}_c^{\multimap}$ iff β is derivable in $\mathsf{FDE}_c^{\multimap}$ from $\{\alpha\}$. A sequent $\varphi \vdash \psi$ is provable in $\mathsf{FDE}_c^{\multimap}$ ($\varphi \vdash_{\mathsf{FDE}_c^{\multimap}} \psi$) iff it is derivable from $\emptyset$. A formula φ is provable in $\mathsf{FDE}_c^{\multimap}$ ($\vdash_{\mathsf{FDE}_c^{\multimap}} \varphi$) iff $\top \vdash \varphi$ is provable in $\mathsf{FDE}_c^{\multimap}$.

It is well known (see e.g. [14]) that the double negation laws and contraposition allow to derive the identity sequent $\varphi \vdash \varphi$ as well as the De Morgan laws $\neg(\varphi \wedge \psi) \dashv\vdash \neg\varphi \vee \neg\psi$ and $\neg(\varphi \vee \psi) \dashv\vdash \neg\varphi \wedge \neg\psi$.

Proposition 1 (Provable Sequents and Derivable Rules) *The following sequents (rules of inference) are provable (derivable) in* $\mathsf{FDE}_c^{\multimap}$*:*

[5] We note that with a second "backward looking" strict implication that would residuate with & from the left side we would be able to derive (I3). We opted for paying the price of adding one extra inference rule instead of adding another operator.

1. $\varphi \vdash \psi \mathrel{-\!\!3} (\varphi \,\&\, \psi)$;
2. $(\varphi \mathrel{-\!\!3} \psi) \,\&\, \varphi \vdash \psi$;
3. $\varphi \,\&\, \psi \vdash \psi$
4. $\varphi \vdash \top$;
5. $\bot \vdash \varphi$;
6. $(\varphi \mathrel{-\!\!3} \psi) \wedge (\varphi \mathrel{-\!\!3} \chi) \vdash \varphi \mathrel{-\!\!3} (\psi \wedge \chi)$;
7. $(\varphi \vee \psi) \,\&\, \chi \vdash (\varphi \,\&\, \chi) \vee (\psi \,\&\, \chi)$;
8. $\varphi_1 \vdash \varphi_2,\ \psi_1 \vdash \psi_2 \Rrightarrow \varphi_1 \,\&\, \psi_1 \vdash \varphi_2 \,\&\, \psi_2$;
9. $\varphi \wedge \psi \vdash \chi \Rrightarrow (\theta \mathrel{-\!\!3} \varphi) \vdash (\theta \wedge \psi) \mathrel{-\!\!3} \chi$.

Proof Items 1 and 2 are easy consequences of (I1). Item 3 is established using (I2) and Item 8. Item 4 is an easy consequence of Item 3, and Item 5 follows from Item 4 and contraposition. Item 6 is established using (FDE1)-(FDE3) and (I1), and Item 7 is established similarly using (FDE4)-(FDE6) and (I1). Items 8 and 9 are established as follows (note that the proof of the former is independent of Item 3):

$$\dfrac{\dfrac{\varphi_1 \vdash \varphi_2 \qquad \dfrac{\varphi_2 \,\&\, \psi_2 \vdash \varphi_2 \,\&\, \psi_2}{\varphi_2 \vdash \psi_2 \mathrel{-\!\!3} (\varphi_2 \,\&\, \psi_2)}\,(\mathsf{I1})}{\varphi_1 \vdash \psi_2 \mathrel{-\!\!3} (\varphi_2 \,\&\, \psi_2)}\,(\mathsf{FDE10}) \qquad \dfrac{\psi_1 \vdash \psi_2}{\psi_2 \mathrel{-\!\!3} (\varphi_2 \,\&\, \psi_2) \vdash \psi_1 \mathrel{-\!\!3} (\varphi_2 \,\&\, \psi_2)}\,(\mathsf{I3})}{\dfrac{\varphi_1 \vdash \psi_1 \mathrel{-\!\!3} (\varphi_2 \,\&\, \psi_2)}{\varphi_1 \,\&\, \psi_1 \vdash \varphi_2 \,\&\, \psi_2}\,(\mathsf{I1})}\,(\mathsf{FDE10})$$

$$\dfrac{\dfrac{\dfrac{\dfrac{\theta \mathrel{-\!\!3} \varphi \vdash \theta \mathrel{-\!\!3} \varphi}{(\theta \mathrel{-\!\!3} \varphi) \,\&\, \theta \vdash \varphi}\,(\mathsf{I1})}{((\theta \mathrel{-\!\!3} \varphi) \,\&\, \theta) \wedge \psi \vdash \varphi \wedge \psi} \qquad \varphi \wedge \psi \vdash \chi}{((\theta \mathrel{-\!\!3} \varphi) \,\&\, \theta) \wedge \psi \vdash \chi}\,(\mathsf{FDE10})}{\dfrac{(\theta \mathrel{-\!\!3} \varphi) \,\&\, (\theta \wedge \psi) \vdash \chi}{\theta \mathrel{-\!\!3} \varphi \vdash \theta \wedge \psi \mathrel{-\!\!3} \chi}\,(\mathsf{I1})}\,(\mathsf{I2})$$

□

For any frame property X from Table 1, let $\mathsf{FDE}_c^{-\!3}$.X denote the axiom system resulting from adding the sequents corresponding to X to $\mathsf{FDE}_c^{-\!3}$ and $\mathbf{FDE}_c^{-\!3}$.**X** the corresponding class of $\mathsf{FDE}_c^{-\!3}$-models satisfying the frame property X. (Equivalently, a sequent $\varphi \vdash \psi$ is considered provable in $\mathsf{FDE}_c^{-\!3}$.X iff it is derivable in $\mathsf{FDE}_c^{-\!3}$ from the set of all substitution instances of sequents corresponding to X.)

Theorem 1 **($\mathsf{FDE}_c^{-\!3}$-soundness)** *For all $\varphi, \psi \in Fm_C$,*

$$\varphi \vdash_{\mathsf{FDE}_c^{-\!3}} \psi \textit{ only if } \varphi \vdash \psi \in Log(\mathsf{FDE}_c^{-\!3}).$$

Proof Proof by induction on the length of $\mathsf{FDE}_c^{-\!3}$ proofs. The base case and the induction step for the axioms and rules of FDE are standard, hence we focus on the remaining cases. For (I1), if $s \models \varphi \,\&\, \psi$, then there is t such that Rts, $t \models \varphi$ and $s \models \psi$. If $\varphi \models_M \psi \mathrel{-\!\!3} \chi$, then $t \models \psi \mathrel{-\!\!3} \chi$, and so $s \models \chi$. Hence, $\varphi \,\&\, \psi \models_M \chi$. Conversely, assume that $s \models \varphi$ and there is t such that Rst and $t \models \psi$. Hence, $t \models \varphi \,\&\, \psi$. Hence, if $\varphi \,\&\, \psi \models_M \chi$, then $t \models \chi$. This means that $\varphi \models_M \psi \mathrel{-\!\!3} \chi$. For (I2), direction from left to right, assume $s \models \varphi \,\&\, (\psi \wedge \chi)$. This means that $s \models \psi$ and $s \models \chi$ and there is t such that Rts and $t \models \varphi$. Then $s \models \varphi \,\&\, \psi$ and

$s \models \chi$, which measn that $s \models (\varphi \,\&\, \psi) \wedge \chi$. The converse direction is established similarly. For (I3), assume $\varphi \models \psi$, $s \models \psi \dashv\!\!\!\!\!\!\rightarrow \chi$, Rst and $t \models \varphi$. By $\varphi \models \psi$ we have $t \models \psi$ and by $s \models \psi \dashv\!\!\!\!\!\!\rightarrow \chi$ we have $t \models \chi$. Hence we conclude that $s \models \varphi \dashv\!\!\!\!\!\!\rightarrow \chi$. For (C1) and (C2), assuming $s \in C$ we have that $s \models \neg\varphi$ iff $s^* \not\models \varphi$ iff $s \not\models \varphi$. For (C3), take any s and assume that $s \not\models c$. Then $s \notin C$. If it were the case that $s^* \models c$, then $s^* \in C$ and so $s = s^*$, which would contradict $s \notin C$. Hence, $s^* \not\models c$ and so $s \models \neg c$ as desired. □

For all $\mathsf{FDE}_c^{\dashv\!\rightarrow}.\mathsf{X}$ and $\mathsf{L} \in \{\mathsf{FDE}, \mathsf{FDE}_c^{\dashv\!\rightarrow}.\mathsf{X}\}$, an L-*theory* is any set of formulas Γ such that (i) $\varphi \in \Gamma$ and $\psi \in \Gamma$ only if $\varphi \wedge \psi \in \Gamma$; and (ii) if $\varphi \in \Gamma$ and $\varphi \vdash_{\mathsf{L}} \psi$, then $\psi \in \Gamma$. An L-theory Γ is *prime* iff $\varphi \vee \psi \in \Gamma$ only if $\varphi \in \Gamma$ or $\psi \in \Gamma$. An L-theory Γ is *proper* iff Γ does not contain all formulas. A pair of sets of formulas (Γ, Δ) is L-*independent* iff there are no finite non-empty sets $\Gamma' \subseteq \Gamma$ and $\Delta' \subseteq \Delta$ such that $\bigwedge \Gamma' \vdash_{\mathsf{L}} \bigvee \Delta'$.

Lemma 3 (Extension Lemma) *For all* $\mathsf{FDE}_c^{\dashv\!\rightarrow}.\mathsf{X}$ *and* $\mathsf{L} \in \{\mathsf{FDE}, \mathsf{FDE}_c^{\dashv\!\rightarrow}.\mathsf{X}\}$, *if* (Γ, Δ) *is* L-*independent, then there is a prime* L-*theory* Σ *such that* $\Gamma \subseteq \Sigma$ *and* $\Delta \cap \Sigma = \emptyset$.

Proof [30, p .94]. □

Definition 5 ($\mathsf{FDE}_c^{\dashv\!\rightarrow}$ **Canonical Model)** The canonical $\mathsf{FDE}_c^{\dashv\!\rightarrow}$-model is $(S, \leq, C, {}^*, R, V)$ where

- S is the set of all proper non-empty prime $\mathsf{FDE}_c^{\dashv\!\rightarrow}$-theories;
- $\leq$ is set inclusion;
- C is the set of all elements of S containing c;
- $s^* = \{\varphi \mid \neg\varphi \notin s\}$;
- Rst iff, for all $\varphi, \psi \in Fm_C$, if $\varphi \dashv\!\!\!\!\!\!\rightarrow \psi \in s$ and $\varphi \in t$, then $\psi \in t$;
- $V(p) = \{s \mid p \in s\}$.

Lemma 4 (R-equivalence) *In the canonical* $\mathsf{FDE}_c^{\dashv\!\rightarrow}$-*model the following are equivalent:*

1. Rst;
2. *for all* $\varphi \in Fm_C$, $\top \dashv\!\!\!\!\!\!\rightarrow \varphi \in s \Rightarrow \varphi \in t$;
3. *for all* $\varphi \in Fm_C$, $\varphi \in s \Rightarrow \varphi \,\&\, \top \in t$.

Proof

(1) implies (2): Assume Rst and $\top \dashv\!\!\!\!\!\!\rightarrow \varphi \in s$. Since t is non-empty, by Proposition 1 (item 4) we have $\top \in t$. Hence, by the assumption Rst, $\varphi \in t$ which establishes (2).

(2) implies (3): Assume (2) and $\varphi \in s$. By Proposition 1 (item 1), $\top \dashv\!\!\!\!\!\!\rightarrow (\varphi \,\&\, \top) \in s$, and so $\varphi \,\&\, \top \in t$ using (2), which establishes (3).

(3) implies (1): Assume (3) and let $\varphi \dashv\!\!\!\!\!\!\rightarrow \psi \in s$ and $\varphi \in t$. We obtain $((\varphi \dashv\!\!\!\!\!\!\rightarrow \psi) \,\&\, \top) \wedge \varphi \in t$ which by (I2) means that $(\varphi \dashv\!\!\!\!\!\!\rightarrow \psi) \,\&\, (\top \wedge \varphi) \in t$. Hence,

by Proposition 1 (item 8) we obtain $(\varphi \dashv\!3\, \psi)\,\&\,\varphi \in t$, which entails $\psi \in t$ by Proposition 1 (item 2).

□

Lemma 5 ($\mathsf{FDE}_c^{\dashv 3}$ **Canonicity**) *The canonical* $\mathsf{FDE}_c^{\dashv 3}$*-model is a* $\mathsf{FDE}_c^{\dashv 3}$*-model.*

Proof Clearly, $(S, \subseteq)$ is a partially ordered set and $s \subseteq t$, $p \in s$ only if $p \in t$. Moreover, that s^* is a non-empty $\mathsf{FDE}_c^{\dashv 3}$-theory follows by s being a prime $\mathsf{FDE}_c^{\dashv 3}$-theory and $\neg(\varphi \wedge \psi) \dashv\vdash_{\mathsf{FDE}_c^{\dashv 3}} \neg\varphi \vee \neg\psi$, while that s^* is prime follows by $\neg(\varphi \vee \psi) \dashv\vdash_{\mathsf{FDE}_c^{\dashv 3}} \neg\varphi \wedge \neg\psi$. Hence, it is sufficient to show that $(F1)$-$(F4)$ hold in the canonical $\mathsf{FDE}_c^{\dashv 3}$-model. For $(F1)$, $\varphi \in t^*$ iff $\neg\varphi \notin t$ only if by $s \subseteq t$ $\neg\varphi \notin s$ iff $\varphi \in s^*$. For $(F2)$, $\varphi \in s^{**}$ iff $\neg\varphi \notin s^*$ iff $\neg\neg\varphi \in s$ iff $\varphi \in s$ by (FDE7). For $(F3)$, assume that $s' \subseteq s$, $t \subseteq t'$ and Rst. By Lemma 4, it is sufficient to derive $\varphi \in t'$ from $\top \dashv\!3\, \varphi \in s'$. The latter implies by $s' \subseteq s$ that $\top \dashv\!3\, \varphi \in s$, which implies by Rst that $\varphi \in t$, from which we conclude $\varphi \in t'$ by $t \subseteq t'$. For $(F4)$, we need to show that, assuming $c \in s$, $\varphi \in s$ iff $\neg\varphi \notin s$. If $\varphi \in s$ and $\neg\varphi \in s$, then $\bot \in s$ by (C2), hence by Proposition 1 (item 5) $\varphi \in s$ for all φ, which contradicts the assumption that s is proper. Conversely, assume $\neg\varphi \notin s$. By (C1), we have $\varphi \vee \neg\varphi \in s$, by which we conclude that $\varphi \in s$. □

Lemma 6 ($\mathsf{FDE}_c^{\dashv 3}$ **Truth**) *For all* $\varphi \in Fm_C$ *and all* $s \in S$ *in the canonical* $\mathsf{FDE}_c^{\dashv 3}$*-model* M*,* $\varphi \in s$ *iff* $(M, s) \models \varphi$*.*

Proof By induction on the complexity of φ. The base case and the case where $\varphi := c$ are trivial and the induction steps where the main connective is $\wedge, \vee, \neg$ are established by standard arguments. The only interesting cases are the induction step for $\varphi := \psi \dashv\!3\, \chi$ and $\varphi := \psi\,\&\,\chi$. The former case is established as follows. For one direction, assume $\psi \dashv\!3\, \chi \in s$, Rst and $\psi \in t$. Hence, by definition of R, $\chi \in t$. For the other direction, assume $\psi \dashv\!3\, \chi \notin s$ and consider the pair $t_0 = (\{\xi \mid \top \dashv\!3\, \xi \in s\} \cup \{\psi\}, \{\chi\})$. The pair is $\mathsf{FDE}_c^{\dashv 3}$-independent since otherwise we could assume that there is a conjunction $\bigwedge \xi$ of some finite subset of formulas in $\{\xi \mid \top \dashv\!3\, \xi \in s\}$ such that $(\bigwedge \xi) \wedge \psi \vdash_{\mathsf{FDE}_c^{\dashv 3}} \chi$, and we could reason using Proposition 1 as follows:

$$(\textstyle\bigwedge \xi) \wedge \psi \vdash_{\mathsf{FDE}_c^{\dashv 3}} \chi \implies \top \dashv\!3\, (\textstyle\bigwedge \xi) \vdash_{\mathsf{FDE}_c^{\dashv 3}} (\top \wedge \psi) \dashv\!3\, \chi \quad \text{(item 9)}$$

$$\implies \textstyle\bigwedge(\top \dashv\!3\, \xi) \vdash_{\mathsf{FDE}_c^{\dashv 3}} (\top \wedge \psi) \dashv\!3\, \chi \quad \text{(item 6)}$$

$$\implies \textstyle\bigwedge(\top \dashv\!3\, \xi) \vdash_{\mathsf{FDE}_c^{\dashv 3}} \psi \dashv\!3\, \chi \quad \text{(item 4 and (I3))}$$

But since $\bigwedge(\top \dashv\!3\, \xi) \in s$, this would mean that $\psi \dashv\!3\, \chi \in s$, which is a contradiction. Hence, by Lemma 3 and 4 there is $t \in S^c$ such that $t_0 \subseteq t$, $R^c st$, $\psi \in t$ and $\chi \notin t$, by which we conclude that $s \not\models \psi \dashv\!3\, \chi$.

The case of the induction step for $\varphi := \psi\,\&\,\chi$ is established as follows. For one direction, assume that Rst, $\psi \in s$ and $\chi \in t$. By Proposition 1 (item 1), $\chi \dashv\!3\, (\psi\,\&\,\chi) \in s$ and so, by the previous case of the induction step, $\psi\,\&\,\chi \in t$.

Conversely, assume that $\psi \,\&\, \chi \in t$. By Lemma 4, it is sufficient to show that $\chi \in t$ and there is s such that $\psi \in s$ and, for all $\varphi \in s$, $\varphi \,\&\, \top \in t$. The first part is easy: $\psi \,\&\, \chi \in t$ implies $\chi \in t$ by Proposition 1 (item 3). To prove the second part, we show that the pair $s_0 = (\{\psi\}, \{\xi \mid \xi \,\&\, \top \notin t\})$ is independent. If it were not, then there would be a disjunction $\bigvee \xi$ of elements of some finite subset of $\{\xi \mid \xi \,\&\, \top \notin t\}$ such that $\psi \vdash_{\mathsf{FDE}_c^{-3}} (\bigvee \xi)$. But then we could reason using Proposition 1 as follows:

$$\psi \vdash_{\mathsf{FDE}_c^{-3}} (\bigvee \xi) \implies \psi \,\&\, \top \vdash_{\mathsf{FDE}_c^{-3}} (\bigvee \xi) \,\&\, \top \qquad \text{(item 8)}$$

$$\implies \psi \,\&\, \top \vdash_{\mathsf{FDE}_c^{-3}} \bigvee (\xi \,\&\, \top) \qquad \text{(item 7)}$$

Since we know that $\bigvee(\xi \,\&\, \top) \notin t$, we may infer that $\psi \,\&\, \top \notin t$, and so $\psi \,\&\, \chi \notin t$ (thanks to items 4 and 8 of Proposition 1). But this is a contradiction with our assumption. Hence, by Lemmas 3 and 4 there is $s \in S^c$ such that s extends s_0, $R^c st$, $\psi \in s$ and $\chi \in t$, by which we conclude that $t \models \psi \,\&\, \chi$. □

We note that the job of the strict conjunction operator is to allow for a more concise and elegant axiomatisation and to enable, via Lemma 4, a smoother proof of the truth lemma. As the reader will realize, the use of the original definition of R^c in the above proof would require a lengthier argument.

Lemma 7 (FDE_c^{-3} **Extension Canonicity)** *For any frame property* X *of Table 1, the canonical* FDE_c^{-3}.X*-model satisfies* X.

Proof We rely on Lemma 4. For ($\Box_e$1), assume $\Box_e\varphi \in s$, hence by the provability of the corresponding sequent that $\varphi \in s$, by which we conclude that Rss. For ($\Box_e$2), assume Rst, Rtu. If $\Box_e\varphi \in s$, by the provability of the corresponding sequent $\Box_e\Box_e\varphi \in s$, hence by Rst $\Box_e\varphi \in t$ and by Rtu $\varphi \in u$, by which we conclude that Rsu. For ($\Box_e$3), assume Rs^*t, Rsu. If $\Box_e\varphi \in u^*$, $\neg\Box_e\varphi \notin u$, hence by Rsu $\Box_e\neg\Box_e\varphi \notin s$, by which by the provability of the ($\Box_e$3)-sequent $\neg\Box_e\varphi \notin s$. Hence, $\Box_e\varphi \in s^*$ and by Rs^*t $\varphi \in t$, by which we conclude that Ru^*t. For ($\Box_e$4)-($\Box_e$6), note that if $s \in C$, $\varphi \in s$ and $c \vdash_{\mathsf{FDE}_c^{-3}} \varphi \supset \psi$, then we have $\psi \in s$, and so these cases can be argued for similarly as ($\Box_e$1)-($\Box_e$3). For (Lprec), if not Rst, then there are φ, ψ such that $\varphi \multimap \psi \in s$, $\varphi \in t$ and $\psi \notin t$. By the corresponding provable sequent we have that $\neg(\varphi \multimap \psi) \notin s$, and so $\varphi \multimap \psi \in s^*$. Hence, we conclude that not Rs^*t. For (Rprec), if not Rs^*t, then there are φ, ψ such that $\varphi \multimap \psi \in s^*$, $\varphi \in t$ and $\psi \notin t$. By $\varphi \multimap \psi \in s^*$ we have that $\neg(\varphi \multimap \psi) \notin s$, and by the corresponding provable sequent we have that $\varphi \multimap \psi \in s$, which. Hence, we conclude that not Rst. The proofs for (Lpostc) and (Rpostc) are similar to those for (Lprec) and (Rprec). □

Theorem 2 (FDE_c^{-3} **Soundness and Completeness)** *For all* $\varphi, \psi \in Fm_C$,

$$\varphi \vdash_{\mathsf{FDE}_c^{-3}.\mathsf{X}} \psi \textit{ iff } \varphi \vdash \psi \in Log(\mathsf{FDE}_c^{-3}.X).$$

Proof The implication from left to right follows from Theorem 1 and Lemma 1. The converse implication follows from Lemmas 5–7. □

Clearly, the usual soundness and completeness result for formulas, i.e. $\varphi \in Log(\mathbf{FDE}_c^{\dashv}.\mathbf{X})$ iff $\vdash_{\mathsf{FDE}_c^{\dashv}.\mathsf{X}} \varphi$, follows from Theorem 2 as a special case.

3 Explicit and Implicit Belief

In this section, we clarify the distinction between explicit and implicit belief, we introduce in our framework based on $\mathsf{FDE}_c^{\dashv}$-models explicit and implicit beliefs as derived modalities, and we discuss the implications of adopting Australian-style semantic for negation.

According to the standard possible worlds account of belief, agents' beliefs are closed under logical consequence, i.e. agents believe all (classical) consequences of what they believe. This feature of standard models of belief was famously addressed by Levesque [20] as unrealistic for the ordinary concept of explicit belief. However, in [20] it is argued that the possible worlds account realistically characterises a distinct yet related concept, that of implicit belief. Roughly speaking, implicit beliefs correspond to what the world would be like if the agent's explicit beliefs were to be true, without the agent necessarily realising it. Accordingly, it is plausible that implicit belief, as opposed to explicit belief, is closed under classical logical consequence. Implicit belief is in general understood as the closure of explicit belief under some stronger logic than that under which explicit belief is closed. While we remain agnostic as to which logic exactly should provide implicit belief's stronger closure principles, we remark that Levesque's semantics models implicit belief as closed under classical consequence (see Sect. 4). In line with Levesque's model, we also adopt classical logic as the distinguished stronger logic.

3.1 Two Approaches to Implicit Belief

It is commonly held in the philosophical and psychological literature that the difference between explicit and implicit attitudes is that agents are aware of the former, which are conscious, but not of the latter.[6] Also in the logical literature the distinction between explicit and implicit belief is intertwined with awareness logic, a framework for dealing with logical omniscience which identifies awareness as the missing ingredient linking implicit belief and explicit belief. For example, in [23] it is argued that 'by definition, one has to be aware of explicit belief', while Fagin

[6] Implicit attitudes, in turn, are characterised in various ways, e.g. in terms of dispositions, measures from tests like the so-called 'implicit-association test', map-like mental representations or degrees of accessibility of explicit attitudes (see the discussion in [6]).

and Halpern [15] read Levesque's logic of explicit and implicit belief as capturing the lack of logical omniscience that arises through lack of awareness.[7] It is worth noting the generality of the concept of awareness employed in [15]: first, awareness is modeled by an arbitrary syntactic function; second, no precise meaning is attached to being aware of φ, which can indicate that the agents take φ into account, hold a mental representation of the concept expressed by φ [24], that φ is included in the agents' vocabulary [17], or simply that the agents have immediate, computationally-free, access to φ [15].[8]

At any rate, the distinction between explicit and implicit belief needs not make reference to the concept of awareness. In most logical frameworks, the working intuition is to understand explicit belief as a conceptual primitive, as belief actually held by agents or belief 'at hand'. Then, implicit belief is what agents would believe explicitly in some ideal state. Formally, this usually results in double-barreled analysis of explicit belief (in terms of implicit belief plus e.g. awareness, justification [3], acknowledgement [40], etc.), where implicit belief is assumed as a logical primitive. However, the opposite route is just as viable: for example, in syntactic approaches like that of [1], a set of explicitly believed formulas is assumed, and its closure under logical consequence models implicit beliefs. In [39] explicit knowledge is modeled by means of neighborhood semantics and implicit knowledge is defined as explicit knowledge with respect to a neighborhood model extending the original one, which behaves like a relational Kripke model. Kripke models represent, then, ideal states agents arrive at after having performed all relevant deductive steps. Levesque's model of explicit and implicit belief can be regarded in the same fashion, as a semantic framework in which implicit belief is defined as explicit belief in some ideal state. This idea is made precise by the semantics of relevant logic: propositions are conceived of as sets of situations, a generalisation of possible worlds in which a formula, its negation, both or none can be supported.

Employing (possibly incomplete and inconsistent) situations as the main semantic object in [20] allows to model the information relevant only to what agents actually, explicitly, believe. Then, possible worlds constitute the limit, ideal, case of complete and consistent situations, and implicit belief is modeled as what is true in all possible worlds compatible with some doxastic alternative. Levesque's framework employs the resources of non-classical, relevant, logic as a formalisation

---

[7] (For the sake of simplicity, in this footnote we re-use our symbols for modal operators without assuming our definitions of the associated formulas.) In [15], a modal operator for awareness in Levesque's logic is defined as $A\varphi := \bigwedge\{\Box_e(p \vee \neg p) \mid p \in Pr(\varphi)\}$. This is not the only possible definition of awareness, e.g. [24] define $A\varphi := \Box_i\varphi \vee \Box_i\neg\Box_i\varphi$. Fagin and Halpern propose their own logic of general awareness, with a primitive awareness operator A such that $\Box_e\varphi := \Box_i\varphi \wedge A\varphi$. Again, this is not the only possible definition of explicit knowledge in terms of awareness, e.g. [37] define $\Box_e\varphi := \Box_i(\varphi \wedge A\varphi)$.

[8] A recent contribution to awareness logic is [38], where awareness is modeled by means of partial valuations and weakly reflexive accessibility relations, so as to account for agents that do not take into consideration at all the truth value of a proposition and possibly differ in what proposition they consider the truth value of.

of agents' beliefs and classical logic as a formalisation of propositional logic. This is accomplished by (i) defining possible worlds as a special subset of situations, (ii) defining validity with respect to possible worlds and (iii) defining agents' belief bases as sets of situations.

The presence of incomplete situations in Levesque's logic allows agents to fail to believe valid formulas, since valid formulas may not be supported by all situations, while the presence of inconsistent situations allows for the invalidity of e.g. $\Box_e\varphi \wedge \Box_e(\varphi \supset \psi) \supset \Box_e\psi$.[9]

3.2 Explicit and Implicit Belief in FDE-models

We do not take a stance on the nature of explicit and implicit beliefs. However, in $\mathsf{FDE}_c^{\dashv}$-models it is possible to express the distinction between explicit and implicit belief by means of strict implication. By defining explicit belief as $\Box_e\varphi := \top \dashv\!3\, \varphi$ (with its adjoint $\diamondminus_e\varphi := \varphi \,\&\, \top$) and implicit belief as $\Box_i\varphi := c \dashv\!3\, \varphi$, we obtain the following derived semantic clauses:

- $(M, s) \models \Box_e\varphi$ iff for all t such that Rst, $(M, t) \models \varphi$;
- $(M, s) \models \diamondminus_e\varphi$ iff there is t such that Rts and $(M, t) \models \varphi$;
- $(M, s) \models \Box_i\varphi$ iff for all t such that Rst and $(t) \in C$, $(M, t) \models \varphi$.

It is interesting to note that more epistemically perspicuous sequents and rules for $\Box_e$, $\diamondminus_e$, $\Box_i$ can be derived from $\mathsf{FDE}_c^{\dashv}$, as the following proposition clarifies.

Proposition 2 (Epistemic Sequents) *The following sequents (rules of inference) are provable (derivable) in* $\mathsf{FDE}_c^{\dashv}$*:*

1. $\Box_e\varphi \wedge \Box_e\psi \vdash \Box_e(\varphi \wedge \psi)$;
2. $\varphi \vdash \Box_e\diamondminus_e\varphi$;
3. $\diamondminus_e\Box_e\varphi \vdash \varphi$;
4. $\Box_i\varphi \wedge \Box_i\psi \vdash \Box_i(\varphi \wedge \psi)$;
5. $\Box_e\varphi \vdash \Box_i\varphi$;
6. $\varphi \vdash \psi \dashv\!3\, \chi \Rightarrow \diamondminus_e\varphi \wedge \psi \vdash \chi$;
7. $\varphi \vdash \psi \Rightarrow \Box_e\varphi \vdash \Box_e\psi$;
8. $\varphi \wedge \psi \vdash \chi \Rightarrow \Box_e\varphi \vdash \psi \dashv\!3\, \chi$;
9. $\varphi \vdash \psi \Rightarrow \Box_i\varphi \vdash \Box_i\psi$.

Proof Items 1 and 4 are easy consequences of Proposition 1 (item 6), while Item 2 and 3 are instances of Proposition 1 (items 1 and 2, respectively). Item 5 is an easy consequence of (I3) and Proposition 1 (item 4). Item 6 is established as follows:

---

[9] It is noted in [15] that in Levesque's logic a related principle is valid, namely, $\Box_e\varphi \wedge \Box_e(\varphi \supset \psi) \supset \Box_e(\psi \vee (\varphi \wedge \neg\varphi))$. The latter is just as problematic as the former, since it says that beliefs fail to be closed under $\supset$ only if some situation the agents believe possible is in fact inconsistent. Despite the fact that situations may support both a formula and its negation, it seems problematic to assume that agents may consider doxastically possible some explicitly inconsistent situation. However, in a framework based on FDE such problem is avoided altogether since material conditional does not provide the intended semantics of implication.

$$\dfrac{\varphi \vdash \psi \strictif \chi \qquad \dfrac{\top \wedge \psi \vdash \psi}{\psi \strictif \chi \vdash \top \wedge \psi \strictif \chi}\,(\mathsf{I3})}{\dfrac{\dfrac{\varphi \vdash \top \wedge \psi \strictif \chi}{\varphi \,\&\, (\top \wedge \psi) \vdash \chi}\,(\mathsf{I1})}{\Diamondminus_e \varphi \wedge \psi \vdash \chi}\,(\mathsf{I2})}\,(\mathsf{FDE10})$$

Finally, items 7–9 are easy consequences of Proposition 1 (item 9). □

Recall now the two approaches to the logical formalisation of the distinction between explicit and implicit belief outlined above. According to the first, implicit belief is a primitive modality providing only a necessary condition on explicit belief, which is in turn defined in terms of implicit belief and a further modality. According to the second, explicit belief is taken as a primitive, whereas implicit belief is defined as the closure of explicit belief under some logic the agent would adopt in some ideal epistemic state. Our framework falls under the second approach: implicit beliefs, which are true in all accessible classical information states, represent the classical closure of explicit beliefs, which are true in all accessible information states.

Proposition 3 makes the above observation precise, where CL is a standard sequent axiomatisation of classical logic, CL-models and provability of a formula in CL are defined as usual and completeness of CL with respect to CL-models is assumed.

Proposition 3 (Classical Closure) *For all $\varphi_1, \ldots, \varphi_n, \psi \in Fm$, the following are equivalent:*

1. $\Box_e \varphi_1 \wedge \cdots \wedge \Box_e \varphi_n \vdash_{\mathsf{FDE}_c^{\strictif}} \Box_i \psi$;
2. $\varphi_1 \wedge \cdots \wedge \varphi_n \vdash_{\mathsf{CL}} \psi$.

Proof By contraposition, assume (2) does not hold with counter-CL-model $(\{s\}, V)$. Then, take the $\mathsf{FDE}_c^{\strictif}$-model $M = (\{s\}, \leq, C, *, R, V)$ such that $C = \{s\}$ and $R = \{(s, s)\}$. Clearly, we have that $(M, s) \models \Box_e \varphi_1 \wedge \cdots \wedge \Box_e \varphi_n$, since by assumption we have that $(M, s) \models \varphi_1 \wedge \cdots \wedge \varphi_n$. However, $(M, s) \not\models \Box_i \psi$, since by assumption $(M, s) \not\models \psi$ (note that this is possible only because $\varphi_1, \ldots, \varphi_n, \psi \in Fm$). Hence, we conclude by Theorem 2 that (1) does not hold. Conversely, assume (2). Hence, $\Box_i \varphi_1 \wedge \ldots \wedge \Box_i \varphi_n \vdash_{\mathsf{FDE}_c^{\strictif}} \Box_i \psi$ (otherwise from the counter-$\mathsf{FDE}_c^{\strictif}$-model $(S, \leq, C, *, R, V)$ to the latter we can extract the counter-CL-model (C, V) to the former). By Proposition 2 (item 5), $\Box_e \varphi_1 \wedge \ldots \wedge \Box_e \varphi_n \vdash_{\mathsf{FDE}_c^{\strictif}} \Box_i \varphi_1 \wedge \ldots \wedge \Box_i \varphi_n$. Hence, we conclude that (1) holds. □

While we can force classical behavior of propositional variables, this does not apply to formulas containing strict implication, as the following counter-example to a modal version of Proposition 3 shows. Note that since we consider formulas in Fm_C, we state the result with respect to the basic normal modal logic K, hence we assume familiarity with Kripke semantics for K, as well completeness of K with respect to K-Kripke models.

Proposition 4 (Modal Closure Failure) *The following hold:*

1. $\vdash_{\mathsf{K}} \top \supset (\Box_e \neg(p \wedge \neg p))$;
2. $\Box_e \top \nvdash_{\mathsf{FDE}_c^{\dashv}} \Box_i \Box_e \neg(p \wedge \neg p)$.

Proof It is immediate to verify that $\top \supset (\Box_e \neg(p \wedge \neg p))$ is valid in every Kripke model, as $p \wedge \neg p$ is true at no state in the model. Hence, by completeness of K with respect to K-Kripke models we have (1). Then, consider the $\mathsf{FDE}_c^{\dashv}$-model $M = (S, \leq, C, *, R, V)$ such that $S = \{s, t, u\}$, $\leq = id_S$, $C = \{t\}$ $R = \{(s, t), (t, u)\}$, $s^* = u$ and $t^* = t$, $V(p) = \{s, t\}$. Clearly we have that $(M, s) \models \Box_e \top$. However, $(M, s) \not\models \Box_i \Box_e \neg(p \wedge \neg p)$, since we have Rst, $t \in C$ and $(M, t) \not\models \Box_e \neg(p \wedge \neg p)$ (as Rtu, $u^* = s^{**} = s \in V(p)$ and $u \notin V(p)$). Hence, by Theorem 2 we conclude that (2) holds. □

On the one hand, the above proposition can be regarded as a drawback, in that it shows that implicit belief in $\mathsf{FDE}_c^{\dashv}$ cannot be defined as the closure of explicit belief under the classical modal logic K. On the other hand however, the above result is in line with our understanding of agents' explicit beliefs as being regimented by non-classical logics, thus not being closed under K. Therefore, it is to be expected that implicit belief is not the classical closure of explicit belief when epistemic formulas are considered. What is more, when considering explicit beliefs one may note that failure of classical closure is a desirable property: $\mathsf{FDE}_c^{\dashv}$ correctly prescribes that agents can hold explicit beliefs about some classical tautology without explicitly believing all of them, that is e.g. $\Box_e \top \nvdash_{\mathsf{FDE}_c^{\dashv}} \Box_e \neg(\varphi \wedge \neg \varphi)$. Then, as an anonymous reviewer correctly points out, if $\Box_e \top \vdash_{\mathsf{FDE}_c^{\dashv}} \Box_i \Box_e \neg(p \wedge \neg p)$ was the case then we would assume that agents implicitly believe that they explicitly believe all classical tautologies of the form $\neg(p \wedge \neg p)$, thereby committing agents to hold an infinite number of false implicit beliefs. In fact, a similar argument can be put forward for any classical tautology, not just $\neg(p \wedge \neg p)$.

3.3 Closure Under Contraposition and the Australian Plan

In this section we discuss the American and Australian plans for negation, relating $\mathsf{FDE}_c^{\dashv}$ with the existing literature. We stress that our motivation for adopting the Australian plan does not come primarily from considerations concerning epistemic modalities. Rather, we wish to provide an appealing framework to epistemic logicians who prefer the Australian plan over the American plan, and who discuss the logical properties and the philosophical assumptions which may favour one over the other.

As noted in [26, Remark 8], American and Australian-style semantics are equivalent for FDE. However, things change (i) when we add further constraints or (ii) when we extend FDE with additional vocabulary. In particular, from the perspective of the present paper it is noteworthy that as a result of either (i) or (ii), logics may fail to be closed under the rule of contraposition ($\mathsf{FDE8}$), i.e.

$\varphi \vdash \psi \Rrightarrow \neg\psi \vdash \neg\varphi$. For an example of (i), consider Priest's logic of paradox LP (which can be seen as an extension of FDE where paracompleteness is dropped).[10] For an example of (ii), consider Priest's logic of intentionality extending FDE with a strict implication operator [27], which is also not closed under contraposition. A similar problem affects Lakemeyer's generalisation of Levesque's logic [19] to arbitrary formulas $B\varphi$ for explicit belief and $L\varphi$ for implicit belief, in which (contrary to Levesque's original logic of Sect. 4) $B\varphi$ entails $L\varphi$ but $\neg L\varphi$ does not entail $\neg B\varphi$.

A typical reason why logics based on American-style semantic fail to be closed under contraposition is that positive and negative semantic clauses can be formulated independently one from the other. The American plan for negation has been motivated by an account of $\neg$ as a contradictory-forming operator, i.e. $\neg\varphi$ is true iff φ is false and $\neg\varphi$ is false iff φ is true. On this account, truth and falsity are taken as primitives and negation is taken to be a sentential operator that flip-flops truth and falsity [9]. On the other hand, Australian-style negation is understood as a modality, whose distinctive accessibility relation is compatibility [12] and whose functional role is to express exclusion. On this account, compatibility is taken as a primitive and fundamental notion, and negation is grounded in compatibility, so that s supports $\neg\varphi$ iff any state compatible with s does not support φ.[11]

Grounding negation in compatibility has the effect that the inferential properties of negation are dictated by its semantic clause and the properties of compatibility. For example, a common assumption on compatibility, defended also in [7], is symmetry, from which it follows that negation satisfies double negation introduction. Another example of a typical feature of negation in Australian-style semantic is that it satisfies contraposition by design.[12] From the perspective of the present paper, we take contraposition as underlying a rather weak inferential feature of negation.[13] Moreover, contraposition is a rule of inference which is admissible in Levesque's logic.

Therefore, although we do not take a position in the debate between the American and Australian plan for negation, FDE_c^{-3} provides an Australian-style model, alternative to the American-style ones of [19, 20], for the formalisation of the

[10] In [9, 25], this fact is taken as one reason to favour the American over the Australian-style semantic, since the validity of contraposition is built into the latter but not the former.

[11] It is well-known that $*$ can be defined in terms of a symmetric, serial and convergent compatibility relation (see e.g. [7]) as maximal compatibility with respect to the partial order $\leq$. Hence, our framework falls under the Australian plan.

[12] Note that grounding negation in compatibility is problematic for a number of paraconsistent negations for which contraposition fails, like Da Costa's negation and Nelson's N4 negation. Note also that contraposition is invalid for the Australian-style semantic of LP based on an extension of pointed models of [32] for FDE, as shown in [25].

[13] [25] observe that the validity of contraposition has some limitations. In particular, in the presence of contraposition it is impossible to keep non-paraconsistency apart from non-paracompleteness, fine-tuning which part of a theory behave classically or semi-classically. However, in the context of FDE_c^{-3} (where we locally require classicality for elements of C), the fact that shrieking truth implies shrugging it (and viceversa) is harmless.

distinction between explicit and implicit belief. This motivates the study, in Sect. 4 below, of the embedding of Levesque's logic of explicit and implicit belief into FDE_c^{-3}.

4 Levesque's Logic Within FDE

In this section we show that Levesque's logic of explicit and implicit belief embeds into our extension of FDE. We do this in a series of steps. First, we introduce Levesque's original axiomatic system and define a generalised version of its semantics generating the same logic BL (set of valid formulas) in the language $Fm_{(BL)}$, consisting of all formulas of Fm_{BL} that do not contain nested instances of B and L. Second, we define a translation t from $Fm_{(BL)}$ to Fm_C and we show how to construct from the canonical generalised Levesque model G a FDE_c^{-3}-model G^* such that a formula $\varphi \in Fm_{(BL)}$ is valid in G if and only if $c \vdash t(\varphi)$ is valid in G^*. This, together with the completeness results for BL and FDE_c^{-3}, provides an embedding result of BL into FDE_c^{-3}.

The embedding of BL into FDE_c^{-3} shows that our framework is a generalisation of Levesque's original logic. Moreover, in FDE_c^{-3} we are able to express the distinction between explicit and implicit beliefs not only for formulas without nested beliefs, but for formulas where $\Box_e$, $\Box_i$ occur at any depth.

4.1 BL *and Generalised Levesque's Models*

In this section we first outline Levesque's logic of explicit and implicit belief BL, as introduced in [20]. Then we formulate a generalised version of his semantics, with respect to which BL is sound and complete. The generalized form of Levesque's semantics will be useful in the next subsection establishing the embedding result.

Definition 6 (L-axiomatisation, Provability) Let an axiomatisation of the logic BL consist of an axiom system for classical logic (CPC) plus the following axioms:

(BL1) $B(\varphi \wedge \psi) \equiv B(\psi \wedge \varphi)$

(BL2) $B(\varphi \vee \psi) \equiv B(\psi \vee \varphi)$

(BL3) $B(\varphi \wedge (\psi \wedge \chi)) \equiv B((\varphi \wedge \psi) \wedge \chi)$

(BL4) $B(\varphi \vee (\psi \vee \chi)) \equiv B((\varphi \vee \psi) \vee \chi)$

(BL5) $B(\varphi \wedge (\psi \vee \chi)) \equiv B((\varphi \wedge \psi) \vee (\varphi \wedge \chi))$

(BL6) $B\neg(\varphi \vee \psi) \equiv B(\neg\varphi \wedge \neg\psi)$

(BL8) $B\neg\neg\varphi \equiv B\varphi$

(BL9) $B\varphi \wedge B\psi \equiv B(\varphi \wedge \psi)$

(BL10) $B\varphi \vee B\psi \supset B(\varphi \vee \psi)$

(BL11) $B\varphi \supset L\varphi$

(BL12) $L\varphi \wedge L(\varphi \supset \psi) \supset L\psi$

(BL13) $\varphi \Rrightarrow L\varphi$

(BL7) $B\neg(\varphi \wedge \psi) \equiv B(\neg\varphi \vee \neg\psi)$

(BL14) $(B\varphi \vee B\psi) \supset B\chi \Rightarrow B(\varphi \vee \psi) \supset B\chi$

The formula φ is provable in BL ($\vdash_{BL} \varphi$) iff there is a sequence of formulas ending with φ, where each element is either an axiom of BL or follows from previous elements by means of a rule of inference of BL.

Note that the logic BL is defined as a set of formulas, not sequents. The logic extends classical propositional logic CL, so the material conditional $\supset$ can play the role of the turnstile $\vdash$ in sequents.

Lemma 8 (FDE-BL) *If* $\varphi \vdash_{FDE} \psi$, *then* $\vdash_{BL} B\varphi \supset B\psi$

Proof To prove this lemma, it is useful to switch to an equivalent formula-formula system for FDE that does not have the contraposition rule (see e.g. Dunn's $\mathbf{R}_{\text{fde}}$ [13]). The lemma is then established by a straightforward induction on the length of proofs. □

Definition 7 (GL-models) A generalised Levesque model, or GL-model, is a structure $G = (S, W, B, T, F)$ where $W, B \subseteq S$, and $T, F : Pr \to \mathscr{P}(S)$ such that:

(G1) $s \in W \Rightarrow (s \in F(p) \text{ iff } s \notin T(p))$;
(G2) $s \in B, s \sqsubseteq t \Rightarrow t \in B$.

We define $s \sqsubseteq t$ iff $s \in X(p) \Rightarrow t \in X(p)$ for $X \in \{T, F\}$ and for all $p \in Pr$.

Following Levesque's [20], GL-models consist of a set of situations S together with a set B of situations which are epistemically accessible to the agent. The situations in B express the belief state of the agent; these are situations that support the information the agent believes. $T(p)$ and $F(p)$ comprise situations supporting the truth, respectively falsity, of p. The relation $\sqsubseteq$ expresses an information ordering on situations: $s \sqsubseteq t$ can be read as saying that all the "atomic" information supported by s is supported by t. Finally, W is a set consisting of complete and consistent situations. The difference between the original models of [20] and GL-models is that, in the latter, (i) the set of states representing the belief state of the agent, B, is assumed to be upwards closed under the "information order" $\sqsubseteq$, and (ii) the set of "possible worlds" W is not necessarily the set of all complete and consistent situations. We will see that this difference is immaterial since Levesque's logic BL is sound and complete with respect to GL models.

Definition 8 (GL-evaluation and Logic) Let $\models^+$ and $\models^-$ be two binary relations between states in GL-models and formulas in $Fm_{(BL)}$ defined by double recursion as follows:

- $(G, s) \models^+ p$ iff $s \in T(p)$;
- $(G, s) \models^- p$ iff $s \in F(p)$;
- $(G, s) \models^+ \neg\varphi$ iff $(G, s) \models^- \varphi$;

- $(G, s) \models^- \neg\varphi$ iff $(G, s) \models^+ \varphi$;
- $(G, s) \models^+ \varphi \wedge \psi$ iff $\models^+ \varphi$ and $(G, s) \models^+ \psi$;
- $(G, s) \models^- \varphi \wedge \psi$ iff $(G, s) \models^- \varphi$ or $(G, s) \models^- \psi$;
- $(G, s) \models^+ \varphi \vee \psi$ iff $(G, s) \models^+ \varphi$ or $(G, s) \models^+ \psi$;
- $(G, s) \models^- \varphi \vee \psi$ iff $(G, s) \models^- \varphi$ and $(G, s) \models^- \psi$;
- $(G, s) \models^+ B\varphi$ iff for all $t \in B((G, t) \models^+ \varphi)$;
- $(G, s) \models^- B\varphi$ iff $(G, s) \not\models^+ B\varphi$;
- $(G, s) \models^+ L\varphi$ iff for all $t \in B \cap W((G, t) \models^+ \varphi)$;
- $(G, s) \models^- L\varphi$ iff $(G, s) \not\models^+ L\varphi$.

Let a formula φ be valid in a GL-model G ($G \models^+ \varphi$) iff for all $w \in W$ $G, w \models^+ \varphi$. **GL** is the class of **GL**-models, $Log(G)$ is the set of formulas valid in G and $Log(\mathbf{GL})$ is the set of formulas valid in **GL**. $Log(\mathbf{GL})$ is the set of formulas valid in **GL**.

It is easily seen that $(G, s) \models^+ \varphi$ and $s \sqsubseteq t$ only if $(G, t) \models^+ \varphi$. In GL-models, $B\varphi \models^+_{\mathbf{GL}} L\varphi$ does imply that $\neg L\varphi \models^+_{\mathbf{GL}} \neg B\varphi$, and in fact it is easy to observe that, differently than in Lakemayer's system, contraposition is an admissible rule of inference in BL. We now show that the logic BL is sound and complete with respect to generalised Levesque models, where the completeness proof adapts the canonical model construction of [21, p.201].

Theorem 3 **(BL-soundness)** *For all* $\varphi \in Fm_{(BL)}$,

$$\vdash_{\mathsf{BL}} \varphi \textit{ only if } \varphi \in Log(\mathbf{GL}).$$

Proof By induction on the length of BL-proofs, where the only interesting cases are those involving L, (BL11)-(BL14). For (BL11), assume for all $w \in W$ that $w \models^+ B\varphi$ iff for all $t \in B$ $t \models \varphi$. Hence for all $t \in B \cap W$ we have that $t \models \varphi$, by which we conclude that $w \models L\varphi$. For (BL12), assume for all $w \in W$ that $w \models^+ L\varphi$ and $w \models L(\varphi \supset \psi)$. Equivalently for all $t \in B \cap W$, $t \models \varphi$ and $t \models \varphi \supset \psi$. Hence for all $t \in B \cap W$ we have $t \models \psi$, by which we conclude that $w \models L\psi$. For (BL13), assume $w \models^+ \varphi$ for all $w \in W$. Hence for all $w \in W \cap B$ we have that $w \models^+ \varphi$, by which we conclude that $\models^+ L\varphi$. To show that (BL14) preserves validity, we first claim the following.

Claim 1 Let $G_1 = (S, W, B_1, T, F)$ and $G_2 = (S, W, B_2, T, F)$ and $G = (S, W, B_1 \cup B_2, T, F)$ be GL-models. If $(G_1, w) \models^+ B\chi$ and $(G_2, w) \models^+ B\chi$, then $(G, w) \models B\chi$.

The proof of the claim is immediate: χ is a non-modal formula and $(G_1, s) \models^+ \chi$ iff $(G_2, s) \models^+ \chi$ iff $(G, s) \models^+ \chi$ for non-modal χ. Then, assume that $B\varphi \vee B\psi \supset B\chi$ is valid in all models. Take a model $G = (S, W, B, T, F)$, and assume that $(G, w) \models^+ B(\varphi \vee \psi)$ for some $w \in W$, i.e.

$$B \subseteq \underbrace{\{s \in S \mid (G, s) \models^+ \varphi\}}_{X_1} \cup \underbrace{\{s \in S \mid (G, s) \models^+ \psi\}}_{X_2}.$$

Define $G_1 = (S, W, B_1, T, F)$ and $G_2 = (S, W, B_2, T, F)$ where $B_1 = B \cap X_1$ and $B_2 = B \cap X_2$. Both B_1 and B_2 are clearly upwards-closed and so G_1 and G_2 are GL-models. By our assumption, $G_1 \models^+ B\varphi \vee B\psi \supset B\chi$ and $G_2 \models^+ B\varphi \vee B\psi \supset B\chi$. Importantly, $(G_1, w) \models^+ B\varphi$ and $(G_2, w) \models^+ B\psi$ by the respective definitions of B_1 and B_2. Hence, $(G_i, w) \models^+ B\varphi \vee B\psi$ and so $(G_i, w) \models^+ B\chi$ for $i \in \{1, 2\}$ by the assumption. Moreover, clearly $B = B_1 \cup B_2$ (since $B \subseteq X_1 \cup X_2$). Hence, our claim implies $(G, w) \models^+ B\chi$ and we are done. □

A BL-theory is any set of formulas $\Gamma \subseteq Fm_{(BL)}$ such that (i) if $\varphi, \psi \in \Gamma$, then $\varphi \wedge \psi \in \Gamma$ and (ii) if $\varphi \in \Gamma$ and $\vdash_{\mathsf{BL}} \varphi \supset \psi$, then $\psi \in \Gamma$. A maximally consistent BL-theory $U \subseteq Fm_{(BL)}$ is a BL-theory $U \subseteq Fm_{(BL)}$ such that (i) for all $\varphi \in Fm_{((BL)}$, $\varphi \notin U$ only if $\neg\varphi \in U$ and (ii) for all $U' \supset U$ there is $\varphi \in Fm_{((BL)}$ such that $\varphi, \neg\varphi \in U'$. To prove completeness, note that if $\nvdash_{\mathsf{BL}} \varphi_0$, then there is a maximally consistent BL-theory $U \subseteq Fm_{(BL)}$ such that $\varphi_0 \notin U$. Hence, we define the canonical BL-model as parametrised on the maximally consistent BL-theory U not containing φ_0. Our definitions and proofs expand on those given in [21, pp. 199–202].

Definition 9 (Canonical *U*-model) Let U be a maximal BL-consistent theory. The canonical U-model is $G_U = (S, W, B, T, F)$ where:

- S is the set of all prime FDE-theories over Fm;
- W is the set of all complete and consistent prime FDE-theories over Fm;
- $B_0 \subseteq S$ is the union of:
  - $\{s \in S \mid \exists\chi(\chi \in s \;\&\; \neg\chi \in s) \;\&\; \forall\psi(B\psi \in U \implies \psi \in s)\}$;
  - $\{w \in W \mid \forall\psi : L\psi \in U \implies \psi \in w\}$;
- $B = \{s \in S \mid \exists t \in B_0 : t \subseteq s\}$;
- $T(p) = \{s \in S \mid p \in s\}$;
- $F(p) = \{s \in S \mid \neg p \in s\}$.

Since U is a maximal BL-consistent theory, $U \cap Fm$ is a complete and consistent prime FDE-theory over Fm, hence $u = U \cap Fm \in W$.

Lemma 9 (BL Canonicity) *The canonical U-model G_U is a BL-model.*

Proof Clearly, $B, W \subseteq S$ and $(G2)$ hold by definition of G_U. Finally, $(G1)$ holds since every complete and consistent FDE-theory u is such that $\neg\varphi \in u$ iff $\varphi \notin u$. □

Lemma 10 (BL Truth) *The following hold for all G_U:*

1. *For all $\varphi \in Fm$: $\varphi \in s$ iff $(G_U, s) \models^+ \varphi$ and $\neg\varphi \in s$ iff $(G_U, s) \models^- \varphi$;*
2. *For all $\varphi \in Fm_{(BL)}$: $\varphi \in U$ iff $(G_U, u) \models^+ \varphi$ where $u = U \cap Fm \in W$.*

Proof Item 1 is established by double induction on the complexity of φ as usual, where we show only the case where $\varphi := \neg\psi$ as an illustration. If $\varphi := \neg\psi$, then $(G_U, s) \models^+ \neg\psi$ iff $(G_U, s) \models^- \psi$ iff (by the IH) $\neg\psi \in s$; and $(G_U, s) \models^- \neg\psi$ iff $(G_U, s) \models^+ \psi$ iff (by the IH) $\psi \in s$ iff (by (FDE7)) $\neg\neg\psi \in s$. To prove

Item 2, we reason by cases. First, if φ is a propositional formula, then we need to prove $\varphi \in u$ iff $(G_U, u) \models^+ \varphi$. This is an instance of Item 1. If $\varphi := B\chi$, then we reason as follows. If $B\chi \in U$, then $\chi \in t$ for all $t \in B_0$ by (BL11). Hence, $\chi \in t$ for all $t \in B$ and so $(G_U, t) \models^+ \chi$ for all $t \in B$ by Item 1, which means that $(G_U, u) \models^+ B\chi$. Conversely, if $B\chi \notin U$, then we reason as follows. We claim that there is $s \in B_0$ such that $\chi \notin s$. $(G_U, u) \not\models^+ B\chi$ follows from this claim by Item 1. The claim is established as follows. It can be shown that the pair $(\{\psi \mid B\psi \in U\} \cup \{p \wedge \neg p\}, \{\chi\})$ is FDE-independent if p is any variable not occurring in χ. (If it were not, then there would be $\psi_1, \ldots, \psi_n$ such that $B\psi_1, \ldots, B\psi_n \in U$ and $\psi_1 \wedge \ldots \wedge \psi_n \wedge p \wedge \neg p \vdash_{\mathsf{FDE}} \chi$. But then $\psi_1 \wedge \ldots \wedge \psi_n \vdash_{\mathsf{FDE}} \chi$.[14] It follows that $\vdash_{\mathsf{BL}} B\psi_1 \wedge \ldots \wedge B\psi_n \supset B\chi$, which contradicts the assumption that $B\chi \notin U$.) It follows that there is an inconsistent $s \in S$ such that $\{\psi \mid B\psi \in U\} \subseteq s$, but $\chi \notin s$. If $\varphi := L\chi$, then we reason as follows. If $L\chi \in U$, then $\chi \in t$ for all $t \in B_0 \cap W$ by the definition of B_0. But then $\chi \in t$ for all $t \in B \cap W$ since obviously $B \cap W \subseteq B_0 \cap W$. (If $t \in B \cap W$, then there is $u \subseteq t$ such that $u \in B_0$. Clearly, u cannot be inconsistent, so $u \in W$. But this means that $u = t$ and so $t \in B_0 \cap W$.) Hence, $(G_U, u) \models^+ L\chi$ by the induction hypothesis. Conversely, if $L\chi \notin U$, then we reason as follows. We claim that there is a $w \in B_0 \cap W$ such that $\chi \notin w$. $(G_U, u) \not\models^+ L\chi$ follows from this claim by Item 1. The claim is established as follows. If $\chi \in w$ for all $w \in B_0 \cap W$, then χ is entailed in classical logic by the set $\{\psi \mid L\psi \in U\}$ (since W contains all classical theories over Fm). By compactness of classical logic, $\psi_1 \wedge \ldots \wedge \psi_n \supset \chi$ is a theorem of classical logic for some $\{\psi_1, \ldots, \psi_n\} \subseteq \{\psi \mid L\psi \in U\}$. Hence, $\vdash_{\mathsf{BL}} L\psi_1 \wedge \ldots \wedge L\psi_n \supset L\chi$ by (BL12)-(BL13). But this contradicts the assumption that $L\chi \notin U$, which is a BL-theory. The final case is the case when φ is obtained using the propositional connectives $\neg, \wedge, \vee$ from some $\{p_1, \ldots, p_n\} \cup \{B\psi_1, \ldots, B\psi_m\} \cup \{L\chi_1, \ldots, L\chi_k\}$. But then the claim is established easily by induction on the complexity of φ. The base cases where φ is a propositional variable or a modal formula of the form $B\psi$ or $L\chi$ were established in the previous claims. The rest is straightforward. (If $\varphi := \neg\psi$, then the argument makes use of the fact that u is a complete and consistent situation in the model G_U and so $(G_U, u) \not\models^+ \chi$ iff $(G_U, u) \models^- \chi$ for all $\chi \in Fm_{(BL)}$.) □

Theorem 4 (**BL Soundness and Completeness**) *For all* $\varphi, \psi \in Fm_{(BL)}$,

$$\vdash_{\mathsf{BL}} \varphi \textit{ iff } \varphi \in Log(\mathbf{GL}).$$

Proof The implication from left to right is Theorem 3, while the converse follows from Lemmas 9 and 10. □

[14] In general, if p does not occur in χ, then $\psi \wedge (p \wedge \neg p) \vdash_{\mathsf{FDE}} \chi$ entails $\psi \vdash_{\mathsf{FDE}} \chi$. If the latter fails for some ψ, χ, then completeness of FDE entails that there is a FDE-model s such that $s \models^+ \psi$ and $s \not\models^+ \chi$. Now change s to s' by setting $s' \models^+ p$ and $s' \models^- p$. An easy proof by induction on complexity of formulas shows that $s' \models^+ \psi$ and $s' \not\models^+ \chi$.

4.2 Embedding Levesque's Logic in FDE with Strict Implication

In this section we show that each canonical U-model G can be transformed into an $\mathsf{FDE}_c^{\dashv}$-model G^* such that $\varphi \in Fm_{(BL)}$ is valid in G iff $t(\varphi) \in Fm_C$ is valid in G^*, for a natural translation function t that takes $B\varphi$ into $\Box_e t(\varphi)$ and $L\varphi$ into $\Box_i t(\varphi)$.

Definition 10 (Embedding Model) For every canonical U-model $G = (S, B, W, T, F)$, define $G^* = (S, \sqsubseteq, C, ^*, R, V)$ as follows:

- $\sqsubseteq$ is set inclusion;
- $C = W$;
- $s^* = \{\varphi \mid \neg\varphi \notin s\}$;
- Rst iff $t \in B$;
- $s \in V(p)$ iff $p \in s$.

In effect, G^* is the canonical FDE-model extended with R and C. It can also be seen as the canonical U-model G where the negative valuation F is replaced by the Routley star operation *. Since FDE is closed under contraposition, we can define * and show that the construction gives rise to a $\mathsf{FDE}_c^{\dashv}$-model. Note that by definition of $\sqsubseteq$ as set inclusion, for all $s, t \in S$ we have that $s \sqsubseteq t$ iff $s \in X(p) \Rightarrow t \in X(p)$ for $X \in \{T, B\}$ as in Definition 8.

Lemma 11 (Embedding Canonicity) *For all canonical U-models G, G^* is a $\mathsf{FDE}_c^{\dashv}$-model.*

Proof Clearly, $(S, \sqsubseteq)$ is a partially ordered set and $C \subseteq S$. As S consists of all FDE-theories and s^* is defined as in the canonical $\mathsf{FDE}_c^{\dashv}$-model, $(F1)$, $(F2)$ and $(F4)$ are established as in Lemma 5. For $(F3)$, assuming $s' \sqsubseteq s$ and $t \sqsubseteq t'$, we have Rst iff $t \in B$ only if $t' \in B$ (by $(G2)$) iff Rst' iff $Rs't'$. □

Definition 11 (c-Translation) Define the c-translation function $t : Fm_{(BL)} \to Fm_{(C)}$ as follows, where $Fm_{(C)}$ is the subset of Fm_C without & where $c, \top$ only occur as antecedents of $\dashv$-formulas:

$$t(p) = p$$

$$t(\neg\varphi) = \neg t(\varphi)$$

$$t(\varphi \wedge \psi) = t(\varphi) \wedge t(\psi)$$

$$t(\varphi \vee \psi) = t(\varphi) \vee t(\psi)$$

$$t(B\varphi) = \top \dashv t(\varphi)$$

$$t(L\varphi) = c \dashv t(\varphi)$$

Lemma 12 (Embedding Equivalence) *For all canonical U-models G:*

$$\varphi \in Log(G) \text{ iff } c \vdash t(\varphi) \text{ is valid in } G^*.$$

Proof The result follows immediately by $W = C$ and the following claim, which holds for all $\varphi \in Fm_{(BL)}$ and is established by double induction on the complexity of φ.

$$(G^*, s) \models t(\varphi) \iff (G, s) \models^+ \varphi \qquad (G^*, s^*) \not\models t(\varphi) \iff (G, s) \models^- \varphi.$$

The base case holds since $s \in T(p)$ iff (by Definition 11) $s \in T(t(p))$ iff (by Definition 9) $t(p) \in s$ iff (by Definition 10) $s \in V(t(p))$; and $s \in F(p)$ (by Definition 9) $\neg p \in s$ iff (by Definition 11) $\neg t(p) \in s$ iff (by definition of $*$ in G^*) $t(p) \notin s^*$ iff (by Definition 10) $s^* \notin V(t(p))$. The induction step is straightforward, hence we show only the cases $\varphi := \neg\psi$ and $\varphi := B\psi$ as an illustration. If $\varphi := \neg\psi$, then $(G, s) \models^+ \neg\psi$ iff $(G, s) \models^- \psi$ iff (by the IH) $(G^*, s^*) \not\models t(\psi)$ iff $(G^*, s) \models \neg t(\psi)$ iff (by Definition 11) $(G^*, s) \models t(\neg\psi)$; and $(G, s) \models^- \neg\psi$ iff $(G, s) \models^+ \psi$ iff (by the IH) $(G^*, s) \models t(\psi)$ iff $(G^*, s^*) \not\models \neg t(\psi)$ iff (by Definition 11) $(G^*, s^*) \not\models t(\neg\psi)$. If $\varphi := B\psi$, then $(G, s) \models^+ B\psi$ iff $(G, t) \models^+ \psi$ for all $t \in B$ iff (by Definition 10) $(G, t) \models^+ \psi$ for all t such that Rst iff (by the IH) $(G^*, t) \models t(\psi)$ for all t such that Rst iff $(G^*, s) \models \top \dashv\!3\, t(\psi)$ iff (by Definition 11) $(G^*, s) \models t(B\psi)$; and $(G, s) \models^- B\psi$ iff $(G, s) \not\models^+ B\psi$ iff there is $t \in B$ such that $(G, t) \not\models^+ \psi$ iff (by the IH) there is $t \in B$ such that $(G^*, t) \not\models t(\psi)$ iff (by Definition 10) there is t such that Rs^*t and $(G^*, t) \not\models t(\psi)$ iff $(G^*, s^*) \not\models \top \dashv\!3\, t(\psi)$ iff (by Definition 11) $(G^*, s^*) \not\models t(B\psi)$. □

Finally, Theorem 5 below shows that not only is $\mathsf{FDE}_c^{\dashv 3}$ able to express natural variants of the informal Levesque-style notions of explicit and implicit belief (as discussed in Sects. 2 and 3), but it also possible to embed Levesque's logic (a specific formal model of the intuitive notions) into it.

Theorem 5 (Embedding) *For all $\varphi \in Fm_{(BL)}$, the following are equivalent*

$$\vdash_{\mathsf{BL}} \varphi \textit{ iff } c \vdash_{\mathsf{FDE}_c^{\dashv 3}} t(\varphi).$$

Proof One direction is established, thanks to Theorems 4 and 2, by induction on the length of BL-proofs, by showing that the translation of each BL-axiom (rule of inference) is c-valid (preserves c-validity) in $\mathsf{FDE}_c^{\dashv 3}$-models. The cases for (BL1) – (BL10) follow from the fact that $\varphi \vdash \psi$ is valid in FDE only if $c \vdash \Box_e\varphi \supset \Box_e\psi$ is valid in $\mathsf{FDE}_c^{\dashv 3}$. For (BL11)-(BL14), we reason as in Theorem 3. Conversely, assume $\not\vdash_{\mathsf{BL}} \varphi$. By Theorem 4 we have that $\varphi \notin Log(G)$, where G is the canonical U-model such that $\varphi \notin U$. By Lemmas 11 and 12, G^* is a $\mathsf{FDE}_c^{\dashv 3}$-model such that $c \vdash t(\varphi)$ is not valid in G^*. We conclude by Theorem 2 that $c \not\vdash_{\mathsf{FDE}_c^{\dashv 3}} t(\varphi)$. □

5 Conclusion

In this paper we introduced $\mathsf{FDE}_c^{\dashv 3}$, an extension of FDE with strict implication $\dashv\!3$ (and its adjoint $\&$) and the classicality constant c. We have argued that the logic is able to express the distinction between explicit belief, formalised as $\top \dashv\!3\, \varphi$, and implicit belief, formalised as $c \multimap \varphi$. We have shown that implicit belief expresses classical closure of explicit belief, akin to Levesque's formalization of the distinction in [20]. In fact, we proved that Levesque's logic of explicit and implicit belief embeds into $\mathsf{FDE}_c^{\multimap}$. The embedding result shows that $\mathsf{FDE}_c^{\multimap}$ can be seen as a generalization of Levesque's logic where nesting of epistemic operators is allowed and where belief may lack some strong introspection properties.

A similar generalization was provided by Lakemeyer [19]. His logic assumes some strong introspection properties and it is based on a four-valued version of the semantics for FDE (the American plan). In view of the ongoing debates concerning the respective merits of four-valued and relational semantics (the Australian plan) for FDE and its extensions [7, 9, 25], our relational framework offers an alternative to Lakemeyer's approach potentially attractive to proponents of the Australian plan.

We conclude by listing some natural topics for future research. First, it is interesting to extend our framework with familiar notions of group belief such as common and distributed belief. The formalization of common belief could build on an Australian-style alternative to Belnapian Propositional Dynamic Logic discussed in [34]. Second, it is natural to consider extensions of our framework with various mechanisms of information update such as public announcements or various kinds of "soft" information update. One approach, building on [28], could use relevant implication as an operator expressing non-monotonic information update.

Acknowledgments The authors thank Adam Přenosil for his advice on proof systems for FDE, and the anonymous reviewers for their useful comments on earlier versions of this paper. Work of the first author was supported by the grant 22-01137S of the Czech Science Foundation and by the long-term strategic development financing of the Institute of Computer Science (RVO:67985807).

References

1. Alchourrón, C. E., Gärdenfors, P., & Makinson, D. (1985). On the logic of theory change: Partial meet contraction and revision functions. *Journal of Symbolic Logic, 50*(2), 510–530.
2. Anderson, A. R., & Belnap, N. D. (1975). *Entailment: The logic of relevance and necessity, Volume I*. Princeton: Princeton University Press.
3. Artemov, S. (2008). The logic of justification. *The Review of Symbolic Logic, 1*(4), 477–513.
4. Belnap, N. (1977). A useful four-valued logic. In J. M. Dunn & G. Epstein (Eds.), *Modern uses of multiple-valued logic* (pp. 5–37). Dordrecht: Springer Netherlands.
5. Belnap, N. (1977). How a computer should think. In G. Ryle (Ed.), *Contemporary aspects of philosophy* (pp. 30–55). Stocksfield: Oriel Press Ltd.
6. Bendana, J. (2022). Implicit beliefs. In J. R. Thompson (Ed.), *The Routledge handbook of philosophy and implicit cognition* (12). New York: Routledge.
7. Berto, F., & Restall, G. (2019). Negation on the Australian plan. *Journal of Philosophical Logic, 48*(6), 1119–1144.
8. da Costa, N. C. A. (1974). On the theory of inconsistent formal systems. *Notre Dame Journal of Formal Logic, 15*(4), 497–510.
9. De, M., & Omori, H. (2018). There is more to negation than modality. *Journal of Philosophical Logic, 47*(2), 281–299.

10. Dunn, J. M. (1976). Intuitive semantics for first-degree entailments and "coupled trees". *Philosophical Studies, 29*(3), 149–168.
11. Dunn, J. M. (1993). Star and perp: Two treatments of negation. *Philosophical Perspectives, 7*, 331–357.
12. Dunn, J. M. (1999). A comparative study of various model-theoretic treatments of negation: A history of formal negation. In D. Gabbay & H. Wansing (Eds.), *What is negation?*. Applied logic series (Vol. 13, pp. 23–51). Dordrecht: Springer.
13. Dunn, J. M. (2000). Partiality and its dual. *Studia Logica, 66*(1), 5–40.
14. Dunn, J. M., & Zhou, C. (2005). Negation in the context of gaggle theory. *Studia Logica: An International Journal for Symbolic Logic, 80*(2/3), 235–264.
15. Fagin, R., & Halpern, J. Y. (1987). Belief, awareness, and limited reasoning. *Artificial Intelligence, 34*(1), 39–76.
16. Fuhrmann, A. (1990). Models for relevant modal logics. *Studia Logica, 49*(4), 501–514.
17. Halpern, J. (2000). Alternative semantics for unawareness. *Games and Economic Behavior, 37*, 321–339.
18. Hintikka, J. (1962). *Knowledge and belief.* Ithaca: Cornell University Press.
19. Lakemeyer, G. (1987). Tractable meta-reasoning in propositional logics of belief. In *IJCAI 1987* (pp. 401–408). Burlington: Morgan Kaufmann Publishers Inc.
20. Levesque, H. (1984). A logic of implicit and explicit belief. In *Proceedings of AAAI 1984* (pp. 198–202).
21. Levesque, H. J., & Lakemeyer, G. (2001). *The logic of knowledge bases*. Cambridge: MIT Press.
22. Lorini, E. (2020). Rethinking epistemic logic with belief bases. *Artificial Intelligence, 282*(C), 103233.
23. Meyer, J.-J. C., & Van Der Hoek, W. (1995). *Epistemic logic for AI and computer science*. Cambridge: Cambridge University Press.
24. Modica, S., & Rustichini, A. (1999). Unawareness and partitional information structures. *Games and Economic Behavior, 27*(2), 265–298.
25. Omori, H., & De, M. (2022). Shrieking, shrugging, and the Australian plan. *Notre Dame Journal of Formal Logic, 63*(2), 137–141.
26. Omori, H., & Wansing, H. (2017). 40 years of FDE: An introductory overview. *Studia Logica, 105*(6), 1021–1049.
27. Priest, G. (2016). *Towards non-being*. Oxford: Oxford University Press.
28. Punčochář, V., Sedlár, I., & Tedder, A. (2023). Relevant epistemic logic with public announcements and common knowledge. *Journal of Logic and Computation, 33*, 436–461.
29. Restall, G. (1999). Negation in relevant logics (How I stopped worrying and learned to love the Routley star). In D. M. Gabbay & H. Wansing (Eds.), *What is negation?* (pp. 53–76). Berlin: Springer.
30. Restall, G. (2000). *An introduction to substructural logics*. London: Routledge.
31. Routley, R., & Meyer, R. K. (1973). Semantics of entailment. In H. Leblanc (Ed.), *Truth, syntax and modality* (pp. 194–243). Amsterdam: North Holland.
32. Routley, R., & Routley, V. (1972). The semantics of first degree entailment. *Noûs, 6*(4), 335–359.
33. Routley, R., Plumwood, V., Meyer, R. K., & Brady, R. T. (1982). *Relevant logics and their rivals* (Vol. 1). Atascadero: Ridgeview.
34. Sedlár, I. (2016). Propositional dynamic logic with Belnapian truth values. In L. Beklemishev, S. Demri, & A. Máté (Eds.), *Advances in modal logic* (Vol. 11, pp. 503–519). London: College Publications.
35. Sedlár, I., & Vigiani, P. (2022). Relevant reasoners in a classical world. In D. Fernández Duque, A. Palmigiano, & S. Pichinat (Eds.), *Advances in modal logic* (Vol. 14, pp. 697–718). London: College Publications.
36. Standefer, S., Shear, T., & French, R. (2023). Getting some (non-classical) closure with justification logic. *Asian Journal of Philosophy, 2*(2), 1–25.

37. van Benthem, J., & Velázquez-Quesada, F. R. (2010). The dynamics of awareness. *Synthese, 177*, 5–27.
38. van den Berg, L., Atencia, M., & Euzenat, J. (2023). Raising awareness without disclosing truth. *Annals of Mathematics and Artificial Intelligence, 91*(4).
39. Velázquez-Quesada, F. R. (2013). Explicit and implicit knowledge in neighbourhood models. In *Proceedings of the 4th International Workshop on Logic, Rationality, and Interaction - Volume 8196*, LORI 2013 (pp. 239–252). Berlin: Springer.
40. Velazquez-Quesada, F. (2014). Dynamic epistemic logic for implicit and explicit beliefs. *Journal of Logic Language and Information, 23*, 107–140.

Relevant Rational Arithmetic

John Slaney

Abstract The arithmetic $\mathbf{R}^\sharp$ is obtained by postulating the Peano axioms on the basis of the relevant logic **R**. $\mathbf{R}^\sharp$ is a remarkable arithmetic, not least in that it has finite models. In this paper we examine the options for extending $\mathbf{R}^\sharp$ from natural numbers to rational numbers, as this is the essential next step towards providing a relevant basis for mathematics and for applications. Relevant rational number theory is problematic in that the most obvious approaches lead to non-conservative extensions of $\mathbf{R}^\sharp$. We consider three ways in which relevant theories of rational arithmetic can be formulated, and note in particular how these fare in the finite models of $\mathbf{R}^\sharp$.

1 Introduction

R. K. Meyer observed [8, 9] that relevant logic has been misconstrued, even by the authors of its canonical exposition [1], as a logic of relevance. It is, he noted, no such thing: on the contrary, like classical logic, it is best seen not as a "logic of" anything, but as a generic theory of valid inference. It embodies a paradigm of reasoning, characterised by keeping the logical principle of monotonicity (weakening) firmly in its place, requiring that nothing should count as a derivation of a conclusion *from* premises unless it *uses* those premises to obtain that conclusion. As is familiar, what counts as a "use" gets to be a delicate matter, especially where uses of the same premise need to be counted or where the order of premises needs to be shuffled, so there is actually a range of substructural logics motivated more or less by paying attention to these details of the abstract nature of proof. Half a century of research into substructural proof systems and their semantics has clarified our view of proof structure and at least begun to make a systematic theory possible. However, while it is important to remember that these variants of the idea exist, in the present paper

J. Slaney (✉)
Australian National University, Canberra, ACT, Australia
e-mail: John.Slaney@anu.edu.au

I. Sedlár et al. (eds.), *New Directions in Relevant Logic*, Trends in Logic 63,
https://doi.org/10.1007/978-3-031-69940-5_18

we follow standard usage in taking “relevant logic” to be the Anderson-Belnap system **R**, of which more shortly. Relevant logic is designed to *exhibit* the feature of relevance, just as intuitionistic logic is designed to be constructive, but in each case the *topic* of the theory is the validity of argument forms, not relevance or constructivity as such.

A logic for generic application must be judged at least in part by reference to the theories it allows. In the case of intuitionistic logic, this point is familiar: while indeed we may argue in the abstract for or against constructivity in reasoning, constructive mathematics is what it is, so in the end we may know the logic by its fruits. Something similar should be true of relevant logic: if it is to take its place as the logical basis of a reasoning paradigm on a par with its more famous cousins, we need examples of relevant theories which will bear deep and sustained investigation. To date, the best candidate for such a theory is the relevant arithmetic $\mathbf{R}^\sharp$ [6, 11] first proposed half a century ago, but still little known outside a small circle of specialists.

$\mathbf{R}^\sharp$ is obtained by adding some very natural versions of the Peano-Dedekind axioms for natural number theory to a basis not of classical logic but of the relevant logic **R**. The resulting theory has a great many fascinating properties, not the least of which is that it has *finite* models, based on the integers modulo n for any chosen n. Since in any such model 0 is the successor of $n-1$, while one of the axioms says that 0 is not a successor, the theories of these models are of course inconsistent; but in relevant logic, inconsistency is not the worst of crimes against rationality, so they are nonetheless nontrivial in the sense that not all formulae hold in them. This means, among other things, that Gödel’s second incompleteness theorem has to be stated carefully for $\mathbf{R}^\sharp$, since there is a finitary proof of the theory’s nontriviality, and indeed of its *reliability* in the sense that it never gives the wrong result for a calculation: it contains no ground equation that is not true of the natural numbers.

$\mathbf{R}^\sharp$, however, is a theory of the natural numbers only. If relevant logical theories are to get further than that, they must extend at least to other parts of mathematics standardly treated in primary school, which means most obviously to rational arithmetic. It is the purpose of the present paper to take it in that direction.

By way of background, we first review $\mathbf{R}^\sharp$. For that, we begin with the logic **R**, presented as usual as a Hilbert system, in terms of axioms and the expected rules of inference. The following axioms suffice:

A1. $A \to A$
A2. $(A \to (A \to B)) \to (A \to B)$
A3. $(A \to B) \to ((B \to C) \to (A \to C))$
A4. $(A \to (B \to C)) \to (B \to (A \to C))$
A5. $(A \wedge B) \to A$
A6. $(A \wedge B) \to B$
A7. $((A \to B) \wedge (A \to C)) \to (A \to (B \wedge C))$
A8. $A \to (A \vee B)$
A9. $B \to (A \vee B)$
A10. $((A \to C) \wedge (B \to C)) \to ((A \vee B) \to C)$
A11. $(A \wedge (B \vee C)) \to ((A \wedge B) \vee C)$

A12. $\neg\neg A \to A$
A13. $(A \to \neg B) \to (B \to \neg A)$

To these are applied two rules:

R1. $A \to B,\ A \ \Rightarrow\ B$
R2. $A,\ B \ \Rightarrow\ A \wedge B$

Axioms A1–A4 with R1 generate the theorems of the pure implication fragment, axioms A1–A7 with both rules the implication-conjunction fragment and A1–A11 the positive logic. It is usual to add some defined operations:

D1. $A \leftrightarrow B \ =_{\text{df.}}\ (A \to B) \wedge (B \to A)$
D2. $A \circ B \ =_{\text{df.}}\ \neg(A \to \neg B)$

The algebraic structures corresponding to **R** are the De Morgan monoids introduced by Dunn [4]. These serve to model the logic, and theories such as $\mathbf{R}^\sharp$ based upon it, in the same sense that boolean algebras model classical logic and theories.

The quantified logic **RQ** is obtained by adding the axioms:

A14. $\forall x A \to A_{[x \leftarrow t]}$	t free for x in A
A15. $\forall x(A \to B) \to (\forall x A \to \forall x B)$	
A16. $(\forall x A \wedge \forall x B) \to \forall x(A \wedge B)$	
A17. $\forall x(A \vee B) \to (A \vee \forall x B)$	x not free in A
A18. $A \to \forall x A$	x not free in A
A19. $\forall x A$	A an axiom

Here $A_{[x \leftarrow t]}$ is the formula A with all and only free occurrences of the variable x replaced by the term t. The existential quantifier is defined:

D3. $\exists x A \ =_{\text{df.}}\ \neg\forall x \neg A$

See the standard presentations [1, 3, 7] for fuller accounts of this well-known relevant logic and its semantics.

To obtain a theory of the natural numbers, we take the language to have just four function symbols: '0' (nullary), ''' (unary), '+' and '·' (binary). The only predicate symbol is the binary '='. For reasons of familiarity, we write the binary operations in infix. The axioms are the universal closures of:

P1. $x = x$
P2. $x = y \to (x = z \to y = z)$
P3. $x = y \to x' = y'$
P4. $x' = y' \to x = y$
P5. $\neg(x' = 0)$
P6. $x + 0 = x$
P7. $x + y' = (x + y)'$
P8. $x \cdot 0 = 0$
P9. $x \cdot y' = (x \cdot y) + x$
P10. $(A_{[x \leftarrow 0]} \wedge \forall x(A \to A_{[x \leftarrow x']})) \to A$

P1–P3 suffice to make the equality relation a congruence with respect to the primitive operators; P4 and P5 are the expected axioms for the successor function; P6–P9 recursively define addition and multiplication. Note that P10, the induction axiom, is an axiom *schema* with infinitely many instances. As usual in first order arithmetic, while it permits proofs by induction, in models of the theory it forces induction to hold only of those properties expressible in the language.

Ordering is readily defined in ways that make sense for the natural numbers:

D4. $a \leq b =_{\text{df}} \exists x(a + x = b)$
D5. $a < b =_{\text{df}} \exists x(a + x' = b)$

For the most part, $\mathbf{R}^\sharp$ has the same theorems as any other arithmetic, though Meyer and Friedman [6] established that 'for the most part' does not extend all the way to classical Peano arithmetic. At any rate, it is easy to show

$$a = b + c \to (a = b \leftrightarrow 0 = c),$$

whence every equation is equivalent to one of the form $0 = c$. The implication order between such equations exhibits a nice regularity: $0 = a \to 0 = b$ is a theorem iff a divides b. In particular, $0 = 1$ implies every equation, and every equation implies $0 = 0$. It is quite easy to show that $0 = 0$ plays the role of the identity proposition: $0 = 0 \to A$ is equivalent to A. It follows that any equation $a = b$ implies all of the theorems, including anything of the form $A \to A$, and this fact allows a proof by induction on the structure of A that for any terms t and u, replacement of equals holds generally in the form of the theorem schema

$$t = u \to (A_{[x \leftarrow t]} \to A_{[x \leftarrow u]})$$

It is worth rehearsing the construction of finite models of this arithmetic $\mathbf{R}^\sharp$. As domain of discourse, we take an initial segment of the natural number sequence; suppose for instance, we take the set $\{0, \ldots, 9\}$. The operations of successor, addition and multiplication we define as for the integers modulo 10. This has the consequence that 0 is the successor of 9, despite P5 that says it isn't. The structure of propositions used to evaluate logical compounds had therefore better not be the classical boolean one. In fact, many **R** propositional structures (or De Morgan monoids) will suffice,[1] the simplest consisting of three propositions or values: T (just true), F (just false) and c (both true and false, or confused). The matrices for the connectives are those of the logic RM3:

A	$\neg A$
T	F
c	c
F	T

$\wedge$	T	c	F
T	T	c	F
c	c	c	F
F	F	F	F

$\to$	T	c	F
T	T	F	F
c	T	c	F
F	T	T	T

[1] Even in the simple case where the modulus is a prime, there are infinitely many subdirectly irreducible De Morgan monoids capable of supporting models. The propositional structure of arithmetic based on the stronger logic RM is under better control, though still not trivial [14].

The "designated values" are T and c. That is, the formulae which hold on an interpretation are those which are (at least) true, whether or not they are also false. Equations $t = u$ are assigned the value c if t and u are the same element, and F if they are different. Evidently, this construction may be repeated with any other modulus, showing that no false ground equations are provable in $\mathbf{R}^\sharp$.

The key observation is that for any n greater than 1 there is a modular model like the one above in which $0 = n$ has the value c while $0 = 1$ is assigned F. That means that although the inference from $a = b$ to $a \cdot x = b \cdot x$ is relevantly fine,[2] there is no relevant warrant for the converse inference from $a \cdot x = b \cdot x$ (for $x > 0$) to $a = b$.

2 How Rational Is Rational?

$\mathbf{R}^\sharp$ is reasonably established [6, 12, 15] but if relevant logic is to support a full-blown number theory then it needs to be extended at least to deal with rational numbers as well as integers. We restrict attention here to the non-negative rationals.

We might hope to develop such an account with little trouble, since it is standard to think of the theory of the rationals as in a good sense just a fragment of that of the naturals (or integers). A classical construal of the rational numbers is as equivalence classes of vulgar fractions, which are essentially just pairs of integers (excluding the case of denominator zero). An attractive idea is then to reduce the language of rational number theory to that of integer theory, or in the case of the non-negative rationals to natural number theory, using a recursive set of rewriting rules to give contextual definitions. First, if there are quantifiers over rationals, they can be replaced by pairs of quantifiers over naturals, and any rational variable q by a pair of natural variables $\frac{x}{y'}$ (rewriting bound variables as necessary to avoid side effects). Then we can reduce terms to the form $\frac{a}{b}$ where a and b do not involve the pairing operator. This is a matter of pushing the addition and multiplication operators inside pairing, as is familiar. Where a, b, c, d are terms not involving the pairing operator:

$$\frac{a}{b} + \frac{c}{d} \rhd \frac{ad + bc}{bd}$$

$$\frac{a}{b} \cdot \frac{c}{d} \rhd \frac{ac}{bd}$$

[2] This inference essentially involves the contraction axiom A2. In arithmetic based on linear logic and its close relatives, it fails too [15].

Then equality between fractions may be reduced to equality between natural numbers:

$$\frac{a}{b} = \frac{c}{d} \rhd a \cdot d = b \cdot c$$

These definitions are easily extended to a more generous vocabulary, for example allowing a natural number and a fraction to be added by coercing n to $\frac{n}{1}$ and the like. The reduction of rational equations to natural ones also applies to the other standard comparison relation symbols such as '$<$' and '$\leq$'. Details are omitted, as they are well known.

Unfortunately, as is also well known, this definitional reduction of rational number theory to the theory of integers (or naturals) does not quite work. For the theories of *positive* numbers it is fine, but as soon as we extend to the *non-negative* ones it stumbles on the traditional awkwardness of avoiding division by zero.

To remedy that, there is a simple trick. Instead of introducing the notation of vulgar fractions directly, we introduce a binary operator ρ with the intended reading

$$\rho(a, b) = \frac{a}{b+1}$$

In terms of this, the definitional reduction can be re-worked more carefully:

$$\rho(a, b) + \rho(c, d) \rhd \rho(ad + bc + a + c, bd + b + d)$$

$$\rho(a, b) \cdot \rho(c, d) \rhd \rho(ac, bd + b + d)$$

$$\rho(a, b) = \rho(c, d) \rhd ad + a = bc + c$$

Again it is straightforward to extend this scheme to allow terms of the forms $\rho(\rho(a, b), c)$ etc. or to allow natural and rational numbers to be mixed in terms as usual, but for the purposes of the present paper it is more convenient to keep types separate, so we assume that the 'ρ' symbol applies only to expressions for natural numbers, and that there is no such expression as '$3 + \rho(1, 2)$', although '$\rho(3, 0) + \rho(1, 2)$' is entirely in order.

The usual notation '$\frac{a}{b}$' can then be regarded as a *façon de parler*—a loose way of referring to $\rho(a, b - 1)$ where $b - 1$ exists. In the remainder of this paper, we shall feel free to switch between the official 'ρ' notation and the more familiar fractions, and to use the latter for readability wherever no confusion is likely.

The definitional reduction allows all standard variable-free equations of rational arithmetic to be proved unproblematically. For example, $\frac{5}{7} + \frac{11}{14} = \frac{3}{2}$ is the same thing as $\rho(5, 6) + \rho(11, 13) = \rho(3, 1)$ which reduces as follows:

$$\rho(5, 6) + \rho(11, 13) = \rho(3, 1)$$

$$\rhd \quad \rho(5 \cdot 13 + 5 + 6 \cdot 11 + 11, 6 \cdot 13 + 6 + 13) = \rho(3, 1)$$

$$\text{i.e.} \quad \rho(147, 97) = \rho(3, 1)$$

$$\rhd \quad 147 \cdot 1 + 147 = 97 \cdot 3 + 3$$

$$\text{i.e.} \quad 294 = 294$$

Here a little natural number arithmetic has been performed inline in the interests of brevity. This is unproblematic.

There is, however, a problem. Consider the following chain of reasoning, where $\mathcal{N}$ is a theory of the natural numbers and $\mathcal{Q}$ a corresponding theory of the rationals:

1.	$\vdash_{\mathcal{N}} 3 \cdot 4 = 2 \cdot 6$	
2.	$\vdash_{\mathcal{Q}} \frac{2}{3} = \frac{4}{6}$	from 1, cross-dividing
3.	$\vdash_{\mathcal{Q}} (\frac{4}{6} = \frac{1}{1}) \to (\frac{2}{3} = \frac{1}{1})$	from 2, transitivity of '='
4.	$\vdash_{\mathcal{N}} (4 \cdot 1 = 1 \cdot 6) \to (2 \cdot 1 = 1 \cdot 3)$	from 3, cross-multiplying
5.	$\vdash_{\mathcal{N}} 4 = 6 \to 2 = 3$	from 4, calculation

If $\mathcal{N}$ is $\mathbf{R}^\sharp$, then the inference from (4) to (5) is good, and (1) holds, but (4) and (5) do not. (2) must also hold, both because it is definitionally equivalent to (1) and because if it were to fail, rational arithmetic would go down with it. If there is to be a rational number theory corresponding to $\mathbf{R}^\sharp$, therefore, the above argument must be broken at one of the two steps involved in getting from (2) to (4).

The step from (2) to (3) uses a transitivity principle reminiscent of the rules of affixing in weak relevant logics. Reversing the order of terms in (2) would make it an instance of the *rule* form of axiom P2, applied to rationals. This inference is clearly in the spirit of $\mathbf{R}^\sharp$, and indeed of weaker relevant arithmetics. Moreover, if it fails, then relevant rational arithmetic is indeed a strange mathematical theory, in that its "identity" relation either is not even transitive or else does not appear to mesh with its implication connective even in the most obvious way.

That leaves the step from (3) to (4), which is perhaps less immediately obvious because of the nesting of the inference inside an arrow. The warrant for it is the same thought as for the step from (1) to (2): that what the equation between rationals means just is the same as the equation between naturals got by cross-multiplying. Again the rule form of that inference, which would go from (2) to (1), for example, seems clear, but that it applies in the same way in arbitrary implicational contexts is less clear.

Let us desist from theorising about that inference for the moment, and re-examine the remaining fact, that (5) fails in the case of $\mathbf{R}^\sharp$. The reason for its failure is the existence of the modular models, and in particular in this case the model in the integers modulo 2. How does rational number theory fare in such a model?

At first glance, it does not fare well. Glaringly, since 0 is the same thing as 2, and there are many theorems about rationals such as $\frac{1}{2}$, division by zero has to make sense. Indeed, whatever is true of $\frac{2}{4}$ must also be true of $\frac{0}{0}$, since both 2 and 4 *are* 0 in the integers modulo 2. What we should make of formulae involving the expression $\rho(a, b)$ in a model in which b' is 0 is not intuitively clear.

3 Relevant Number Theories

What *is* clear is that we cannot secure all desiderata in a single theory. No conservative extension of $\mathbf{R}^\sharp$ can simultaneously maintain the expected definitional relationship between rational and natural arithmetic and the transitivity of rational equality. This leaves at least three options[3] for relevant theories of rational numbers: extend $\mathbf{R}^\sharp$ to a stronger theory of natural numbers, abandon the idea of numerical equality as a transitive relation between ratios, or formulate a *sui generis* rational number theory whose ties to relevant Peano arithmetic are looser than expected. All three options are possible. Without developing any of them in detail here, it is worthwhile noting the responses they embody to the models in the integers modulo n.

3.1 *Extending* $\mathbf{R}^\sharp$

Since $\mathbf{R}^\sharp$ is insufficient to make rational equality transitive, and since transitivity is an obvious property of any usable approximation to an equality relation, a reasonable response is to strengthen the underlying natural arithmetic to the point where transitivity *does* become provable. As just noted, this requires that at least some of the finite models in the integers modulo n be lost. Well, those were wildly unintended models anyway, and the lack of a cancellation principle was never seen by anyone except the True Relevantist[4] as a strength of the system, so perhaps the loss is not so great.

[3] There is in a sense a fourth option in the form of an extension of $\mathbf{R}^\sharp$ pointed out by Meyer and Mortensen [5, 13]: the axiom $\forall x(x = 0 \vee \exists y(xy = 1))$ can be added to $\mathbf{R}^\sharp$, apparently asserting the existence of multiplicative inverses and allowing the fraction $\frac{a}{b}$ to be defined as $\frac{1}{b} \cdot a$. It is bizarre, and of course false on any reasonable account including $\mathbf{R}^\sharp$ itself, to suggest that the rationals are actually natural numbers, but the theory does have models because the extra axiom holds in any finite model of $\mathbf{R}^\sharp$ modulo a prime. In any such model modulo a specific prime p, however, the inverse $\frac{1}{p}$ does not exist, so to secure a model in which *all* rationals exist, the "prime" has to be bigger than every finite p, and thus a nonstandard number. The associated number theory has as a theorem the formula $0 = n \rightarrow 0 = 1$ where n is any positive natural number, making it a serious departure from $\mathbf{R}^\sharp$, even ignoring its false theorems. However, it cannot contain the universal $\forall x(0 = x' \rightarrow 0 = 1)$ without collapsing arithmetic by making all equations true. It is therefore a fascinating curiosity, but hardly a satisfactory basis for relevant number theory.

[4] The "True Relevantist" is a (largely) fictional character introduced by Belnap and Dunn [2] who adheres unswervingly to the doctrine that **R** or something like it is the One True Logic. Meyer [8] famously claimed not to be one.

The transitivity of rational equality is equivalent in context to the cancellation principle

$$a \cdot x' = b \cdot x' \rightarrow a = b$$

which of course is a statement about natural numbers, so we may take the arithmetic in question, which we hereby name $\mathbf{R}^\natural$, to be the result of adding the universal closure of the above formula as an axiom to $\mathbf{R}^\sharp$ and introducing notation for the rationals by means of equations and biconditionals corresponding to the definitional reduction. Cancellation is plausibly a desirable property anyway, even without considering that it may be needed for rational arithmetic. Consider that $\mathbf{R}^\sharp$ provides full relevant support for three of the four most basic operations on equations: we can add

$$a = b \rightarrow a + x = b + x$$

subtract

$$a + x = b + x \rightarrow a = b$$

and multiply

$$a = b \rightarrow a \cdot x = b \cdot x$$

but the operation of division is missing, unless the cancellation axiom is added.

Proof that cancellation is equivalent in context to transitivity is very easy. To show that transitivity implies cancellation, note that since $x' \cdot y' = x' \cdot y'$ is a theorem, by replacement of equivalents so is $\rho(x', x) = \rho(y', y)$. In particular, $\rho(x', x) = \rho(1, 0)$. If transitivity holds for equality of ratios, therefore, $\rho(x', x) = \rho(a, b) \rightarrow \rho(1, 0) = \rho(a, b)$ is a theorem. Replacing provable equivalents again, that means $a \cdot x' = b' \cdot x' \rightarrow a \cdot 1 = b' \cdot 1$ is a theorem, giving cancellation for a and b' and by parity of reasoning for a' and b, so it holds where either of the numbers involved is not 0. The case where they are both 0 holds in $\mathbf{R}^\sharp$ anyway, so cancellation holds in general.

To show that cancellation implies transitivity, note that $\rho(a, b) = \rho(c, d)$ amounts to $a \cdot d' = b' \cdot c$, which implies $a \cdot d' \cdot f' = b' \cdot c \cdot f'$. Similarly, note that $\rho(c, d) = \rho(e, f)$ implies $b' \cdot c \cdot f' = b' \cdot d' \cdot e$. By the transitivity of equality for naturals (postulate P2 of $\mathbf{R}^\sharp$) therefore, $\rho(a, b) = \rho(c, d)$ and $\rho(c, d) = \rho(e, f)$ taken together imply $a \cdot d' \cdot f' = b' \cdot d' \cdot e$. But by cancellation, this implies $a \cdot f' = b' \cdot e$, which amounts to $\rho(a, b) = \rho(e, f)$ as required.

So cancellation and the transitivity of rational equality march together in $\mathbf{R}^\sharp$ extended by the addition of the ρ operator governed by the definitional reductions suggested above. Hence the theory $\mathbf{R}^\natural$, defined as suggested above, features a suitably transitive rational equality relation. Since cancellation holds, $\mathbf{R}^\natural$ does not permit the inconsistent models in the integers modulo n—with one exception. There remains the extreme case of the integers modulo 1, where there is only one number and all equations are true. It is amusing that this model, the wildest and least intended of all, survives. Since all atomic formulae evaluate to the Ackermann constant t, in this model, and the quantifiers all amount to the identity operator ($\forall x A$ is effectively A) the **R** propositional structures available for this domain are all constant (generated from t alone). The 3-valued structure exhibited above is not one of them, since in that structure the "confused" element c, which is the value of t, is a fixed point for all connectives. There is, in fact, only one non-trivial, inconsistent, constant De Morgan monoid.[5] It has four elements:

A	$\neg A$
T	F
f	t
t	f
F	T

$\wedge$	T	f	t	F
T	T	f	t	F
f	f	f	t	F
t	t	t	t	F
F	F	F	F	F

$\to$	T	f	t	F
T	T	F	F	F
f	T	t	F	F
t	T	f	t	F
F	T	T	T	T

The "designated values" are all of T, f and t. This is the *only* finite model of relevant rational arithmetic satisfying both the definitional equivalences and the transitivity of rational equality. Although it does not falsify any equations, it does still provide a finitary proof that the arithmetic is non-trivial in that, for example, $0 \neq 0 \to 0 = 0$ is not a theorem of it.

3.2 *Staying Within* $\mathbf{R}^\sharp$

While the extension of $\mathbf{R}^\sharp$ to accommodate such standard numerical reasoning as dividing by a constant leads to a theory in the same style as everybody else's arithmetic, to the True Relevantist or indeed to a less fanatical $\mathbf{R}^\sharp$ enthusiast such as Meyer, it may well appear to concede too much to the classical world view. In particular, with the loss of the models in the integers modulo n the finitary guarantee of reliability is also lost. Moreover, the fact that reasoning in accordance with **R** suffices for the derivation from the Peano axioms of the monotonicity of multiplication but not its converse is an *interesting* fact. Classically, $ax' = bx'$ is equivalent to $a = b$; from a substructural logical point of view, one direction of this equivalence is strongly linked to contraction while the other direction requires weakening. This suggests that moves which occur in ordinary *numerical* reasoning may have a *logical* character which only the study of substructural arithmetics will uncover.

[5] Why is there only one? Well, the four elements are t, f, $f \to t$ and f^2. If the De Morgan monoid is inconsistent, either $t = f$ and the algebra is trivial, or else $t < f$ and the order of the four elements is the chain as shown. Since it is a chain, the lattice operations $\wedge$ and $\vee$ cannot generate anything new, and it is not hard to show that the set of four elements is also closed under $\to$. A simple proof is to observe that the structure is generated from $\{f\}$ using $\to$ alone, and therefore results by imposing a congruence on the 6-element algebra for $\mathbf{R}^1_\to$ described by Meyer [10].

The definitional reduction of rational number theory to that of the naturals is standard practice, and a feature of the classical account as well as of the $\mathbf{R}^\sharp$ extension canvassed above. There is an obvious way to formulate such a theory of the rationals while staying firmly within the confines of $\mathbf{R}^\sharp$ itself: simply impose the definitions as stated, without changing the underlying arithmetic in any way, and let the theorems fall where they will. This will, of course, result in a conservative extension of $\mathbf{R}^\sharp$, since no postulates have been added—not even hidden ones such as the assumption that the defined relation of equality behaves like identity.

On a minimal reading, the resulting theory carries no new ontological weight. That is, it can be viewed as a purely syntactic construct, adding nothing to $\mathbf{R}^\sharp$ beyond a little vocabulary for describing certain relationships between natural numbers. On such a view, when we say that $\frac{2}{3} < \frac{3}{4}$ we literally mean that $2 \cdot 4 < 3 \cdot 3$, or in other words that 8 is smaller than 9. Models of the theory, then, are just models of $\mathbf{R}^\sharp$: the domain of such a model contains only natural numbers, and there are no such things as ratios between them. A question like "What is $\frac{2}{3}$?" is an ontological mistake, as all reference to $\frac{2}{3}$ is to be eliminated by a series of contextual definitions, so that there is nothing in any model of the theory with which it can be identified.

Even a theorem like "The rationals are dense" is not literally about objects called rational numbers. It appears to assert the existence of a rational between a given pair, but on unpacking the definitions, it really asserts the existence of two *natural* numbers related in a certain way to four others:

$$\tfrac{a}{b'} < \tfrac{c}{d'} \rightarrow \exists x \exists y (\tfrac{a}{b'} < \tfrac{x}{y'} \wedge \tfrac{x}{y'} < \tfrac{c}{d'})$$

That is, unpacking the definition of order for fractions:

$$ad' < cb' \rightarrow \exists x \exists y (ay' < xb' \wedge xd' < cy')$$

This is purely a fact about natural numbers, and is easily proved in $\mathbf{R}^\sharp$ in much the same way as it would be classically: the mean of two rational numbers is rational, and the mean of $\frac{a}{b'}$ and $\frac{c}{d'}$ is $\frac{ad'+cb'}{2b'd'}$, so we may let x be $ad' + cb'$ and y' be $2b'd'$. Hence we need only prove in $\mathbf{R}^\sharp$ the two relevant conditionals

$$ad' < cb' \rightarrow 2ab'd' < b'(ad' + cb')$$

$$ad' < cb' \rightarrow d'(ad' + cb') < 2cb'd'$$

These are similar, and follow in $\mathbf{R}^\sharp$ from the three principles

$$x < y \rightarrow xz < yz$$

$$x(y + z) = xy + xz$$

$$x < y \rightarrow 2x < x + y$$

all of which are available. Note once again that the first of these is *not* valid in contraction-free relatives of $\mathbf{R}^\sharp$, so the proof in Peano arithmetic that the rational numbers are dense also appears to be contraction-related.

The suggested construction does result in a theory, which we may read as a description of "the rational numbers" if we are prepared to adopt an ontology on which there is no such thing as a rational number. It is, however, difficult to square this reading of the theory with the actual practice which motivates the exercise. Consider addition, for example. We have two sheep; we add three goats; now we have five animals. Natural number theory provides a simple account of this operation. A model of the theory provides some abstract objects: the number of sheep, the number of goats and the number of animals. It is utterly natural to model addition as a function taking two of these numbers to the third. The arithmetic of fractions arises from a similar need to model *amounts* of things without the need to discretise: we have a rod measuring half a metre, and a second rod measuring 50% more than the first; together they measure a metre and a quarter. Although we *can* base the calculation on an ontology of integers alone, in fact that is not how we think of it. The natural mathematical model again features (fractional) numbers representing the lengths in metres, and again the intuitive picture of adding the lengths is abstracted from that of concatenating the objects. On this ground, therefore, rational number theory should at least sustain interpretations on which the existential quantifiers mean what they appear to mean and on which the referring expressions do genuinely refer.

Nothing, of course, prevents such a naturalistic construal of the canvassed definitional extension of $\mathbf{R}^\sharp$. It needs a domain of objects, which could be some sets of ordered pairs of natural numbers if we wish, such that every expression of the form $\rho(a, b)$ refers to one of them, and the symbol '=' applied to such expressions is interpreted as a relation between these objects. Unfortunately, some models—including crucially the finite models in the integers modulo n—are not at all what we might expect. Classically, the object modelling an expression like '$\frac{3}{4}$' is the relation between natural numbers a and b such that $4a = 3b$. Relevantly, however, it cannot be that relation, because '$\frac{3}{4}$' and '$\frac{6}{8}$' have different properties: substituting one for the other may result in equations which are not relevantly equivalent (e.g. $\frac{6}{8} = \frac{1}{1}$ is true in the integers modulo 2 whereas $\frac{3}{4} = \frac{1}{1}$ is not).

The difficulty of understanding "ordinary" models of $\mathbf{R}^\sharp$ as a first order theory of rational numbers is nothing but the difficulty of understanding the theory, brought into sharp focus. There are provable equations, but the inferences we would expect to make with these, such as substituting equals for equals, are restricted or unavailable. Even the definitions themselves are not quite as obvious as they seem. Addition in particular should perhaps be re-thought: is the result of adding $\frac{2}{7}$ to $\frac{3}{7}$ really $\frac{35}{49}$, or is it rather $\frac{5}{7}$? It cannot be both, as these are "equal" but different numbers in models of the theory. Denying the mathematician the right to simplify equations or to string them together is worse than denying a boxer the use of his fists: it is closer to denying him the right to breathe!

3.3 Cutting the Knot

If relevant arithmetic is to retain the principle that equations between ratios are reducible to those between natural numbers, we seem to be forced into a choice between unpalatable options: either the lack of a properly functioning relation of numerical equality threatens to render the theory unusable, or the relevant Peano arithmetic $\mathbf{R}^\sharp$ is abandoned in favour of a more classical account. There is, however, a third way. In order to allow the theory to remain true to its roots by conservatively extending $\mathbf{R}^\sharp$ while also supporting basic equational reasoning, it is necessary to break, or at least to loosen, the ties between rational equations and the underlying natural ones. The biconditional

$$\tfrac{a}{b} = \tfrac{c}{d} \leftrightarrow ad = bc$$

records two very different arithmetical inferences. The right to left conditional

$$ad = bc \rightarrow \tfrac{a}{b} = \tfrac{c}{d}$$

is the means whereby an antecedently given theory of natural arithmetic generates the true equations between rational numbers. It serves to justify and explain the core theory of reasoning about ratios and proportions. The left to right conditional,

$$\tfrac{a}{b} = \tfrac{c}{d} \rightarrow ad = bc$$

on the other hand, not only serves to specify the cases in which rational equations are to be held false, but also transfers equality between fractions which may hold in certain models back in the other direction, causing naturals which might otherwise be distinct to become conflated. If something has to give in order to secure a rational arithmetic consonant with $\mathbf{R}^\sharp$, this intrusive inference back from rational to natural equations is the obvious candidate.

There is no unique approach to using $\mathbf{R}^\sharp$ to inform a rational number theory if the latter is to be axiomatised on its own terms, but the following is suggested as a minimum under the constraints that its equality relation should behave like identity between rationals just as that of $\mathbf{R}^\sharp$ does on the naturals.

The suggested theory is two-sorted: we quantify over natural and rational numbers separately, and models of the theory will have domains in which the two sets of objects are kept distinct. We thus begin by adopting two kinds of variable: we use 'a', 'b', 'c' and 'x', 'y', 'z', etc. to refer to natural numbers, and the fraktur (gothic) counterparts '$\mathfrak{a}$', '$\mathfrak{b}$', '$\mathfrak{c}$,' and '$\mathfrak{x}$', '$\mathfrak{y}$', '$\mathfrak{z}$', ... as notation for rational numbers. It is convenient to keep the types clear, so recall that ρ is strictly a function from (pairs of) natural numbers to rational numbers. The addition and multiplication symbols are overloaded, as it would be too painful to separate the notations for addition and multiplication, but types must be consistent, so we cannot add or multiply a natural number and a rational. The identity symbol is similarly overloaded, but again makes sense only if the types match.

The axiomatisation of rational number theory begins with the axioms P1–P10 of $\mathbf{R}^\sharp$. To these are added axioms for the rationals. Axioms Q2 and Q3 ensure that equality is an equivalence relation, arising as advertised from the analogous relation of natural arithmetic, and axioms Q4–Q7 make it a congruence for the axiomatised functions:

Q1. $\exists x \exists y (\mathfrak{x} = \rho(x, y))$
Q2. $wx' = yz' \rightarrow \rho(w, z) = \rho(y, x)$
Q3. $\mathfrak{x} = \mathfrak{y} \rightarrow (x = z \rightarrow \mathfrak{y} = \mathfrak{z})$

Q4. $x = y \rightarrow \rho(x, z) = \rho(y, z)$
Q5. $x = y \rightarrow \rho(z, x) = \rho(z, y)$
Q6. $\mathfrak{x} = \mathfrak{y} \rightarrow \mathfrak{x} + \mathfrak{z} = \mathfrak{y} + \mathfrak{z}$
Q7. $\mathfrak{x} = \mathfrak{y} \rightarrow \mathfrak{x} \cdot \mathfrak{z} = \mathfrak{y} \cdot \mathfrak{z}$

Addition and multiplication are defined as expected:

Q8. $\rho(w, x) + \rho(y, z) = \rho(wz + xy + w + y,\ xz + x + z)$
Q9. $\rho(w, x) \cdot \rho(y, z) = \rho(wy,\ xz + x + z)$

The analogue of P5, specifying which equations are false, is the inference rule

R3. $ad' \neq b'c \ \Rightarrow\ \rho(a, b) \neq \rho(c, d)$

Axiomatic extensions of this minimal theory could be entertained. For example, in the absence of the converse of Q2, there is no obvious way to prove important conditionals with equations as antecedents, such as the principle that addition is cancellative:

$$\mathfrak{a} + \mathfrak{c} = \mathfrak{b} + \mathfrak{c} \rightarrow \mathfrak{a} = \mathfrak{b}$$

Whether or not this is *required*, it is at least *useful* for basic numerical reasoning, and is clearly in the spirit of $\mathbf{R}^\sharp$, so plausibly it should be added if it does not flow from the given basis. Such piecemeal extension of the theory is rather unsatisfactory, however. Unless there is a clearly defined end to the list of plausible additions, extensions will always have an *ad hoc* character.

Leaving extensions aside, we may note that the theory does contain all the correct rational equations, does allow reasoning using standard properties of equality (including its transitivity) in arbitrary contexts, and does feature such ordinary classical theorems as trichotomy:

$$\forall \mathfrak{x} \forall \mathfrak{y}(\mathfrak{x} < \mathfrak{y} \vee \mathfrak{x} = \mathfrak{y} \vee \mathfrak{x} > \mathfrak{y})$$

The lack of the converse of Q2 is vital where the underlying model of $\mathbf{R}^\sharp$ is in one of the finite domains. The model of rational arithmetic in such a case is extremely simple: if zero is the successor of a natural number, then there is only one rational number, making all equations in the rational vocabulary true.

Proof of this fact is easy. Consider a model in which $0 = n$ for some positive n. Then by Q4, $\frac{0}{n} = \frac{n}{n}$. As theorems of arithmetic, $\frac{0}{n} = \frac{0}{1}$ and $\frac{n}{n} = \frac{1}{1}$, so by transitivity, $\frac{0}{1} = \frac{1}{1}$. That is, in the domain of rational numbers, the zero element and the unit element are the same. These are, respectively, the fixed point and the identity for multiplication: that is, for any a and b, $\frac{0}{1} \cdot \frac{a}{b} = \frac{0}{1}$ and $\frac{1}{1} \cdot \frac{a}{b} = \frac{a}{b}$. By Q7 and Q3, therefore, $\frac{a}{b} = \frac{0}{1}$. Since all rationals are equal to $\frac{0}{1}$, they are all equal to each other.

The collapse of rational arithmetic in the finite models is in keeping with the overall character of this number theory, which exhibits strikingly classical features without upsetting the relevant base in $\mathbf{R}^\sharp$. The result may be strengthened, in fact: where $\mathfrak{a}$ is any positive rational, we have the theorem

$$\rho(0, 0) = \mathfrak{a} \ \rightarrow\ \forall \mathfrak{x} \forall \mathfrak{y}(\mathfrak{x} = \mathfrak{y})$$

Proof is again simple. Let a and b be two natural numbers such that $\mathfrak{a} = \rho(a', b)$, and consider any other number $\mathfrak{x}$. Let c and d be naturals such that $\mathfrak{x} = \rho(c, d)$. As an instance of Q7,

$$\rho(0,0) = \mathfrak{a} \rightarrow \rho(0,0) \cdot \tfrac{b'c}{a'd'} = \tfrac{a'}{b'} \cdot \tfrac{b'c}{a'd'}$$

But $\rho(0,0) \cdot \frac{b'c}{a'd'} = \rho(0,0)$, and by Q9 $\frac{a'}{b'} \cdot \frac{b'c}{a'd'} = \frac{a'b'c}{a'b'd'}$. By Q2, $\frac{a'b'c}{a'b'd'} = \frac{c}{d'}$, so $\rho(0,0) = \mathfrak{x}$. Since $\mathfrak{x}$ was any number, on supposition that $\rho(0,0) = \mathfrak{a}$, we have that all numbers are equal to $\rho(0,0)$ and therefore to each other.

4 Conclusion

Extending relevant number theory from the natural numbers to the rationals is not simple. Developing rational arithmetic by definition within $\mathbf{R}^\sharp$ is possible, but leads to a theory in which many of the inferences central to numerical reasoning are not present in the form of relevant implications. Even such fundamental features as the transitivity of equality are not derivable. The correct equations are present, but they are not usable for much of mathematical reasoning without stepping outside the set of inferences warranted according to the theory.

Adding some of the missing principles such as transitivity leads directly to $\mathbf{R}^\natural$, a non-conservative extension of the base theory $\mathbf{R}^\sharp$. That may not be so bad in the end, but it lies outside not only the letter but the spirit of relevant arithmetic as set forth by its founders. It also leads to the loss of the finitary proof of reliability which is one of the main attractions of $\mathbf{R}^\sharp$.

The attempt to steer a middle path by removing the inferences which cause non-conservativeness by leading back from the rationals to the naturals, while explicitly postulating transitivity and other useful inference forms, leads to a range of theories all of which are subsets of $\mathbf{R}^\natural$ but which it is hoped (without proof at this stage) will exactly contain $\mathbf{R}^\sharp$ as their account of the natural numbers. What emerges clearly from a preliminary investigation of this idea is that rational number theory in a relevant logical context rapidly gains a very classical flavour. Even with very modest postulates added to $\mathbf{R}^\sharp$, the inconsistent assumption that zero is equal to a positive number relevantly implies that all rational numbers are equal and hence leads to the collapse of rational number theory. This is in stark contrast to the relevant theory of the natural numbers, where such inconsistent assumptions may be entertained quite readily.

No one theory stands out as *the* relevant account of rational numbers. All are perfectly ordinary relevant theories, of course, and all arise out of $\mathbf{R}^\sharp$ in reasonably well motivated ways. All are worthy of further investigation. Problems arising naturally include the questions of whether the rule γ (material detachment) is admissible in $\mathbf{R}^\natural$, and what axiomatic extensions in the style of the last section are really conservative extensions of $\mathbf{R}^\sharp$. These questions we leave for another time, the groundwork sufficing for the present paper.

References

1. Anderson, A. R., & Belnap, N. D. (1975). *Entailmant: The logic of relevance and necessity, Vol. 1*. Princeton: Princeton University Press.
2. Belnap, N. D., & Dunn, J. M. (1981). Entailment and the disjunctive syllogism. In G. Fløistad & G. H. von Wright (Eds.), *Tome 1 Philosophie du langage, Logique philosophique* (pp. 337–366). Dordrecht: Springer.
3. Dunn, J. M. (1985). Relevance logic and entailmant. In D. Gabbay & F. Guenthner (Eds.), *Handbook of philosophical logic, Vol. 3* (pp. 117–224). Dordrecht: Reidel.
4. Dunn, J. M. (2019). *The algebra of intensional logics*. London: College Publications.
5. Ferguson, T. M., & Ramirez-Camara, E. (2021). Episodes in model-theoretic xenology: Rationals as positive integers in $R^\sharp$. *Australasian Journal of Logic*, *18*, 428.
6. Friedman, H., & Meyer, R. K. (1992). Whither relevant arithmetic. *Journal of Symbolic Logic*, *57*, 824–831.
7. Mares, E. (2004). *Relevant logic: A philosophical interpretation*. Cambridge: Cambridge University Press.
8. Meyer, R. K. (1978). *Why I am not a relevantist*. Technical report, Australian National University.
9. Meyer, R. K. (1985). A farewell to entailment. In G. Dorn & P. Weingartner (Eds.), *Foundations of logic and linguistics: Problems and solutions* (pp. 577–636). New York: Plenum Press.
10. Meyer, R. K. (1986). Sentential constants in R and $R^\neg$. *Studia Logica*, *45*, 301–327.
11. Meyer, R. K. (2021). Arithmetic formulated relevantly. *Australasian Journal of Logic*, *18*, 154–288.
12. Meyer, R. K., & Mortensen, C. (1984). Inconsistent models for relevant arithmetics. *Journal of Symbolic Logic*, *49*, 917–929.
13. Meyer, R. K., & Mortensen, C. (2021). Alien intruders in relevant arithmetic. *Australasian Journal of Logic*, *18*, 401.
14. Slaney, J. (2022). Models of relevant arithmetic. *Australasian Journal of Logic*, *19*(1), 32–50.
15. Slaney, J., Meyer, R. K., & Restall, G. (1998). Linear arithmetic desecsed. *Logique et Analyse*, *39*, 379–388.

MIX
Papier aus verantwortungsvollen Quellen
Paper from responsible sources
FSC® C105338

If you have any concerns about our products,
you can contact us on
ProductSafety@springernature.com

In case Publisher is established outside the EU,
the EU authorized representative is:
Springer Nature Customer Service Center GmbH
Europaplatz 3, 69115 Heidelberg, Germany

Printed by Libri Plureos GmbH
in Hamburg, Germany